REVIEWS in MINERALOGY
Volume 25

OXIDE MINERALS:
PETROLOGIC AND MAGNETIC SIGNIFICANCE

DONALD H. LINDSLEY, Editor

INTRODUCTION TO OXYGEN FUGACITY AND ITS PETROLOGIC IMPORTANCE	B. R. FROST
CRYSTAL CHEMISTRY OF OXIDES AND OXYHYDROXIDES	G. A. WAYCHUNAS
EXPERIMENTAL STUDIES OF OXIDE MINERALS	D. H. LINDSLEY
MAGNETIC PROPERTIES OF FE-TI OXIDES	S. K. BANERJEE
OXIDE TEXTURES -- A MINI-ATLAS	S. E. HAGGERTY
THERMOCHEMISTRY OF THE OXIDE MINERALS	M. S. GHIORSO & R. O. SACK
MACROSCOPIC AND MICROSCOPIC THERMODYNAMIC PROPERTIES OF OXIDES	B. J. WOOD, J. NELL & A. B. WOODLAND
INTERPLAY OF CHEMICAL AND MAGNETIC ORDERING	B. P. BURTON
CHROMITE AS A PETROGENETIC INDICATOR	R. O. SACK & M. S. GHIORSO
OXIDE MINERALOGY OF THE UPPER MANTLE	S. E. HAGGERTY
OXYGEN BAROMETRY OF SPINEL PERIDOTITES	B. J. WOOD
OCCURRENCE OF IRON-TITANIUM OXIDES IN IGNEOUS ROCKS	B. R. FROST & D. H. LINDSLEY
STABILITY OF OXIDE MINERALS IN METAMORPHIC ROCKS	B. R. FROST
MAGNETIC PETROLOGY: FACTORS THAT CONTROL THE OCCURRENCE OF MAGNETITE IN CRUSTAL ROCKS	B. R. FROST

Series Editor: Paul H. Ribbe

MINERALOGICAL SOCIETY OF AMERICA

REVIEWS in MINERALOGY   Volume 25

OXIDE MINERALS:
PETROLOGIC AND
MAGNETIC SIGNIFICANCE
DONALD H. LINDSLEY, EDITOR

The authors:

Subir Banerjee
Dept. of Geology & Geophysics
University of Minnesota
Minneapolis, Minnesota 55455

Benjamin P. Burton
National Institute of Standards and
  Technology, B150/223, Div. 450
Gaithersburg, Maryland 20899

B. Ronald Frost
Dept. of Geology & Geophysics
University of Wyoming
Laramie, Wyoming 82071

Mark S. Ghiorso
Dept. of Geological Sciences
University of Washington
Seattle, Washington 98195

Stephen E. Haggerty
Dept. of Geology & Geography
University of Massachusetts
Amherst, Massachusetts 01003

Donald H. Lindsley
Dept. of Earth & Space Sciences
State University of New York
Stony Brook, New York 11794

J. Nell
Dept. of Geological Sciences
Bristol University
Queen's Road
Bristol BS8 1RJ England

Richard O. Sack
Dept. of Earth &
  Atmospheric Sciences
Purdue University
E. Lafayette, Indiana 47907

Glenn A. Waychunas
Center for Materials Research
Stanford University
Stanford, California 94305

Bernard J. Wood
Dept. of Geological Sciences
Bristol University
Queen's Road
Bristol BS8 1RJ England

A. B. Woodland
Bayerisches Geoinstitut
Bayreuth
Federal Republic of Germany

Series Editor:

PAUL H. RIBBE
DEPARTMENT OF GEOLOGICAL SCIENCES
Virginia Polytechnic Institute and State University
Blacksburg, Virginia 24061

COPYRIGHT: 1991

MINERALOGICAL SOCIETY of AMERICA

Printed by BookCrafters, Inc., Chelsea, Michigan 48118

REVIEWS in MINERALOGY

Formerly: *SHORT COURSE NOTES*
ISSN 0275-0279
Volume 25
OXIDE MINERALS: THEIR PETROLOGIC
AND MAGNETIC SIGNIFICANCE
ISBN 0-939950-30-8

ADDITIONAL COPIES of this volume as well as those listed below
may be obtained from the MINERALOGICAL SOCIETY OF AMERICA,
1130 Seventeenth Street, N.W., Suite 330, Washington, D.C. 20036 U.S.A.

Vol.	Year	Pages	Editor(s)	Title
1	1974	284	P. H. Ribbe	SULFIDE MINERALOGY
2	1983	362	P. H. Ribbe	FELDSPAR MINERALOGY (2nd edition)
3	out of print			OXIDE MINERALS
4	1977	232	F. A. Mumpton	MINERALOGY AND GEOLOGY OF NATURAL ZEOLITES
5	1982	450	P. H. Ribbe	ORTHOSILICATES (2nd edition)
6	1979	380	R. G. Burns	MARINE MINERALS
7	1980	525	C. T. Prewitt	PYROXENES
8	1981	398	A. C. Lasaga R. J. Kirkpatrick	KINETICS OF GEOCHEMICAL PROCESSES
9A	1981	372	D. R. Veblen	AMPHIBOLES AND OTHER HYDROUS PYRIBOLES—MINERALOGY
9B	1982	390	D. R. Veblen P. H. Ribbe	AMPHIBOLES: PETROLOGY AND EXPERIMENTAL PHASE RELATIONS
10	1982	397	J. M. Ferry	CHARACTERIZATION OF METAMORPHISM THROUGH MINERAL EQUILIBRIA
11	1983	394	R. J. Reeder	CARBONATES: MINERALOGY AND CHEMISTRY
12	1983	644	E. Roedder	FLUID INCLUSIONS (Monograph)
13	1984	584	S. W. Bailey	MICAS
14	1985	428	S. W. Kieffer A. Navrotsky	MICROSCOPIC TO MACROSCOPIC: ATOMIC ENVIRONMENTS TO MINERAL THERMODYNAMICS
15	1990	406	M. B. Boisen, Jr. G. V. Gibbs	MATHEMATICAL CRYSTALLOGRAPHY (Revised)
16	1986	570	J. W. Valley H. P. Taylor, Jr. J. R. O'Neil	STABLE ISOTOPES IN HIGH TEMPERATURE GEOLOGICAL PROCESSES
17	1987	500	H. P. Eugster I. S. E. Carmichael	THERMODYNAMIC MODELLING OF GEOLOGICAL MATERIALS: MINERALS, FLUIDS, MELTS
18	1988	698	F. C. Hawthorne	SPECTROSCOPIC METHODS IN MINERALOGY AND GEOLOGY
19	1988	698	S. W. Bailey	HYDROUS PHYLLOSILICATES (EXCLUSIVE OF MICAS)
20	1989	369	D. L. Bish J. E. Post	MODERN POWDER DIFFRACTION
21	1989	348	B. R. Lipin G. A. McKay	GEOCHEMISTRY AND MINERALOGY OF RARE EARTH ELEMENTS
22	1990	406	D. M. Kerrick	THE Al_2SiO_5 POLYMORPHS (Monograph)
23	1990	603	M. F. Hochella, Jr. A. F. White	MINERAL-WATER INTERFACE GEOCHEMISTRY
24	1990	314	J. Nicholls J. K. Russell	MODERN METHODS OF IGNEOUS PETROLOGY—UNDERSTANDING MAGMATIC PROCESSES

OXIDE MINERALS: PETROLOGIC AND MAGNETIC SIGNIFICANCE

FOREWORD

The Mineralogical Society of America has been publishing *Reviews in Mineralogy* since 1974. See opposite page for a list of available volumes. This is the first of 25 volumes that has been redone and expanded under a new title and volume number. "Oxide Minerals: Their Petrologic and Magnetic Significance" was edited by Don Lindsley and represents contributions of 11 authors.

As editor of *Reviews*, I thank Don Lindsley for a great job of pressuring authors and helping with final copy editing - far above the call of duty. Marianne Stern did most of the paste-up for camera-ready copy. Her tireless efforts with a tedious job are greatly appreciated.

Paul H. Ribbe
Blacksburg, VA
April 2, 1991

EDITOR'S INTRODUCTION

This volume is scheduled to be published in time to be used as the textbook for the Short Course on Fe-Ti Oxides: Their Petrologic and Magnetic Significance, to be held May 24-27, 1991, organized by B.R. Frost, D.H. Lindsley, and S.K. Banerjee and jointly sponsored by the Mineralogical Society of America and the American Geophysical Union.

It has been fourteen and a half years since the last MSA Short Course on Oxide Minerals and the appearance of Volume 3 of *Reviews in Mineralogy*. Much progress has been made in the interim. This is particularly evident in the coverage of the thermodynamic properties of oxide minerals: nothing in Volume 3, while in contrast, Volume 25 has three chapters (6, 7, and 8) presenting various aspects of the thermodynamics of oxide minerals; and other chapters (9, 11, 12) build extensively on thermodynamic models. The coverage of magnetic properties has also been considerably expanded (Chapters 4, 8, and 14). Finally, the interaction of oxides and silicates is emphasized in Chapters 9, 11, 12, 13, and 14. One of the prime benefits of *Reviews in Mineralogy* has been that any scientist can afford to have it at his or her fingertips. Because Volume 3 is out of print and will not be readily available to newcomers to our science, as much as possible we have tried to make Volume 25 a replacement for, rather than a supplement to, the earlier volume. Chapters on crystal chemistry, phase equilibria, and oxide minerals in both igneous and metamorphic rocks have been rewritten or extensively revised. The well received photographs of oxide textures in Volume 3 have been collected and expanded into a "Mini-Atlas" In Volume 25. Topics that receive less attention than in the earlier volume are oxides in lunar rocks and meteorites, and the manganese minerals. We hope that the new volume will turn out to be as useful as the previous one was.

This volume has been made possible by the efforts of many people. The authors have produced manuscripts of high quality. However, it is Series Editor Paul Ribbe and his staff who have converted those manuscripts, in less than two months, into the finished volume, and the authors join me in offering a heartfelt thanks to them for this remarkable achievement.

Donald H. Lindsley
Stony Brook, N. Y.
April 2, 1991

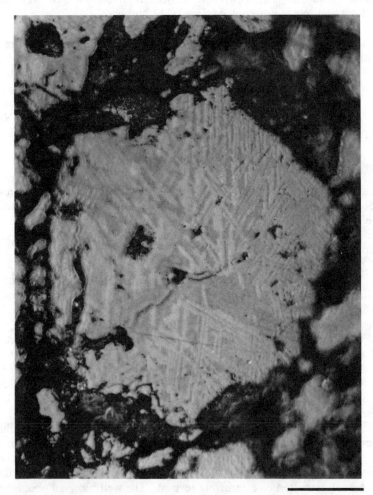

10 microns

DEDICATION

In the years since the last Oxides Volume appeared (1976), two giants in oxide mineralogy have passed away: Paul Ramdohr, the consummate ore mineralogist whose ability to identify and interpret silicate minerals in reflected light was also truly astounding, and Arthur F. Buddington, an eminent silicate petrologist who came to cherish "opaques" only late in his career, but whose impact on the understanding of Fe-Ti oxide minerals was enormous.

We dedicate this book with great respect and affection to the memory of those pioneers of oxide mineralogy, A. F. Buddington and Paul Ramdohr.

The Photograph

In 1962 I was a young postdoctoral fellow at the Geophysical Laboratory studying Fe-Ti oxides experimentally. I had started out accepting the gospel that there was extensive solubility of ilmenite in magnetite - the assumption underlying the 1955 magnetite-ilmenite thermometer of Buddington, Fahey, and Vlisidis. A few workers like Knut Heier and E. A. Vincent, however, were promoting the heresy (first promulgated by V. M. Goldschmidt even before ulvöspinel had been found in nature or synthesized in the laboratory), that the classic trellis "exsolution" texture of ilmenite lamellae in the (111) planes of magnetite might actually result from the oxidation of Fe_2TiO_4 component in the magnetite. It turned out to be remarkably easy for me to produce that trellis texture by controlled oxidation of ulvöspinel-rich solid solutions under hydrothermal conditions at 1-2 kbar and 600-1000°C. One of the best examples resulted from oxidizing pure Fe_2TiO_4 at 1000°C and 670 bar for 1.5 hours, with the oxygen fugacity imposed by the walls of the stellite pressure vessel (approximately NNO). Paul Ramdohr was visiting the Geophysical Laboratory at the time and very kindly took a photograph of it (facing page).

Meanwhile, Buddington had written a new manuscript on oxide thermometry, and had sent it to me for comments. Two things were very clear: the manuscript was fatally flawed by the assumption of extensive $FeTiO_3$ solubility in magnetite, and I in turn needed Bud's vast array of analytical data in order to apply the new magnetite-ilmenite thermometer/oxybarometer that was emerging from my experiments. I suggested to Bud that we combine forces - all he had to do was to forswear exsolution and accept the radical notion that ilmenomagnetites formed by oxidation! Bud was already in his seventies, and it would have been understandable had he refused. But he was A. F. Buddington, and he wrote back, "Sure - if you can convince me of the oxidation hypothesis." I sent him the photograph, and he responded, "You win; let's write the paper."

It is poignantly ironic that Ramdohr, whose then state-of-the-art photograph made the Buddington-Lindsley collaboration possible, never accepted the importance in nature of the oxidation process for producing the trellis ilmenomagnetite texture. I cherish a reprint of a paper that the Professor, then past 90, wrote defending the exsolution origin of ilmenomagnetite. On the cover he inscribed "Dr. Lindsley mit bestem Gruss. Was sagen Sie jetzt [What do you have to say now]? Paul Ramdohr". May we all be so active and feisty in our nineties!

D. H. Lindsley

Table of Contents

Page

ii Copyright; Additional volumes of *Reviews in Mineralogy*
iii Foreword and Editor's Introduction
iv Dedication

Chapter 1 — B. Ronald Frost
INTRODUCTION TO OXYGEN FUGACITY AND ITS PETROLOGIC IMPORTANCE

1 THE IMPORTANCE OF OXYGEN FUGACITY
3 CALIBRATION OF OXYGEN FUGACITY
4 AN OVERVIEW OF OXYGEN FUGACITY SPACE
6 SOME COMMON MISCONCEPTIONS
6   Buffers
7   Oxygen as fluid species
8   Ferrous/ferric ratio
8 REFERENCES

Chapter 2 — Glenn A. Waychunas
CRYSTAL CHEMISTRY OF OXIDES AND OXYHYDROXIDES

11 INTRODUCTION
11   What is crystal chemistry?
12   Solid solubility
14   Order-disorder
14   Element site partitioning
14   Electronic properties peculiar to oxides
15 DEFECTS AND NONSTOICHIOMETRY
16 BASIC STRUCTURAL TOPOLOGIES
16   HCP and CCP O and OH arrangements
18   NaCl structure
18   Perovskite structure
18   Fluorite and rutile structures
18   Spinel structure
22   Rhombohedral oxide structures
22 IRON OXIDES
22   Wüstite, magnesiowüstite and other solid solutions
23   Hematite, hydrohematite and protohematite
24 SPINELS
24   Predictions and determinations of cation site preference in spinels
25   Magnetite, ulvöspinel and titanomagnetites
30   Maghemite, titanomaghemite
32   Distorted spinel-like phases
33 IRON OXYHYDROXIDES
33   Goethite
34   Akaganeite
34   Lepidocrocite
37   Feroxyhyte
37   Ferrihydrite and protoferrihydrite

38	TITANIUM AND IRON-TITANIUM OXIDES
40	Rutile
40	Pseudorutile
40	Anatase
42	Brookite
43	Ilmenite, solid solutions and related phases
43	Pseudobrookite solid solutions
46	MANGANESE OXIDES AND OXYHYDROXIDES
47	Tetravalent manganese oxides and oxyhydroxides
47	Pyrolusite
47	Ramsdellite
47	Nsutite
47	Hollandite and related phases
51	Romanechite
51	Todorokite
52	LAYER STRUCTURE MANGANESE OXYHYDROXIDES (PHYLLOMANGANATES)
52	Chalcophanite
53	Birnessite
54	Buserite
54	Asbolane
54	Vernadite
55	Lithiophorite
55	TRIVALENT MANGANESE OXIDES AND OXYHYDROXIDES
55	Bixbyite
55	Groutite and feitknechtite
55	Manganosite
56	ALUMINUM OXIDES AND OXYHYDROXIDES
56	Corundum
56	Boehmite and diaspore
56	COMPLEX OXIDE GROUPS
56	Pyrochlore group
58	Columbite-tantalite group
59	OTHER SIGNIFICANT OXIDES
59	Perovskites
59	Uraninite
61	BIBLIOGRAPHY AND REFERENCES

Chapter 3   Donald H. Lindsley
EXPERIMENTAL STUDIES OF OXIDE MINERALS

69	INTRODUCTION
69	Synthesis vs. phase equilibrium
70	Control of experimental conditions
70	*Oxygen fugacity*
70	*Container problem*
71	Minerals and phases considered
72	Fe-O join
72	*Wüstite*
72	*Thermodynamically impossible diagram for wüstite*
73	*Effects of high pressures on wüstite*
73	*Magnetite*
73	*Fe_2O_3 in magnetite*
73	*Hematite*
74	*Melting of iron oxides in the presence of C-O vapor*
74	TiO_2

74	FeO-TiO$_2$ join
79	Fe$_2$O$_3$-TiO$_2$ join
79	FeO-Fe$_2$O$_3$-TiO$_2$(-Ti$_2$O$_3$) join
79	*Magnetite-ulvöspinel join*
79	*Hematite-ilmenite join*
80	*Pseudobrookite-"ferropseudobrookite" join*
80	*1300 °C isotherm*
86	*Phase relations at other temperatures*
87	*Titanomaghemites*
87	*Reduction of Fe-Ti oxides*
87	Fe-O-MgO-TiO$_2$ SYSTEM
87	FeO-MgO join
87	FeO-MgO-Fe$_2$O$_3$ join
88	FeO-MgO-TiO$_2$-(Ti$_2$O$_3$) join
90	FeO-Fe$_2$O$_3$-MgO-TiO$_2$ join
90	Fe-O-MnO-TiO$_2$ SYSTEM
90	FeO-Fe$_2$O$_3$-MnO join
92	Mn-O-TiO$_2$ join
93	FeO-MnO-TiO$_2$ join
93	FeO-Fe$_2$O$_3$-MnO-TiO$_2$ join
94	Fe-O-Al$_2$O$_3$-TiO$_2$ AND Fe-O-MgO-Al$_2$O$_3$ SYSTEMS
94	Fe$_2$O$_3$-Al$_2$O$_3$ join
95	FeO-Fe$_2$O$_3$-Al$_2$O$_3$ join
95	MgO-Al$_2$O$_3$ join
95	FeO-MgO-Fe$_2$O$_3$-Al$_2$O$_3$ systems
95	FeO-Al$_2$O$_3$-TiO$_2$ join
98	FeO-Fe$_2$O$_3$-Al$_2$O$_3$ systems
98	Cr$_2$O$_3$-BEARING SYSTEMS
98	FeCr$_2$O$_4$-Fe$_3$O$_4$-FeAl$_2$O$_4$ join
98	MISCELLANEOUS SYSTEMS
100	ACKNOWLEDGMENTS
100	REFERENCES

Chapter 4   Subir K. Banerjee

MAGNETIC PROPERTIES OF FE-TI OXIDES

107	INTRODUCTION
107	CRYSTAL STRUCTURE AND COMPOSITION
109	Rhombohedral oxides
110	Spinel oxides
111	Cation-deficient spinel
113	INTRODUCTION TO THEORIES OF MAGNETIC SUPEREXCHANGE
114	Antiferromagnetism
114	Ferrimagnetism
117	Non-collinear structures
117	SATURATION MAGNETIZATION AND CURIE (AND NEEL) POINTS
119	Rhombohedral oxides
120	Spinel oxides
121	Cation-deficient spinels
122	MAGNETOCRYSTALLINE ANISOTROPY AND MAGNETOSTRICTION
122	Basic theory
123	Isotropic points
126	OVERVIEW AND CONCLUSION
127	ACKNOWLEDGMENTS
127	REFERENCES

Chapter 5 Stephen E. Haggerty
OXIDE TEXTURES – A MINI-ATLAS

129	INTRODUCTION
130	Mini-Atlas
130	OXIDATION OF Fe-Ti SPINELS AND ILMENITES
130	Oxidation of titanomagnetite
131	*Trellis types*
132	*Sandwich types*
132	*Composite types*
132	*Oxidation of titanomagnetite-ilmenite intergrowths*
135	Oxidation of discrete primary ilmenite
137	REFERENCES

Chapter 6 M. S. Ghiorso & R. O. Sack
THERMOCHEMISTRY OF THE OXIDE MINERALS

221	INTRODUCTION
221	CRYSTAL CHEMICAL CONSTRAINTS ON THERMODYNAMIC MODELS
221	Cation substitution and size mismatch
223	Cation-ordering phenomena
226	*Convergent ordering*
233	*Non-convergent ordering*
236	Magnetic contributions
240	Crystalline defects
242	SOLUTION MODELS FOR THE OXIDE MINERALS
243	Cubic oxides: spinels
252	Rhombohedral oxides
256	Orthorhombic oxides
258	RECOMMENDED STANDARD STATE PROPERTIES
260	OXIDE GEOTHERMOMETERS
260	DIRECTIONS FOR FUTURE RESEARCH
262	REFERENCES

Chapter 7 B. J. Wood, J. Nell & A.B. Woodland
MACROSCOPIC AND MICROSCOPIC THERMODYNAMIC PROPERTIES OF OXIDES

265	INTRODUCTION
266	ACTIVITY MEASUREMENTS
266	Equilibration with metal
268	Equilibration with noble metals
268	Interphase partitioning
273	ENTHALPY MEASUREMENTS
273	SPECIFIC SYSTEMS
273	FeO-MgO
274	Other rocksalt structure oxides
275	Spinel structure oxides
276	Fe_3O_4-Fe_2TiO_4
278	$MgAl_2O_4$-Cr_2O_4
278	$FeAl_2O_4$-$FeCr_2O_4$
278	Fe_3O_4-$FeCr_2O_4$
279	Fe_3O_4-$FeAl_2O_4$
279	$FeAl_2O_4$; $MgAl_2$-Mg_2TiO_4

279	RHOMBOHEDRAL OXIDES
279	Al_2O_3-Cr_2O_3
280	Al_2O_3-Fe_2O_3
280	Fe_2O_3-Mn_2O_3
280	$FeTiO_3$-$MnTiO_3$
280	$FeTiO_3$-$MgTiO_3$
281	Fe_2O_3-$FeTiO_3$
283	MICROSCOPIC PROPERTIES
284	Electrical conduction in Fe_3O_4
285	Measurement technique
287	Fe_3O_4-$MgFe_2O_4$
289	Fe_3O_4-$FeAl_2O_4$
289	Fe_3O_4-$FeCr_2O_4$
293	Fe_3O_4-$Fe_2Ti_2O_4$
295	More complex spinels
295	Integrating order-disorder and activity relations
299	SUMMARY
299	REFERENCES

Chapter 8   Benjamin P. Burton

THE INTERPLAY OF CHEMICAL AND MAGNETIC ORDERING

303	TERMINOLOGY
304	INTRODUCTION
304	MAGNETIC HEAT CAPACITIES
308	THEORETICAL METHODS
308	Method (1): Meijering's "regular pseudo-ternary model"
310	Method (2): The phenomenological $C_m(T)$ approach
312	EFFECTS ON PHASE DIAGRAM TOPOLOGIES
312	Metallurgical systems
312	Ceramic and mineral systems
313	Rock salt structure systems
313	Hematite-ilmenite
315	Magnetite-ulvöspinel
318	SUMMARY
318	OPPORTUNITIES FOR FUTURE RESEARCH
319	REFERENCES

Chapter 9   R. O. Sack & M. S. Ghiorso

CHROMITE AS A PETROGENETIC INDICATOR

323	INTRODUCTION
323	Nomenclature and chemical species
324	IRON-MAGNESIUM EXCHANGE REACTIONS
325	Fe-Mg nonideality
328	Cation ordering and the olivine-spinel geothermometer
335	PETROLOGICAL APPLICATIONS
338	Chromian spinel in MORB-type lavas
340	Chromian spinels in HED meteorites
343	Miscibility gaps in the spinel prism
348	A PETROLOGICAL CONCERN (PASTICHE)
350	REFERENCES

Chapter 10 — Stephen E. Haggerty
OXIDE MINERALOGY OF THE UPPER MANTLE

355 INTRODUCTION
356 SPINEL MINERAL GROUP
357   Spinel-bearing xenoliths in alkali basalts
357   Alpine-type peridotites, abyssal peridotites and ophiolites
365   Spinel-bearing xenoliths in kimberlites and lamproites
370   Discussion
372 ILMENITE MINERAL GROUP
372   Ilmenite in alkali volcanics
376   Ilmenite xenoliths in kimberlites
382   Ilmenite-pyroxene intergrowth
382   Large ion lithophile element (LILE) and high field strength element (HFSE) oxide minerals
384   Crichtonite mineral group
387   Magnetoplumbite mineral group
392   Armalcolite mineral group
392   Rutile
397   Ilmenite and spinel in LILE and HFSE mineral oxide assemblages
399   Priderite and related minerals
399   Baddeleyite
400   Zirconolite
401   Freudenbergite
402   Petrogenetic implications
404 WÜSTITE AND PERICLASE
404 UPPER MANTLE OXIDE STRATIGRAPHY AND REDOX STATE
407 REFERENCES

Chapter 11 — Bernard J. Wood
OXYGEN BAROMETRY OF SPINEL PERIDOTITES

417 INTRODUCTION
419 SPINEL PERIDOTITE OXYGEN BAROMETRY
422   Determination of Fe^{3+} contents of spinel
423 OXYGEN BAROMETRY OF CONTINENTAL SPINEL LHERZOLITE XENOLITHS
425 SUBOCEANIC PERIDOTITES
426 MASSIF PERIDOTITES
426   Possible role of carbon in controlling fO_2
428 SUMMARY
429 REFERENCES

Chapter 12 — B. R. Frost & D. H. Lindsley
OCCURRENCE OF IRON-TITANIUM OXIDES IN IGNEOUS ROCKS

433 INTRODUCTION
433 THE COMPOSITION OF Fe-Ti OXIDES IN IGNEOUS ROCKS
433   Fe-Ti spinel in volcanic rocks
434   *Tholeiitic rocks*
434   *Calc-alkalic volcanics*
434   *Alkalic rocks*
435   *Silicic volcanics*

435	Fe-Ti spinel in plutonic rocks
436	*Mafic plutonic rocks*
436	*Alkalic plutonic rocks*
437	*Felsic plutonic rocks*
437	Ilmenite-hematite in volcanic rocks
437	*Tholeiitic volcanics*
437	*Calc-alkalic lavas*
438	*Ilmenite in alkalic lavas*
438	*Ilmenite in silicic lavas*
439	Ilmenite from plutonic rocks
439	*Ilmenite from mafic plutons*
439	*Ilmenite from alkalic plutons*
439	*Ilmenite in felsic plutons*
440	Pseudobrookite in igneous rocks
441	Armalcolite in terrestrial rocks
441	OXIDE-SILICATE REACTIONS IN MAGMATIC SYSTEMS
441	System Fe-Ti-Si-O
442	Topologic relations in orthopyroxene-bearing systems
442	*System Fe-Mg-Si-O*
445	*System Fe-Mg-Ti-Si-O*
447	*System Ca-Fe-Mg-Ti-Si-O*
447	*QUIF-type equilibria with biotite and hornblende*
449	*Petrologic significance*
452	CRYSTALLIZATION CONDITIONS OF COMMON IGNEOUS ROCKS
452	Volcanic rocks
452	*Tholeiitic and alkalic rocks*
452	*Calc-alkalic rocks*
452	Mafic plutonic rocks
453	*Tholeiitic plutons*
455	*Anorthositic complexes*
456	*Mafic plutons of calc-alkalic affinity*
456	Alkalic plutons
458	Pyroxene-bearing felsic rocks
458	DISCUSSION
460	Crystallization trends for tholeiitic rocks
461	Differentiation trends in calc-alkalic magmas
462	ACKNOWLEDGMENTS
462	REFERENCES

Chapter 13   B. Ronald Frost

STABILITY OF OXIDE MINERALS IN METAMORPHIC ROCKS

469	INTRODUCTION
469	NONMAGNETIC OXIDES
469	Corundum
469	"Green spinel"
470	*Impure dolomitic marbles*
470	*Metapelitic rocks*
474	*Metabasites*
474	*Other protoliths*
476	Chrome spinels
476	*Metaperidotite*
477	*Other protoliths*

478	OCCURRENCE AND CHEMISTRY OF Fe-Ti OXIDES
480	Rutile
481	Hematite
481	Ilmenite
482	Magnetite
483	REFERENCES

Chapter 14   B. Ronald Frost

MAGNETIC PETROLOGY: FACTORS THAT CONTROL THE OCCURRENCE OF MAGNETTE IN CRUSTAL ROCKS

489	INTRODUCTION
490	CONTROLS OF MAGNETITE STABILITY DURING COOLING OF PLUTONIC ROCKS
490	Oxide-silicate re-equilibration
491	Oxide-oxide re-equilibration
491	Intraoxide re-equilibration
494	MINERAL EQUILIBRIA EFFECTING MAGNETITE STABILITY DURING METAMORPHISM
494	Silicate-oxide and silicate-oxide-carbonate equilibria
494	*Oxidation-reduction equilibria*
497	*Equilibria involving ferric iron in silicates*
497	Graphite-fluid equilibria
499	Oxide-sulfide equilibria
500	EFFECT OF PROGRADE METAMORPHISM ON VARIOUS PROTOLITHS
500	Metaperidotite and iron-formation
502	Metabasites
504	Graphite-free rocks with Fe-Ti oxides
504	Pelitic schists
506	Quartzofeldspathic gneisses
506	CONCLUSIONS AND SUGGESTIONS FOR FUTURE WORK
506	REFERENCES

Chapter 1 B. Ronald Frost

INTRODUCTION TO OXYGEN FUGACITY AND ITS PETROLOGIC IMPORTANCE

When discussing the stability of iron oxides, one must invariably deal with oxygen fugacity as a variable. To many geophysicists the importance of this variable in determining the stability and composition of oxides is not immediately apparent. Indeed, even within the petrologic community there are many misconceptions about what oxygen fugacity is and how it applies to rocks. The purpose of this chapter is to introduce the concept of oxygen fugacity so that later discussions within this book that use oxygen fugacity as a variable will be more comprehensible.

THE IMPORTANCE OF OXYGEN FUGACITY IN PETROLOGY

Because iron, which is the fourth most abundant element in the Earth's crust, can exist in three oxidation states, some variable must be used in geologic systems to indicate the potential for iron to occur in a more oxidized or more reduced state. We use oxygen fugacity as this variable. At very low oxygen fugacities, such as are found in the Earth's core and in meteorites, iron (Fe^0) is present as a metal. At higher oxygen fugacities and in silica-bearing systems, iron occurs as a divalant cation and is incorporated mostly into silicates. The reaction governing this change can be modelled as following (by convention reactions are written so that the high entropy side, that is the side with oxygen, is on the right):

$$Fe_2SiO_4 = 2\ Fe^0 + SiO_2 + O_2 \qquad \text{(QIF)}$$
$$\text{fayalite} \qquad \text{iron} \qquad \text{quartz}$$

At still higher oxygen fugacities, iron is present both in the ferrous (divalent) and the ferric (trivalent) states and is mostly incorporated into magnetite. This change is described by the reaction:

$$2\ Fe_3O_4 + 3\ SiO_2 = 3\ Fe_2SiO_4 + O_2 \qquad \text{(FMQ)}$$
$$\text{magnetite} \qquad \text{quartz} \qquad \text{fayalite}$$

At very high oxygen fugacities iron occurs in the ferric state in hematite. The formation of hematite from magnetite occurs according to the reaction:

$$6\ Fe_2O_3 = 4\ Fe_3O_4 + O_2 \qquad \text{(MH)}$$
$$\text{hematite} \qquad \text{magnetite}$$

For the system $Fe-O-SiO_2$ the reaction FMQ and QIF mark the upper and lower oxygen fugacity limit for fayalite, whereas MH is the upper oxygen fugacity limit for the stability of magnetite (Fig. 1). As a first approximation, therefore, oxygen fugacity is a

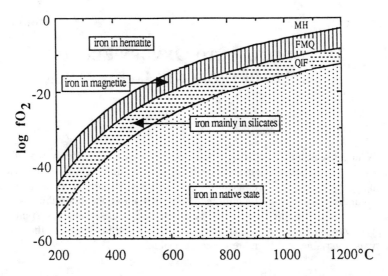

Figure 1: Log oxygen fugacity-T diagram showing the relative stabilities of the various oxidation states of iron in the system Fe-Si-O.

variable that tells us whether iron is likely to be found in its native state, as a divalent ion in a silicate or as a divalent or trivalent ion in an oxide. In natural systems, however, the presence of magnesium (the fifth most abundant element in the crust) and titanium (the tenth most abundant element in the crust) also play an important role in determining the relative stabilities of silicates, magnetite, and ilmenite. Magnesium and ferrous iron substitute for each other in a variety of silicates. The solution of Mg into iron silicates stabilizes them to higher oxygen fugacity, with the result that Fe^{2+}-bearing silicates can be stable even in the presence of hematite (i.e. at relatively high oxygen fugacity) if they have incorporated enough Mg. In addition, titanium and ferrous iron will substitute for ferric iron in both magnetite and hematite, stabilizing these oxides relative to silicates. As a result in an assemblage containing both silicates and oxides, the Fe/Mg ratio of the silicates (a very important petrologic variable), the titanium contents and the ferrous/ferric ratios of the oxides, and oxygen fugacities are all interrelated. In many rocks it is appropriate to say that oxygen fugacity is a function of the Fe/Mg ratio of the silicates and the Ti content of the oxides, for it is more likely that oxygen fugacity is a variable that is governed by the mineral assemblages in the rock, rather than one that is imposed from the environment. The question as to whether oxygen fugacity is a dependent or independent variable will surface many times throughout the remainder of this book.

From the above discussion it is evident that the occurrence and composition of the Fe-Ti oxides and the composition of coexisting silicates is a function of oxygen fugacity (or more properly a monitor of oxygen fugacity). In addition, oxygen fugacity plays an important role in determining the composition of any fluid phase that is associated with igneous and metamorphic rocks. Although this is not a topic we will consider in depth here, it is important to realize that oxygen fugacity plays an important role in determining not only the species present in such fluids, but it also strongly affects the stability of graphite and sulfides. Although in most rocks these minerals are usually present in minor concentrations, they may have significant petrologic or geophysical effects (Frost et al., 1989).

CALIBRATION OF OXYGEN FUGACITY

The concept of oxygen fugacity was introduced to petrology by Hans Eugster who, in 1956, developed a method to control the oxidation potential in experimental runs (Eugster, 1957). He was attempting to synthesize the iron biotite, annite, and found that traditional hydrothermal runs always produced the oxidized equivalent magnetite + sanidine. He discovered that he could eliminate this problem by surrounding the capsule containing the starting materials with a second charge that contained an assemblage which could fix the oxygen potential at a low enough value to stabilize annite. Such oxygen-controlling equilibria, for example the reactions IQF, FMQ, MH listed above, are called oxygen buffers. (See Table 1 for a list of the commonly used buffers).

In an oxygen-controlled experiment a platinum (or a Ag-Pd alloy) capsule containing the experimental starting material (for example FeO, K_2O, Al_2O_3, and SiO_2 in the proportion to produce annite along with excess H_2O) is placed into a gold capsule containing the buffering assemblage and a small quantity of H_2O. When the charge is subjected to experimental T and P the buffer assemblage equilibrates with the H_2O, producing a small amount of H_2 due to the dissociation of H_2O. Since platinum is permeable to hydrogen, H_2 produced by this reaction diffuses through the walls of the inner capsule and combines with available oxygen in the fluid of the inner capsule producing H_2O and reducing the oxygen fugacity in the central charge. After a short time the oxygen fugacity of the inner capsule becomes reduced to that of the oxygen buffer.

Shortly after his invention of oxygen buffers Eugster wrote a paper showing how variations in oxygen potential apply to petrology (Eugster, 1959). In his 1957 and 1959 papers, Eugster referred to the variable controlling oxidation potential as the partial pressure of oxygen (PO_2). This was soon replaced with oxygen fugacity (Eugster and Wones, 1962) because fugacity is a more rigorous thermodynamic term. The reason for this change is subtle but important. Free oxygen in petrologic environments is never present in great quantities. For example the oxygen fugacity of the FMQ buffer at 500°C is 10^{-22} bars. This is the pressure that would be applied by 1 molecule of oxygen in a cubic meter of space. In a rock with a fluid phase the partial pressure of oxygen can be defined, even if it is vanishingly low. In many rocks there may have been no fluid phase at the time of formation (Lamb and Valley, 1984). In such a condensed system the partial pressure of oxygen has no physical meaning. In contrast, oxygen fugacity can be used to describe a condensed system because fugacity monitors the chemical potential.

For example, to obtain oxygen fugacity for the FMQ buffer, one takes the equilibrium constant for the FMQ reaction:

$$K_{FMQ} = \frac{(a_{fay})^3}{(a_{SiO_2})^3 * (a_{mt})} * (a_{O_2}) \qquad (1)$$

Where a_i = activity of component i. For pure substances activity is unity. Thus, for a system with pure quartz, magnetite, and fayalite, reaction (1) reduces to:

$$K_{FMQ} = a_{O_2} = \frac{f_{O_2}}{f^o_{O_2}} \qquad (2)$$

Where fO_2 = fugacity of oxygen at a given T and P and $f°O_2$ = fugacity of oxygen in the standard state. If we adopt a standard state of 1 bar and temperature of interest, the standard state fugacity will be unity and the activity of oxygen will equal its fugacity at the P and T of interest. Equation (2) then becomes:

$$K_{FMQ} = f_{O_2} \qquad (3)$$

Since:

$$\log K = \frac{-\Delta G°_{FMQ}}{2.303 * R * T} \qquad (4)$$

Then:

$$\log f_{O_2} = \frac{-\Delta G°_{FMQ}}{2.303 * R * T} \qquad (5)$$

Where $\Delta G°_{FMQ}$ is the free energy change for the FMQ reaction in its standard state and R is the gas constant.

In a natural system the solid phases of the FMQ reaction do not usually have the pure end-member composition. To determine the oxygen fugacity at which a natural assemblage of olivine, magnetite, and quartz formed one must adjust the oxygen fugacity of the FMQ buffer using equation (1). To use equation (1) we must know how the compositions of the phases (which can be measured by microprobe) relate to activity. This is a reason why the thermodynamics of solutions (for example see the chapters by Ghiorso and Sack and Wood et al.) is so important to petrology.

AN OVERVIEW OF OXYGEN FUGACITY SPACE

From the above discussion it is evident that oxygen fugacity is "measured" by the free energy change between the "oxidized" and the "reduced" portions of an assemblage in a rock or in the buffer capsule (see Table 1). The location of the common buffers in fO_2 space is shown on Figure 2. An important feature to note on Figure 2 is that the oxygen fugacity of all buffers increases with increasing T. Over the range of geologic temperatures, for example, the oxygen fugacity defined by the FMQ buffer changes by 60 orders of magnitude. This change in oxygen fugacity is merely a reflection of the fact that with increasing T most reactions undergo devolatilization. For an oxygen buffer, devolatilization means release of oxygen and hence an associated increase in oxygen fugacity. Another important point to note is that all oxygen buffers have approximately the same slope in $\log fO_2$ - T space. This is due to the fact that the enthalpy change associated with oxidation is approximately the same for all these reactions.

As we will discuss in detail later, most igneous rocks have equilibrated within a few log units of oxygen fugacity of the FMQ buffer (Haggerty, 1976). A few mafic rocks seem to have equilibrated at oxygen fugacities nearly two log units below FMQ, and there are rare mafic rocks with native iron that probably crystallized more than four log units below FMQ (Ulff-Møller, 1986). Most felsic rocks crystallized at oxygen fugacities one

TABLE 1

SOME COMMON PETROLOGIC BUFFERS

Equilibria

$Fe_2SiO_4 = 2 Fe + SiO_2 + O_2$	(QIF)
$2 Fe_xO = 2x Fe + O_2$	(IW)
$2x/(4x-3) Fe_3O_4 = 6/(4x-3) Fe_xO + O_2$	(WM)
$0.5 Fe_3O_4 = 1.5 Fe + O_2$	(IM)
$2 CoO = 2 Co + O_2$	(CoCoO)
$2 Fe_3O_4 + 3 SiO_2 = 3 Fe_2SiO_4 + O_2$	(FMQ)
$2 NiO = 2 Ni + O_2$	(NiNiO)
$6 Fe_2O_3 = 4 Fe_3O_4 + O_2$	(MH)

Equilibrium Expressions*

buffer	A	B	C	Temp. range (°C)	Ref
αQIF	-29,435.7	7.391	.044	150 - 573	1
βQIF	-29,520.8	7.492	.050	573 - 1200	1
IW	-27,489	6.702	.055	565 - 1200	1,2
WM	-32,807	13.012	.083	565 - 1200	1,2
IM	-28,690.6	8.13	.056	300 - 565	3
CoCoO	-24,332.6	7.295	.052	600 - 1200	4
FMαQ	-26,455.3	10.344	.092	400 - 573	1
FMβQ	-25,096.3	8.735	.110	573 - 1200	1
NiNiO	-24,930	9.36	.046	600 - 1200	5
MH	-25,497.5	14.330	.019	300 - 573	1
MH	-26,452.6	15.455	.019	573 - 682	1
MH	-25,700.6	14.558	.019	682 - 1100	1

* where $\log fO_2 = A/T + B + C(P-1)/T$ (T in Kelvins)
Note: temperature of the alpha - beta quartz transition is approximated by the expression: T(°C) = 573 + 0.025P(bars).
Caution: The buffers may be strongly curved at low pressures and it is advised not to extrapolate these expressions below the limits indicated.
Reference: 1: linear fit to Hass (personal comm.); 2. Huebner (1971); 3. $\log K_{IM} = 1/4(\log K_{FMQ} + 3 \log K_{QIF})$; 4. Holmes, et al., (1986); 5. Huebner and Sato (1970)

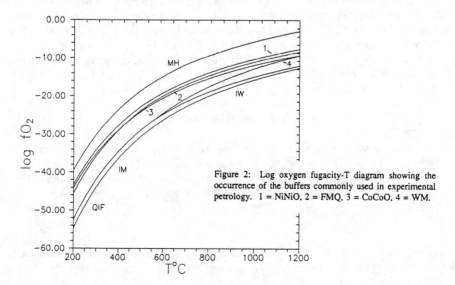

Figure 2: Log oxygen fugacity-T diagram showing the occurrence of the buffers commonly used in experimental petrology. 1 = NiNiO, 2 = FMQ, 3 = CoCoO, 4 = WM.

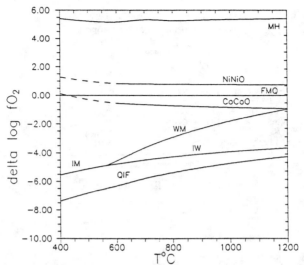

Figure 3: Δ log fO$_2$ - T diagram showing the common petrologic buffers as normalized to the oxygen fugacity of the FMQ buffer. Portions of the NiNiO and CoCoO buffers that lie outside of the range of experimental calibration are dashed.

to two log units above FMQ, although a few silicic rocks formed more than three log units above FMQ (Haggerty, 1976). Metamorphic rocks seem to have a wider range in oxygen fugacity than do igneous rocks, partially as a result of the wider range of bulk compositions found in metamorphic rocks (Frost, 1988). Oxygen fugacities of metamorphic rocks range from those of the magnetite-hematite buffer (5-6 log units above FMQ) to those of the iron-magnetite buffer (5-6 log units below FMQ). The relative range of oxygen fugacities found in rocks, therefore is around 10 log units.

Because the oxygen fugacity for a given buffer changes dramatically with temperature yet the range of oxygen fugacity between the different buffers at any given temperature is relatively small, it is useful to normalize oxygen fugacity to that of a given buffer. This allows one to display changes in oxygen fugacity on a larger scale, and removes the strong temperature-dependence from the buffer curves. The most common normalizing buffer is FMQ (Ohmoto and Kerrick, 1977; Frost, 1985; Frost et al., 1988; Lindsley et al., 1990), although WM (Sato and Valenza, 1980), NiNiO (Mathez et al., 1989), and HM (Anovitz et al., 1985) have been used. The relation between the common oxygen buffers when normalized to oxygen fugacity of the FMQ buffer (this variable is known as Δlog fO$_2$) is shown in Figure 3.

SOME COMMON MISCONCEPTIONS

There are some common misconceptions about oxygen fugacity and its petrologic implications that deserve comment. Briefly stated these are: (1) that buffers actually exist in nature; (2) that oxygen can be treated as a major species in the fluid phase; and (3) that the Fe^{2+}/Fe^{3+} ratio of rocks can be used as a rigorous monitor of oxygen fugacity.

<u>Buffers</u>

As we have noted above, the concept of an oxygen buffer was introduced to provide a way to control oxygen fugacity in experimental charges. Although the various buffers are an important means by which we can orient ourselves in oxygen fugacity space, the buffers listed on Table 1 themselves rarely exist in nature. Thus when one says "Rock A equilibrated on the NiNiO buffer" one really means that "Rock A equilibrated at oxygen fugacities equivalent to those of the NiNiO buffer." This seems to be a trivial point, but it is important to understand the difference between the behavior of oxygen fugacity in an experimental charge and its behavior in geologic systems.

In an experimental charge, oxygen fugacity is imposed upon the system by the isothermally, isobarically invariant reaction in the surrounding buffer. In such conditions, equilibration involves changes in the composition of the solid phases to accomodate the imposed oxygen fugacity. In natural systems oxygen fugacity is usually internally controlled by multivariant equilibria involving the silicates, the oxides and, in igneous rocks, the melt. Although it is still valid in these systems to say that the mineral compositions buffer the oxygen fugacity, the type of buffering is markedly different from that found in experiments. Even under isobaric, isothermal conditions natural equilibria are usually multivariant (the only truely isobarically isothermally invariant assemblage I know of is the association of pure hematite and pure magnetite often found in iron-formations). Thus, in such systems, oxygen fugacity is not fixed, but rather it varies to accommodate the dictates of the mineral compositions. Such variation is, of course, not random, but rather follows T-log fO_2 trends that are sub-parallel to those of the buffers (see Carmichael, 1967; also the chapter by Frost and Lindsley). When one finds that a suite of rocks defines a T-fO_2 trend that lies on that of the NiNiO buffer, for instance, it is important to note that it is entirely a coincidence that has nothing to do with the NiNiO buffer (certainly there wasn't a huge chunk of Ni and NiO adjacent to the rock that imposed the oxygen fugacity upon it).

Oxygen as a fluid species

Because the oxygen fugacity of an experimental charge is controlled by a surrounding buffer petrologists naturally think of oxygen fugacity as a parameter that is imposed upon a rock by the environment. The imposition of oxygen fugacity can readily be induced in experiments because the volume of the charge is usually small relative to that of the buffer, the amount of fluid in the system is relatively high, and the distance hydrogen has to diffuse is very small. However, such conditions rarely hold in natural systems. In most rocks, therefore, oxygen fugacity is more likely to be a function of the composition of the primary melt (if the rock is igneous) and of the mineral reactions that occurred during the formation of the rock than to be a parameter that was imposed from the outside.

Because oxygen is likely to be an internally controlled parameter and because free oxygen is so rare in most geologic environments, it is important to make a distinction as to how one writes natural oxidation-reduction reactions. For example, if one found a texture in which quartz and magnetite formed rims around fayalite one may be tempted to model that reaction using the FMQ reaction given above. That reaction may accurately represent the energy balance of the system, but because free oxygen is essentially absent in petrologic environments, it would not correctly describe the actual physical process by which the texture formed. A reaction that correctly models the process must also contain a source for oxygen. It may be the dissociation of H_2O, CO_2, or SO_2, the reaction of N_2 to NH_3 in aqueous fluids, the precipitation of graphite, the reduction of ferric iron, or anything else the author may dream up, but it cannot be simply free oxygen in the fluid. A discussion of the kind of oxygen-conserving reactions that can produce magnetite in metamorphic reactions is given in the chapter on magnetite petrology.

Ferrous/ferric ratio

Although the ferrous/ferric ratio has long been used as a monitor of oxygen fugacity in rocks, this ratio is not a simple function of oxygen fugacity. It it is true that this ratio monitors oxygen fugacity in melts or glasses, and it may be used as in indicator of oxygen fugacity in volcanic rocks that can be safely assumed to be free of cumulate minerals. In most rocks, however, oxygen fugacity is monitored by mineral equilibria and these equilibria are dependent more on mineral composition, rather than on simply on ferrous/ferric ratio. This can be understood merely by pointing out that oxygen fugacity is an intensive variable and intensive variables are independent of the relative proportion of phases in a rock. In contrast, the ferrous/ferric ratio is strongly dependent on the relative proportion of phases in a rock. Two rocks with the assemblage fayalite-magnetite-quartz may have crystallized at the same oxygen fugacity but can have vastly different ferric/ferrous ratios depending on the relative proportion of fayalite to magnetite. Another way to illustrate this argument consider two examples of iron-formation. Rock (1) has 49% quartz, 50% orthopyroxene (X_{Fe} = 0.20), and 1% hematite. It clearly formed at oxygen fugacities above those of the MH buffer. Rock (2) has 49% quartz, 50% magnetite, and 1% fayalite (X_{Fe} = 1.0). It formed on the FMQ buffer. The ferrous/ferric ratio of rock (1) is 6.3 whereas that of rock (2) is 0.47. Although rock (1) formed at oxygen fugacities at least 4 log units above those of rock (2), it contains far more ferrous iron; the ferrous iron is stabilized in the silicate by Mg. This is admittedly an extreme example, but it clearly shows that, although ferrous/ferric ratios have been used for a long time to indicate the relative "oxidation state" of rocks, it is often difficult to equate such ratios to the oxygen fugacity at which the rocks formed.

REFERENCES

Anovitz, L.M., Treiman, A.H., Essene, E.J., Hemingway, B.S., Westrum, E.F., Wall, V.J., Burriel, R., and Bohlen, S.R. (1985) The heat-capacity of ilmenite and phase equilibria in the system Fe-Ti-O. Geochim. Cosmochim. Acta 49, 2027-2040.

Carmichael, I.S.E. (1967) The iron-titanium oxides of salic volcanic rocks and their associated ferromagnesian silicates. Contrib. Mineral. Petrol. 14, 36-64.

Eugster, H.P. (1957) Heterogeneous reactions involving oxidation and reduction at high pressures and temperatures. J. Chem. Phys. 26, 1760.

Eugster, H.P. (1959) Oxidation and reduction in metamorphism. In Ableson, P. H., ed. Researches In Geochemistry: New York, John Wiley & Sons, 397-426.

Eugster, H.P. and Wones, D.R. (1963) Stability relations of the ferruginous biotite, annite. J. Petrol. 3, 82-125.

Frost, B.R. (1985) On the stability of sulfides, oxides, and native metals in serpentinite. J. Petrol. 26, 31-63.

Frost, B.R. (1988) A review of graphite-sulfide-oxide-silicate equilibria in metamorphic rocks. Rendiconti Societa Italiana Mineral. Petrol. 43, 25-40.

Frost, B.R., Fyfe, W.S., Tazaki, K., and Chan, T. (1989) Grain-boundary graphite in rocks from the Laramie Anorthosite Complex: Implications for lower crustal conductivity. Nature. 340, 134-136.

Frost, B.R., Lindsley, D.H., and Andersen, D.J. (1988) Fe-Ti oxide - silicate equilibria: Assemblages with fayalitic olivine. Am. Mineral. 73, 727-740.

Haggerty, S.E. (1976) Opaque mineral oxides in terrestrial igneous rocks. Rev. Mineral. 3, Hg101-Hg300.
Holmes, R.D., O'Neill, H.St.C., and Arculus, R.J. (1986) Standard Gibbs free energy of formation for Cu_2O, NiO, CoO, and Fe_xO: High resolution electrochemical measurements using zirconia solid electrolytes from 900-1400K. Geochim. Cosmochim. Acta 50, 2439-2452.
Huebner, J.S. (1971) Buffering techniques for hydrostatic systems at elevated pressures. In Ulmer, G. C. Research Techniques for High Pressures and High Temperatures. Springer-Verlag, New York, 123-177.
Huebner, J.S. and Sato, M. (1970) The oxygen fugacity-temperature relationships of manganese oxide and nickel oxide buffers Am. Mineral. 55, 934-952.
Lamb, W.M. and Valley, J.W. (1985) Metamorphism of reduced granulites in low-CO_2 vapour-free environment. Nature 312, 56-58.
Mathez, E.A., Dietrich, V.J., Holloway, J.R., and Boudreau, A.E. (1989) Carbon distribution in the Stillwater Complex and evolution of vapor during crystallization of the Stillwater and Bushveld magmas. J. Petrol. 30, 153-173.
Ohmoto, H. and Kerrick, D.M. (1977) Devolatilization equilibria in graphitic schists. Am. J. Sci. 277, 1013-1044.
Sato, M. and Valenza, M. (1980) Oxygen fugacities of the layered series of the Skaergaard intrusion, East Greenland. Am. J. Sci. 280-A, 134-158.
Ulff-Møller, F. (1985) Solidification history of the Kidlit Lens: Immiscible metal and sulfide liquids from a basaltic dyke on Disko, Central West Greenland. J. Petrol. 26, 64-91.

Chapter 2 Glenn A. Waychunas

CRYSTAL CHEMISTRY OF OXIDES AND OXYHYDROXIDES

INTRODUCTION

Fleischer (1987) lists over 200 distinct oxide-oxyhydroxide minerals. Of these, perhaps twenty are common enough in crystalline rocks to be of great interest to petrologists. A second group, limited largely to pegmatites and selected geochemical provinces, are important for their concentration of rare earths and metals. A third group, largely hydrous, is extremely important in soil and surface enviromental chemistry and geochemistry. The remaining rarer phases will probably never be observed by the nonspecialist, and are relatively unimportant from the geochemical standpoint. However, even these uncommon minerals may have interesting and varied chemistry, and can give insight into aspects of crystal chemistry not afforded by the more common phases.

In the present chapter the common phases, and especially the Fe-Ti phases treated in detail in later chapters, are emphasized. Representative examples of the other important oxide-oxyhydroxide groups are also included. The coverage is admittedly biased by the interests of the author, but the intention is to provide a survey wide enough to be of general utility. The emphasis is also on material published in the period since Volume 3 of this series, entitled *OXIDE MINERALS*.

<u>What is crystal chemistry?</u>

During the writing of this chapter, the author frequently asked himself the question: What comprises the subject area that we call crystal chemistry? Inspection of many texts whose title included the words "crystal chemistry" revealed few attempts at a lucid definition (e.g., Evans, 1966). Earlier texts were mainly concerned with a description of the possible chemistries and structures of particular natural phases. Later this was enlarged to consider where specific elements were located in mineral structures, how much of which element could be located in which site, possible ordering schemes, the existence of other defects, solid solubility, and so on.

But, even lacking a formal definition, there is clearly an operational definition that can be extracted from these works. Specifically, crystal chemistry is the combination of the rules of crystallography and solid state physics and chemistry to allow interpretations, rationalizations and predictions of atomic arrangements in crystals. Inherent in this process is the continuing question: Why is one particular arrangement or structure stable over other possibilities? Although, ultimately, the answer to this question must come from calculation of the total energy of the arrangement or phase in question and all possible alternative structures[1], geochemists and mineralogists have been using comparative arguments based on particular structural details (bond angles, ion or atom radii, valence sums, etc.) to successfully answer this question (at least in more "ionic" structures) for decades. More recently, calculations of the energy of small clusters of atoms which are characteristic parts of a larger mineral structure have bridged the traditional and the ultimate methodologies (Gibbs et al., 1981; Burnham, 1990).

---

[1] The following comment from Pearson (1973) is probably still accurate today: "No really certain answer can be given to the question: 'Will two components form an intermediate phase, and if so, what will its structure be?' We are still concerned with *a posteriori* explanations, and by the time that these give understanding, all possible phases are likely to have been found, so that our understanding cannot be tested by prediction."

From the opposite direction, modern methodologies (Burdett and McLarnan, 1984) have also been used to explain (if not justify) such traditional crystal chemical rules as those of Pauling (1929) and Zacharison (1954; 1963; Baur, 1970), and the validity of ion radii comparisons (Burdett et al., 1981). Hence modern crystal chemistry is achieving practical unification of old and new rules in the characterization of crystal architecture.

Besides structure, a second major aspect of crystal chemistry, noted at least as early as Stillwell (1938), is the influence of the atomic and electronic structure of crystals on their physical properties and chemical reactivity. This area, like that of structural analysis, has also progressed with the evolution of schemes for calculating and modeling band structures, vibrational spectra, heat capacity, occupancy and composition of molecular orbitals, and other structure-dependent properties. Much of the present volume is concerned with magnetic properties, which can vary dramatically even with subtle structural changes.

The third aspect of crystal chemistry, and one that is evolving and expanding perhaps more quickly than the other areas, is its encompassing of non-crystalline or non-equilibrium details of mineral structures. These include: stacking faults, clustered defects, surface rearrangements or reconstructions, adjustments in surface chemistry, grain boundary structure, epitaxial geometries and twinning relationships. Such areas are increasingly important in modern geochemical research because of their significance in determining chemical reactivity and physical properties. In some cases, especially in terms of reactivity, these noncrystalline details may have more significance that the bulk crystal structure.

In passing it should also be noted that there is an increasing incidence of "crystal chemical rules" being applied to amorphous materials, e.g., coordination number deduction based on radius ratios, valence sums, relative electronegativity, and "site" size. In these cases the strength of crystallographic constraints is lost, but chemical and physical characteristics seem to be maintained. The success of this approach may, in fact, be a confirmation of the validity of modern quantum mechanical small-cluster calculations in predicting and rationalizing structural motifs.

This chapter continues in the tradition of descriptive and rationalized crystal chemistry, and, where possible, considers the problems of relative stability and prediction of atom arrangements with references to energy (and related) calculation schemes. Significant aspects of crystal chemistry that cannot be treated in any detail due to space limitations, or specificity, are referenced.

Solid solubility

Solid solutions tend to be thought of as being formed from isostructural end-member compositions (Laves, 1980). Hence a traditional condition for solid solubility is that such end members with appropriate composition exist. A second condition for solubility is reasonably similar unit cell parameters, and thus atomic sizes. However, it is not necessary that end member compositions actually exist as stable phases in all cases. In particular, for certain types of defect solid solutions, one of the endmembers may be structurally unstable. Also, there are many examples of considerable miscibility, where one of the end members exists only in a markedly different structure than the solid solution. A third important consideration in predicting miscibility is bonding character. For example, isostructural sulfides and oxides do not ordinarily form solid solutions at moderate pressures and temperatures. Rationalizations for miscibility in the solid state have been detailed by many authors including Goldschmidt (1937), Wells (1975), Evans (1966), Henderson (1982), and Jaffe (1988) and Kitaigorodsky (1984).

Oxides form solid solutions mainly on the basis of homovalent and heterovalent cation substitution, and by interstitial (normally unoccupied site) occupation. In homovalent substitution the solubility is affected by ionic radius differences between the substituting species, specific electronic peculiarities of a particular ion (such as the tendency to produce a Jahn-Teller distortion), and the strength of the M-O bond for a particular M cation. Simulations of homovalent substitutions have shown how a given structure responds to ion size variations (Dollase, 1980). As expected, most of the structural relaxation is in the first anion neighbors, with smaller but significant adjustments in the next nearest neighbor cations. The degree of structural relaxation, or compliance, is related to the coordination number of the anions. Hence some structures can accomodate large ionic radii differences, while other structures are relatively intolerant. An interesting result of such analysis is that site compliance may also be a function of composition in a solid solution. This would produce nonlinearity in the unit cell volume-composition function, i.e., non-Vegard behavior, and could be an explanation for the "volume of mixing" in solutions.

For ions like Cu^{2+} ($3d^9$) or Mn^{3+} ($3d^4$) that have strong Jahn-Teller (J-T) distortions in octahedral crystal fields, a structure with regular octahedral sites may not be able to tolerate significant substitution, even replacing a similarly sized ion. [As a consequence, these J-T ions also tend to form distorted structures when present as major constituents.] Hence ions that prefer unusual site geometries are very selectively dissolved in many phases. Finally, ions like Si^{4+} also may not show much oxide solubility (in substitution for Ge^{4+}, Al^{3+} or Fe^{3+}) because of the tendency to form more stable silicates.

In heterovalent substitution there is always some type of charge-coupled substitution (CCS). The CCS may involve cation sites of the same crystallographic type, or several differing types, as in Tschermak substitutions. Cation and anion vacancies and other types of point defects are also possible charge-balancing entities, allowing many complex substitution possibilities. Solutions involving interstitial sites can be viewed as a kind of heterovalent substitution, e.g., the solution of Fe_3O_4 in FeO where the ideally unoccupied tetrahedral sites in FeO are occupied by Fe^{3+} ions. In oxide-oxyhydroxides, CCS can involve protonation-deprotonation reactions. For example, Ge^{4+} dissolves in goethite by substituting for a Fe^{3+} ion and one of the protons attached to a coordinating oxygen: $Ge^{4+} = Fe^{3+} + H^+$ (Bernstein and Waychunas, 1987). This sort of CCS is in part responsible for the flexibility of substitutions in the hydrous Fe and Mn oxyhydroxides.

One of the most obvious crystal-chemical questions concerns the localization of charge-coupled substitutions. Are the substituting ions located as close as they can be in a structure? Or, are the substituting ions distributed more randomly, such that the local charge imbalance in the structure is compensated on a statistical basis? Spectroscopic and many structure refinement studies seem generally to indicate the former situation, but there is also clear evidence for the latter case. These questions are intimately tied to the nature of cation ordering processes. For example, at low temperatures there is reduced solubility of hematite in ilmenite, and the structure is that of the long range ordered phase with alternating (Fe^{2+}, Fe^{3+}) and (Ti^{4+}, Fe^{3+}) planes along [0001]. For small hematite component we can raise the temperature until the solution disorders, so that alternating planes have identical chemistry on average. Even in this state there is still short range order and probable localization of Fe^{2+}-Ti^{4+} ion pairs within planes or close to one another on adjacent planes. As temperature increases further this short range order must decrease, until ultimately, charge balancing is statistical over the whole structure, or the phase decomposes into other more stable alternates. But, at this point, how does the structure "know" not to dissolve too much Ti^{4+}, given the available Fe^{2+}, if the charge balancing is non-localized?

Order-disorder

Second order phase transitions can be classified as: order-disorder, displacive, and a combination of both (Franzen, 1986). The identifying characteristic of this type of transition is its continuous nature. The structure can change gradually from one having a given symmetry, to another with a distinctly different space group. The symmetry changes can be analyzed with Landau theory (Landau and Lifshitz, 1959; Haas, 1965), which can be used to test the order of the transition. For example, structures related by a second order phase transition must have space groups that are in the relation of a group and a subgroup.

The order-disorder transitions in question are all Long Range Order (LRO) phase transitions, and reflect the increasing tendency of particular crystallographic sites in a structure to have certain atomic occupants as the degree of LRO increases. There are many mineralogical examples, e.g., hematite-ilmenite, calcite-dolomite and rutile-trirutile. LRO can often be quenched easily because many atoms must change positions to effect significant LRO variation. Hence the degree of LRO can be used, in principle, as a tracer of geologic history.

Minerals may also have Short Range Order (SRO), variations in which do not change the structure symmetry or lead to phase transitions. SRO refers to the tendency to have particular arrangements of atoms among particular sites (identical or crystallographically distinct) without any LRO. CCS, assuming close cation association, is a form of SRO. Clustering or ordering of cations at temperatures above a miscibility gap, are other examples. SRO states are relatively difficult to quench. SRO increases with decreasing temperature and is altered by relatively few diffusive jumps.

LRO and SRO often occur in the same solid solutions, with SRO at high temperatures being an indication of low temperature LRO. Many oxide systems display LRO transitions. Studies of SRO are difficult and SRO has not been well characterized in many mineral systems, let alone oxides and oxyhydroxides.

Element site partitioning

An extensive part of the crystal chemical literature has addressed the topic of element partitioning, including mineral intersite, mineral-mineral, mineral-melt and mineral-aqueous solution partitioning. In this chapter only intersite partitioning within specific phases or solid solutions is considered. In general, the rationales applied to cation substitutions in solid solutions can also be applied to intersite partitioning. For example, the larger ions will tend to favor the larger or more distorted sites, the more highly charged ions will favor more highly coordinated sites, electronic properties of ions will create preferences for octahedral or other geometry sites, and so forth. If specific energies for site substitution can be calculated the partitioning can be determined by Boltzmann statistics.

Electronic properties peculiar to oxides

The oxide and oxyhydroxide minerals are distinguished by their usually high transition metal content, and by the electronic interactions of these metal ions. Coupling of the magnetic moments of individual metal ions, either directly or via intervening oxygens, leads to strong bulk magnetism not observed in silicate phases. This strong coupling, along with strong charge-transfer absorption bands, leads to intense coloration (Rossman, 1975; Sherman, 1985a). Indeed, most of the colors observed on the surface of the earth are due to the more common oxide and oxyhydroxide minerals.

Transition metal-containing oxides and oxyhydroxides usually have large band gaps (Sherman, 1985b), but the presence of many impurities can produce electronic levels within the gaps, leading to semiconducting properties. In several Fe and Mn oxides (notably magnetite) there is a large amount of electron hopping (or small polarons) which produces near-metallic conductivity. Some oxide systems also exhibit relatively high ionic conductivity, particularly the hollandites and related phases. In contrast, most silicates are either insulating or very poor semiconductors. Sulfides can have high conductivity, but less magnetism due to the greater likelihood of paired electronic spins.

DEFECTS AND NONSTOICHIOMETRY

The most common defects in mineral structures are point defects. These are equilibrium defects whose energy of formation is compensated by entropic considerations, and consist of: cation or anion vacancies, interstitial cations or anions, and simple impurity atoms. Clustered point defects are possible and widespread (see e.g., wüstite, below), but become unstable with respect to the nucleation and growth of a new phase as the cluster size increases. Unstable defects, which persist in structures after formation due to sluggish kinetics, include: dislocations, stacking faults, grain boundaries and crystal surfaces. Finally there are extended defects, which generally represent stacking variations in a structure that have nearly the same energy as the regular arrangements. Such defect structure can lead to nearly continuous variation in overall composition.

Dislocations, whether caused by growth anomalies or mechanical-induced deformation, can diffuse out of a structure if given sufficient time. Because of high strain energy near the dislocation core, this region is a sink for impurities, and dislocations in natural phases can be "decorated" by solute atoms which either decrease local strain or are simply trapped in the vicinity of the defect. Screw dislocations are important considerations in crystal growth and crystal surface reactivity.

Stacking faults are mistakes in the packing order of a crystal. For example, in the ABC cubic close packing (see below) the sequence ABCABCABCBABC might occur. Locally the ABCBA sequence looks like a twin, and twinning may result in some cases by this sort of mistake in the sequence of crystal growth. Stacking faults can provide nuclei for the growth of new phases, for phase transformations, and for the beginnings of melting.

Grain boundaries and crystal surfaces are not generally thought of as defects, yet they are thermodynamically unstable entities. In an attempt to lower the bulk energy of the crystal, crystal faces of high energy will tend to be reduced in area during growth. Similarly, the shape of grain boundaries is modified to produce the lowest energy contact region with adjacent grains. Grain boundaries and surfaces may be reconstructed or even have chemistries different from the bulk crystal, all in an attempt to reduce the overall energy due to the defect.

Intermediate between grain boundaries and stacking faults are antiphase domain boundaries (APB). These are planar defects separating parts of a crystal with identical bulk composition, but with broken translational symmetry. APBs consist of some combination of stacking faults and dislocations.

Oxide minerals are often nonstoichiometric. This means that the ratio of elements in the basic structure of the mineral is indefinite, and can vary over a finite range. Nonstoichiometry is associated with defects in the oxide structure, and such phases can be looked at as defect solid solutions. In many cases the defect component has no existence except within the framework of the host phase, e.g., in wüstite and uraninite. In other cases the defect component is a separate stable phase, e.g., γ-Fe_2O_3 (maghemite) component in magnetite.

Most nonstoichiometry in oxides is associated with multi-valent transition or actinide metal ions. Higher valences of the metal ion may allow fewer cations to charge balance a given set of anions. Thus wüstite, $Fe_{1-x}O$, with Fe^{3+} replacing Fe^{2+}, is metal-deficient. On the other hand, uraninite, UO_{2+x}, with U^{5+} and U^{6+} replacing U^{4+}, has all metal sites filled, and requires excess (interstitial) oxygens for charge balancing. In highly defective oxides with large deviations from stoichiometry there can be ambiguity in the proper basis for a description of their crystal structures (Greenwood, 1968; Catlow and Mackrodt, 1986).

Another cause of nonstoichiometry is variation in chemical composition at grain surfaces or grain boundaries. Within the vicinity of the surface there can be ion rearrangements and high vacancy or interstitial content that stabilizes the interface. This has been observed in spinel ceramics (Chiang and Peng, 1986). Nonstoichiometry reaches massive proportions in certain families of oxides where crystallographic adjustments, such as crystallographic shear and planar intergrowths, can occur. Much of this occurs in oxide systems (Tilley, 1987; Wadsley, 1963).

BASIC STRUCTURAL TOPOLOGIES

A large fraction of oxide and oxyhydroxide structures can be described through the use of a relative few structural motifs. Those described below will be referenced throughout the rest of the chapter.

HCP and CCP O and OH arrangements

A close packed layer of oxygen atoms forms a hexagonal pattern with six atoms around any given atom. Between the atoms are interstices, twice as many as atoms in the layer. A second layer will pack to the highest density on top of the first if the atoms are aligned with one set of half of the interstices of the first layer. This can be done by shifts of the second layer in any of the six possible directions shown in Figure 1. By convention, these two layers are now called A and B, and further layers can continue the pattern of positioning to form an ABABAB structure called Hexagonal Close Packing or HCP. The coordination of any given atom is 12, three below it, three above it, and six within its layer.

We can elect also to place another kind of layer on top of the original AB pairing. This layer can have a displacement relative to the A layer in some direction 60° away from the B layer displacement. The new atoms thus have a position over the A layer interstices not taken up by the B layer. We call this third layer C and the motif created by repeated layering is -ABCABC-; it is called Cubic Close Packing or CCP. In the CCP structure each atom is also 12-coordinated, but the arrangement differs inasmuch as, considering atoms in the B layer, the "top" layer atoms are rotated with respect to the "bottom" layer. This has the consequence of changing the site symmetry, and also yields an alternative view, that of an atom in the center of a cube with atoms at each of the edge-center positions (Fig. 2e). In this perspective one sees that there are eight tetrahedral and four octahedral interstices within one such cube.

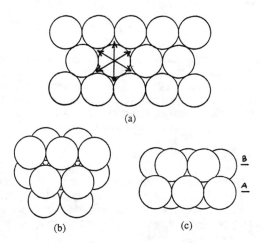

Figure 1. (a) Close packed layer of anions. The arrows indicate the directions and displacement necessary for an overlayer to snuggle down in the interstices. (b) A piece of such a double layer created by a displacement of the second layer toward the top of the page. (c) Side view of the bilayer with the planes designated A and B. After Clark (1972).

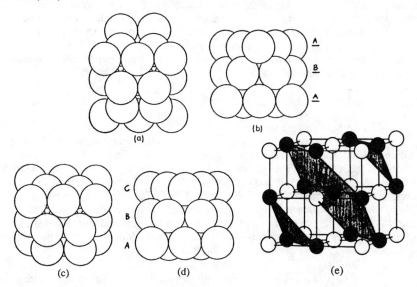

Figure 2. HCP layering created by addition of a third layer with identical positioning to first, creating an ABA sequence. (a) Top view. (b) Side view. CCP layering created by displacement of a third layer into an inequivalent position from the other two, producing the sequence ABC. (c) Top view. (d) Side view. (e) View of CCP packing (solid circles) located on a NaCl-structure unit cell. Note that each layer has a different orientation. The open circles show the octahedral interstices in CCP.

NaCl structure

The NaCl structure is derived from the CCP arrangement by filling up all of the octahedral interstices with cations. Each cation and anion has six nearest neighbors. Viewed as cation-centered octahedra, all such units share edges. All tetrahedral interstices are empty in stoichiometric structures (Fig. 3).

Perovskite structure

The perovskite structure is derived from the CCP oxygen arrangement by removing one fourth of the oxygen atoms, e.g., the central one in the cube in Figure 2e, filling in the missing oxygen with a large A cation, and filling up one quarter of the octahedral interstices (the ones on the cube corners) with B cations. This yields the stoichiometry ABO_3, with A in 12-coordination (Fig. 4). In this structure the cation-centered octahedra share only apices, but share faces with the A cation-centered dodecahedron.

Fluorite and rutile structures

The fluorite structure is derived from the CCP oxygen arrangement by inverting the atom identities and placing A cations in each of the oxygen CCP positions. Oxygens are then placed into all of the tetrahedral interstices, producing the stoichiometry AO_2. The A cation is 8-coordinated by oxygen, oxygens are 4-coordinated by A cations (Fig. 5).

The rutile structure is based on the HCP packing with considerable distortion which lowers the symmetry to tetragonal. Ti cations fill one half of the octahedral interstices in an alternating pattern (Fig. 6). The tetragonal symmetry is not a consequence of this cation arrangement, which would result in an orthorhombic structure, but rather appears due to an undulation of the oxygen layers to produce a more energetically stable structure (Sahl, 1965). In the ideal rutile structure, without the undulation, the rows of empty octahedral interstices along <001> do not provide much space for diffusing ions. In the real rutile structure, however, these interstices have opened into square cross-section "tunnels" which allow considerable cation diffusion.

Spinel structure

The spinel structure is derived from the CCP oxygen packing by cation occupation of one quarter of the tetrahedral and one half of the octahedral interstices. The occupation is ordered such that the overall cubic symmetry is retained (Fig. 7) if the unit cell is doubled in size. If the tetrahedral sites are called A, and the octahedral sites B, the stoichiometry is AB_2O_4. All spinels contain two differing cations, or at least two different valences of the same cation, in the ratio of 2:1. Spinels are classified as normal or inverse, depending on where the more abundant of these cations resides. If it occupies the octahedral sites, the spinel is *normal*. If it is split evenly between the octahedral and tetrahedral sites, the spinel is *inverse*. Many spinels have mixed character by these definitions, and solid solutions can have changing character over compositional space.

If the CCP oxygens in the ideal spinel structure are assumed to have a radius of 1.28 Å, the octahedral and tetrahedral cation sites will have radii of 0.53 and 0.29 Å, respectively. No spinel structures with cations having these radii are known, but the physical structure can distort to accommodate a large range of cation sizes. The distortion consists of an expansion of the anion packing to produce greater O-O distances than found in CCP arrays, and a displacement of anions common to both octahedral and tetrahedral

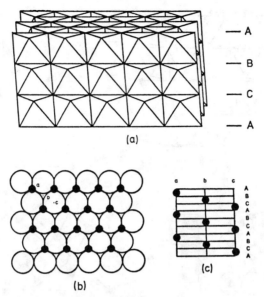

Figure 3. Packing scheme for NaCl structures. (a) The edge sharing layers of anion octahedra produced from the CCP arrangement, each with a central Na atom. (b) The arrangement of metal cations on a given layer. All octahedral sites are filled. (c) Cation occupation plan in terms of interstitial anion positions (a,b,c) and stacking layers (-ABCABC-).

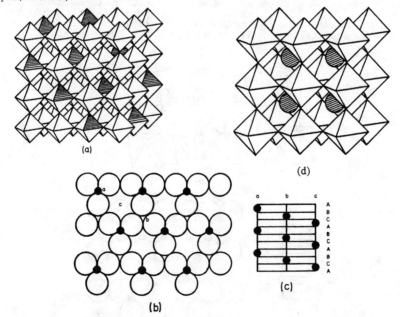

Figure 4. Perovskite structure and its close cousin, the ReO$_3$ structure. (a) Layers of anion octahedra sharing only vertices. This is produced by the CCP anion layer scheme in (b), where one quarter of the anions are missing, and the remaining anion octahedral sites are filled with cations. This is the ReO$_3$ structure. (c) Cation distribution scheme. Note that the anion layers are stacked on [111] relative to the perovskite cubic structure. After Clark (1972). (d) Ideal perovskite structure. Large cations fill the large interstices between the connected octahedra.

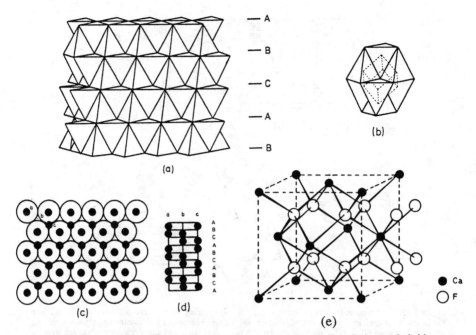

Figure 5. Nontraditional and traditional views of the Fluorite structure. (a) Face sharing tetrahedral layers consisting of FCa_4 tetrahedra. Layering acheme of the cations is not CCP. (b) Cubic coordination (outline) within a unit of CCP Ca^{2+} cations. (c) CP layer of Ca^{2+} anions with superimposed (dark circles) fluorite anions. (d) Anion occupation scheme. After Clark (1972). (e) Traditional view of the fluorite structure. Compare Ca arrangement with white circles in structure of Figure 2 bottom. After Greenwood (1968).

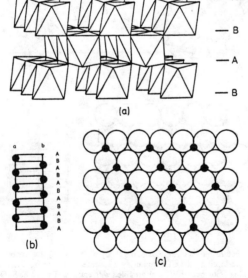

Figure 6. Rutile structure. (a) Array of chains of edge-sharing octahedra (into the page), running normal to the HCP stacking direction. (b) Cation occupation scheme. (c) One anion layer, showing positions of rows of octahedral Ti^{4+} ions.

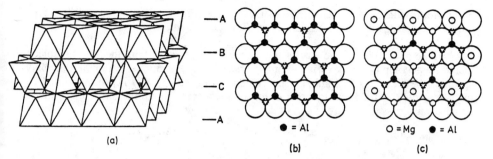

Figure 7. Spinel structure. (a) Alternating layers of octahedral and octahedral-tetrahedral polyhedra based on the CCP anion arrangement. (b) CP layer of anions with octahedral cations superimposed (Al for MgAl$_2$O$_4$). (c) CP layer of anions with octahedral (filled circles) and tetrahedral (open circles) superimposed. The dashed circles indicate the cations in the adjacent cation layer. After Clark (1972).

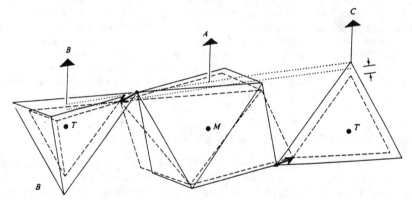

Figure 8. Effect of displacement of anions on void spaces in spinel. Solid line represents M/T (octahedral/tetrahedral) cation radius ratio of 1.84. Displacement of anions (increasing u) shared between polyhedra (arrows) results in no overall structure symmetry change, but changes void size and acceptable M/T ratio dramatically (dashed lines) to about 0.92. After Zoltai and Stout (1984).

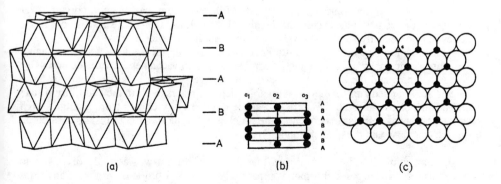

Figure 9. Corundum-hematite structure. (a) Layering of HCP anions and filled octahedra to produce edge and face-sharing polyhedra. (b) Cation distribution scheme on stacked anion layers. After Clark (1972). (c) Arrangement of cations on a given anion layer.

polyhedra to change the relative size of these voids (Fig. 8). The anion displacements introduce a corrugation into the anion layers (Zoltai and Stout, 1984).

Due to the high symmetry of the spinel space group, $Fd\bar{3}m$, the anion displacements are accomplished by variation of a single structural positional parameter, u, which specifies the location of all anions (see Table L-3 in Lindsley, 1976). In the ideal CCP oxygen spinel u is 0.25, but is larger in most known spinels. An increase in u produces larger tetrahedral sites at the expense of decreased octahedral sites.

Rhombohedral oxide structures

The trigonal (rhombohedral) oxides have ideal structures based on the HCP oxygen packing scheme. The trivalent cations occupy 2/3 of the octahedral interstices to produce the A_2O_3 stoichiometry. The arrangement of the occupied cation sites is in the form of connected six-fold rings (honeycomb) within each cation layer (parallel to (0001)), but shifted in adjacent upper and lower neighboring layers (Fig. 9). The repeat layer distance comprises six anion and cation layers. This cation occupation scheme results in each occupied octahedron sharing one octahedral face and three edges with adjacent occupied octahedra. The shared face introduces distortion in the real oxide structures, as the cations in face sharing octahedra are displaced away from one another along the normal to their (0001) cation planes. This distortion is traditionally attributed to cation-cation repulsion (Lindsley, 1976). There is also expansion of the unoccupied octahedra toward the occupied octahedral sites. The latter displacement results in "triplets" of "touching" oxygens at the shared face. In hematite, for example, the shared face O-O distance is 2.669 Å, while the O-O distance for unshared octahedral faces is 3.035 Å.

As a result of their common basis in the HCP oxygen array, the idealized structures of rutile and trigonal A_2O_3 have similar oxygen planes parallel to (010) and (0001), respectively. These can lead to exsolution or epitaxial growth between these structures (Armbruster, 1981).

IRON OXIDES

Wüstite, magnesiowüstite and other solid solutions

Wüstite is a rare mineral on the surface of the earth, but may be an important lower mantle or core constituent (Hazen and Jeanloz, 1984). It has the NaCl structure but is not believed to be exactly stoichiometric under any conditions (McCammon and Liu, 1984). The nonstoichiometry is due to the presence of Fe^{3+} species on octahedral sites replacing Fe^{2+}, and on normally unoccupied tetrahedral sites. Cation vacancies provide the necessary charge balance, and there have been years of controversy about how these various point defects are arranged and condensed into larger defect clusters, both in the equilibrium state and in quenched metastable material. Part of the difficulty comes from wüstite's high concentration of defects, inasmuch as the number and type of defect clusters formed on quench can vary with the quench conditions.

For some time it was believed that the wüstite field was actually divided into subfields of differing stoichiometry separated by second-order or higher transitions (Gavarri et al., 1976). In fact, quenches from these regions do result in differing defect structures (Andersson and Sletnes, 1977), but no unchallenged evidence seems to support the equilibrium subfield idea (Giddings and Gordon, 1974; Hayakawa et al., 1977). A recent summary of defect cluster geometries in wüstite is given by Gartstein et al. (1986).

Wüstite is unstable below 570°C, but rapidly quenched material will disproportionate to form a near stoichiometric and a less stoichiometric metastable wüstite upon suitable anneal (Hentschel, 1970). The near stoichiometric wüstite has a lattice parameter of 4.332 Å, and yields a singlet Mössbauer spectrum, consistent with a very low defect population and near perfect local cubic symmetry.

The most important solid solution is with periclase, MgO, which also has the NaCl structure. Magnesiowüstite also has a large defect concentration, with point and clustered defects similar to wüstite (Waychunas, 1983). EXAFS studies have explored the possibility of SRO in quenched low-defect magnesiowüstite solid solutions, but at synthesis temperatures above 1080°C the series appears to have randomly mixed Fe^{2+}-Mg cations (Waychunas et al., 1986). Other solid solutions are with bunsenite (NiO), manganosite (MnO) and, probably rather limited, with monteponite (CdO), all having the NaCl structure. The cell parameters for these oxides and representative wüstite and magnesiowüstite compositions appear in Table 1.

Hematite, hydrohematite and protohematite

Hematite has the trigonal oxide structure noted above. The space group is $R\bar{3}c$, (no. 167). Oxygen anions are located in special positions along the twofold axes at $\pm[u,0,1/4; 0,u,1/4; -u,-u,1/4]$ with $u = 0.3059$. The Fe^{3+} ions reside on threefold axes at $\pm[0,0,w; 0,0,w+1/2]$ with $w = 0.3553$ (Blake et al., 1966) [note that the cell also includes ions on positions related to these by R translations: (2/3,1/3,1/3) and (1/3,2/3,2/3)]. Hematite forms partial to complete solid solutions with other $R\bar{3}c$ oxides corundum (Al_2O_3), eskolaite (Cr_2O_3) and karelianite (V_2O_3). The solid solution with corundum is interesting inasmuch as it is limited at moderate temperatures according to the phase relations of Turnock and Eugster (1962), with solubility of only a few mol % Al_2O_3 in the hematite-rich end. However, aluminous hematites with up to about 15 mol % Al_2O_3 occur in soil and clay environments (Fysh and Clark, 1982a). These hematites are probably formed from the dehydration of highly aluminous goethite, and are thus metastable. Table 1 includes cell parameters for the $R\bar{3}c$ oxides. Solutions with ordered $R\bar{3}$ oxides are considered below with ilmenite.

Hydrohematite is a defect solid solution where OH^- ions replace oxygens, and charge balancing is achieved by octahedral Fe^{3+} ion vacancies. This yields the formula: $Fe_{2-x/3}(OH)_xO_{3-x}$ (Wolska and Szajda, 1985), where x can be as large as about 0.5 in alumina-free hydrohematite, and up to 0.66 in alumina-containing hydrohematite. Formation is by the low-temperature dehydrogenation of goethite. Very little change from hematite cell dimensions is noted, as the cell size is mainly determined by the close packed anion layers which remain complete. However, the increase in cation vacancies produces appropriate intensity variations in powder XRD structure factors, and the structural hydroxyl has been well substantiated. There are no observations suggesting ordering of the cation vacancies or hydroxyls, but a complete refinement of the structure has apparently not been done. Wolska and Schwertmann (1989) suggest that hydro-hematites with small x values may be more stable than anhydrous hematites.

Protohematite, as its name implies, is an early-formed hematite-like product of goethite dehydrogenation. It is characterized by a defective structure with large quantities of structural water, and broadened x-ray diffraction patterns (Wolska and Schwertmann, 1989), and has chemistry like the hydrohematite solid solution. It is probably only a transitional phase, but appears to be metastable at low temperatures and in the compositional

Table 1. NaCl and Trigonal A_2O_3 oxides

NaCl structure		$Fm\overline{3}m$ O^5_h #225		Cations in 4a; anions in 4b.
			ao	reference
Wüstite	$Fe_{1-x}O$		4.332	Hentschel (1970)
			4.334	McCammon and Liu (1984)
Manganosite	$Mn_{1-x}O$		4.4448	Landolt-Bornstein (LB)
Periclase	MgO		4.2117	LB
Bunsenite	NiO		4.1946	LB
Monteponite	CdO		4.6951	LB
Magnesiowüstite	$(Fe,Mg)O$			Waychunas (1979)
$Fe_{.05}Mg_{.95}$			4.2183	
$Fe^{3+}_{.002}Fe_{.097}Mg_{.90}$			4.2246	
$Fe^{3+}_{.004}Fe_{.196}Mg_{.80}$			4.2369	
$Fe^{3+}_{.008}Fe_{.292}Mg_{.70}$			4.2493	

hematite structure	$R\overline{3}c$ D^6_{3d} #167	Cations in 12c; O in 18e. See text for all atom positions.

		a	c	u	w	reference
Hematite	Fe_2O_3	5.034	13.750	0.3059	0.3553	Blake (1966)
Corundum	Al_2O_3	4.7592	12.9918			Chatterjee et al.(1982)
Ruby	$(Al,Cr)_2O_3$	4.7606	12.994	0.3064	0.3622 (Cr) 0.3520 (Al)	Tsirel'son et al. (1985)
Eskolaite	Cr_2O_3	4.9585	13.5952			Chaterjee et al.(1982)
	Cr_2O_3	4.9576	13.5874			LB
Karelianite	V_2O_3	4.9515	14.003	0.3118	0.34629	LB

range ca. $1.0 \geq x \geq 0.55$. The infrared absorption spectra of protohematite and hydrohematite appear distinctive.

SPINELS

Predictions and determinations of cation site preference in spinels

Site preferences for ions in spinels have been a subject of considerable research. Such preferences provide a good test bed for crystal chemical or bonding models applicable to many other oxide and silicate phases. Early success with relatively simple theories, e.g., calculation of Madelung energies for (A)[B] vs (B)[A] occupations, has not led to satisfactory general application. Somewhat later, the crystal field theory approach (Dunitz and Orgel, 1957; McClure, 1957), though apparently useful in many cases, has been shown to have numerous theoretical and practical difficulties. The most serious of these is that the crystal field stabilization energy (CFSE) is usually only a small fraction of the total lattice energy associated with ion site exchange. Hence, unless this energy is quite small, i.e., in cases of very equivocal site preference, CFSE is not an important determining factor. Thus many of the successes of CFSE calculations are probably fortuitous.

An entirely different approach in determining site preference is that proposed by Price et al. (1982), who use pseudopotential radii (Zunger, 1980) to sort the site preferences. The ability to successfully predict normal vs. inverse character using this method is considerably better than using CFSE comparisons, and the applicability extends to ions without unfilled 3d or 4d states.

Neither the crystal field or pseudopotential radii approach can deal well with the temperature dependence of site distribution, the mixed normal-inverse character, or the change in site distribution with composition in many spinels. Instead, experimentally derived distributions can be used with thermochemical data to develop thermodynamic models which can then be used for predictive purposes (see e.g., Navrotsky, 1986; O'Neill and Navrotsky, 1984; Mason, 1987; Nell et al., 1989).

A particularly interesting study of cation arrangements is that in the magnetite-chromite-franklinite-$ZnCr_2O_4$ system (Marshall and Dollase, 1984). These authors demonstrate that much of the site preferences in this system seem to agree with traditional size and crystal field stabilization energy values, except where electron hopping is statistically likely among Fe^{2+} and Fe^{3+} ions on octahedral sites. In that case the hopping appears to be the most important factor in stabilizing the inverse distribution.

In Table 2 and Figure 10 the cation distributions for several spinels and spinel solid solution series are assembled. Table 2 also lists structural parameters for known spinel end-member compositions and selected natural phases. Most of the data come from the compilation by Hill et al. (1979), with other sources referenced separately. Also included are data for the tetragonal spinel-like structures

Magnetite, ulvöspinel and titanomagnetites

Magnetite has the inverse spinel structure and is one of the most studied of all minerals because of its abundance, magnetism, remarkable conductivity and interesting phase transitions. The electrical conductivity is distinctive, approaching that of a metal, and is dominantly electronic. The high electronic conductivity makes thermoelectric measurements to determine cation distribution feasible in magnetite solid solutions. The conduction results from fast electron transfer between the octahedral Fe^{3+} and Fe^{2+} ions by a hopping mechanism (small polaron model) with a relatively small activation energy (0.065 eV). Averaged over any period of many hops, the individual Fe ions appear to have an intermediate valence of 2.5. As magnetite is cooled the hopping rate slows, until at about 120 K the conductivity decreases abruptly. This is the onset of the Verwey transition, and signals the ordering of the octahedral cations. Below about 115 K the magnetite structure is orthorhombic (or perhaps pseudo-orthorhombic). Much of the interest in this transition arises from its metal—>insulator nature.

Magnetite occurs naturally as a solid solution with many spinel components, but probably the most important is the magnetite-ulvöspinel solution $Ti_xFe_{3-x}O_4$ (known as titanomagnetites), and its oxidation products, the titanomaghemites. Ulvöspinel is an inverse spinel with Ti^{4+} and Fe^{2+} randomly filling the octahedral sites and Fe^{2+} filling the tetrahedral sites. This sort of mixing on the octahedral site is sometimes referred to as "valency disorder" inasmuch as the charge coupled substitution $2L^{3+} = M^{4+} + N^{2+}$ occurs, relative to a hypothetical $Fe^{2+}L^{3+}_2O_4$ normal spinel. If one assumes that cations do not exchange sites as mixing occurs, then Ti^{4+} should replace octahedral Fe^{3+} and tetrahedral Fe^{2+} should replace tetrahedral Fe^{3+} across this join. This model of cation occupation was suggested by Akimoto (1954), but it does not predict saturation moments

Table 2. Spinel and spinelloid minerals

$Fd\bar{3}m$ O^7_h # 227 Tetragonal cations in 8a; Octahedral cations in 16d; O in 32e; Cell origin at center of symmetry $\bar{3}m$ (origin choice 2 International Tables Volume A) B ions in octahedral sites in completely normal spinels; half of B and all A in octahedral sites in completely inverse spinels. Cations in square brackets are on octahedral sites.

name	composition	a_o	u	inversion	reference
	A B$_2$ O$_4$				
Brunogeierite	$Ge^{2+}Fe^{3+}_2O_4$	8.411	0.250	0.00	Hill et al. (1979)
Chromite	$Fe^{2+}Cr^{3+}_2O_4$	8.393		0.00	Hill et al. (1979)
Cochromite	$Co^{2+}Cr^{3+}_2O_4$	8.332		0.00	Hill et al. (1979)
Coulsonite	$Fe^{2+}V^{3+}_2O_4$	8.4530	0.2610	0.00	Hill et al. (1979)
Cuprospinel	$Cu^{2+}Fe^{3+}_2O_4$	8.369	0.255	1.00	Hill et al. (1979)
Franklinite	$Zn^{2+}Fe^{3+}_2O_4$	8.4432	0.2615	0.00	Hill et al. (1979)
Franklinite	$Zn^{2+}_{.61}Mn^{2+}_{.39}[Mn^{3+}_{.06}Fe^{3+}_{1.94}]O_4$				
		8.466		0.00	Shirakashi and Kubo (1979)
Gahnite	$Zn^{2+}Al^{3+}_2O_4$	8.086	0.2636	0.03	Hill et al. (1979)
Galaxite	$Mn^{2+}Al^{3+}_2O_4$	8.2410	0.2650	0.29	Hill et al. (1979)
Hercynite	$Fe^{2+}Al^{3+}_2O_4$	8.1490	0.2650	0.00	Hill et al. (1979)
Hercynite (post 850°C anneal)		8.15579	0.2633	0.163	Hill (1984)
Jacobsite	$Mn^{2+}Fe^{3+}_2O_4$	8.511	0.2615	0.15	Hill et al. (1979)
Magnesiochromite					
	$Mg^{2+}Cr^{3+}_2O_4$	8.333	0.2612	0.00	Hill et al. (1979)
Magnesioferrite					
	$Mg^{2+}Fe^{3+}_2O_4$	8.360	0.2570	0.90	Hill et al. (1979)
Magnetite	$Fe^{2+}Fe^{3+}_2O_4$	8.3958	0.25470	1.00	Wechsler et al. (1984)
Magnetite		8.3940	0.2548	1.00	Hill et al. (1979)
Magnetite	$Fe_{3.005}O_4$	8.3969	0.25468	1.00	Fleet (1984)
Magnetite	$Fe_{2.96}Mg_{0.04}O_4$	8.3975	0.25491	1.00	Fleet (1984)
Manganochromite					
	$Mn^{2+}Cr^{3+}_2O_4$	8.437	0.2641	0.00	Hill et al. (1979)
Nichromite	$Ni^{2+}Cr^{3+}_2O_4$	8.305	0.260	0.00	Hill et al. (1979)
Qandilite	$Ti^{4+}Mg^{2+}_2O_4$	8.4376	0.26049	1.00	Wechsler and Von Dreele (1989)
(high T form synthesized at 973 K)					
Spinel	$Mg^{2+}Al^{3+}_2O_4$	8.0832	0.2624	0.07	Hill et al. (1979)
Spinel		8.0806	0.2623	0.00	Yamanaka and Takeuchi (1983)
Spinel (700°C)		8.0834		0.21	Wood et al. (1986)
Spinel (1050°C)		8.0855		0.38	
Spinel (1300°C)		8.0842		0.37	
Trevorite	$Ni^{2+}Fe^{3+}_2O_4$	8.325	0.2573	1.00	Hill et al. (1979)
Ulvöspinel	$Ti^{4+}Fe^{2+}_2O_4$	8.5348	0.26038	1.00	Wechsler et al. (1989)
Ulvöspinel		8.536	0.265	1.00	Lindsley (1976)
Ulvöspinel			0.261	1.00	Forster and Hall (1965)
Ulvöspinel		8.530	0.265	1.00	Hill et al. (1979)
Vuorelainenite	$Mn^{2+}V^{3+}_2O_4$	8.520	0.2633	0.00	Hill et al. (1979)

Tetragonal or lower symmetry Cubic spinels

SG	name	composition	a	c	reference
P4/nnm	Donathite	$(Fe_{.79}Mg_{.14}Zn_{.08})[Cr_{1.28}Fe_{0.70}Al_{.02}]O_4$			
			8.342	8.305	Seeliger and Mucke (1969)
$I4_1/amd$	Hausmannite	Mn_3O_4	5.7621	9.4696	JCPDS card 24-734
$I4_1/amd$	Hetaerolite	$ZnMn^{3+}{}_2O_4$	5.7204	9.245	JCPDS card 24-1133
$P4_2/nnm$	Iwakiite	$(Mn_{.983}Mg_{.035})[Fe_{1.277}Mn_{.660}Si_{.018}Al_{.014}Ti_{.009}]O_4$			
			8.519	8.540	Matsubara et al. (1979)
P cubic	Maghemite	$Fe^{3+}[Fe^{3+}{}_{1.67}[\ ]_{.33}]O_4$			
			8.3505		Goss (1988)
$P4_1$	Maghemite		8.330	24.990	Van Oosterhout and Rooijmans (1958)
$P4_32_12$	Maghemite		8.3396	24.996	Greaves (1983)
$P4_122$	Qandilite	$Ti^{4+}Mg^{2+}{}_2O_4$	5.9738	8.414	Wechsler and Von Dreele (1989)
	(low T form synthesized at 773 K)				
$F4\bar{3}m$	Spinel	$Mg^{2+}Al^{3+}{}_2O_4$	8.0858		Grimes et al. (1983)
$P4_132$	Titanomaghemite	$(Fe^{3+}{}_{.96}[\ ]_{.04})[Fe^{2+}{}_{.23}Fe^{3+}{}_{.99}Ti^{4+}{}_{.42}[\ ]_{.37}]O_4$			
			8.341		Collyer et al. (1988)
not given	Titanomaghemite	$(Fe_{.77}Ti_{.22}Zn_{.01})[Fe_{1.19}Ti_{.26}Mn_{.02}Al_{.04}[\ ]_{.49}]O_4$			
			8.348		Allan et al. (1989)
not given	Titanomaghemite	$(Fe^{2+}{}_{.247}Fe^{3+}{}_{.653}[\ ]_{.100})[Fe^{3+}{}_{.750}[\ ]_{.425}Ti^{4+}{}_{.825}]O_4$			
			not given		Schmidbauer (1987)

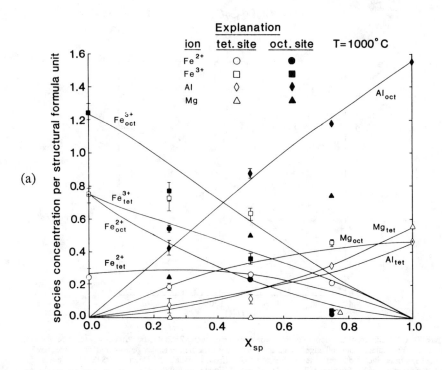

Figure 10. (a) Calculated (curves) compared with observed intersite cation distributions at 1000°C in magnetite-spinel solid solutions (from Nell et al., 1989).

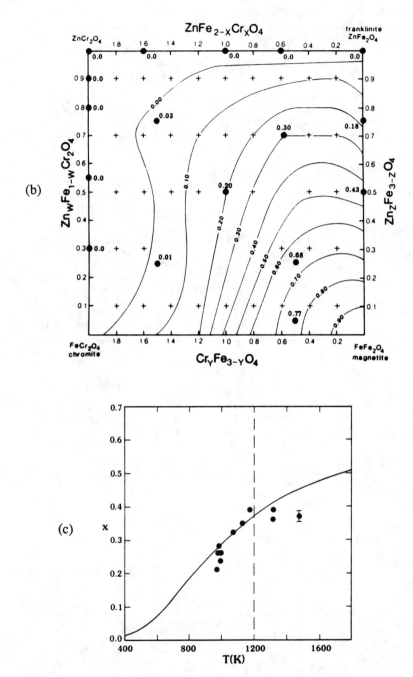

Figure 10. (b) Contours of the amount of inversion over the chromite-ZnCr$_2$O$_4$-franklinite-magnetite compositional field, determined from cell edge values and Mössbauer measurements. Large dots are directly measured unit cells, crosses indicate interpolated cell edge values (from Marshall and Dollase, 1984). (c) Values of inversion parameter (x) as function of T from NMR data of Wood et al. (1986).

accurately. Neel (1955) and Chevallier et al. (1955) suggested a model where Ti^{4+} also replaces octahedral Fe^{3+}, but Fe^{2+} substitutes initially for octahedral Fe^{3+} until all sites contain Fe^{2+}, then additional Fe^{2+} substitutes only for tetrahedral Fe^{3+}. This model also fails to predict saturation moments accurately, but in the opposite sense compared to the Akimoto model.

A more accurate model which fits the available magnetic data quite well was proposed by O'Reilly and Banerjee (1965). In this model the cation substitution is the same as the Neel model from $0 \geq x \leq 0.2$ and $0.8 < x < 1.0$, but has substitution like the Akimoto model in the intermediate region. This "patchwork" model may seem strange since there are no compelling crystal chemical reasons for changing the substitution process at these x values. However, as Lindsley (1976) details, there are numerous experimental observations which indicate physical property changes in the solution at precisely these values.

Recent studies have considered the effects of both temperature of formation and quench rate on the cation distributions. O'Donovan and O'Reilly (1980) examined the effects of annealing at 600° and 1000°C on samples synthesized at 1350° and 1425°C. No effect on cation distribution was observed as measured by cell parameters, saturation magnetization or Curie temperature. Wechsler et al. (1984) utilized both x-ray and neutron diffraction to obtain high precision cell and oxygen position parameters for a suite of samples along the join. These workers found that varied sample synthesis temperature or later lower temperature sample anneal had no effect on the cation distribution, and furthermore, that the distribution was consistent with the Akimoto model. All samples had Ti^{4+} present only on octahedral sites, as expected on general crystal chemical grounds. However, the Akimoto model appears to be inconsistent with the spinel space group $Fd\bar{3}m$, which requires that all octahedral sites be equivalent. The model sets the tetrahedral and octahedral Fe^{3+} occupations exactly equal, and futher, requires that exactly half of the octahedral sites must be occupied by Fe^{2+}. These restrictions are most easily achieved by symmetry-breaking, e.g., lowering the symmetry to produce inequivalent octahedral sites. But Wechsler et al. (1984) suggested that crystal chemical constraints alone may be strong enough to produce the Akimoto distribution. The suggested constraints are: (1) Ti^{4+} has a strong octahedral site preference. (2) The octahedral sublattice may not tolerate more than one ($Ti^{4+} + Fe^{3+}$) per formula unit. and (3) that equal numbers of Fe^{3+} and Fe^{2+} may be favored on the octahedral sites. (1) is certainly substantiated by many observations. (3) appears likely due to the tendency for electron delocalization among identical symmetry sites. This delocalization can be expected to stabilize the structure slightly. (2) is weaker, but might be necessary for localized charge balancing and electrostatic stability.

Trestman-Matts et al. (1983) examined magnetite-ulvöspinel cation distributions via thermoelectric measurements over the temperature range 600°-1300°C. The cation distribution relation which they developed from these data produced a remarkable result. Their higher temperature results agreed with none of the previous models, but their extrapolated results for 300°, 20°, and 0°C agreed strikingly well with the Akimoto, O'Reilly and Banerjee, and Neel-Chevallier models, respectively. This result could only be explained by a strong dependence of the cation distribution on temperature, which was observed directly, and an inability to quench-in the high temperature distribution. Hence, the samples studied by Wechsler et al. (1984) were probably quenched more rapidly than those of earlier workers, thus preserving the highest temperature distribution. The failure of both Wechsler et al. (1984) and O'Donovan and O'Reilly (1980) to detect any temperature effects on the cation distribution is then due to constancy of quench rate.

The results of Trestman-Matts et al. (1983) were, however, consistent with the thermodynamic model of O'Neill and Navrotsky (1984), and pointed out the relative speed of purely electronic redistributions as compared to actual ion exchanges. For example, Rezlescu et al. (1972) had determined that electronic redistribution occurs in $MgFe_2O_4$ spinel within tens of minutes at 200°-300°C. Thus normal synthesis quench rates will only preserve electronic distributions up to perhaps 400°C.

Maghemite, titanomaghemite

Lindsley (1976) noted that the structure and even the existence of maghemite were debated subjects. The existence is no longer in any doubt, but the structure remains problematical, and may have many variants. Much work supports the interpretation that maghemite is a defect spinel with incomplete spinel cation site occupancy. The appropriate number of Fe^{3+} ions for charge balance are statistically distributed on the spinel unit cell in 21-1/3 of the 16 octahedral and 8 tetrahedral (normally filled) sites. The structure can be thought of as derived from magnetite, by replacing the eight Fe^{2+} ions on octahedral sites by the charge equivalent of 5-1/3 Fe^{3+} ions plus 2-2/3 cation vacancies. This picture has been verified by Mössbauer spectra, and by structure refinements which show that all tetrahedral spinel sites are filled, and by the fact that the vacancies occur on the octahedral sublattice. Unfortunately, the exact positioning of the vacancies on the octahedral sites seems to be variable, and this has led to ordering schemes resulting in a tetragonal symmetry superlattice with c = 3a (Boudeulle et al., 1983; Greaves, 1983), or to a primitive cubic cell with the ordered $LiFe_5O_8$ spinel structure (Braun, 1952; Smith, 1979).

Maghemite is metastable with respect to hematite, and historically has been difficult to synthesize under anhydrous conditions. These facts suggest that the structure may require bonded water or H^+ for stabilization. One suggested structure is that of a phase related to $(Fe^{3+})_8[Fe_{12}{}^{3+} \square_4](OH)_4O_{28}$, the hydrous analog of $LiFe_5O_8$, which can be rewritten $(Fe^{3+})_8[Fe_{12}{}^{3+}Li_4{}^+]O_{32}$ (van Oosterhout and Rooijmans, 1958). This structure has a primitive lattice. These workers also prepared maghemites with tetragonal symmetry (space group $P4_1$ or $P4_3$) and a tripling of the spinel cell. However the neutron diffraction study of Greaves (1983), on just such a tetragonal maghemite, gave no indication of any structural H^+, and other types of studies have not confirmed the requirement of structural hydrogen. Lindsley (1976) details other maghemite observations between 1936 and 1971.

More recently, Goss (1988) examined maghemite formed by the oxidation of structurally and magnetically well characterized collodial magnetites. Goss found that initial oxidation at 200°C produced a magnetite-maghemite solution with FCC symmetry. Oxidation at 500°C produced a maghemite with a primitive cell and a lattice parameter of 8.3505 Å. The cation distributions consistent with this cell and the observed saturation magnetization have increased tetrahedral Fe^{3+} (8.4) and octahedral vacancies (3) compared to the "ideal" defect maghemite structure (8.0 and 2-2/3, respectively). Recent structural information on maghemites is collected in Table 2.

Maghemite-magnetite solid solutions appear to be complete, and are formed by varying degrees of oxidation of magnetite. The solution can be considered as a defect magnetite with excess cation vacancies having the stoichiometry: $(Fe^{2+})[Fe^{3+}{}_{1+2/3y} Fe^{2+}{}_{1-y} \square_{y/3}]O_4$. Molar volumes and cell parameters in the $MgAl_2O_4$-Fe_3O_4-Fe_2O_3 series have been measured recently by Mattioli et al. (1987).

Titanomaghemite has been defined by Lindsley (1976) to be any Fe-Ti spinel lying off the Fe_3O_4-Fe_2TiO_4 join, i.e., magnetite-ulvöspinel solutions with additional Fe_2O_3 or

FeTiO$_3$ component, making them cation deficient nonstoichiometric spinels. The origin of these spinels is predominantly by oxidation of precursor titanomagnetites, and much of the analysis of their structure, composition and properties has been in this context. Since Fe/Ti ratio ideally does not change with oxidation degree (Akimoto and Katsura, 1959), this ratio and a suitable oxidation parameter provide independent variables which describe titanomaghemite composition. Lindsley (1976) has reviewed the various compositional parameters and cation distribution schemes in detail, thus the topic is only briefly explained here.

The oxidation process must occur with the addition of oxygen to the crystal boundary, as cations are being preserved and raised in total valence. The creation of a uniform sample then requires ample diffusion of cations, and particularly Fe^{2+}. Models which begin with a cation distribution in the parent spinel and then modify this during oxidation include:

(1) O'Reilly and Banerjee (1967) which assumes these authors' (1965) cation distribution model for the parent spinel, and that only octahedral Fe^{2+} is oxidized initially, and that only these Fe^{2+} ions diffuse. This model has all other ions immobile, including original octahedral Fe^{3+}, and hence only octahedral vacancies can be formed. There are several crystal chemical difficulties with this model, perhaps the most serious being that it would require the creation of a maghemite rim around an oxidized titanomaghemite core since Ti^{4+} cannot diffuse out of the core. Another questionable requirement is the constancy of the Fe^{3+}/Fe^{2+} ratio among tetrahedral sites of parent and oxidized product spinel.

(2) Readman and O'Reilly (1971) developed two revised models with less serious difficulties. Their model 3 allows diffusion of Ti^{4+} into newly created octahedral sites and transport of electrons to establish overall, if not localized, charge balance. Their model 4 further allows diffusion of tetrahedral cations from original positions. These authors' analysis includes an availability parameter, which is a measure of how many tetrahedral Fe^{2+} ions diffuse and are oxidized relative to octahedral Fe^{2+} ions. The reliability of these models can thus in principle be tested by examining octahedral and tetrahedral diffusion rates in well characterized spinels.

Zeller et al. (1967) determined the cation distribution of titanomaghemites by solving a set of eight simultaneous linear equations for the eight distribution variables: three kinds of cations and a vacancy on two kinds of sites. The equations were novel inasmuch as they included expressions for the dependence of the Curie temperature, saturation magnetization and paramagnetic susceptibility on the constituent distributions. The results agreed well with the O'Reilly and Banerjee (1967) model assumptions, affirming that no Ti^{4+} or cation vacancies occur on tetrahedral sites.

Freer and O'Reilly (1980) measured diffusion coefficients in several spinels to test the availability concept noted in the Readman and O'Reilly (1971) model. They found activation energies of 0.27 and 0.71 eV for Fe^{2+} in octahedral and tetrahedral sites, respectively, supporting their earlier model's assumption of preferential octahedral diffusion and oxidation.

More recently, Schmidbauer (1987) analyzed magnetization data and Mössbauer spectra for a titanomaghemite of composition Fe$_{1.650}\square_{0.525}$Ti$_{0.825}$O$_4$. The use of the Readman and O'Reilly model gave only fair agreement with the observed magnetization. Agreement was improved by the assumption of tetrahedral vacancies and spin canting of the tetrahedral Fe moments, the latter due to octahedral vacancies and the presence of nonmagnetic Ti^{4+}. The cation distribution with these assumptions was:

$$Fe^{2+}_{0.247}Fe^{3+}_{0.653}\square_{0.100}(Fe^{3+}_{0.750}\square_{0.425}Ti^{4+}_{0.825})O_4 \; .$$

Collyer et al. (1988) performed an x-ray diffraction structure refinement of a titanomaghemite from Pretoria, South Africa. They determined a primitive lattice with a = 8.341 Å and space group $P4_332$ (or $P4_132$). The combination of Mössbauer and refinement results gave the site distribution as:

$$Fe^{3+}_{0.96}\square_{0.04}(Fe^{2+}_{0.23}Fe^{3+}_{0.99}Ti^{4+}_{0.42}\square_{0.37})O_4 \; .$$

This distribution was consistent with the Chevallier et al. (1955) model of precursor titanomagnetite spinel cation distribution, and, except for the small amout of tetrahedral vacancies, with the model of Readman and O'Reilly (1971) for titano-magnetite oxidation. No confirming magnetization measurements were attempted.

In contrast, Allan et al. (1989) determined the distribution of cations and vacancies in a fully-oxidized titanomaghemite from chemical and Mössbauer analyses. The lattice parameter was 8.348 Å. Their results yielded the distribution:

$$Fe_{0.77}Ti_{0.22}Zn_{0.01}(Fe_{1.19}Ti_{0.26}Mn_{0.02}Al_{0.04}\square_{0.49})O_4 \; .$$

Canting of the spin structure was noted, predominantly on the octahedral sublattice. This implies major tetrahedral Ti^{4+}, in conflict with most other results or crystal chemical arguments.

It is clear from all of these studies that there is still much work to do to sort out the detailed structure and crystal chemistry of titanomaghemites. Certainly the basic precursor models must be reevaluated in the light of the results of Trestman-Matts et al. (1983).

<u>Distorted spinel-like phases</u>

Hausmannite (Mn_3O_4), hetaerolite ($ZnMn_2O_4$) and iwakiite ($Mn(Fe,Mn)_2O_4$) have major Mn^{3+} on octahedral sites, an ion that is prone to produce large Jahn-Teller distortions on such sites. In these structures this distortion lowers the symmetry to tetragonal, and both hausmannite and hetaerolite have space group $I4_1amd$ and a cell with z = 4. Referred to a spinel FCC cell (FCC cubic a = 2 BC tetragonal a), the tetragonal distortion is c/a = 1.162. Iwakiite (Matsubara et al., 1979) appears to be only slightly distorted relative to the spinel structure, perhaps because of its large octahedral Fe^{3+} content, and has space group $P4_2/nnm$ with z = 8. This means that the compositional series $MnMn_2O_4$-$MnFe_2O_4$ has two symmetry changes: $I4_1amd \Rightarrow P4_2/nnm \Rightarrow Fd\bar{3}m$. By application of the Landau rules (Haas, 1965) there must be at least one first order phase transition in this series with a miscibility gap, and another second or possibly first order transition. Tetragonal distortions in synthetic manganate spinelloids have been examined by Buhl (1969).

Wechsler and Von Dreele (1989) found that qandilite (Mg_2TiO_4) is a completely inverse $Fd\bar{3}m$ spinel at high temperatures but undergoes a cubic $\Rightarrow$ tetragonal transition in the vicinity of 933 K. The tetragonal structure (space group $P4_122$) is a very slight distortion of the cubic one, and there is substantial ordering of Mg and Ti between two distinct octahedral sites which form rows of alternating Mg-rich and Ti-rich sites along [100]. The symmetry change requires a first order phase transition (Haas, 1965).

Another curious spinelloid with reported space group P4$_2$/nnm is donathite, ideally FeCr$_2$O$_4$ (Seeliger and Mücke, 1969), which is claimed to be a dimorph of chromite. This structure is reported from powder material, and may possibly be due to inhomogeneity in a cubic material rather than true tetragonality. Magnetic ordering and magnetostriction could conceivably reduce the physical symmetry at very low temperatures.

Several Ni^{2+} spinels are reported to be tetragonal. This may be possible as Ni^{2+} is also a Jahn-Teller ion in tetrahedral sites. Nichromite (NiCr$_2$O$_4$) is also reported with cubic symmetry (Hill et al.,1979), and the cubic->tetragonal transition occurs near 60°C.

An interesting variant from the spinel structure is that of högbomite, a hexagonal Al oxide-hydroxide with considerable Fe, Mg, Ti and Zn (Gatehouse and Grey, 1982). The structure is derived from a close packed anion framework with an eight layer mixed stacking sequence (-ABCABACB-), with ordered cations in 6 tetrahedral and 16 octahedral sites per unit cell. Slabs of the structure similar to the aluminate spinels hercynite, spinel and gahnite, alternate in the [001] direction with layers of nolanite-structure. Polytypism in högbomites has been discussed by McKie (1963) and more generally for spinels by Price (1983).

IRON OXYHYDROXIDES

The FeOOH polymorphs are abundant near-surface phases that have important roles in aqueous geochemistry. Due to usual small grain sizes, they tend to have high reactivity for surface adsorption and complexing, and thus can act as strong scavengers for heavy metals and other aqueous complexes in the environment.

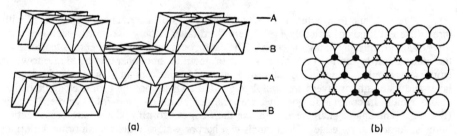

Figure 11. Structure of goethite (α-FeOOH). (a) Anion packing in HCP arrangement with double rows of filled octahedral sites. The octahedra share edges within the chains, but only vertices between chains. (b) Cation distribution on a close packed anion layer. Dashed circles represent cations in next higher or lower layer. After Clark (1972).

Goethite

Goethite (α-FeOOH) is the most abundant of the iron oxyhydroxides and the most stable. The structure is based on HCP packing of oxygen and hydroxide ions with Fe^{+3} ions arranged in the octahedral interstices in a series of double rows separated by vacant double rows (Fig. 11). Each double row with associated anions constitutes a double chain of edge-sharing octahedra, which extend along [001]. The double chains are connected to one another by shared vertices at the chain edges. In this structure, the close packed anion stacking is along [100], and the oxygen and hydroxide ions alternate as shown within the anion planes. There are also channels in the structure along [001] provided by the vacant rows of octahedral sites, and these presumably allow enhanced diffusion in this direction. The surface expression of these channels are "grooves" which serve as excellent attachment sites for adsorbed species (Sposito, 1984). Goethite tends to have an acicular habit, with crystals elongated down [001], reflecting the dioctahedral chain and channel extensions.

Goethite has the same structural topology as several other oxide and oxyhydroxide minerals, and forms limited to extensive solid solutions with most of these. In particular, there is considerable solubility toward diaspore, the direct aluminum analog of goethite. Goethites with aluminum substitution of as much as 33% have been reported (Fysh and Clark, 1982b). It is intriguing to speculate whether charge-coupled substitution of Mn^{4+} or other tetravalent ions into goethite gives rise to local domains of the very similar ramsdellite (γ-MnO_2) structure, or whether tetravalent ions remain separated in such a defect solid solution. Bernstein and Waychunas (1987) found that Ge^{4+} in goethite was not clustered and was probably associated with hydroxyl sites that had been deprotonated. Si^{4+} also appears to dissolve in goethite, but this has not been well characterized. Table 3 lists structural parameters for many goethite-like species.

Progressive dehydrogenation of goethite appears to result in the protohematite phase when the compositional parameter x in $Fe_{2-x/3}(OH)_xO_{3-x}$ falls to ca. 1 or below. Further dehydrogenation leads to the formation of hydrohematite at ca. $x = 0.5$ (Wolska and Schwertmann, 1989).

<u>Akaganeite</u>

Akaganeite (β-FeOOH) has a tunnel structure closely similar to the hollandite minerals, $(Ba,Pb,Na,K)_{2-x}(Fe,Mn)_8[O(OH)]_{16}$ (or ideally, α-MnO_2), but with a variable amount of a large anion (mainly Cl^- or F^-) in the alkali site (Fig. 12). It is probably not a proper oxyhydroxide of iron, as there is evidence that it is not stable, or at least cannot be formed successfully, without the presence of these extra species (Feitknecht et al., 1973; Chambaere and DeGrave, 1984).

The akaganeite structure is based on a defect close packed anion arrangement with three different kinds of anion layers. Every third layer is only two-thirds occupied with rows of anions missing along the c-axis. The cation occupation of octahedral sites between the other anion layers is in double rows, as in goethite, but separated by single rows of empty sites along c. Finally, the octahedral cation sites that remain between the third anion layer and its neighbor layers are completely filled. This topology produces dioctahedral chains which, in contrast to goethite, are arranged about the four-fold symmetry c-axis (Fig. 12). The chains share vertices along their edges, forming the square-cross section tunnels, some 5 Å on edge. The tunnels can harbor water, hydroxyl, fluorine, chlorine, sulfate and nitrate ions, but it is unclear how much of these can be taken up by the structure or how the structure compensates for the presence of additional anionic species. Although the tunnels seem large it must be noted that only a single row of anions is missing. Hence only species with sizes similar to O^{2-} ions can be readily accomodated.

Akaganeite has a remarkable crystal habit, described by Murray (1979). The crystals are needle shaped groupings of 5 x 5 x n unit cells where n refers to replication down the c axis. By some observations, these spindle crystals have empty cores, that is, a 3 x 3 x n cell hole runs down the center of the crystal, producing a square channel about 31.4 Å on a side. Other workers have determined the individual crystals to be solid. The needle-like crystals form a bunch called a somatoid. Crystallographic parameters of akaganeite and related phases appear in Table 3.

<u>Lepidocrocite</u>

The lepidocrocite (γ-FeOOH) structure is based on the CCP packing of anions. Each layer of cations fills double rows of octahedral sites with intervening double rows of

Table 3. Fe and Al oxyhydroxides

space group #62 D^{16}_{2h}

Goethite	FeOOH	Pnma	Szytula et al. (1968).		all atoms in 4c positions.	
	cell dimensions	coordinate	Fe	O_1	O_2	H
a	9.95	x	0.145	-0.199	-0.053	-0.08
b	3.01	y	1/4	1/4	1/4	1/4
c	4.62	z	-0.045	0.288	-0.198	-0.38

Diaspore	AlOOH	Pbnm	(space group #62 but structure refined in different setting than Goethite.) Hyde and Andersson (1989). all atoms in 4c.			
	cell dimensions	coordinate	Al	O_1	O_2	H
a	4.396	x	-0.0451	0.2880	-0.1970	-0.4095
b	9.426	y	0.1446	-0.1989	-0.0532	-0.0876
c	2.844	z	1/4	1/4	1/4	1/4

space group #63 D^{17}_{2h}

Lepidocrocite	FeOOH	Cmcm	Christensen et al. (1982)	Fe, O in 4c.	H in 8f.	
	cell dimensions	coordinate	Fe	O_1	O_2	H
a	3.07	x	0.	0.	0.	0.
b	12.53	y	-0.3137	0.2842	0.0724	0.514
c	3.876	z	1/4	1/4	1/4	0.366

Boehmite	AlOOH	Cmcm	Christensen et al. (1982)	Al,O in 4c.	H in 8f.	
	cell dimensions	coordinate	Al	O_1	O_2	H
a	2.876	x	0.	0.	0.	0.
b	12.24	y	-0.3172	0.2902	0.0820	0.519
c	3.709	z	1/4	1/4	1/4	0.392

Akaganeite	FeOOH		space group C^5_{4h} I4/m #87 Szytula et al. (1970) Fe, O and H in 8h. Cl- or H$_2$O in 2b.				
	cell dimensions	coordinate	Fe	O1	O2	H	Cl
a	10.44	x	0.332	0.168	0.527	0.214	0.
c	3.01	y	0.135	0.204	0.190	0.151	0.
		z	0.	0.	0.	0.	1/2

Feroxyhyte	FeOOH		space group D^3_{3d} P$\bar{3}$m1 #164 Patrat et al. (1983) Two O and one H on 2d. One Fe divided between 1a and 1b.			
	cell dimensions	coordinate	Fe_1	Fe_2	O	H
a	2.95	x	0.	0.	1/3	1/3
c	4.56	y	0.	0.	2/3	2/3
		z	0.	1/2	0.246	0.510

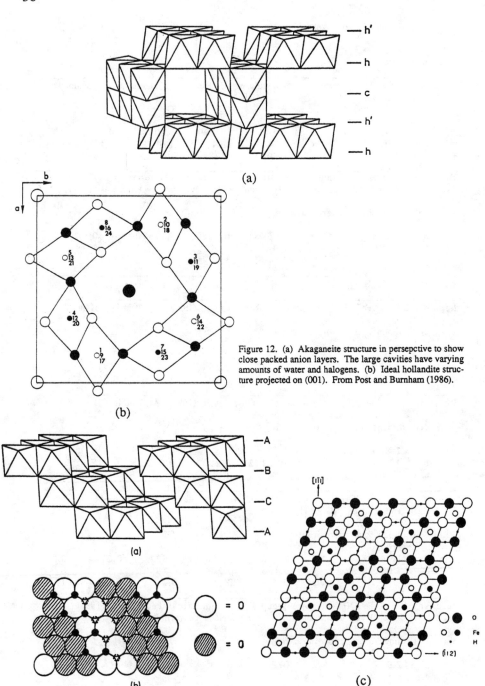

Figure 12. (a) Akaganeite structure in persepctive to show close packed anion layers. The large cavities have varying amounts of water and halogens. (b) Ideal hollandite structure projected on (001). From Post and Burnham (1986).

Figure 13. Lepidocrocite structure. (a) CCP anion packing scheme and layering effect of offset dioctahedral chains. (b) Arrangement of cations on a close packed anion layer. (Both open and shaded oxygens are equivalent) After Clark (1972). (c) Projection of the structure on (110). Shaded and open circles indicate atoms at different elevations. After Misewa et al. (1974).

empty sites, but the cation layers are offset by one row as one proceeds through the packing layers (Fig. 13). This arrangement produces stepped layers of dioctahedral Fe(O,OH)$_2$ chains. The actual structure has orthorhombic symmetry, and the CCP stacking sequence is along [051] (Fasiska, 1967; Murray, 1979; Oles et al., 1970). Well-crystallized maghemite can be made by the direct dehydration of lepidocrocite, a consequence of the similar CCP anion stacking along [111] in maghemite. Unlike the other FeOOH polymorphs, there are no analogous Mn-containing mineral structures.

At this point the reader is reminded that the idealized packing structures represented in this chapter are intended to show the structural topology, rather than correct individual details. For the hydrogen bonded oxyhydroxides the anion layers can be corrugated or otherwise altered by the particular proton positions.

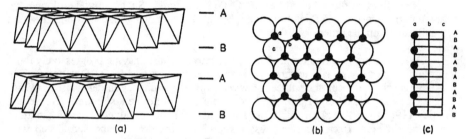

Figure 14. Feroxyhyte idealized structure. (a) HCP close packed anion layers, and sets of edge-sharing octahedra arranged in sheets. (b) Cation arrangement on close packed anion planes. Alternate cation layers are vacant. (c) Cation stacking scheme. After Clark (1972).

Feroxyhyte

The structure of feroxyhyte, (δ'-FeOOH), is in need of clarification. Murray (1979) reported that the structure is thought to be a disordered cation version of synthetic δ-FeOOH, which itself has a CdI$_2$ structure. The CdI$_2$ is a layer structure based on HCP anion packing, with alternating octahedral cation layers completely filled or vacant. This produces sheets of Fe^{3+}(O,OH) octahedra that share six edges with six surrounding such octahedra (Fig. 14). In δ-FeOOH the anions are apparently disordered instead of having regular alternation of OH$^-$ and O^{2-} within layers. In feroxyhyte the cations are also disordered, or only slightly ordered, and this can be judged from electron diffraction patterns (Chukhrov et al., 1976). Patrat et al. (1983) examined the local ordering in synthetic δ-FeOOH by diffuse powder x-ray diffraction methods, finding evidence for a regular periodicity of the Fe^{3+} occupation in the [001] direction. This sequence consisted of two Fe^{3+} occupied sites then two vacancies as one proceeds along [001]. In hematite the observed occupation sequence is two filled sites, one vacant site, etc.

Feroxyhyte is not an important phase from the standpoint of magnetic behavior, but its existence in nature is unusual due to its relative instability. Found in manganese nodules, it is unstable in air and alters to goethite.

Ferrihydrite and protoferrihydrite

Ferrihydrite is the mineral name identifying poorly crystalline to amorphous hydrous iron oxides with chemical composition near 5Fe$_2$O$_3$·9H$_2$O. It is formed abundantly as the initial product of the hydrolysis and precipitation of dissolved iron in natural geologic systems. Like goethite, it is a strong scavenger of metal ions from aqueous solutions. Because of its usual small grain size and poor x-ray diffraction patterns, the structure of ferrihydrite is a subject of some dispute.

Ferrihydrite that is effectively amorphous, yielding only two broad "peaks" in a powder x-ray pattern, has often been called "protoferrihydrite" or "two-line ferrihydrite." Feitknecht et al. (1973) proposed a simple structure for this material consisting of small sheets of edge-sharing $Fe^{3+}(O,OH)$ octahedra without any three dimensional organization, and assumed diffraction due only to (11) and (30) type planes in these platelets. More recently, Manceau and Drits (pers. comm., 1990) have suggested that a better explanation of the structure might be a disordered mixture of CCP and HCP anion domains within three dimensional particles. In a study of adsorbed arsenate on two-line ferrihydrite via EXAFS, Waychunas et al. (1990) reported that highly arsenated ferrihydrite was effectively broken down into individual dioctahedral $Fe^{3+}(O,OH)_2$ chain units. Such material still gave the characteristic two-line XRD pattern, though with broader lines than normal reflecting crystallite sizes on the order of 8-10 Å.

With ageing, two-line ferrihydrite orders into structures having progressively more long range order, giving rise to three, four, five and six line powder XRD patterns. The six-line material has been the subject of several investigations. Van der Giessen (1966) suggested a cubic unit cell of 8.37 Å on edge, but did not define a structure. Towe and Bradley (1967) used very similarly prepared material, but obtained a six-line pattern somewhat different from van der Giessen's, and suggested a rhombo-hedral cell with similarities to the hematite structure. Towe and Bradley found no evidence for structural OH^-, and concluded all H^+ was present as H_2O. However, Russell (1979) showed conclusively by IR measurements with D_2O exchange that about half of the H^+ is present as OH^-.

In detail the Towe and Bradley (1967) structure differs from hematite becaue of the large number of vacant octahedral sites and concomitant ferric ion disorder in ferrihydrite, and the four layer octahedral repeat rather than six in hematite. Thus hematite and six line ferrihydrite have related cell dimensions according to this model. Objections to the Towe and Bradley (1967) structure have been made by Atkinson et al. (1968) and Chukhrov et al. (1973). Recently a somewhat different model has been proposed by Eggleton and Fitzpatrick (1988). Their model consists of alternating bi-layers of edge sharing $Fe^{3+}(O,OH)$ octahedra and corner sharing $Fe^{3+}(O,OH)$ tetrahedra and octahedra. The overall motif is similar to a β-alumina structure. However XANES studies of six-line ferrihydrites by Manceau et al. (1990) conclusively refute the possibility of practically any tetrahedral Fe^{3+} in the ferrihydrite structure.

As with some of the other iron oxyhydroxides, this phase is in need of more exacting characterization work. This will no doubt be forthcoming in view of the significance of ferrihydrite to environmental metal-pollutant cycling.

TITANIUM AND IRON-TITANIUM OXIDES

The three polymorphs of TiO_2 have structures based on the HCP, CCP and mixed HCP arrangements of anions. The rutile structure was described earlier (Fig. 6), and is also observed in cassiterite (SnO_2), pyrolusite (β-MnO_2), and plattnerite (PbO_2). The actual rutile structure departs considerably from a close packed arrangement, but is still the densest of the three polymorphs. Its stability on a molecular orbital basis has been examined by Burdett (1984). Rutile-structure minerals are compared in Table 4.

Table 4. Ti oxides and related phases.

space group D^{14}_{4h} $P4_2mnm$ #136

Rutile	TiO_2	X-ray data		Abrahams and Bernstein (1971)		
	Ti in 2a. O in 4f.					
	cell dimensions		coordinates	Ti	O	Ti-O distances
a	4.593659		x	0.	0.30479	
c	2.958682		y	0.	0.30479	
			z	0.	0.	
	Neutron diffraction data			Gonschorek and Feld (1982)		
a	4.5937		x	0.	0.3048	4@1.9485
c	2.9587		y	0.	0.3048	2@1.982
			z	0.	0.	

Cassiterite SnO_2		Baur (1956)	Sn in 2a. O in 4f.		
	cell dimensions	coordinates	Sn	O	Sn-O distnaces
a	4.737	x	0.	0.307	4@2.052
c	3.185	y	0.	0.307	2@2.056
		z	0.	0.	

Pyrolusite MnO_2			Mn in 2a. O in 4f.		
	cell dimensions	coordinates	Mn	O	Mn-O distances
a	4.398	x	0.	0.302	4@1.880
c	2.873	y	0.	0.302	2@1.898
		z	0.	0.	

Plattnerite PbO_2			Pb in 2a. O in 4f.	Hill (1982)	
	cell dimensions	coordinates	Pb	O	Pb-O distances
a	4.9578	x	0.	0.3067	4@2.169
c	3.3878	y	0.	0.3067	2@2.149
		z	0.	0.	

Tapiolite $FeTa_2O_6$ Von Heidenstam (1968)   2 (Fe,Ta) in 2a. 4 (Fe,Ta) in 4e.
4O in 4f. 8O in 8j. Neutron diffraction data.

	cell dimensions	coordinates	2(Fe,Ta)	4(Fe,Ta)	4O1	4O2
a	4.7515	x	0.	0.	0.3030	0.2986
c	9.254	y	0.	0.	0.3030	0.2986
		z	0.	0.3329	0.	0.3298

Anatase	TiO_2	space group D^{19}_{4h} $I4_1/amd$ #141		Horn et al. (1972)	
	Ti in 4a. O in 8e.				
	cell dimensions	coordinates	Ti	O	Ti-O distances
a	3.7842	x	0.	0.	4@1.9338
c	9.5146	y	0.	0.	2@1.9797
		z	0.	0.2081	

Brookite	TiO_2	space group D^{15}_{2h} Pbca #61		Meagher and Lager (1979)	
	All atoms in 8c.				
	cell dimensions	coordinates	Ti	O	Ti-O distances 1 each @
a	9.174	x	0.1289	0.0095	1.990 2.052
b	5.449	y	0.0972	0.1491	1.863 1.930
c	5.138	z	0.8628	0.1835	1.999 1.923

Rutile

Rutile has limited solid solution with other isostructural minerals at low temperatures, and forms many other solutions by virtue of charge-coupled substitutions. Chief among these are the solutions with: $2 M^{5+} + N^{2+} = 3 Ti^{4+}$, where M can be V, Ta, Nb and Sb, while N is Fe^{2+} or Mn^{2+}; and the solutions with $K^{3+} + M^{5+} = 2 Ti^{4+}$, where K is Fe^{3+}, Mn^{3+} or Al^{3+}. The latter set apparently never results in any ordered rutile phase, even for the end-member cases; but the former substitution scheme at high concentrations can either produce a valency disordered rutile structure or ordered rutile variants known as trirutiles. As the name suggests, these have a tripling of the rutile unit cell in the c direction, with ordered cations in the sequence $(-M^{5+}-M^{5+}-N^{2+}-)_n$ along the edge-sharing octahedral chains. Tapiolite ($FeTa_2O_6$, but often extensively substituted) is the prime example of a trirutile (Clark and Fejer, 1978; Von Heidenstam, 1968). Tapiolite can form solid solutions with valency disordered rutile-structures like $Fe^{3+}TaO_4$ (Schmidbauer and Lebkuchner-Neugebauer, 1987), setting up a series that should show interesting long and short range order transitions as a function of composition and temperature.

Rutile formed at moderate temperatures also may contain a small amount of Fe^{3+}, which is observed to exsolve as hematite (Putnis, 1978). The nature of this defect solution is unclear, but as elevated temperatures are required for significant Fe^{3+} solubility (3% at 1400°C), the presence of charge-compensating stacking faults and cation or anion vacancies is likely. The Fe^{3+} may selectively occupy planes in rutile crystallographically similar to those in hematite. During anneal at lower temperatures these regions would serve as nucleation points for the exsolution development.

Pseudorutile

Pseudorutile is an unusual, yet apparently abundant, mineral formed from the hydrothermal alteration of ilmenite (Grey et al., 1983; Wort and Jones, 1980). Since goethite and rutile have modified HCP anion arrangements it is possible to construct unit-cell scale intergrowths by the creation of antiphase boundaries separating rutile and goethite slabs. This topology is shown in Figure 15, where the structures of tivanite ($VTiO_3OH$) and pseudorutile (approximately $Fe_2Ti_3O_8(OH)_2$) are depicted. The pseudorutile structure may actually be comprised of domains of tivanite-structure twins of various types. The individual microtwin domains are about 40-50 Å in size with coherency of the close packed anion arrangement over 3 or 4 such domains. Variation in the thickness of slabs affect overall stoichiometry in a manner similar to structures having crystallographic shear. South African pseudorutile has the composition

$$Fe^{3+}_{1.81}Fe^{2+}_{0.07}Mn^{2+}_{0.03}Ti_{3.09}O_{8.33}(OH)_{1.33}(H_2O)_{0.07}$$

with an anion/metal ratio of 1.93, which is close to the 2.00 predicted for an ordered goethite-rutile intergrowth. The cell dimensions are a = 2.868 Å, c = 4.607 Å.

Anatase

Anatase has a tetragonal structure based on CCP anions (Fig. 16) (Cromer and Herrington, 1955; Horn et al., 1972). From a topological standpoint it differs from the other TiO_2 polymorphs by having TiO_6 octahedra which share four edges with other such octahedra. In the rutile structure each octahedron shares two such edges, while in brookite three such edges are shared. Since shared edges should lead to cation-cation repulsion and structural destabilization by traditional reasoning, one might expect brookite to be more stable than anatase. However the reverse appears true as brookite is rarer in nature. Post

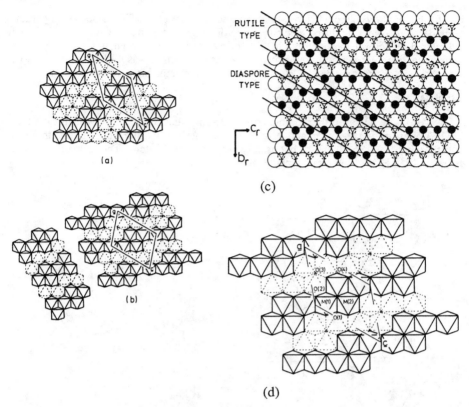

Figure 15. Goethite-rutile intergrowths $(FeOOH)_{2p}(TiO_2)_q$. (a) $p = 2, q = 1$. (b) $p = 1, q = 3$ (both from Grey et al., 1983). [Small unit on lower left is $Fe_2Ti_3O_9$ motif.] (c) Structure of tivanite seen along b showing cation distribution on a close packed anion layer, and slabs of diaspore (goethite) and rutile structure (see Figs. 6 and 11). (d) Polyhedral representation of tivanite structure. Dashed polyhedra are in level below solid polyhedra (from Grey and Nickel, 1981).

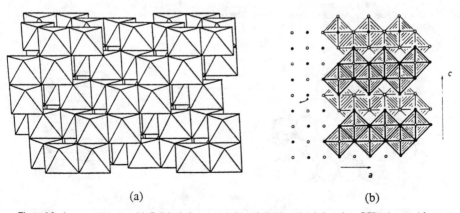

Figure 16. Anatase structure. (a) Polyhedral representation. Anion layering is based on CCP scheme. After Clark (1972). (b) Idealized structure as seen projected on (010). Filled and open circles refer to alternate elevations. After Hyde and Andersson (1989).

and Burnham (1986a) modeled the structures and energies of the TiO_2 oxides with parameters derived from a modified electron gas (MEG) approximation. They found that the calculated structural parameters closely followed those observed, though polyhedra distortions were not perfectly replicated. The rutile structure was 4 kJ/mole more stable than the anatase structure, and 20 kJ/mole more stable than the brookite structure.

Brookite

Brookite has an orthorhombic structure somewhat more complex than the other TiO_2 polymorphs (Meagher and Lager, 1979). As with rutile and anatase there is substitution of Nb, Sb, Sn, Fe^{3+}, but usually only to a small extent. The structure is based on a modified HCP anion arrangement (Fig. 17).

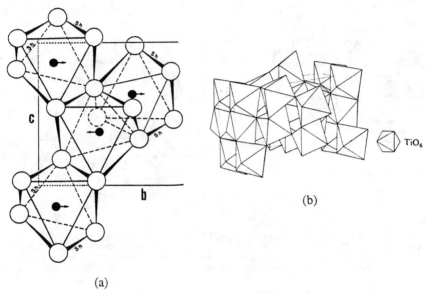

Figure 17. Brookite structure. (a) Part of the structure projected on (100). Sh indicates shared edge. The arrows indicate Ti displacement as temperature is increased. From Meagher and Lager (1979). (b) Polyhedral representation. The c axis is vertical. After Zoltai and Stout (1984).

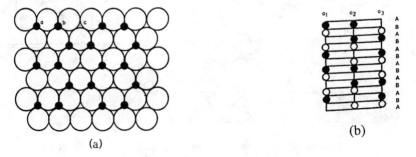

Figure 18. Ilmenite cation distribution. (a) Cation arrangement on close packed anion layer. (b) Stacking sequence of cations. Open circles: titanium, solid circles: Fe^{2+}. After Clark (1972).

Ilmenite, solid solutions and related phases

Ilmenite is an ordered version of hematite where the honeycomb-arranged cation sites within alternating (0001) planes are occupied with Fe^{2+} or Ti^{4+} instead of Fe^{3+} (Fig. 18). This doubles the number of crystallographically inequivalent cation sites and reduces the space group symmetry to $R\bar{3}$, relative to hematite's $R\bar{3}c$. Down the [0001] direction the cation stacking repeat is $(-Fe-Ti-\square-Ti-Fe-\square-)_n$ such that a complete unit cell has six layers. The ordering results in adjustments of the anions, such that they move away from the layers with larger cations and toward the layers with smaller cations. The basic structure is shared by several other important phases, among them pyrophanite ($MnTiO_3$), geikielite ($MgTiO_3$), and ecandrewsite ($ZnTiO_3$). There appears to be considerable solubility between these phases (Gaspar and Wyllie, 1983).

Ilmenite forms solid solutions with disordered $R\bar{3}c$ trigonal oxides, especially hematite, with large miscibility gaps at low temperatures. The solution with hematite is of particular interest since it combines aspects of an order-disorder transition in the cation sublattice, and phase separation in a miscibility gap, with a magnetic order-disorder transition (Nord and Lawson, 1989; Burton, 1984; Burton, 1991). Although at some temperature ilmenite probably disorders to $R\bar{3}c$ symmetry, this form has not been observed upon quench. However, intermediate solution compositions can be quenched in the disordered state, reflecting the strongly asymmetric nature of the phase relations. The difficulty in quenching disordered ilmenite (and other $R\bar{3}$ trigonal oxides, Wechsler and Von Dreele, 1989) probably arises from the presence of considerable short range order among the Ti^{4+} and Fe^{2+} ions, serving as protonuclei for growth of domains of the ordered phase. This SRO may be disrupted by addition of Fe^{3+}, hence changing the quenchability and phase relations.

The magnetic properies of ilmenite-hematite solutions are complex, and have been reviewed by Lindsley (1976), Nord and Lawson (1989), Ishikawa and Akimoto (1957), Ishikawa and Syono (1963), and many other workers. Specifically unusual is the existence of two compositional ranges wherein reverse thermoremanent magnetization (TRM) can be produced. The individual magnetic properties of hematite and ilmenite are considered in Chapters 4 and 8 in this volume. Crystallographic data for the Ilmenite-type minerals appear in Table 5.

Pseudobrookite solid solutions

Pseudobrookite (ideally Fe_2TiO_5) and armalcolite (ideally $Fe_{0.5}Mg_{0.5}Ti_2O_5$), are the accepted mineral phases within the pseudobrookite solid solution (Bowles, 1988). Solubility toward the magnesium rich composition $MgTi_2O_5$ ("karrooite") is observed, but no natural composition with such major Mg has been found. However, solubility toward $FeTi_2O_5$, sometimes called "ferropseudobrookite", is complete. Lunar armalcolite has substantial Ti^{3+}, but all titanium appears to be tetravalent in terrestrial pseudobrookites. Although the pseudobrookite structure has somewhat distorted octahedra, it can be described in terms of an ideal CCP anion arrangement, from which it is based, though this may be misleading. The actual structure (Fig. 19) has two types of occupied metal ion octahedral sites, M1 and M2, in the ratio of 1:2, respectively. The metal octahedra share edges with one another and form ribbons along [010]. Specifically, M1 shares edges with two M2 octahedra, and M2 shares edges with one M1 and one M2 octahedron. In the completely ordered forms of these minerals all of the Ti occupies M2 following the stoichiometry, but significant and possibly complete disorder can be produced at elevated

Table 5. Ilmenite structure minerals.

space group R3 C^2_{3i} #148

Ilmenite	FeTiO3		Wechsler and Prewitt (1984)				
	Fe and Ti in 6c. O in 18f.						
						Metal-oxygen distances	
	cell dimensions		Fe	Ti	O	Fe-O	Ti-O
	a	5.0884	x 0.	0.	0.31743	3@2.0784	3@1.8742
	c	14.0855	y 0.	0.	0.02332	3@2.2005	3@2.0871
			z 0.35537	0.14640	0.24506		

Pyrophanite	MnTiO3				
	Mn and Ti in 6c. O in 18f.				
	cell dimensions				
	a	5.14 (Jaffe, 1988)	5.126	(Posnjak and Barth, 1934)	
	c	14.36	14.33		

Geikielite	MgTiO3		Wechsler and Von Dreele (1989)				
	Mg and Ti in 6c. O in 18f.						
						Metal-oxygen distances	
	cell dimensions		Mg	Ti	O	Mg-O	Ti-O
	a	5.05478	x 0.	0.	0.31591	3@2.047	3@1.867
	c	13.8992	y 0.	0.	0.02146	3@2.168	3@2.090
			z 0.35570	0.14510	0.24635		

Ecandrewsite	ZnTiO3		Bartram and Slepetys (1961)		
	Zn and Ti in 6c. O in 18f.				
	cell dimensions		Zn	Ti	O
	a	5.077	x 0.	0.	0.550
	c	13.92	y 0.	0.	-0.050
			z 0.359	0.150	0.250

Comment: The R factor for the powder refinement was 13.5%, which is not very good. Calculation of the interatomic distances based on these positional parameters yields unreasonable values. However Bartram and Slepetys (1961) give Zn-O and Ti-O average distances of 2.15 and 2.06 Å, respectively.

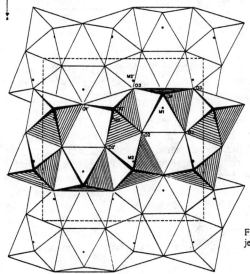

Figure 19. Pseudobrookite (karrooite) structure projected on (001). From Brown and Navrotsky (1989).

Table 6. Pseudobrookite minerals.

D^{17}_{2h} #63 (traditional space group is Bbmm b-centered; but c-centered Cmcm is also used. Note monoclinic refinement C^3_2 #5 for ferropseudobrookite.)

Karooite MgTi$_2$O$_5$ Bbmm Wechsler and Von Dreele (1989)
M1 and O1 on 4c. M2, O2 and O3 on 8f.

Sample quenched from 973 K.

cell dimensions			M1	M2	O1	O2	O3
a	9.7289	x	0.19521	0.13255	0.77696	0.04606	0.31332
b	10.0057	y	1/4	0.56474	1/4	0.11407	0.06499
c	3.7416	z	0	0	0	0	0

Ti fraction   0.213   0.894
M1-O distances 2@2.033; 2@1.989; 2@ 2.179
M2-O distances 2.052; 1.806; 1.977; 2@1.944; 2.186

Sample quenched from 1773 K.

a	9.7492	x	0.19482	0.13298	0.77021	0.04653	0.31235
b	9.9896	y	1/4	0.56108	1/4	0.11521	0.06803
c	3.7460	z	0	0	0	0	0

Ti fraction   0.400   0.800
M1-O distances 2@2.012; 2@1.976; 2@2.149
M2-O distances 2.110; 1.832; 1.952; 2@1.949; 2.173

Pseudobrookite Fe$^{3+}_2$TiO$_5$ Cmcm Tiedemann and Müller-Buschbaum (1982)

cell dimensions			M1	M2	O1	O2	O3
a	3.739	x	0	0	0	0	0
b	9.779	y	0.1890	0.1360	0.766	0.048	0.311
c	9.978	z	1/4	0.5642	1/4	0.117	0.070

Ti fraction   0.333   0.333
M1-O distances 2@2.01; 2@1.92; 2@2.15
M2-O distances 2.09; 1.87; 2.00; 2@1.94; 2.00

Pseudobrookite Fe$^{3+}_2$TiO$_5$ C^3_2 C2 #5 Shiojiri et al. (1984)
All atoms in 4c in this space group.

cell dimensions			Fe1	Fe2	Fe3	Fe4	Ti1	Ti2	O1	O2	O3	O4
a	22.23	x	.475	.624	.970	.774	.131	.283	.035	.027	.198	.206
b	3.73	y	0	0	0	0	0	0	0	0	0	0
c	9.80	z	.615	.812	.107	.927	.313	.430	.086	.329	.248	.522
β	116.2°											

	O5	O6	O7	O8	O9	O10
	.384	.875	.710	.699	.547	.555
	0	0	0	0	0	0
	.653	.148	.031	.755	.889	.600

Armalcolite Fe$^{2+}_{.5}$Mg$_{.5}$Ti$_2$O$_5$ Bbmm Wechsler (1977)
Sample Arm-E synthesized and rapidly quenched from 1200°C; refinement at 24° C before annealing; A second armalcolite sample was also refined; Cation distribution evaluated at several temperatures.

Table 6 (continued).

cell dimensions			M1	M2	O1	O2	O3
a	9.7762	x	0.80743	0.13464	0.2243	0.0464	0.3133
b	10.0214	y	1/4	0.43528	1/4	0.8846	0.9348
c	3.7504	z	0	0	0	0	0
Ti fraction			.335	.832			
Mg fraction			.332	.084			
Fe fraction			.332	.084			
M1-O distances			2@2.044; 2@1.966; 2@2.198				
M2-O distances			2.055; 2.002; 1.841; 2@1.943;2.180				

Note: Although refined in the same space group and setting, some karooite and armalcolite coordinates differ because of the arbitrary use of (1-u) or (u) values in the refinements. Either coordinate set is crystallographically equivalent.

temperatures (Wechsler, 1977; Brown and Navrotsky, 1989; Wechsler and Von Dreele, 1989).

The pseudobrookite minerals have cation distributions that are a function of both synthesis temperature and cooling rate. Wechsler and Von Dreele (1989) found that karrooite synthesized at 1673 K but annealed at 973 K had 0.213 Ti in M1 and 1.788 Ti in the two M2 sites, while similarly synthesized material annealed at 1773 K had the distribution 0.400 Ti in M1 and 1.600 Ti in the two M2 sites. These distributions probably include some reordering during quench as lattice parameters were found to remain constant in samples quenched from any temperatures above 1400 K, even though it is known that the disordering continues up to about 1800 K with a continuous change in lattice parameter (Brown and Navrotsky, 1989).

The temperature dependence of the metal ion distribution in armalcolite was studied by Wechsler (1977). Material quenched from 1200°C had substantial cation disorder with nearly equal Mg, Fe and Ti occupation of M(1) and substantial Mg and Fe in M(2). This sample ordered almost completely after one day of annealing at 400°C. Another sample quenched from 1100°C had slight disorder.

Strong ordering of Ti^{4+} into M2 and Fe^{3+} into M1 is usually observed in Fe_2TiO_5, although Tiedemann and Müller-Buschbaum refined a disordered cation distribution. In $FeTi_2O_5$, Ti^{4+} prefers M2 and Fe^{2+} prefers M1 (Bowles, 1988; Grey and Ward, 1973). Shiojiri et al. (1984) found that Fe_2TiO_5 could be refined in a completely ordered monoclinic cell based on a doubled orthorhombic cell with four distinct Fe positions and two distinct Ti positions. Table 6 lists the crystallographic parameters of several of these pseudobrookites.

MANGANESE OXIDES AND OXYHYDROXIDES

Manganese is found in three valence states in its minerals, and this is in part responsible for the wide variety of interesting and complex manganese phases. However the most notable feature of manganese mineral crystal chemistry is the striking tunnel and layer structures found in these materials. There is such an wide range of structural motifs and varieties, many of which are incompletely described and characterized, that justice to their novelty and scope cannot be done in a short review such as this one. Other surveys have been done by Burns and Burns (1979), Burns and Burns (1977) and Huebner (1976). Here the major phases are described along with the results of work since the last oxide RIM volume.

Tetravalent manganese oxides and oxyhydroxides

The dominantly tetravalent oxides are composed mainly of edge-sharing octahedral $Mn^{4+}O_6$ units, arranged in single chains and multioctahedral slabs, which share vertices to form linear arrays, tunnel and layer structures. There are many similarities to the Fe oxyhydroxide structures, as replacement of the OH^- and Fe^{3+} with O^{2-} and Mn^{4+} is a balanced valence substitution, and these species have similar sizes.

Pyrolusite

Pyrolusite (β-MnO_2) has the tetragonal rutile structure (Fig. 6) with chains of edge sharing MnO_6 octahedra along [001]. Its crystallographic parameters appear in Table 4. It can be formed as a primary phase, or by dehydration and oxidation of manganite, MnOOH, (Rask and Buseck, 1986). An orthorhombic variety of pyrolusite apparently occurs (Potter and Rossman, 1979), but is not well characterized.

Ramsdellite

Ramsdellite (γ-MnO_2), has a structure with topology like goethite, with double chains of edge sharing MnO_6 octahedra (= dioctahedral MnO_3 slabs) along [001] connected at their edges by vertices (Fig. 11). Like goethite, ramsdellite is orthorhombic and usually has an acicular habit. Although generally thought to be anhydrous and stoichiometric, Potter and Rossman (1979) found a single well-ordered type of structural water in the ramsdellites they studied. This suggests that water may be an integral part of the ramsdellite structure. Any such water is probably associated with the open channels along [001] rather than occupying other defect sites. Crystallographic parameters are listed in Table 7.

Nsutite

Nsutite (nominally γ-MnO_2, but substitution of hydroxyl suggests the formula $Mn^{4+}_{1-x}Mn^{3+}_xO_{2-x}OH_x$) has a structure that combines the single chains of octahedra in pyrolusite with the dioctahedral chains of ramsdellite in a random fashion (deWolff, 1959) (Fig. 20). All chains run parallel to [001], but have minimal order in the [010] or [100] directions, though Turner and Buseck (1983) observe some tendency for regular alternation of single and double chains in some nsutite grains. The structure can be considered as having large scale stacking faults, and the packing disorder is associated with nonstoichiometry, cation vacancies and charge-compensating species. Heat treatment of nsutite leads to an increase in the population of pyrolusite microdomains and gradual ordering.

Hollandite and related phases

Hollandite (generically $A_{0-2}(B^+Mn^{4+})_8O_{16}$) has a structure consisting of dioctahedral chains sharing vertices along [001] to form a square tunnel arrangement. This is the motif also found in akaganeite (Fig. 12). The tunnels are partially filled with large A cations which may be mono- or divalent. Charge compensation is achieved by substitution for Mn^{4+} in the tunnel walls by tri- and divalent B cations, and also some OH^- may replace O^{2-}. Ba is the major A cation in hollandite *sensu stricto*, while K takes the A position in

Table 7.   Manganese oxides and hydroxides.

Ramsdellite   MnO$_2$ Pbnm   D$^{16}_{2h}$ #62   Byström (1949)
All atoms in 4c positions.

cell dimensions		Mn	O1	O2
a	4.533	x 0.022	0.17	-0.21
b	9.27	y 0.136	-0.23	-0.033
c	2.866	z 1/4	1/4	1/4

Hollandite   (Ba,Pb,Na,K)(Mn,Fe,Al)$_8$(O,OH)$_{16}$ I2/m (C2/m)   C$^3_{2h}$ #12   Post et al. (1982)

cell dimensions		Mn1	Mn2	O1	O2	O3	O4	Pb	Ba
a	10.026	x .85180	.33670	.6583	.6552	.2940	.0415	0	0
b	2.8782	y 0	0	0	0	0	0	.202	0
c	9.729	z .33266	.15345	.3022	.0414	.3502	.3222	0	0
b	91.03°								

Mn1-O distances: 1.958; 2@1.949;2@1.892;1.907
Mn2-O distances: 1.899;1.969;2@1.946;2@1.899

Coronadite   (Pb,Ba)(Mn,V,Al)$_8$O$_{16}$ I2/m   C$^3_{2h}$ #12   Post and Bish (1989)

cell dimensions		Mn1	Mn2	O1	O2	O3	O4	Pb	Ba
a	9.938	x .852	.333	.637	.657	.284	.034	0	0
b	2.8678	y 0	0	0	0	0	0	.214	0
c	9.834	z .333	.156	.294	.051	.327	.329	0	0
b	91.39°								

Mn1-O distances: 2.14; 2@1.94;2@1.88;1.85
Mn2-O distances: 1.97;1.93;2@1.91;2@1.94

Romanechite   (Ba,H$_2$O)$_2$Mn$_5$O$_{10}$ C2/m   C$^3_{2h}$ #12   Turner and Post (1988)

cell dimensions		M-O distances	subcell (supercell b'=3b also refined)
a	13.929	Mn1-O:	2@1.918;4@1.904
b	2.8459	Mn2-O:	2@1.935;1.933;2@1.896;2.118
c	9.678	Mn3-O:	2@1.870;1.921;2@1.933;1.899
β	92.39°	Ba-O:	2@2.943;3.640;2@2.992;3.149;2@2.942;3.005

	Mn1	Mn2	Mn3	O1	O2	O3	O4	O5	O6	Ba
x	0	.99978	.3401	.5669	.0976	.7663	.4248	.9263	.250	.2466
y	0.5	0	0	0	0	0	0	0	0	0
z	0	.2682	.4841	.1773	.4244	.3958	.3354	.0721	.131	.1242

Todorokite   (Na,Ca,K)(Mn,Mg)$_6$O$_{12}$·nH$_2$O   P2/m   C$^1_{2h}$ #10   Post and Bish (1988)

cell dimensions		mean M-O distances			Mn1	Mn2	Mn3	Mn4
a	9.764	Mn1-O:	1.80	x	1/2	0.764	0	0.974
b	2.8416	Mn2-O:	1.94	y	1/2	0	0	1/2
c	9.551	Mn3-O:	1.87	z	0	0.002	1/2	0.765
b	94.06°	Mn4-O:	1.96					

	O1	O2	O3	O4	O5	O6	H$_2$O1	H$_2$O2	H$_2$O3
	0.178	0.418	0.665	0.917	0.913	0.880	0.363	0.696	1/2
	1/2	0	1/2	0	1/2	0	0	1/2	1/2
	0.119	0.079	0.090	0.150	0.407	0.649	0.353	0.380	1/2
occupancy:							0.92	0.42	0.48

Table 7 (continued).

Chalcophanite	$ZnMn_3O_7 \cdot 3H_2O$	$R\bar{3}$	C^2_{3i}	#148	Post and Appleman (1988)
	Bisbee sample (Sterling Hill sample also refined)				

cell dimensions		Mn	Zn	O1	O2	O3	O4
a	7.533	x	0.71869	0	0.52784	0.26077	0	0.17900
c	20.794	y	0.57771	0	0.62297	0.20655	0	0.93107
		z	0.99948	0.09997	0.04721	0.05048	0.71250	0.16435

Mn-O distances: 1.967;1.891;1.913;1.857;1.869;1.937
Zn-O distances: 3@2.069;3@2.140

Birnessite	$NaMn_2O_4 \cdot 1.5H_2O$	C2/m (arbitrary, could also be Cm or C2)	C^3_{2h} #12
	Post and Veblen (1990) (Mg and K birnessites also refined)		

cell dimensions		Mn	O1	O2	O3	
a	5.175	x	0	0.376	0.595	0
b	2.850	y	0	0	0	0
c	7.337	z	0	0.133	0.500	1/2
β	103.18°	occupation:		2.00	0.3	(atoms/cell)

Mn-O distances: 4@1.92;2@1.97

Manganite	MnOOH	$B2_1m$	C^2_{2h}	#11	(Dachs, 1963; Buerger, 1936)
	All atoms on 4f positions.				

cell dimensions		Mn	O1	O2	H
a	8.88	x	0.0047	0.1218	0.1239	0.2252
b	5.25	y	-0.0119	0.1250	0.1250	0.0254
c	5.71	z	0.2401	0.0011	0.5016	0.5031
b	ca. 90o					

Mn-O distances: 1.878;1.868;2.199;1.981;1.965;2.333

Bixbyite	Mn_2O_3	Pcab	D^{15}_{2h}	#61	Geller (1971)

a=9.4157 b=9.4233 c=9.4047   5 different Mn positions; 6 different O positions

$(Mn_{0.983}Fe_{0.017})_2O_3$ Ia3 T^7_h #206
metal atoms on 8a and 24d; O on 48e.

			M1	M2	O	M-O distances
a	9.4146	x	0	0.28508	0.12913	M1-O: 6@2.009
		y	0	0	0.14708	M2-O: 2@2.147;
		z	0	1/4	-0.08347	2@1.930;2@2.033

$(Mn_{0.37}Fe_{.63})_2O_3$ Ia3

a	9.4126	x	0	0.28473	0.13527	M1-O: 6@2.003
		y	0	0	0.13950	M2-O: 2@2.242;
		z	0	1/4	-0.08830	2@1.898;2@1.987

Groutite	MnOOH	Pbnm	D^{16}_{2h}	#62	Dent-Glasser and Ingram (1968)
	All atoms in 4c.				

cell dimensions		Mn	O1	O2	Mn-O distances
a		x	-0.0501	0.2987	-0.1945	Mn-O1: 2.178;1.896
b		y	0.1401	-0.1868	-0.0697	Mn-O2: 2.340;1.968
c		z	1/4	1/4	1/4	

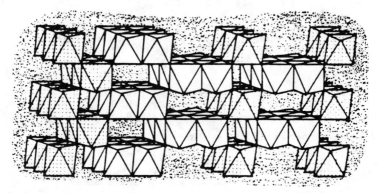

Figure 20. "Representative" nsutite as drafted by Potter and Rossman (1979). The stippled octahedra belong to pyrolusite-like slabs; the clear octahedrons are ramsdellite-like. See Figures 11 and 6.

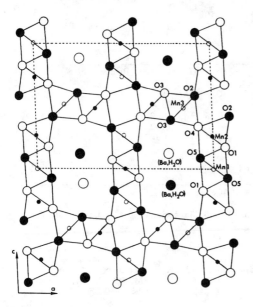

Figure 21. Romanechite structure projected on (010). From Turner and Post (1988).

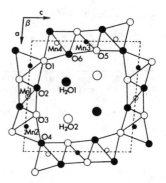

Figure 22. Todorokite structure projected on (010). From Post and Bish (1988).

cryptomelane, Pb^{2+} in coronadite, and Na in manjiroite. Ti^{4+} analogs include priderite (A = K,Ba; B = Ti^{4+}), mannardite (A = Ba,H_2O; B = Ti^{4+},V^{3+},Cr^{3+}) and retledgite (A = Ba; B = Ti^{4+}). Pure natural end-member hollandite has been reported by Meyer et al. (1986).

Hollandite structure minerals have been refined by Post and Bish (1989) (coronadite via Rietveld method), Post et al. (1982) (hollandite, cryptomelane and priderite), Gatehouse et al. (1986) (redledgeite), Szymanski (1986) (mannardite), and Sinclair and McLaughlin (1982) (priderite). Post and Burnham (1986b) have used structure-energy calculations based on MEG theory-based parameters to model the tunnel cation positioning. The ideal hollandite structure has I4/m space group symmetry with A cations at special position 2a at (0,0,0), but the site is always partially vacant. Hollandites with divalent A ions have about half of the A sites filled, while monovalent A ion hollandites may have A site occupancies of 0.67-0.75. Diffraction studies have shown that some of the A ions appear to be displaced from the special position along [001] (down the tunnels). In cryptomelane and priderite, about 20% and 30%, respectively, of the A cations are displaced to a subsidiary site 0.55 Å along the tunnels. Such split site refinements are usually indications of positional disorder. Post and Burnham (1986b) found that the displacements of the A cations was correlated with the positions of the lower valence charge-compensating cations in nearby octahedral sites. Their energy calculations also indicated only a narrow range of energies for a large set of possible configurations of the octahedral ions, suggesting that they are probably disordered in actual hollandites. Hence the A cation positions are also disordered.

Post and Bish (1989) found that coronadite is monoclinic with space group I2/m. Redledgeite also is monoclinic I/2m, with apparent tunnel cation ordering. The tunnel cation ordering may be incommensurate with the framework periodicity (Gatehouse et al., 1986).

Romanechite

Romanechite (also called psilomelane, a name now properly used to describe mixtures of manganese oxide minerals) $(Ba,H_2O)_2Mn_5O_{10}$, has a structure formed from di- and tri-octahedral slabs joined at their vertices to form rectangular tunnels (Fig. 21). The original structure was described by Wadsley (1953), and confirmed by TEM studies (Turner and Buseck, 1979) and in more detail by a recent x-ray structure refinement (Turner and Post, 1988). Romanechite occurs intimately intergrown with other manganese minerals on a unit cell level, hence good single crystals are rare. Barium is the dominant tunnel cation, with traces of other large mono- and divalent species. The Turner and Post (1988) refinement shows that charge compensation in the tunnel walls is achieved mainly by the substitution of Mn^{3+} for Mn^{4+} at the nearest framework octahedral site, Mn2, as originally proposed by Burns et al. (1983). The Mn^{3+} causes distortion of these polyhedra because of its strong Jahn-Teller effect. The basic structure has monoclinic symmetry, with space group C2/m. Superstructure reflections indicate that tripling of the cell length along b, the tunnel axis, occurs as a result of ordering of Ba and H_2O in the tunnels. Interestingly, adjacent tunnels apparently have correlated Ba and H_2O positions in one direction, but not in another, the cause of which is unclear.

Todorokite

Todorokite is a major phase in manganese nodules and has been thought to be a host for important metals residing in these deposits. It has a variable composition, suggested as $(Na,Ca,K,Ba,Sr)_{0.3-0.7}(Mn,Mg,Al)_6O_{12} \cdot 3.2-4.5\ H_2O$ by Post and Bish

(1988), who performed a Rietveld powder x-ray refinement on polycrystalline material from South Africa and Cuba. The true structure has been in question for some time. Because of its fibrous habit, Burns and Burns (1977) proposed a tunnel structure like that of the other manganese oxides, romanechite and hollandite. IR data and occasional platy habit, however, suggested a layer-like structure found in other Mn-minerals, e.g., birnessite (Potter and Rossman, 1979). Giovanoli and Burki (1975) suggested that todorokite was actually an intimate mixture of layer manganese oxides and manganite (MnOOH) and thus was not a valid species; their arguments were contested by Burns et al. (1983).

Post and Bish (1988) verify that todorokite has a tunnel structure with still larger cavities than romanechite, as it is composed of trioctahedral slabs sharing vertices (Fig. 22). There are three well-defined occupied tunnel sites at about (0.36,0,0.35), (0.70,1/2,0.38) and (1/2,1/2,1/2), the first two of which are partially filled with H_2O, and the other partially filled with Na and Ca. The structure was refined in P2/m symmetry which is consistent with the overall framework symmetry, and 2/m Laue symmetry from electron diffraction patterns (Turner, 1982). Similar to the situation with romanechite, Mn^{3+}, substituting for Mn^{4+} to aid in charge compensation, was found to occupy octahedra at the edges of the trioctahedral slabs. These octahedra had mean M-O distances larger than other polyhedra in the structure, and probably contained all of the larger charge balancing ions.

In a related study, Bish and Post (1989) examined the response of tunnel manganese oxides to anneals of up to 1000°C. They found that todorokite lost most of its tunnel water below 400°C, but that romanechite retained most of its water up to over 500°C, and coronadite retained most of its water to 600°C. Thus the water bond is inversely related to the tunnel size. Water in all of these minerals appeared to occupy well defined crystallographic sites.

LAYER STRUCTURE MANGANESE OXYHYDROXIDES (PHYLLOMANGANATES)

<u>Chalcophanite</u>

Chalcophanite, ideally $ZnMn_3O_7 \cdot 3H_2O$, is one of the better characterized phyllomanganates. The structure was originally determined by Wadsley (1955) as triclinic, with a layer structure consisting of sheets of edge-sharing $Mn^{4+}O_6$ octahedra alternating with layers of water molecules and Zn cations in the sequence -Mn-O-Zn-H_2O-Zn-O-Mn- along the c axis. Recently the structure has been refined by Post and Appleman (1988) and found to be trigonal, space group $R\overline{3}$, but topologically very close to Wadsley's (1955) structure.

One out of every seven manganese sites in the manganese sheets is vacant in an ordered arrangement (Fig. 23). The vacant Mn sites coincide directly with Zn sites above and below the manganese sheets. These Zn ions are coordinated to three oxygens in the manganese sheet, and to three water oxygens in the water layer.

Although the particular chalcophanite specimens studied by Post and Appleman (1988) are very close to the ideal composition, there is evidence that the chalcophanite structure accomodates a wide variety of other metal ions including Ag, Ni, Mn^{2+} and Mg. Chalcophanites enriched in these species appear to have the same powder XRD pattern as pure Zn-rich material, and it is likely that these cations substitute for Zn in the structure.

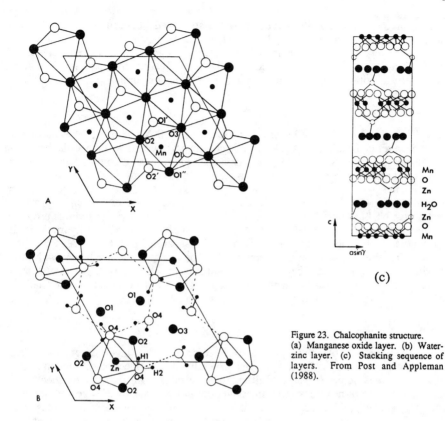

Figure 23. Chalcophanite structure. (a) Manganese oxide layer. (b) Water-zinc layer. (c) Stacking sequence of layers. From Post and Appleman (1988).

Birnessite

Birnessite (sometimes known as "7 Å manganite") is another manganese oxide layer structure that has been subject to many investigations, but has resisted complete characterization. The lack of homogeneous and well crystallized samples has been such a problem that even the crystal system and unit cell parameters have been disputed. The powder XRD patterns of birnessite are similar to those of chalcophanite, and this combined with the tendency for a platy habit, suggest a layer structure though with considerable disorder (Potter and Rossman, 1979). Chukhrov et al. (1985) adapted model structures derived from chalcophanite to fit the powder XRD patterns of natural birnessites with some success. Their model had random Mn vacancies in the manganese layers and Na cations above and below the vacancies, with a hexagonal unit cell, a = 2.86 Å and c = 7 Å. Other possible structures have been suggested by Giovanoli and co-workers, e.g., Giovanoli and Arrhenius (1988).

Most recently, Post and Veblen (1990), refined the structure of Na, Mg and K synthetic birnessites by Rietveld methods assisted by TEM observations. They found a structure very similar to chalcophanite, but with monoclinic, C2/m symmetry (subcell), and with different water molecule and interlayer cation positions in each of the analogs. Each material's XRD pattern also showed different superstructure reflections, indicating that the ordering of the water and interlayer cations also varied between the three types of structures. Post and Veblen (1990) also observed that there are fewer Mn vacancies in the manganese layers than in chalcophanite, and that these might be disordered. The cation position in the interlayer region depended on the type of cation, thus partially explaining why heavily substituted natural birnessites may have such structural disorder.

Buserite

The crystal structure of buserite (also known as "10 Å manganite") has apparently not yet been refined, and there is considerable controversy over its relation to the birnessite and chalcophanite structures and its very existence in the natural environment. What is known is that IR spectra show similarities between the basic manganese layers in buserites and birnessites, with the differences in layer spacing probably due only to interlayer water. Thus Giovanoli (1980) proposed a structure for buserite, based on an earlier proposed structure for Na birnessite (Giovanoli et al., 1970). By this scheme, buserite is proposed to have a birnessite-like layered structure, but with additional water and hydroxyl in the interlayer, such that the manganese oxide layers are separated by 9.6-10.1 Å. One out of six octahedral sites is vacant on an ordered basis in the manganese oxide layers, with Mn^{2+} and Mn^{3+} residing above and below these vacant sites in the interlayer. The hypothetical Na birnessite structure disagrees with that refined by Post and Veblen (1990) in detail, but the basic concepts of the buserite structure may be correct.

Asbolane

Asbolane is another manganese oxide mineral with a layer structure requiring additional characterization. Existing results, generally all from electron diffraction studies, suggest that its structure consists of alternating layers of manganese oxide sheets as in chalcophanite and birnessite with an interlayer comprising islands of edge-sharing metal hydroxide octahedra (Chukhrov et al., 1980). Apparently, considerable quantities of Ni, Cu and Co can be taken up by the asbolane structure. Hence its characterization is important for an understanding of the geochemistry of manganese nodules. Asbolane is suggested to form mixed layer structures with buserite (Chukhrov et al., 1983). EXAFS and XANES studies have been utilized to determine the distribution of metal ions between the hydroxide and manganese oxide layers (Manceau et al., 1987).

Vernadite

Vernadite (disordered δ-MnO_2) exists mainly as hydrous poorly crystallized small particles. It appears to be a scavanger for many metals, including Pb. Its XRD pattern shows only two lines, indexed as hk reflections, suggesting complete lack of any basal direction ordering (somewhat analogous to similar suggestions about protoferrihydrite). IR spectra (Potter and Rossman, 1979), if anything, suggest that vernadite is similar to, but more disordered than, birnessite. Chukhrov et al. (1988) suggest that Fe-containing vernadites consist of sheets of manganese oxide octahedra intermixed with three dimensional chunks of feroxyhyte-like structure. The proposed structure has random domains of CCP and HCP anions, larger than the sheet and chunk units, and with similar metal composition, which supposedly cancels any basal-type x-ray reflections.

Other possibilities may be that the XRD pattern of vernadites is due mainly to small grain size, and the Fe content is due to epitaxial overgrowth during precipitation. Fe and Mn domains may not mix because of somewhat different octahedra sizes and differing

Lithiophorite

Lithiophorite has a layer structure originally determined by Wadsley (1952). The structure consists of layers of manganese oxide octahedra sharing edges, as in chalcophanite but without cation vacancies, alternating with layers of $(Al,Li)(OH)_6$ edge-sharing octahedra (hydrargillite layer). The stacking sequence along [001] is $-O-Mn-O-OH-(Al,Li)-OH-O-Mn-O-$, with the manganese oxide layers separated by about 9.5 Å. There are no cation vacancies in the hydroxide layers. Giovanoli et al. (1973) suggest the formula $(Mn^{4+}{}_5Mn^{2+}O_{12})^{2-}(Al_4Li_2(OH)_{12})^{+2}$ for lithiophorite, and a trigonal unit cell. Pauling and Kamb (1982) reconsidered Wadsley's monoclinic structure and proposed a hexagonal motif which better satisfied the electroneutrality principle. This structure has ideal composition $Al_{14}Li_6(OH)_{42}Mn^{2+}{}_3Mn^{4+}{}_{18}O_{42}$, and is consisted with the observations of Giovanoli et al. (1973) within experimental accuracy. One octahedron in 21 is vacant in the hydrargillite layer, and the layer repeat is six. The space group is $P3_1$ with a = 13.37 Å and c = 28.20 Å.

TRIVALENT MANGANESE OXIDES AND OXYHYDROXIDES

Bixbyite

Bixbyite $(Fe,Mn)_2O_3$ is an example of an anion deficient fluorite structure (Dachs, 1956). Recall that the fluorite structure can be thought of as a CCP array of cations with anions filling all of the tetrahedral interstitial positions. If there are anion vacancies the formula for the hypothetical fluorite structure, $Mn^{4+}O_2$ becomes $Mn^{3+}{}_{2x}Mn^{4+}{}_{2-2x}O_{4-x}[\ ]_x$, where the Mn^{3+} is necessary for charge balancing. If x = 1, we get the formula $Mn^{3+}{}_2O_3$, and a structure with one quarter of its anion sites vacant. This is the bixbyite structure. Pure α-Mn_2O_3 is actually orthorhombic, but all bixbyite is cubic because of the stabilizing effects of even a small amout of Fe^{3+}. Note that Mn^{3+} is not a Jahn-Teller ion in the cubic fluorite-structure crystal field. Geller (1971) has refined the structure of several compositions of bixbyite. The iron analog of bixbyite, β-Fe_2O_3, has been proposed by several workers (e.g., Ben-Dor et al., 1976). It has a cell edge of 9.393 Å.

Groutite, feitknechtite and manganite

Polymorphs groutite (α-MnOOH) (Dent-Glasser and Ingram, 1968) and feitknechtite (β-MnOOH) have structures analogous to goethite (α-FeOOH) and lepidocrocite (β-FeOOH), respectively. Manganite has a structure that is a derivative of the marcasite structure, which itself is similar to that of rutile (Dachs, 1963). In manganite the anion stacking is based on HCP, and half of the octahedral cation sites are filled, forming rows of edge-sharing $Mn(O,OH)_6$ octahedra, as in rutile. However, the octahedra are considerably distorted. Manganite is monoclinic, but pseudo-orthorhombic. The fact that it contains Mn^{3+} in octahedral coordination may be involved with its overall symmetry. The available crystallographic parameters for manganite and groutite appear in Table 7. The structure of feitknechtite has apparently not been refined.

Manganosite

Manganosite ($Mn^{2+}O$), an NaCl-structure manganese oxide which usually contains Mn^{3+} and has a net cation deficiency similar to wüstite, has parameters listed in Table 1.

ALUMINUM OXIDES AND OXYHYDROXIDES

Corundum

Corundum (α-Al_2O_3) has the hematite structure and its parameters are listed in Table 1. Corundum commonly contains trace to minor amounts of Fe^{3+} and Cr^{3+}. These cations are the source of the beautiful colors of sapphire and ruby, two of the gem varieties of corundum.

Boehmite and diaspore

Boehmite and diaspore are AlOOH polymorphs. The boehmite (β-AlOOH) and diaspore (α-AlOOH) structures are closely similar to those of lepidocrocite and goethite, respectively. Goethite and diaspore have considerable solid solution, and up to 33 mol % diaspore can be dissolved in synthetic goethite precipitated under high pH conditions (Franz, 1978). The parameters for these oxyhydroxides appear in Table 3 with the Fe oxyhydroxides.

COMPLEX OXIDE GROUPS

Pyrochlore group

The pyrochlore group comprises the pyrochlore, microlite and betafite subgroups and a total of more than twenty individual species (at last count). All of these are cubic oxides with essential amounts of Nb, Ta or Ti. They crystallize in the space group symmetry $Fd\bar{3}m$, and have general formula $A_{2-m}B_2X_6[(O,OH,F) = Y]_{1-n} \cdot pH_2O$ (Hogarth, 1977), where the A cations can be Na, Ca, K, Sn^{2+}, Ba, Sr, REE, Pb, Mn, Fe^{2+}, Bi, Th and U, and the B cations include Ta, Nb, Ti, Zr, Fe^{3+}, Sn^{4+} and W. The type of B cation determines the subgroup designation. Pyrochlore group oxides have Nb+Ta > 2Ti and Nb > Ta. Microlites have Nb+Ta > 2Ti but Ta ≥ Nb. Betafites have 2Ti ≥ Nb+Ta. Individual species are further categorized by the A cation concentrations. Pyrochlores are frequently recovered as sources of REE, e.g., at the Oka carbonatite in Quebec province, Canada (Perrault, 1968). The structural parameters for several specimens of this pyrochlore appear in Table 8.

The pyrochlore structure (Fig. 24) is not derived from a close packed array of anions, but rather from an interesting packing of octahedral units sharing vertices (Clark, 1972; Hyde and Andersson, 1989). The organization of the layering is a bit complicated, and there is a wide variety of associated polytypes (Yagi and Roth, 1978). All of the atoms in the pyrochlore structure reside on special crystallographic positions in the $Fd\bar{3}m$ unit cell, with the A cations on 16(d) at (5/8,5/8,5/8) and equivalent positions, B cations on 16(c) at (1/8,1/8,1/8) and equivalent, X anions on 48(f) at (x_1,0,0), etc., and the remaining oxygen, OH, F and H_2O on 8(b) at (1/2,1/2,1/2) and equivalents. Given the many valences of the substituting species, there can be a host of charge-compensating schemes in place, and several types of cation vacancies.

Owing to the frequent high concentration of U and Th in pyrochlores, they may have quite damaged crystallinity or be completely metamict. Thus the crystal chemistry of many natural pyrochlore species is not well characterized. Reviews on the crystal chemistry and structure of pyrochlores include Subramanian et al. (1983), Chakoumakos (1984) and Lumpkin and Ewing (1985).

Table 8. Complex Oxides.

Pyrochlore	$(Na,Ca,Ce)_{16}(Nb,Ta)_{16}O_{48}F_8$		$Fd\bar{3}m$	O^7_h	#227	Perrault (1968)	
	Origin 43m. Nb in 16d; F in 8a; Ca in 16c; O in 48f.						
						M-O distances	
cell dimensions		Nb	Ca	F	O	Nb-O	Ca-O
a	10.43	x	5/8	1/8	0	0.316	1.93	2.70
a	10.393	x	5/8	1/8	0	0.313	1.95	2.68
a	10.428	x	5/8	1/8	0	0.322	1.92	2.75
a	10.3947	x	5/8	1/8	0	0.315	1.94	2.70
a	10.4205	x	5/8	1/8	0	0.330	1.90	2.82

Columbite (manganotantalite) $(Mn,Ti,Fe)(Ta,Nb)_2O_6$ Pbcn D^{14}_{2h} #60
   Grice et al. (1976)   A (Mn) on 4c; B(Ta) on 8d; O on 8d.

cell dimensions			A	B	O1	O2	O3
a	14.413	x	0	0.1628	0.099	0.419	0.758
b	5.760	y	0.322	0.1770	0.094	0.112	0.121
c	5.084	z	1/4	0.7367	0.055	0.108	0.099

A(Mn)-O distances: 2@2.18;2@2.20;2@2.16
B(Ta)-O distances: 2.04;1.92;1.81;2.23;2.08;1.98

Wolframite (ferberite) $FeWO_4$ P2/c C^4_{2h} #13
   Cid-Dresdner and Escobar (1968)   W on 2e;Fe on 2f; O on 4g.

cell dimensions			W	Fe	O1	O2	M-O distances
a	4.750	x	0	1/2	0.2158	0.2623	W-O: 2@1.99;
b	5.720	y	0.1808	0.3215	0.1068	0.3850	2@2.11;2@2.08
c	4.970	z	1/4	3/4	0.5833	0.0912	Fe-O: 2@1.95;
b	90.16°						2@2.17;2@2.02

Perovskite $PrFeO_3$ Pbnm D^{16}_{2h} #62   Marezio et al. (1970)
   A (Pr) on 4c; B(Fe) on 4b; 4O on 4c; 8O on 4d.

cell dimensions			Pr	Fe	O1	O2
a	5.482	x	0.99097	0	0.0817	0.7075
b	5.578	y	0.04367	1/2	0.4788	0.2919
c	7.786	z	1/4	0	1/4	0.0437

Pr-O distances: 2.371;2.478;3.130;3.190;2@2.395;2@2.629;2@2.730;2@3.386
Fe-O distances: 2@2.001;2@2.015;2@2.010

$LuFeO_3$

cell dimensions			Lu	Fe	O1	O2
a	5.213	x	0.98003	0	0.1199	0.6893
b	5.547	y	0.07149	1/2	0.4539	0.3071
c	7.565	z	1/4	0	1/4	0.0621

Lu-O distances: 2.185;2.243;3.195;3.502;2@2.225;2@2.455;2@2.687;2@3.599
Fe-O distances: 2@2.008;2@2.024;2@1.997

Table 8 (continued).

Uraninite	$UO_{2.12}$		$Fm\bar{3}m$		O^5_h	#225 Willis (1978; 1963)	
		U on 4a; O mainly on 8c, with some interstitial on 48i and 32f.					
cell dimension		U	O1	O'	O"	U-O distances	
a 5.488	x	0	1/4	1/2	0.61	U-O1:	2.376
	y	0	1/4	0.61	0.61	U-O':	3.401; 3.880
	z	0	1/4	0.61	0.61	U-O":	3.454
occupancy:			1.87	0.13	0.12		

(ideal UO_2 would have 2.00 O1 and no other oxygen positions occupied.)

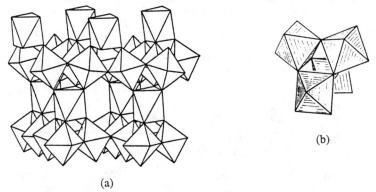

Figure 24. Aspects of the pyrochlore structure, $A_2B_2X_7$. (a) BX_3 framework as a packing of BX_6 octahedra. A atoms and additional X atoms are located in the large voids. (b) Close aspect of the corner-sharing octahedral unit characteristic of the pyrochlore structure. After Clark (1972) and Hyde and Andersson (1989), respectively.

<u>Columbite-tantalite group</u>

The columbite group oxides have basic stoichiometries of AB_2O_6, where usually A is Fe^{2+} or Mn^{2+} and B is Ta^{5+} or Nb^{5+}, but many other substitutions are possible, especially tetravalent cations. The columbite structure (Weitzel, 1976), is derived from the HCP arrangement of anions, with half filled octahedral cation sites set in a zig-zag pattern, and the cation occupation sequence normal to the close packed layers shown in Figure 25. The comparison of the columbite structure with that of a trirutile, e.g., tapiolite (see *Rutile* section) is revealing. In the trirutile ($Fe^{2+}Ta^{5+}_2O_6$) or valency-disordered rutile structure ($Fe^{3+}TaO_4$), two different types of cations are always found in every edge-sharing octahedral chain along [001] and thus also between all pairs of closest packed anion layers. In the columbite structure the cations in any one chain of edge-sharing octahedra are always of the same kind, and thus only one kind of cation is found between any one pair of close packed anion layers.

Columbite-structure minerals may disorder by mixing the A and B cations. This ultimately reduces the cell size to 1/3 the length of the ordered (super)cell, the structure of α-PbO_2. Partial order results in variations in unit cell dimensions (Wise et al., 1985). The crystal chemistry of columbite group minerals and their ordered derivatives has been considered by Graham and Thornber (1974a,b), Barker and Graham (1974), Grice et al. (1976), and Ferguson et al. (1976). Noteworthy among these articles are the differing structures, based on varied cation ordering arrangements, proposed for wodginite. The problem of deciphering the nature of the metamict state in these oxides is discussed by Ewing (1975) and Graham and Thornber (1975).

Wolframite has a structure very closely related to columbite. The anion arrangment is HCP and the cation occupation sites are the same, but the sequence normal to the close packed layers is different (Fig. 25). This yields the stoichiometry ABO_4. The relationship of the $\alpha\text{-}PbO_2$, wolframite and columbite structures is thus of subcell and two types of ordered supercells. Many tungstates and some tantalates and niobates crystallize with the wolframite structure, while most niobates and some tantalates crystallize in the columbite structure. Both structures may show varying degrees of disorder. The parameters for ferberite, $FeWO_4$, are shown in Table 8 (Cid-Dresdner and Escobar, 1968).

OTHER SIGNIFICANT OXIDES

Perovskites

Perovskite has been in the news recently because its structure is related to that of many of the high temperature superconductors. Its silicate analog is also believed to be an important mantle phase (Navrotsky and Weidner, 1989). The ideal cubic structure is shown in Figure 4, but the structure is unique inasmuch as a variety of cations can be accomodated in the central large site by the appropriate tilting and kinking of the relatively rigid corner-linked TiO_6 octahedra. Larger ions prevent the tilting, stabilizing a cubic or near-cubic structure. Smaller species allow considerable tilting, and the symmetry may be as low as monoclinic (Fig. 26). Natural perovskites near $CaTiO_3$ in composition are orthorhombic, space group Pnma. The TiO_6 octahedra are quite regular with Ti-O distances withing a few hundredths of an Ångström of one another. The individual octahedra are tilted about 10° from the ideal structure with the tilt axes in the (001) plane of the orthorhombic cell (equal to the (011) plane of the parent ideal cubic cell) (Hyde and Andersson, 1989). The large cation coordination is altered by the octahedral tilting such that for large tilt angles there is a wide spread of A-O distances. For example, in synthetic $YFeO_3$ the Fe-O distances range from 2.00 to 2.04 Å, but the Y-O distances vary between 2.24 and 3.58 Å (Marezio et al., 1970). In synthetic lanthanide perovskites the tilt angle of the octahedra varies smoothly with the size of the lanthanide on the A site, from 15° for $LaFeO_3$ to 23° for $LuFeO_3$.

A very large number of compounds have the perovskite structure, and all types of charge-coupled substitutions are possible. $K^+W^{5+}O_3$ is an example of a "tungsten bronze", a name derived from its color. The K ions may completely fill the A site, or there can be a variable number of A vacancies with accompanying hexavalent tungsten. More complicated charge-coupled substitution schemes can lead to larger ordered cells.

Natural perovskites have apparently not been thoroughly characterized. Mössbauer spectra of a couple of samples (Muir et al., 1984; Burns, 1989) show mainly Fe^{3+} and a small component of Fe^{2+} in the octahedral Ti site. The Fe^{3+} is presumably stabilized by lanthanide trivalent ions in the A sites. It is now well known that synthetic perovskite can accept large quantities of Fe_2O_3 as a defect solid solution (Cormack, 1989; Smyth, 1985). The extra Fe^{3+} goes into tetrahedral sites formed by the production of anion vacancies, but little ordering is noticed up to 50 mol % Fe_2O_3. With higher substitution brownmillerite-like ordering is observed. This substitution process has not been observed in natural perovskites. The disposition of other substituting elements in natural material is uncertain.

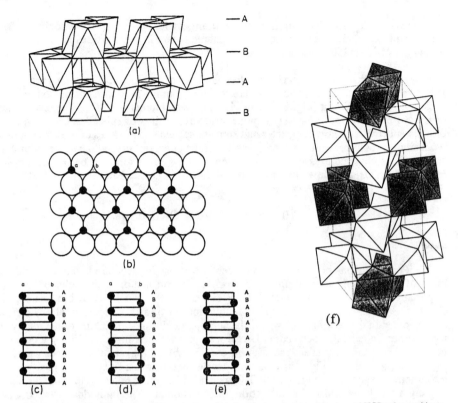

Figure 25. Structures related to α-PbO$_2$. (a) Zig-zag chains of edge-sharing octahedra and HCP anion packing layers. (b) Cation arrangement on close packed anion plane. (c) Stacking diagram for α-PbO$_2$ structure, all cations identical. Equivalent to disordered columbite and wolframite. (d) Stacking sequence for wolframite structure. Each type of cation occupies alternate chains. (e) Stacking sequence for columbite structure. After Clark (1972). (f) Polyhedral drawing for ordered occupation of chains, as in columbite. Vertical axis is b. After Zoltai and Stout (1984).

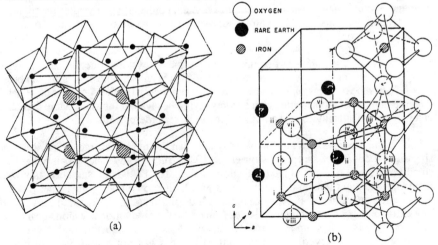

Figure 26. (a) Perovskite structure with a doubled cubic cell showing the tilting of TiO$_6$ octahedra. Large hatched circles are Ca^{2+} ions; small solid circles are Ti^{4+} ions. After Naray-Szabo (1943). (b) Relationship of orthorhombic to ideal cubic perovskite cell in rare earth perovskites, REEFeO$_3$. After Marezio et al. (1970).

Varieties of perovskite include latrappite, $(Ca,Na)(Nb,Ti,Fe)O_3$, loparite-(Ce), $(Ce,Na,Ca)_2(Ti,Nb)_2O_6$, tausonite, $SrTiO_3$, and lueshite, $NaNbO_3$.

Uraninite

Uraninite has a defect fluorite structure noted earlier. It forms a complete solid solution with thorianite (ThO_2), and can contain small amounts of transition metal and heavy metal impurities. At temperatures above about 1150°C there is a complete defect solid solution between UO_2 and $UO_{2.25}$. The arrangement of the oxygen interstitials was once thought to be completely random. However, the defect structures have now been well characterized, and consist of well-defined oxygen clusters of several types (Willis, 1978; Allen et al., 1982; Naito et al., 1989). A well characterized example is shown in Figure 27. At low defect concentrations the oxygen clusters are randomly positioned, but remain intact. The space group is the same as stoichiometric UO_2, $Fm\bar{3}m$. At high concentrations the clusters are ordered into the β-U_4O_9 structure, $I\bar{4}3d$. X-ray diffraction is relatively insensitive to the anion arrangements because of the much stronger scattering from the intact uranium cation array, and hence neutron diffraction and modeling techniques have been particularly important in characterizing uraninite structure. Despite this the low-temperature structure of highly defective U_4O_9 is still unsolved.

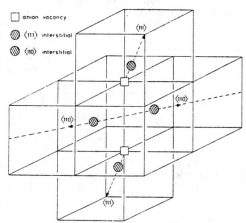

Figure 27. Defect cluster in the uraninite structure. The corners of the cubic outlines indicate the positions of undisplaced anion sites. The cluster contains two anion vacancies, two <111> interstitial oxygens, and two <110> interstitial oxygens. Uranium atoms occupy the cube centers. After Willis (1978).

BIBLIOGRAPHY AND REFERENCES

Abrahams, S.C. and Bernstein, J.L. (1971) Rutile: Normal probability plot analysis and accurate measurements of crystal structure. J. Chem. Phys. 55, 3206-3211.

Akimoto, S. (1954) Thermo-magnetic study of ferromagnetic minerals contained in igneous rocks. J. Geomag. Geoelectricity 6, 1-14.

Akimoto, S. and Katsura, T. (1959) Magneto-chemical study of the generalized titanomagnetite in volcanic rocks. J. Geomag. Geoelectricity 3, 69-90.

Allan, J.E.M., Coey, J.M.D., Sanders, I.S., Schwertmann, U., Friedrich, G. and Wiechowski, A. (1989) An occurrence of a fully-oxidized natural titanomaghemite in basalt. Mineral. Mag. 53, 299-304.

Allen, G.C., Tempest, P.A. and Tyler, J.W. (1982) Coordination model for the defect structure of hyperstoichiometric UO_{2+x} and U_4O_9. Nature (London) 295, 48-49.

Andersson, B. and Sletnes, J.O. (1977) Decomposition and ordering in $Fe_{1-x}O$. Acta Crystallogr. A33, 268-280.

Antonio, P.D. and Santoro, A. (1980) Powder neutron diffraction study of chemically prepared β-lead dioxide. Acta Crystallogr. B36, 2394-2397.

Armbruster, T. (1981) On the origin of sagenites: Structural coherency of rutile with hematite and spinel structure types. N. Jahrb. Mineral. Monat. 7, 329-334.

Atkinson, R.J., Parfitt, R.L. and Smart, R.S.C. (1974) Infrared study of phosphate adsorption on goethite. J. Chem. Soc. Faraday Trans. I, 70, 1472-1479.

Atkinson, R.J., Posner, A.M. and Quirk, J.P. (1968) Crystal nucleation in Fe(III) solutions and hydroxide gels. J. Inorg. Nucl. Chem. 30, 2371-2381.

Barker, W.W. and Graham, J. (1974) The crystal chemistry of complex niobium and tantalum oxides: V. Electrostatic energy calculations. Am. Mineral. 59, 1051-1054.

Bartram, S.F. and Slepetys, R.A. (1961) Compound formation and crystal structure in the system ZnO-TiO_2. J. Amer. Ceram. Soc. 44, 493-499.

Baur, W.H. (1956) Uber die verfeinerung der kristallstrukturbestimmung einiger vertreter des Rutiltyps: TiO_2, SnO_2, GeO_2 und MgF_2. Acta Crystallogr. 9, 515-520.

Baur, W. H. (1970) Bond length variation and distorted coordination polyhedra in inorganic crystals. Trans. Am. Crystallogr. Assoc. 6, 129-155.

Ben-Dur, L., Fischbein, E. and Kalman, Z. (1976) Concerning the β-phase of iron(III) oxide. Acta Crystallogr. B32, 667.

Bernstein. L. R. and Waychunas, G.A. (1987) Germanium crystal chemistry in hematite and goethite from the Apex Mine, Utah, and some new data on germanium in aqueous solution and in stottite. Geochim. Cosmochim. Acta 51, 623-630.

Bish, D.L. and Post, J.E. (1989) Thermal behavior of complex, tunnel-structure manganese oxides. Am. Mineral. 74, 177-186.

Blake, R.L., Hessevick, R.E., Zoltai, T. and Finger, L.W. (1966) Refinement of the hematite structure. Am. Mineral. 51, 123-129.

Boudeulle, M., Batis-Landoulsi, H., Leclercq C.-H., and Vergnon, P. (1983) Structure of γ-Fe_2O_3 microcrystals: vacancy distribution and structure. J. Solid State Chem 48, 21-32.

Bowles, J.F.W. (1988) Definition and range of composition of naturally occurring minerals with the pseudobrookite structure. Am. Mineral. 73, 1377-1383.

Braun, P. (1952) A superstructure in spinels. Nature (London) 170, 1123.

Brown, N.E. and Navrotsky, A. (1989) Structural, thermodynamic, and kinetic aspects of disordering in the pseudobrookite-type compound karrooite, $MgTi_2O_5$. Am. Mineral. 74, 902-912.

Buerger, M.J. (1936) The symmetry and crystal structure of manganite, Mn(OH)O. Zeits. Kristallogr. 95, 163-174.

Buhl, R. (1969) Manganites spinelles purs d'elements de transition preparations et structures cristallographiques. J. Phys. Chem. Solids 30, 805-812.

Burdett, J. K. (1984) Electronic control of the Geometry of Rutile and related structures. Inorg. Chem. 24, 2244-2253.

Burdett, J.K. and McLarnan, T. J. (1984) An orbital interpretation of Pauling's rules. Am. Mineral. 69, 601-621.

Burdett , J. K., Price, G.D. and Price, S.L. (1981) The factors influencing solid state structure. An interpretation using pseudopotential radii structure maps. Phys. Rev. B24, 2903-2912.

Burnham, C.W. (1990) The ionic model: Perceptions and realities in mineralogy. Am. Mineral. 75, 443-463.

Burns, R.G. (1989) Mössbauer spectra of ^{57}Fe in rare earth perovskites: applications to the electronic states of iron in the mantle. In Navrotsky, A. and Weidner, D.J., Eds. Perovskite: A structure of great interest to geophysics and materials science. Geophysical Mono. 45, Am. Geophys. Union, 146 pp.

Burns, R.G. and Burns, V.M. (1977) Mineralogy of Manganese Nodules. In G.P. Glasby (ed.), Marine Manganese Deposits. Elsevier, New York, Chapter 7, 185-248.

Burns, R.G. and Burns, V.M. (1979) Manganese Oxides. Rev. Mineral. 6, 1-46.

Burns, R.G, Burns, V.M., and Stockman, H.W. (1983) A review of the todorokite-buserite problem: Implications to the mineralogy of marine manganese nodules. Am. Mineral. 68, 972-980.

Burton, B. (1984) Thermodynamic analysis of the system Fe_2O_3-$FeTiO_3$. Phys. Chem. Minerals 11,132-139.

Burton, B. (1991) Interplay of chemical and magnetic ordering. Rev. Mineral. 25 (this volume), Chapter 8.

Byström, A.M. (1949) The crystal structure of ramsdellite, an orthorhombic modification of MnO_2. Acta Chem. Scand. 3, 163-173.

Catlow, C.R.A. and Mackrodt, W.C. (1986) Nonstoichiometric Compounds. Advances in Ceramics vol. 23, Am. Ceram. Soc. 726 pp.

Chakoumakos, B.C. (1984) Systematics of the pyrochlore structure type, ideal $A_2B_2X_6Y$. J. S.S. Chem. 53, 120-129.

Chambaere, D.G. and DeGrave, E. (1984) A study of the non-stoichiometrical halogen and water content of β-FeOOH. Physica status solidi 83, 93-102.

Chatterjee, N.D., Leistner, H., Terhart, L., Abraham, K. and Klaska, R. (1982) Thermodynamic mixing properties of corundum-eskolaite, α-$(Al,Cr^{+3})_2O_3$, crystalline solutions at high temperatures and pressures.

Chevallier, R., Bolfa, J. and Mathieu, S. (1955) Titanomagnetites et ilmenites ferromagnetiques. (1) Etude optique, radiocristallographique, chimique. Bull. Soc. Franc. Mineral. Cristallogr. 78, 307-346.
Chiang, Y.M. and Peng, C.J. (1986) Grain-boundary nonstoichiometry in spinels and titanites. in C.R.A. Catlow and W.C. Mackrodt (eds.) Nonstoichiometric Compounds. Advances in Ceramics v. 23, Am. Ceram. Soc. 361-378.
Christensen, A.N., Lehmann, M.S. and P. Convert (1982) Acta Chem. Scand. A36, 303-308.
Chukhrov, F.V., Gorshkov, A.I., Vitovskaya, I.V., Drits, V.A., Sivtsov, A.V. and Rudnitskaya, Y.S. (1980) Crystallochemical nature of Co-Ni asbolan. Akademiya Nauk SSSR Izvestiya, Seriya Geologicheskaya 6, 73-81 [in Russian].
Chukhrov, F.V., Gorshkov, A.I., Drits, V.A., Shterenberg, L. Y., Sivtsov, A.V. and Sakharov, B.A. (1983) Mixed-layer asbolane-buserite minerals and asbolanes in oceanic iron-manganese nodules. Akademiya Nauk SSSR Izvestiya, Seriya Geologicheskaya 5, 91-99 [in Russian].
Chukhrov, F.V., Zvyagin, B.B., Gorshkov, A.I., Ermilova, L.P., Korovushkin, V.V., Rudnitskaya, E.S., and Yu, N. (1976) Feroxyhyte, a new modification of FeO(OH). Izvest. Akad. Nauk SSR, Ser. geol. no. 5, 5-24 [in Russian].
Chukhrov, F.V., Zoyagin, B.B., Gorshkov, A.I., Yermilova, L.P. and Balashova, V.V. (1973) Izvest. Acad. Nauk SSR, Ser. Geol. 23-33 [in Russian].
Chukhrov, F.V., Sakharov, B.A., Gorshkov, A.I., Drits, V.A. and Dikov, H.P. (1985) The structure of birnessite from the Pacific Ocean. Akademiya Nauk SSSR Izvestiya, Seriya Geologicheskaya 8, 66-73 [in Russian].
Chukhrov, F.V., Manceau, A., Sakharov, B.A., Combes, J.-M., Gorshkov, A.I., Salyn, A.L. and Drits, V.A.(1988) Crystal chemistry of oceanic Fe-vernadites. Mineralogicheskii J. 10, 78-92.
Cid-Dresdner, H. and Escobar, C. (1968) The crystal structure of ferberite, $FeWO_4$. Zeits. Kristallogr. 127, 61-72.
Clark, G.M. (1972) The structures of non-molecular solids. John Wiley and Sons, New York. 365 pp.
Clark, A. M. and Fejer, E.E. (1978) Tapiolite, its chemistry and cell dimensions. Mineral. Mag. 42, 477-480.
Collyer, S., Grimes, N.W., Vaughan, D.J. and Longworth, G. (1988) Studies of the crystal structure and crystal chemistry of titanomaghemite. Am. Mineral. 73, 153-160.
Cormack, A.N. (1989) Defect interactions, extended defects and non-stoichiometry in ceramic oxides. In J. Nowotny and W. Weppner (eds.) Non-stoichiometric Compounds. NATO Advanced Workshop. Kluwer, Dordrecht 45-52.
Cromer, D.T. and Herrington, K. (1955) The structures of anatase and rutile. J. Am. Chem. Soc. 77, 4708-4709.
Dachs, H. (1956) Die kristallstruktur des bixbyits $(Fe,Mn)_2O_3$. Zeits. Kristallogr. 107, 370-395.
Dachs, H. (1963) Neutronen- und Röntgenuntersuchungen am Manganit, MnOOH. Zeits. Kristallogr. 118, 303-326.
de Wolff, P.M. (1959) Intrepretation of some γ-MnO_2 diffraction patterns. Acta Crystallogr. 12, 341-345.
Dent-Glasser, L.S. and Ingram, L. (1968) Refinement of the structure of groutite, α-MnOOH. Acta Crystallogr. B24, 1233-1236.
Dollase, W. A. (1980) Optimum distance model of relaxation around substutional defects. Phys. Chem. Minerals 6, 295-304.
Dunitz, J.D. and Orgel, L.E. (1957) Electron properties of transition metal oxides-II. Cation distribution amongst octahedral and tetrahedral sites. J. Phys. Chem. Solids 3, 318-323.
Eggleton, and Fitzpatrick, (1988) New data and a revised structural model for ferrihydrite. Clays and Clay Minerals 36, 111-124.
Evans, R.C. (1966) An introduction to Crystal Chemistry. Cambridge University Press. 410 pp.
Ewing, R.C. (1975) The crystal chemistry of complex niobium and tantalum oxides: IV. The metamict state: Discussion. Am. Mineral 60, 728-733.
Fasiska, E.J. (1967) Structural aspects of the oxides and oxyhydroxides of iron. Corrosion Sci. 7, 833-839.
Feitknecht, W., Giovanoli, R., Michaelis, W., and Müller, M. (1973) Über die hydrolyse von eisen (III) salzlösungen. 1. Die hydrolyse der lösungen von eisen (III) chlorid. Helv. Chim. Acta 56, 2847-2856.
Ferguson, R.B., Hawthorne, F.C. and Grice, J.D. (1976) The crystal structures of tantalite, ixiolite and wodginite from Bernic Lake, Manitoba II. Wodginite. Can. Mineral. 14, 550-560.
Fleet, M.E. (1984) The structure of magnetite: two annealed natural magnetites, $Fe_{3.005}O_4$ and $Fe_{2.96}Mg_{0.04}O_4$. Acta Crystallogr. C40, 1491-1493.
Fleischer, M. (1987) Glossary of Minerals 5th edition.
Forster, R.H. and Hall, E.O. (1965) A neutron and x-ray diffraction study of ulvöspinel, Fe_2TiO_4. Acta Crystallogr. 18, 857-862.
Franz, E.-D. (1978) Synthetic solid solutions between goethite and diaspore. Mineral. Mag. 42, 159.
Franzen, H.F. (1986) Physical Chemistry of Inorganic Crystalline Solids. Springer-Verlag, Berlin. 158 pp.

Freer, R. and O'Reilly, W. (1980) The diffusion of Fe^{2+} ions in spinels with relevance to the process of maghemitization. Mineral. Mag. 43, 889-899.
Fysh, S.A. and Clark, P.E. (1982a) Aluminous Hematite: A Mössbauer study. Phys. Chem. Minerals 8, 257-267.
Fysh, S.A. and Clark, P.E. (1982b) Aluminous Goethite: A Mössbaüer study. Phys. Chem. Minerals 8, 180-187.
Gartstein, E., Mason, T.O. and Cohen, J.B. (1986) The agglomeration of point defects in wüstite at high temperatures. In C.R.A. Catlow and W.C. Mackrodt (eds.) Nonstoichiometric Compounds. Advances in Ceramics v. 23, Am. Ceram. Soc., 699-710.
Gaspar, J.C. and Wyllie, P.J. (1983) Ilmenite (high Mg, Mn, Nb) in the carbonatites from the Jacupiranga Complex, Brazil. Am. Mineral. 68, 960-971.
Gatehouse, B.M. and Grey, I.E. (1982) The crystal structure of högbomite-8H. Am. Mineral. 67, 373-380.
Gatehouse, B.M., Jones, G.C., Pring, A. and Symes, R.F. (1986) The chemistry and structure of redledgeite. Mineral. Mag. 50, 709-715.
Gavarri, J.R., Weigel, D. and Carel, C. (1976) Introduction to description of phase diagram of solid wüstite. II. Structural Review. Mater. Res. Bull. 11, 917
Geller, S. (1971) Structures of α-Mn_2O_3, $(Mn_{0.983}Fe_{0.017})_2O_3$, $(Mn_{0.37}Fe_{0.63})_2O_3$ and relation to magnetic ordering. Acta Crystallogr. B27, 821-828.
Gibbs, G.V., Meagher, E.P., Newton, M.D. and Swanson, D.K. (1981) A comparison of experimental and theoretical bond length and angle variations for minerals, inorganic solids, and molecules. In M. O'Keeffe and A. Navrotsky (eds.) Structure and Bonding in Crystals, vol. 1, p. 195-226. Academic Press, New York.
Giddings, R.A. and Gordon, R.S. (1974) Solid-state coulometric titration: critical analysis and application to wüstite. J. Electrochem. Soc. Solid State Sci. and Tech. 793.
Giovanoli, R. (1980) On natural and synthetic manganese nodules. In I.M. Varentsov, and G. Grasselly, (eds.) Geology and Geochemistry of Manganese, vol. 1, p. 159-202, E. Schweizerbart'sche Verlags-buchhandlung, Stuttgart.
Giovanoli, R. and Arrhenius, G. (1988) Structural chemistry of marine manganese and iron minerals and synthetic model compounds. In P. Halbach, G. Friedrich and U. von Stackelberg, eds., The manganese nodule belt of the Pacific Ocean. Ferdinand Enke Verlag, Stuttgart.
Giovanoli, R., Buhler, H. and Sokolowska, K. (1973) Synthetic lithiophorite: electron microscopy and X-ray diffraction. J. de Microscopie 18, 271-284.
Giovanoli, R. and Bürki, P. (1975) Comparisons of X-ray evidence of marine manganese nodules and non-marine manganese ore deposits. Chimia 29, 114-117.
Giovanoli, R., Stahli, E., and Feitknecht, W. (1970) Über Oxid-hydroxide des vierwertigen Mangans mit Schichtengitter. Helv. Chim. Acta 53, 209-220.
Goldschmidt, V.M. (1937) The principles of distribution of chemical elements in minerals and rocks. J. Chem. Soc. London, 655-673
Gonschorek, W. and Feld, R. (1982) Neutron diffraction study of the thermal and oxygen position parameters in rutile. Zeits. Kristallogr. 161, 1-5.
Goss, C.J. (1988) Saturation magnetisation, coercivity and lattice parameter changes in the system Fe_3O_4-γ-Fe_2O_3, and their relationship to structure. Phys. Chem. Minerals 16, 164-171.
Graham, J. and Thornber, M.R. (1974a) The crystal chemistry of complex niobium and tantalum oxides: I. Structural classification of MO_2 phases. Am. Mineral. 59, 1026-1039.
Graham, J. and Thornber, M.R. (1974b) The crystal chemistry of complex niobium and tantalum oxides: II. Composition and structure of wodginite. Am. Mineral. 59, 1040-1044.
Graham, J. and Thornber, M.R. (1975) The crystal chemistry of complex niobium and tantalum oxides: IV. The metamict state: Reply. Am. Mineral. 60, 734.
Greaves, C. (1983) A powder neutron diffraction investigation of vacancy and covalence in γ-ferric oxide. J. Solid State Chem. 30, 257-263.
Grey, I.E., Li, C. and Watts, J.A. (1983) Hydrothermal synthesis of goethite-rutile intergrowth structures and their relationship to pseudorutile. Am. Mineral. 68, 981-988.
Grey, I.E. and Nickel, E. H. (1981) Tivanite, a new oxyhydroxide mineral from Western Australia, and its structural relationship to diaspore. Am. Mineral. 66, 866-871.
Grey, I.E. and Ward, J. (1973) An X-ray and Mössbauer study of the $FeTi_2O_5$-Ti_3O_5 system. J. Sol. State Chem. 7, 300-307.
Greenwood, N.N. (1968) Ionic crystals, Lattice defects and Nonstoichiometry Butterworths, London 194pp.
Grice, J.D., Ferguson, R.B. and Hawthorne, F.C. (1976) The crystal structures of tantalite, ixiolite and wodginite from Bernic Lake, Manitoba I. Tantalite and ixiolite. Can. Mineral. 14, 540-549.
Grimes, N.W. Thompson, P. and Kay, H.F. (1983) New symmetry and structure for spinel. Proc. Roy. Soc. A 386, 333-345.

Haas, C. (1965) Phase transitions in crystals with the spinel structure. J. Phys. Chem. Solids 26, 1225-1232.
Hayakawa, M., Wagner, J.B., Jr., and Cohen, J.B. (1977) Comments on "Phase diagram of solid wüstite--part I." Mater. Res. Bull. 12, 429-430.
Hazen, R. and Jeanloz, R. (1984) Wüstite ($Fe_{1-x}O$): A review of its defect structure and physical properties. Rev. Geophys. Space Phys. 22, 37-46.
Henderson, P. (1982) Inorganic Geochemistry. Pergamon Press, Oxford. 353 pp.
Hentschel, B. (1970) Stoichiometric FeO as metastable intermediate of the decomposition of wüstite at 225°C. Zeit. Naturforschung 25, 1996-1997.
Hill, R.J. (1984) X-ray powder diffraction profile refinement of synthetic hercynite. Am. Mineral. 69, 937-942.
Hill, R.J. (1982) The crystal structures of lead dioxides from the positive side of the lead/acid battery. Mater. Res. Bull. 17, 769-784.
Hill, R.J. , Craig, J.R., and Gibbs, G.V. (1979) Systematics of the spinel structure type. Phys. Chem. Minerals 4, 317-339.
Hogarth, D.D. (1977) Classification and nomenclature of the pyrochlore group. Am. Mineral. 62, 403-410.
Horn, M., Schwerdtfeger, C.F. and Meagher, E.P. (1972) Refinement of the structure of anatase at several temperatures. Zeits. Kristallogr. 136, 273-281.
Huebner, J.S. (1976) The manganese oxides-a bibliographic commentary. Rev. Mineral. 3, SH1-SH17.
Hyde, B.G. and Andersson, S. (1989) Inorganic Crystal Structures. John Wiley & Sons, New York. 430 pp.
Ishikawa, Y. and Akimoto, S. (1957) Magnetic properties of the $FeTiO_3$-Fe_2O_3 solid solution series. J. Phys. Soc. Japan 12, 1083-1098.
Ishikawa, Y. and Syono, Y. (1963) Order-disorder transformation and reverse thermoremanent magnetism in the $FeTiO_3$-Fe_2O_3 system. J. Phys. Chem. Solids 24, 517-528.
Jaffe, H.W. (1988) Crystal Chemistry and refractivity. Cambridge University Press, Cambridge. 335 pp.
Kitaigorodsky, A.I. (1984) Mixed Crystals. Springer Series in Solid-state Sciences, v. 33, 388 pp.
Landau, L.D, and Lifshitz, E.M. (1959) Statistical Physics. Pergamon Press, Oxford.
Laves, F.H. (1980) Similarity and miscibility of inorganic crystals. Zeits. Kristallogr. 151, 21-29.
Lindsley, D.H. (1976) The crystal chemistry and structure of oxide minerals as exemplified by the Fe-Ti oxides. Rev. Mineral. 3, L1-L60.
Lumpkin, G.R. and Ewing, R.C. (1985) Natural pyrochlores: Analogues for actinide host phases in radioactive waste forms. In C.M. Jantzen, J.A. Stone, and R.C. Ewing (eds.) Scientific basis for nuclear waste management VIII. Materials Res. Soc. Symp. Proc. 44, 647-654.
Manceau, A., Combes, J.-M., and Calas, G. (1990) New data and a revised model for ferrihydrite: a comment on a paper by R.A. Eggleton and R.W. Fitzpatrick. Clays and Clay Minerals 38, 331-334.
Manceau, A., Llorca, S., and Calas, G. (1987) Crystal chemistry of cobalt and nickel in lithiophorite and asbolane from New Caledonia. Geochim. Cosmochim. Acta 51, 105-113.
Marezio, M., Remeika, J.P. and Dernier, P.D. (1970) The crystal chemistry of the rare earth orthoferrites. Acta Crystallogr. B26, 2008-2022.
Marshall, C.P. and Dollase, W.A. (1984) Cation arrangement in iron-zinc-chromium spinel oxides. Am. Mineral. 69, 928-936.
Mason, T.O. (1987) Cation intersite distributions in iron-bearing minerals via electrical conductivity/Seebeck effect. Phys. Chem. Minerals 14, 156-162.
Matsubara, S., Kato, A. and Nagashima, K. (1979) Iwakiite, $Mn^{+2}(Fe^{+3},Mn^{+3})_2O_4$, A new tetragonal spinelloid mineral from the Gozaisho Mine, Fukushima Prefecture, Japan. Mineral. J. (Tokyo) 9, 383-391.
Mattioli, G.S., Wood, B.J. and Carmichael, I.S.E. (1987) Ternary-spinel volumes in the system $MgAl_2O_4$-Fe_3O_4-$\gamma Fe_{8/3}O_4$: Implications for the effect of P on intrinsic fO_2 measurements of mantle-xenolith spinels. Am. Mineral. 72, 468-480.
McCammon, C.A. and Liu, L.-G. (1984) The effects of Pressure and Temperature on Nonstoichiometric wüstite, Fe_xO: The iron-rich phase boundary. Phys. Chem. Minerals 10, 106-113.
McClure, D.S. (1957) Distribution of transition metal cations in spinels. J. Phys. Chem. Solids 3, 311-317.
McKie, D. (1963) The högbomite polytypes. Mineral. Mag. 33, 563-580.
Meagher, E.P. and Lager, G.A. (1979) Polyhedral thermal expansion in the TiO_2 polymorphs: Refinement of the crystal structures of rutile and brookite at high temperatures. Can. Mineral. 17, 77-85.
Meyer, I., Hirdes, W. and Lodziak, J. (1986) A pure barium hollandite from the Nuba Mountains in Sudan. N. Jahrb. Mineral. Monat. 30-36.
Misewa, T., Hashimoto, K. and Shimodaira, S. (1974) The mechanisms of formation of iron oxide and oxyhydroxides in aqueous solutions at room temperature. Corrosion Sci. 14, 131-149.

Muir, I.J., Metson, J.B. and Bancroft, G.M. (1984) ^{57}Fe Mössbauer spectra of perovskite and titanite. Can. Mineral. 22, 689-694.

Murray, J.W. (1979) Iron Oxides. Rev. Mineral. 6, 47-98.

Naito, K., Tsuji, T. and Matsui, T. (1989) Defect structure and the related properties of UO_2 and doped UO_2. In J. Nowotny and W. Weppner, Eds. Non-stoichiometric Compounds. NATO Advanced Workshop. Kluwer, Dordrecht, 27-44.

Naray-Szabo, St. V. (1943) Der Strukturtyp des Perovskits ($CaTiO_3$). Naturwiss. 31, 202-203.

Navrotsky, A. (1986) Cation-distribution energetics and heats of mixing in $MgFe_2O_4$-$MgAl_2O_4$, $ZnFe_2O_4$-$ZnAl_2O_4$, and $NiAl_2O_4$-$ZnAl_2O_4$ spinels: Study by high-temperature calorimetry. Am. Mineral. 71, 1160-1169.

Navrotsky, A. and Weidner, D.J. (1989) Perovskite: A structure of great interest to geophysics and materials science. Geophysical Monograph 45, AGU, 146pp.

Neel, L. (1955) Some theoretical aspects of rock magnetism. Adv. Phys. 4, 191-243.

Nell, J., Wood, B.J. and Mason, T.O. (1989) High-temperature cation distributions in Fe_3O_4-$MgAl_2O_4$-$MgFe_2O_4$-$FeAl_2O_4$ spinels from thermopower and conductivity measurements. Am. Mineral. 74, 339-351.

Newnham, R.E. and deHaan, Y.M. (1962) Refinement of the α-Al_2O_3, Ti_2O_3, V_2O_3 and Cr_2O_3 structures. Zeits. Kristallogr. 117, 235-237.

Nord, G.L., Jr. and Lawson, C. A. (1989) Order-disorder transition-induced twin domains and magnetic properties in ilmenite-hematite. Am. Mineral. 74, 160-176.

O'Donovan, J.B. and O'Reilly, W. (1980) The temperature dependent cation distribution in titanomagnetites: an experimental test. Phys. Chem. Minerals 5, 235-243.

O'Neill, H. St.-C., and Navrotsky, A. (1984) Cation distributions and thermodynamic properties of binary spinel solid solutions. Am. Mineral. 69, 733-755.

O'Reilly, W. and Banerjee, S.K. (1965) Cation distribution in titanomagnetites $(1-x)Fe_3O_4$-xFe_2TiO_4. Phys. Lett. 17, 237-238.

O'Reilly, W. and Banerjee, S.K. (1967) The mechanism of oxidation in titanomagnetites: A magnetic study. Mineral. Mag. 36, 29-37.

Oles, A., Szytula, A. and Wanic, A. (1970) Neutron Diffraction study of γ-FeOOH. Phys. Stat. Solidi 41, 173-177.

Patrat, G, DeBergevin, F., Pernet, M. and Joubert, J.C. (1983) Structure Locale de γ-FeOOH. Acta Crystallogr. B39, 165-170.

Pauling, L. (1929) The principles determining the structure of complex ionic crystals. J. Am. Chem. Soc. 51, 1010-1026.

Pauling, L. and Kamb, B. (1982) The crystal structure of lithiophorite. Am. Mineral. 67, 817-821.

Pearson (1973) Factors controlling the formation and structure of phases. In N. Bruce Hannay (ed.) Treatise on Solid State Chemistry, vol. 1 , 115-174.

Perrault, G. (1968) La composition chimique et la structure cristalline du pyrochlore d'Oka, P.Q. Can. Mineral. 9, 383-402.

Posnjak, E. and Barth, T.F.W. (1934) Notes on some structures of the ilmenite type. Zeits. Kristallogr. 88, 271-280.

Post, J.E. and Appleman, D.E. (1988) Chalcophanite, $ZnMn_3O_7 \cdot 3H_2O$: New crystal-structure determinations. Am. Mineral. 73, 1401-1404.

Post, J.E. and Bish, D.L. (1989) Rietveld refinement of the coronadite structure. Am. Mineral. 74, 913-917.

Post, J.E. and Bish, D.L. (1988) Rietveld refinement of the todorokite structure. Am. Mineral. 73, 861-869.

Post, J.E. and Burnham, C.W. (1986a) Ionic modeling of mineral structures and energies in the electron gas approximation: TiO_2 polymorphs, quartz, forsterite, diopside. Am. Mineral. 71, 142-150.

Post, J.E. and Burnham, C.W. (1986b) Modeling tunnel-cation displacements in hollandites using structure-energy calculations. Am. Mineral. 71, 1178-1185.

Post, J.E. and Veblen, D.R. (1990) Crystal structure determinations of synthetic sodium, magnesium, and potassium birnessite using TEM and the Rietveld method. Am. Mineral. 75, 477-489.

Post, J.E., Von Dreele, R.B., and Buseck, P.R. (1982) Symmetry and cation displacements in hollandites: Structure refinements of hollandite, cryptomelane, and priderite. Acta Crystallogr. B38, 1056-1065.

Potter, R.S. and Rossman, G.R. (1979) The tetravalent manganese oxides: identification, hydration, and structural relationships by infrared spectroscopy. Am. Mineral. 64, 1199-1218.

Price, G.D. (1983) Polytypism and the factors determining the stability of spinelloid structures. Phys. Chem. Minerals 10, 77-83.

Price, G.D., Price, S.L. and Burdett, J.K. (1982) The factors influencing cation site-preferences in spinels. A new Mendelyevian approach. Phys. Chem. Minerals 8, 69-76.

Putnis, A. (1978) The mechanism of exsolution of hematite from iron-bearing rutile. Phys. Chem. Minerals 3, 183-197.
Rask, J.H. and Buseck, P.R. (1986) Topotactic relations among pyrolusite, manganite, and Mn_5O_8: A high-resolution transmission electron microscopy investigation. Am. Mineral. 71, 805-814.
Readman, P.W. and O'Reilly, W. (1971) Oxidation processes in titanomagnetites. Zeits. Geophys. 37, 329-338.
Rezlescu, N., Cuciureanu, E., Ioan, C. and Luca, E. (1972) Time variation of the electrical conductivity in spinel ferrites. Physica status solidi A 11, 351-359.
Rossman, G.R. (1975) Spectroscopic and magnetic studies of ferric iron hydroxy sulfates: Intensification of color in iron(III) clusters bridged by a single hydroxide ion. Am. Mineral. 60, 698-704.
Russell, J.D. (1979) Infrared spectroscopy of ferrihydrites: Evidence for the presence of structural hydroxyl groups. Clay Mineral. 14, 109-114.
Sahl, K. (1965) Gitterenergetische Berechnungen an geometrischen Deformationen der Rutilstruktur. Acta Crystallogr. 19, 1027-1030.
Schmidbauer, E. (1987) ^{57}Fe Mossbauer spectroscopy and magnetization of cation deficient Fe_2TiO_4 and $FeCr_2O_4$. Part II: Magnetization data. Phys. Chem. Minerals 15, 201-207.
Schmidbauer, E. and Lebküchner-Neugebauer, J. (1987) ^{57}Fe Mössbauer study on compositions of the series $Fe^{3+}TaO_4$-$Fe^{2+}Ta_2O_6$. Phys. Chem. Minerals 15, 196-200.
Seeliger, E. and Mücke, A. (1969) Donathit, ein tetragonaler, Zn-reicher mischkristall von magnetit und chromit. N. Jahrb. Mineral. Monat. 49-57.
Sherman, D.M. (1985a) SCF-Xα-SW MO study of Fe-O and Fe-OH chemical bonds; Applications to the Mössbauer spectra and magnetochemistry of hydroxyl-bearing Fe^{3+} oxides and silicates. Phys. Chem. Minerals 12, 311-314.
Sherman, D.M. (1985b) Electronic structures of Fe^{3+} coordination sites in iron oxides; application to spectra, bonding and magnetism. Phys. Chem. Minerals 12 161-175.
Shiojiri, M., sekimoto, S., Maeda, T., Ikeda, Y. and Iwauchi, K. (1984) Crystal structure of Fe_2TiO_5. Physica status solidi A84, 55-64.
Shirakashi, T. and Kubo, T. (1979) Cation distribution in franklinite by nuclear magnetic resonance. Am. Mineral. 64, 599-603.
Sinclair, W. and McLaughlin, G.M. (1982) Structure refinement of priderite. Acta Crystallogr. B38, 245-246.
Smith, P.P.K. (1979) The observation of enantiomorphous domains in a natural maghemite. Contrib. Mineral. Petrol. 69 249-254.
Smyth, D.M. (1985) Defects and order in perovskite-related oxides. Ann. Rev. Mater. Sci. 329-357.
Sposito, G. (1984) The surface chemistry of soils. Oxford University Press, New York, 234 pp.
Stillwell, C.W. (1938) Crystal Chemistry. McGraw-Hill, New York, 431 pp.
Subramanian, M.A., Aravamudan, G. and Subba Rao, G.V. (1983) Oxide Pyrochlores-A review. Progress in Solid State Chemistry 15, 55-143.
Szymanski, J.T. (1986) The crystal structure of mannardite, a new hydrated cryptomelane-group (hollandite) mineral with a doubled short axis. Can. Mineral. 24, 67-78.
Szytula, A., Burewicz, A., Dimitrijevic, Z. , Krasnicki, S., Rzany, H., Todorovic, J., Wanic, A. and Wolski, W. (1980) Neutron diffraction studies of α-FeOOH. Physica status solidi 26, 429-434.
Szytula, A., Balanda, M. and Dimitrijevic, Z. (1970) Neutron diffraction studies of β-FeOOH. Physica status solidi (a) 3, 1033-1037.
Tiedemann, P. and Muller-Buschbaum, Hk. (1982) Zum problem der metallverteilung in pseudobrookiten: $FeAlTiO_5$ und Fe_2TiO_5. Zeits. Anorg. Chem. 494, 98-102.
Tilley, R.J.D. (1987) Defect crystal chemistry and its applications. Chapman and Hall, New York. 236pp.
Towe, K.M. and Bradley, W.F. (1967) Mineralogical constitution of colloidal "hydrous ferric oxides". J. Coll. Interface Sci. 24, 384-392.
Trestman-Matts, A., Dorris, S.E., Kumarakrishnan, S. and Mason, T.O. (1983) Thermoelectric determination of Cation Distributions in Fe_3O_4-Fe_2TiO_4. J. Am. Ceram. Soc. 66, 829-834.
Tsirelson, V.G., Antipin, M.Y., Gerr, R.G., Ozerov, R.P. and Struchkov, J.T. (1985) Ruby structure peculiarities derived from X-ray diffraction data. Physica status solidi A37, 425-433.
Turner, S. (1982) A structural study of tunnel manganese oxides by high-resolution transmission electron microscopy. Ph.D. dissertation, Arizona State University, Tempe, AZ.
Turner, S. and Buseck, P.R. (1979) Manganese oxide tunnel structures and their intergrowths. Science, 203, 456-458.
Turner, S. and Buseck, P.R. (1983) Defects in nsutite (γ-MnO_2) and dry-cell battery efficiency. Nature 304, 143-146.
Turner, S. and Post, J.E. (1988) Refinement of the substructure and superstructure of romanechite. Am. Mineral. 73, 1155-1161.

Turnock, A.C. and Eugster, H.P. (1962) Fe-Al oxides: phase relationships below 1000°C. J. Petrol. 3, 533-565.
van der Giessen, A.A. (1966) The structure of iron (III) oxide-hydrate gels. J. Inorg. Nucl. Chem. 28, 2155-59.
van Oosterhout, G.W. and Rooijmans, C.J.M. (1958) A new superstructure in gamma-ferric oxide. Nature 181, 44.
Von Heidenstam, O. (1968) Neutron and X-ray diffraction studies on Tapiolite and some synthetic substances of trirutile structure. Ark. Kemi. 28, 375-387.
Wadsley, A.D. (1952) The structure of lithiophorite, $(Al,Li)MnO_2(OH)_2$. Acta Crystallogr. 5, 676-680.
Wadsley, A.D. (1953) The crystal structure of psilomelane, $(Ba,H_2O)_2Mn_5O_{10}$. Acta. Crystallogr. 6, 433-438.
Wadsley, A.D. (1955) Crystal structure of chalcophanite, $ZnMn_3O_7.3H_2O$. Acta Crystallogr. 8, 165-72.
Wadsley, A.D. (1963) Nonstoichiometric compounds. L. Mandelcorn (ed.) Academic Press, New York.
Waychunas, G.A. (1979) Spectroscopic study of cation arrangements and defect association in Fm3m oxide solid solutions. Ph.D. dissertation, Univ. of California, Los Angeles, CA.
Waychunas, G.A. (1983) Mössbauer, EXAFS and X-ray diffraction study of Fe^{3+} clusters in MgO:Fe and magnesiowüstite $(Mg,Fe)_{1-x}O$-evidence for specific cluster geometries. J. Mater. Sci. 18, 195-207.
Waychunas, G.A., Dollase, W.A. and Ross, C.R. (1986) Determination of short range order (SRO) parameters from EXAFS pair distribution functions of oxide and silicate solid solutions. J. Physique 47, C8 845-848.
Waychunas, G.A., Rea, B.A., Fuller, C.C, and Davis, J.A. (1990) Fe and As K-edge EXAFS study of arsenate $(AsO_4)^{-3}$ adsorption on "two-line" ferrihydrite. in XAFS VI: X-ray Absorption Fine Structure VI, Ellis Horwood Ltd., Chichester, U.K.
Wechsler, B.A. (1977) Cation distribution and high-temperature crystal chemistry of armalcolite. Am. Mineral. 62, 913-920.
Wechsler, B.A., Lindsley, D.H. and Prewitt, C.T. (1984) Crystal structure and cation distribution in titanomagnetites $(Fe_{3-x}Ti_xO_4)$. Am. Mineral. 69, 754-770.
Wechsler, B.A. and Prewitt, C.T. (1984) Crystal structure of ilmenite $(FeTiO_3)$ at high temperature and at high pressure. Am. Mineral. 69, 176-185.
Wechsler, B.A. and Von Dreele, R. B. (1989) Structure of Mg_2TiO_4, $MgTiO_3$ and $MgTi_2O_5$ by Time-of-Flight neutron powder diffraction. Acta Crystallogr. B45, 542-549.
Weitzel, H. (1976) Kristallstrukturverfeinerung von Wolframiten und Columbiten. Zeits. Kristallogr. 144, 238-258.
Wells, A.F. (1975) Structural Inorganic Chemistry. Oxford University Press, New York.
Willis, B.T.M. (1978) The defect structure of hyper-stoichiometric uranium dioxide. Acta Crystallogr. A34, 88-90.
Willis, B.T.M. (1963) Positions of the atoms in $UO_{2.13}$. Nature 197, 755-756.
Wise, M.A., Turnock, A.C. and Cerny, P. (1985) Improved unit cell dimensions for ordered columbite-tantalite end members. N. Jahrb. Miner. Monat. 372-378.
Wolska, E. (1981) The structure of hydrohematite. Zeits. Kristallogr. 154, 69-75.
Wolska, E. and Szajda, W. (1985) Structural and spectroscopic characteristics of synthetic hydrohematite. J. Mater. Sci. 20, 4407-4412.
Wolska, E. and Schwertmann, U. (1989) Zeits. Kristallogr. 189, 223-237.
Wood, B.J., Kirkpatrick, R.J. and Montez, B. (1986) Order-disorder phenomena in $MgAl_2O_4$ spinel. Am. Mineral. 71, 999-1006.
Wort, M.J. and Jones, M.P. (1980) X-ray diffraction and magnetic studies of altered ilmenite and pseudorutile. Mineral. Mag. 43, 659-663.
Yagi, K. and Roth, R.S. (1978) Electron-microscope study of the crystal structures of mixed oxides in the systems $Rb_2O-Ta_2O_5$, $Rb_2O-Nb_2O_5$ and $K_2O-Ta_2O_5$ with composition ratios near 1:3. I. Stacking characteristics of MO_6 layers. Acta Crystallogr. A34, 765-773.
Yamanaka, T. and Takeuchi, Y. (1983) Order-disorder transition in $MgAl_2O_4$ spinel at high temperatures up to 1700°C. Zeits. Kristallogr. 165, 65-78.
Zachariasen, W.H. (1954) Crystal chemical studies of the 5f-series of elements. XXIII. On the crystal chemistry of uranyl compounds and of related compounds of transuranic elements. Acta Crystallogr. 7, 795-799.
Zachariasen, W.H. (1963) The crystal structure of monoclinic metaboric acid. Acta Crystallogr. 16, 385-9.
Zeller, C., Hubsch, J. and Bolfa, J. (1967) Relations entre les proprietes magnetiques et la structure d'une titanomagnetite au cours son oxydation. C.R. Acad. Sci. B265, 1034-1036.
Zoltai, T. and Stout, J.H. (1984) Mineralogy, Concepts and Principles. Burgess Publishing Co., Minneapolis, MN. 505 pp.
Zunger, A. (1980) Systematization of the stable crystal structure of all AB-type binary compounds: A pseudopotential orbital radii approach. Phys. Rev. B 22, 5839-5872.

Chapter 3   Donald H. Lindsley
EXPERIMENTAL STUDIES OF OXIDE MINERALS

INTRODUCTION

"*Experimental studies of oxide minerals are made for a variety of reasons, including: ore processing, effects on magnetic properties, and determining the conditions under which the oxide minerals form. This review will concentrate mainly on the last topic. The oxide minerals in a rock can, in many cases, be considered as a `subsystem' of the rock, more or less independent of the silicates or other minerals present.*" These opening sentences from the phase equilibrium chapter in the first edition (Lindsley, 1976, p. L-61) serve to highlight important differences in the current volume. Much of the present book emphasizes magnetic properties, and the chapters on phase petrology highlight our progress in understanding oxide-silicate interactions over the past 15 years. Nevertheless, much can be learned from experimental studies of oxides as isolated systems, and that aspect is emphasized here.

In a chapter such as this it is always necessary to choose which minerals and systems to cover and which to omit so as to keep the length within reasonable bounds. There are far too many natural oxide minerals to cover in detail, and there are many more synthetic oxides (including high-temperature superconductors). The goal here is to provide reasonable coverage for phases that most petrologists would recognize as "oxide minerals" in a rock from the crust or uppermost mantle. This definition excludes $(Mg,Fe)_2SiO_4$ spinels and post-spinel phases such as magnesiowüstite and $(Mg,Fe)SiO_3$ "perovskite" (for example, Ito and Takahashi, 1989), which might reasonably be considered oxide minerals. Petrologically, however, all of these but magnesiowüstite clearly belong to the silicates, and there they will stay for the remainder of this chapter. [The arbitrary exclusion of silicates having oxide structures becomes all the more problematical when solid solutions are formed between such silicates and true oxides, as was done for example by Ohtani and Sawamoto (1976).] The emphasis of this chapter therefore is on experiments that give information on the conditions of formation of the oxides or the origin of textures; experiments relating to magnetic properties are reviewed in Chapter 4.

A major change in the past fourteen and one half years is the decrease in the number of studies aimed directly at determining phase diagrams per se, and the concomittant increase in the proportion aimed at determining thermodynamic properties - from which, in favorable cases, improved phase diagrams can be calculated. Because several chapters (6-9, and to a certain extent 11) emphasize thermodynamics and calculated diagrams, I have tried to minimize overlap by directing the reader to the appropriate chapter.

Synthesis vs. phase equilibrium

It is important to bear in mind that many experimental studies that appear to report equilibrium phase relations are in fact *synthesis* studies - that is, the reported phase or phases were grown from oxide mixes or other materials of relatively high free energy. Sometimes synthesis experiments produce equilibrium results, but altogether too often they do not. The reader is urged to treat such results with caution and to rely mainly on phase diagrams that have been verified by *reversed reactions*, that is where phase or assemblage A has been converted to B and then B is converted back to A.

Control of experimental conditions

Since a major purpose of experiments is to reproduce or otherwise identify the conditions under which certain minerals have formed, the experimenter must be able to control or monitor a variety of parameters. Temperature and pressure come immediately to mind; and while there are various technical difficulties in controlling them (see, for example, Holloway and Wood, 1988, and various chapters in Ulmer, 1971, for details) they are relatively simple conceptually and do not need further discussion here. As emphasized in Chapter 1 of this volume, oxygen fugacity (fO_2) is also an important parameter for oxides of Fe, Mn, V, Ti, and Cr, because those elements can exist in more than one oxidation state.

Oxygen fugacity. Rarely in geological environments are oxygen fugacities greater than that of pure air ($fO_2 \sim PO_2 = 0.21$ atm.) encountered. Indeed, for most processes other than weathering and surficial oxidation of volcanic rocks, the appropriate fO_2 values are much lower (see for example, Sato and Wright, 1966; Sato, 1972; Sato et al., 1973). To a limited extent the experimenter can lower the fO_2 simply by decreasing the air pressure and assuming the the proportion of oxygen remains fixed at 0.21. But this method is restricted by the limitations of the vacuum system: $PO_2 \sim 10^{-6}$ to 10^{-8} atm. for good mechanical pumps; and while diffusion pumps can achieve lower pressures, it is unlikely that the proportion of oxygen in the remaining gas remains that of air. Thus other techniques are generally necessary.

Most phase-equilibrium experiments on oxide minerals have used solid-solid buffers (Chapter 1; Huebner, 1971; Chou and Cygan, 1990) or mixtures of gases - usually but not necessarily at a total pressure (P_{tot}) of one atm. In the gas-mixture technique (see, for example, Nafziger et al., 1971) a relatively oxidizing gas such as CO_2 or H_2O is mixed with a reducing gas (CO or H_2) and passed through the furnace in the vicinity of the oxide sample. The gases react, yielding various species, including oxygen. The amount of oxygen (and thus fO_2) is controlled by the mixing ratio of the gases and the temperature of the furnace. Deines et al. (1974) tabulate fO_2 values for a wide range of temperatures and C-H-O gas mixtures. While experimental fO_2 values can be estimated using such tables, virtually all modern studies use an oxygen probe (similar to that described in Chapter 7) to measure fO_2.

Container problem. Perhaps surprisingly for the non-experimentalist, relatively few materials can serve as satisfactory sample containers for oxides during experiments. Many containers commonly used for other phases such as silicates react with the sample, changing its composition. For example, for all but very high values of fO_2, platinum reacts with iron-bearing oxides and silicates (e.g., Merrill and Wyllie, 1973):

$$2 \text{ FeO (in oxide or silicate)} + \text{Pt} = \text{Fe-Pt alloy} + O_2 \qquad (1)$$

Table 1. Names, ideal formulas, and abbreviations of oxide minerals and phases discussed in this chapter.

Name	Ideal Formula	Abbreviation
Spinels		
Magnetite	Fe_3O_4	Mt
Maghemite	γFe_2O_3	Mgh
Ulvospinel	Fe_2TiO_4	Usp
Qandilite	Mg_2TiO_4	Qan
Titanomagnetite	Fe_3O_4-Fe_2TiO_4 (ss)	Ti-Mt
Titanomaghemite	$(Fe^{2+},Fe^{3+},Ti,\square)_3O_4$	Ti-Mgh
Hercynite	$FeAl_2O_4$	Hc
Spinel	$MgAl_2O_4$	Sp
Chromite	$FeCr_2O_4$	Chr
Picrochromite	$MgCr_2O_4$	Mg-Chr
Franklinite	$ZnFe_2O_4$	Frk
Gahnite	$ZnAl_2O_4$	Gah
Rhombohedral series		
Corundum	Al_2O_3	Cor
Hematite	αFe_2O_3	Hem
Ilmenite	$FeTiO_3$	Ilm
Geikielite	$MgTiO_3$	Gk
Pyrophanite	$MnTiO_3$	Prph
Orthorhombic series		
Pseudobrookite	Fe_2TiO_5	Psb
Armalcolite	$Fe_{0.5}Mg_{0.5}Ti_2O_5$	Arm
"Ferropseudobrookite"	$FeTi_2O_5$	Fpb
"Karrooite"	$MgTi_2O_5$	Kar
Others		
Metallic iron	Fe	$Fe°$; I
Wüstite	$Fe_{1-x}O$	Wüs, W
Rutile	TiO_2	Rut, R
Anatase	TiO_2	Anat
Brookite	TiO_2	Brk
αPbO_2 form	TiO_2	"αPbO_2"
Baddeleyite form	TiO_2	"Badd"
Graphite	C	G
Liquid	—	Liq, L
Vapor	—	V

This reaction not only depletes the remaining sample in iron, it also releases oxygen that must be removed or buffered to prevent an unwanted increase in fO_2. Lack of space precludes discussion of the advantages and drawbacks of various container materials that are used, but the reader is cautioned that some experimental studies that appear useful are in fact of dubious validity because inappropriate containers were used. Among the materials that have been found most useful are silver and silver-rich Ag-Pd alloys (Muan, 1963) at relatively low temperatures; pure iron at very low fO_2; loops of very fine platinum wire (Ribaud and Muan, 1962); and Fe-Pt alloys of composition chosen - usually by trial and error - to be inert with respect to the sample (e.g., Huebner, 1973).

Minerals and phases considered

Names, ideal formulas, and abbreviations used in the text and figures are summarized in Table 1 for the minerals and phases considered in this chapter.

Fe-Ti-O SYSTEM

Many oxide minerals fall in or close to the compositional triangle Fe-Ti-O; these include magnetite and its solid solutions with Fe_2TiO_4, hematite, ilmenite, and rutile as well as a number of others.

Fe-O join

The Fe-O join (Fig. 1) includes magnetite and hematite; metallic Fe (Fe^0) and wüstite (both known as minerals but rare in crustal rocks); and maghemite, which does not appear on the equilibrium diagram because it is always metastable with respect to hematite. Figure 1 has been constructed from various sources, the most important being Darken and Gurry (1945, 1946), Greig et al. (1935), Phillips and Muan (1960), and Crouch et al. (1971). It should be noted that fO_2 (and PO_2) vary greatly over Figure 1, ranging from less than 10^{-30} atm in the lower left to more than 10 atm. in the upper right. Muan and Osborn (1965, Fig. 13) show oxygen isobars for part of this diagram.

Wüstite. The rarity of wüstite in crustal rocks stems from several causes: It requires low oxygen fugacity and temperatures above 570°C, but most importantly it reacts with SiO_2 to form silicates in all but the most silica-undersaturated environments. Wüstite apparently is never stoichiometric within its stability field, although some workers (e.g., Hentschel, 1970) have claimed that it disproportionates to FeO and an oxygen rich phase upon cooling. It invariably contains some Fe^{3+} plus vacancies, the amount depending on T and fO_2. Numerous ordering schemes of the Fe^{3+} and vacancies have been reported in quenched samples; at present it is not clear how many (if any) of these have true thermodynamic stability at high temperature. The three possible wüstites shown in Figure 1A have some evidence for existence at high temperatures, but the details are still uncertain (Kleman, 1965; Vallet and Raccah, 1964, 1965; Koch, 1967; Koch and Cohen, 1969; Manenc, 1968; Swaroof and Wagner, 1967; Carel, 1974; Giddings and Gordon, 1974; Carel and Gavarri, 1976; Gavarri et al., 1976; Hayakawa et al., 1977; Knacke, 1983; Hazen and Jeanloz, 1984; Haas, 1987). The existence of polymorphs (or allomorphs) of wüstite at temperature is potentially important because the thermodynamic properties of some minerals have been calculated from their coexistence with magnesiowüstites (Chapter 7).

Thermodynamically impossible diagram for wüstite. There is in the literature an alleged phase diagram for wüstite so outrageously bad that it may be useful to enumerate its faults (Fig. 1B). Hazen and Jeanloz (1984) attribute it to Carel (1978, a reference not available to this author). But it had already appeared in Carel and Gavarri (1976); they attributed it to "P. Vallet (1975)", but neither 1975 paper by Vallet cited in their bibliography (both short notes in Comte Rendu) contains any figure, let alone this one! Thus the ultimate blame for burdening the literature with this diagram cannot now be assigned. Most egregious by far is the offset in the composition of wüstite that coexists with Fe^0 at 911°C (region **a**), the temperature of the α-γ transition in iron. Such an offset violates the laws of thermodynamics. Furthermore, while the intersection of the α-γ transition with the Fe-saturated phase boundary of wüstite can (and must) refract that boundary at **a** in accord with Schreinemakers' Rules, the α-γ boundary can have no meaning within the wüstite

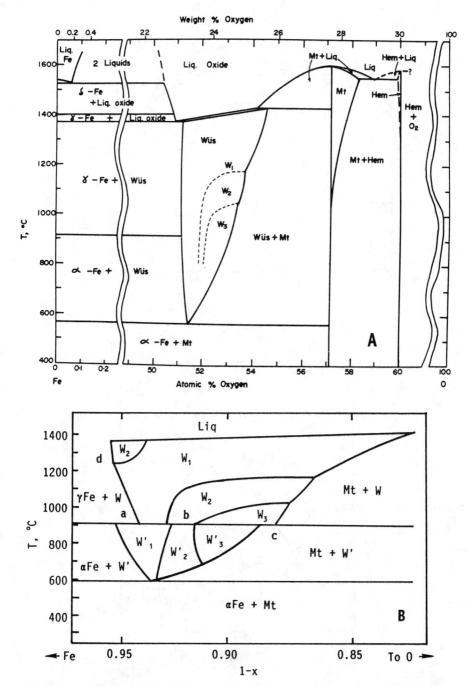

Figure 1. A. Phase diagram for the join Fe-O at low pressures. Note breaks in the compositional scale and also the expanded scale near Fe. Abbreviations as in Table 1. B. Thermodynamically impossible "phase diagram" for the wüstite field, presented as a negative example. Areas labelled a-d highlight violations of the Phase Rule and/or Schreinemakers' Rules. After Hazen and Jeanloz (1984), but see text for its dubious genealogy.

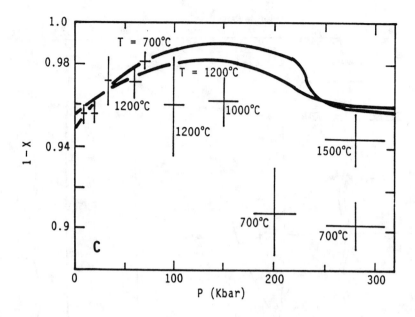

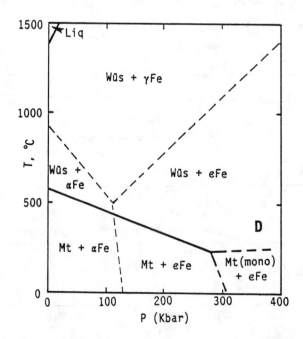

Figure 1. C. Variation of the composition of wüstite (coexisting with Fe°) as a function of pressure for two temperatures. Curves are calculated; crosses show data. After McCammon and Liu (1984). D. Variation, as a function of temperature and pressure, of the phases stable for bulk composition FeO. The heavy curve shows the temperature at which wüstite, magnetite, and Fe° coexist (after Shen et al., 1983). Light dashed lines show phase boundaries for Fe°.

phase field. Thus the indicated offset of the boundary between W_1 and W_2 (area b) at 911°C cannot exist even if the boundary does. The discontinuous offset in the boundary for magnetite-saturated wüstite at 911°C at c is even more ludicrous - and likewise incorrect. Finally, the sharp kink in the Fe°-saturated phase boundary at approximately 1240°C (area d) violates Schreinemakers' Rules; the angle of the boundary to the γFe + W field cannot exceed 180°. I sincerely hope this is the last time this diagram will appear, at least in the geological literature!

Effects of high pressures on wüstite. Because wüstite may be important in the mantle (e.g., Chapter 10) and the core, the effects of pressure on its stability are important. McCammon and Liu (1984) show that the Fe°-saturated phase boundary of wüstite moves to more nearly stoichiometric compositions up to approximately 100 kbar, but moves in the opposite direction at higher pressures (Fig. 1C). As can be seen from Figure 1A, wüstite is unstable below approximately 570°C at one atm, decomposing to magnetite + Fe°. Shen et al. (1983) found that this temperature decreases to 300°C at 200 kbar (Fig. 1D, adapted from Liu and Bassett, 1986).

Magnetite. The unit cell parameter of magnetite has been determined up to 45 kbar by Nakagiri et al. (1986); magnetite is stable to considerably higher pressures, but eventually transforms to a monoclinic(?) form (Fig. 1D) (Mao et al, 1974). Figure 1A shows magnetites in equilibrium with wüstite as stoichiometric; in fact they can have a very small cation excess (up to ~ 10^{-6} Fe^{2+} per three cations) (Dieckmann and Schmalzried, 1977a,b; Aragon and McCallister, 1982).

Fe_2O_3 in magnetite. Only at very high temperatures is there appreciable *stable* solubility of Fe_2O_3 in magnetite. Sosman and Hostetter (1916) had reported complete solubility between hematite and magnetite, but their results were disproved by the very careful study of Greig et al. (1935). Wu and Mason (1981) among many others have determined the distribution of Fe^{2+} and Fe^{3+} between the tetrahedral and octahedral sites of magnetite. Aragon and McCallister (1982) modelled both cation deficiency and cation excess for magnetite at relatively high temperatures; their results are presented in Figure 11 of Chapter 6.

At temperatures below approximately 600°C, non-equilibrium oxidation of magnetite yields metastable, cation-deficient magnetite-maghemite solid solutions, whereas equilibrium oxidation produces hematite. For examples of a vast and contentious literature on the subject, see Kushiro (1960); Columbo et al. (1964, 1965); Feitknecht and Mannweiler (1967); Feitknecht and Gallagher (1970); Heizmann and Baro (1967); Huguenin (1973a, b); Johnson and Merrill (1972); and Johnson and Jensen (1974). Özdemir and Banerjee (1984) reported "high-temperature stability" of maghemite, but the context of their paper indicates that they refer to persistence of maghemite upon heating rather than to true thermodynamic stability.

Hematite. Hematite requires relatively high oxygen fugacities ($fO_2 \geq$ MH, by definition) to be stable as a pure phase. Voigt and Will (1981) studied the system Fe_2O_3-H_2O, and found that the phase boundary between hematite and α FeOOH is described by the relation log PH_2O (kbar) = 3.287 - 1078/T (K), with hematite being stable on the high-temperature (or low PH_2O) side of the boundary.

Melting of iron oxides in the presence of C-O vapor. The melting temperatures in Figure 1 are quite high relative to common magmatic temperatures. Yet the existence of dikes and even lava flows of nearly pure Fe oxides strongly suggest that such liquids can exist in nature. A major step to resolving this conundrum was made by Weidner (1982), who showed that in the presence of graphite and/or C-O vapor, Fe oxides can melt at temperatures below 1000°C if the pressure is at least 0.25 kbar (Fig. 2).

TiO_2

Three polymorphs of TiO_2 - anatase, brookite, and rutile - occur in nature, and two more, one with the columbite (α-PbO_2) structure and named TiO_2-II, (Dachille and Roy, 1962; Bendeliani et al., 1966; McQueen et al., 1967; Jamieson and Olinger, 1968) and another with the baddeleyite (ZrO_2) structure (Sato et al., 1991) have been synthesized at high pressures. Of the natural polymorphs probably only rutile has a true stablity field, both anatase and brookite being metastable with respect to it, for no one has ever successfully converted rutile directly to either of the other polymorphs. Modified-electron-gas calculations (Post and Burnham, 1986) showing that rutile is approximately 4 and 20 kJ more stable than brookite and anatase, while not conclusive, support this interpretation. Anatase and brookite tend to form in low-temperature, hydrothermal environments. Although many studies of the synthesis and inversion of these polymorphs have been published (for example, Osborn, 1953; Glemser and Schwartzmann, 1956; Czanderna et al., 1958; Rao, 1961; Rao et al., 1961; Knoll, 1961, 1963, 1964; Shannon and Pask, 1965; Keesman, 1966; Beard and Foster, 1967; Dachille et al., 1968, 1969; Jamieson and Olinger, 1969; Izumi and Fujiki, 1975; Matthews, 1976; Deelman, 1979), no phase diagram is given for them here because most (perhaps all) published "phase boundaries" between rutile and anatase or brookite reflect kinetics rather than equilibrium. TiO_2-II and the baddeleyite form of TiO_2, on the other hand, may well have true stability fields. Possible fields are shown in Figure 3. See the reference in Chapter 5 to Banfield et al. (in press) for news of a recently discovered natural polymorph.

FeO-TiO_2 join

Minerals within the FeO-TiO_2 join include ilmenite, ulvöspinel, and "ferropseudobrookite", the last two being rare except as components in solid solution series. Figure 4 has been constructed from the data of MacChesney and Muan (1960) and of Taylor (1963, 1964), with the instability of "ferropseudobrookite" at 1140°C from Lindsley (1965). In the presence of $Fe°$, Fe_2TiO_5 takes Ti_3O_5 into solid solution and thereby is stabilized down to 1075°C (Hartzman and Lindsley, 1973; Saha and Biggar, 1974; Lindsley et al., 1974; Lipin and Muan, 1974; Simons, 1974; Simons and Woermann, 1978). Its instability at lower temperatures, combined with its high TiO_2 content, make "ferropseudobrookite" very rare as a mineral. ["Ferropseudobrookite" never had formal standing as a mineral name, and it has been further degraded in a recent review of the pseudobrookite mineral group (Bowles, 1988); it remains a useful if cumbersome term for the $FeTi_2O_5$ endmember.]

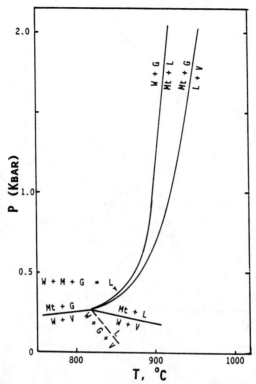

Figure 2. Melting in the system Fe-C-O (after Weidner, 1982). Abbreviations as in Table 1. There is a singular point (not shown, but probably near 0.5 kbar) at which the low-pressure reaction W + Mt + G = L becomes the peritectic W + G = Mt + L at higher pressures.

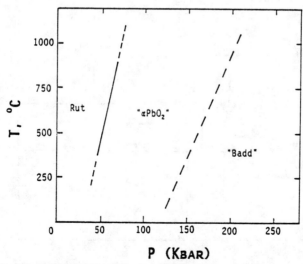

Figure 3. Possible stability fields for TiO_2 polymorphs (after Sato et al., 1991). Anatase and brookite are considered not to have true thermodynamic stability fields, although it is possible that either or both could be stable at low pressures and temperatures, and that the correct relations are obscured by sluggish reaction kinetics in those regions.

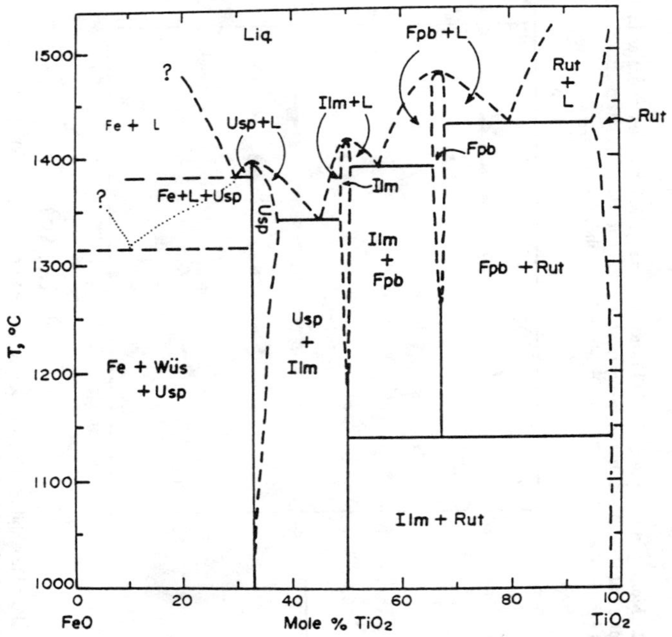

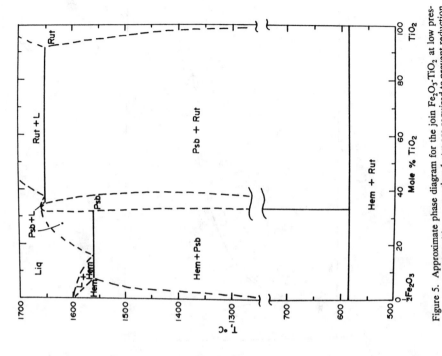

Figure 5. Approximate phase diagram for the join Fe_2O_3-TiO_2 at low pressures. Oxygen fugacities greater than 1 atm are required to prevent reduction of Hem and Psb at high temperatures. Abbreviations as in Table 1.

Figure 4. Approximate phase diagram for the join FeO-TiO_2 at low pressures. Relations near the FeO end are pseudobinary because of the instability of stoichiometric FeO. Abbreviations as in Table 1.

Fe_2O_3-TiO_2 join

In addition to the end members hematite and rutile, the Fe_2O_3-TiO_2 join (Fig. 5) includes pseudobrookite. Figure 5 is based mainly on MacChesney and Muan (1959) and Taylor (1963, 1964). Pure Fe_2TiO_5 is unstable below 585°C (Lindsley, 1965).

FeO-Fe_2O_3-TiO_2(-Ti_2O_3) join

Equilibrium phase relations in the join FeO-Fe_2O_3-TiO_2 have been studied at liquidus temperatures (Taylor, 1963); 1300°C (Taylor, 1964), 1200°C (Webster and Bright, 1961), 1000-1200°C (Katsura et al., 1976), 700-1100°C (Borowiec and Rosenqvist, 1981) and 1000°C (Schmahl et al., 1960) at 1 atm; and under hydrothermal conditions by Brothers et al. (1987), Buddington and Lindsley (1964), Lindh (1972), Lindsley (1962, 1963, 1965, 1973), Lindsley and Lindh (1974), and Matsuoka (1971). Other studies incluse those of Haggerty and Lindsley (1970), Saha and Biggar (1974), Simons (1974), Taylor et al. (1972), Tompkins (1981), and Hammond et al. (1982),. There are three (nominally) binary joins within this system: magnetite-ulvöspinel (Fe_3O_4-Fe_2TiO_4), hematite-ilmenite (Fe_2O_3-$FeTiO_3$), and pseudobrookite-"ferropseudobrookite" (Fe_2TiO_5-$FeTi_2O_5$). Chemically, the substitution is the same along all three joins: Fe^{2+} + Ti^{4+} for 2 Fe^{3+}. At sufficiently high temperatures, solid solution is complete along each join.

Magnetite-ulvöspinel join. Natural intergrowths of magnetite and ulvöspinel along (100) make it clear that there is a miscibility gap along this spinel join. Attempts to determine the miscibility gap experimentally have been made by Vincent et al. (1957), Price (1981), and Lindsley (1981). None of these attempts has been very successful, but it seems clear that the consolute point lies below 600°C (and quite possibly below 500°C), and that it lies closer to Fe_3O_4 than to Fe_2TiO_4. Chapters 6 and 8 show that magnetic ordering in the magnetite component has a strong effect on the location of the miscibility gap. Attempts to determine cation distributions (e.g., Trestmann-Matts et al., 1983; Wechsler et al., 1984) are helpful in developing microscopic thermodynamic models for the join (Chapters 6, 7).

Hematite-ilmenite join. Both hematite and ilmenite are rhombohedral, but at ambient temperatures they have different space groups - R3̄c for hematite and R3̄ for ilmenite (Chapter 2). Nevertheless, they form a complete solid solution series. Intergrowths along (0001) between hematite and ilmenite in natural samples require the existence of a miscibility gap between these rhombohedral phases. Experiments to locate the position of the gap have been made by Carmichael (1961), Lindh (1972), Lindsley (1973), Lindsley and Lindh (1974), and Burton (1982; 1984). These investigations have all been hampered by very slow chemical reaction rates, and they have succeeded mainly in showing that the consolute point lies below ~650°C and is asymmetric towards hematite. Much more progress has been made in locating the second-order R3̄-R3̄c inversion and the magnetic ordering along this join (e.g., Nord and Lawson, 1989), and the miscibility gap calculated including these effects (Fig. 8 of Chapter 8) is doubtless superior to those determined by phase-equilibrium experiments alone.

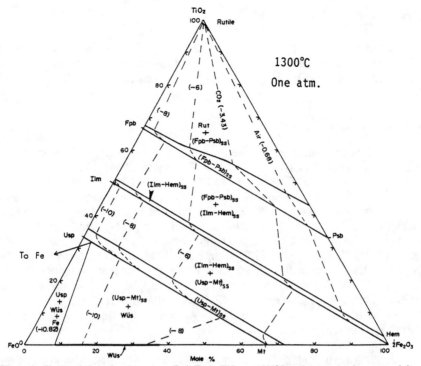

Figure 6. Phase relations for the system $FeO-Fe_2O_3-TiO_2$ at 1300°C and 1 atm total pressure (after Taylor, 1964). Light dashed lines are representative oxygen isobars and are labelled with values of log fO_2 in parentheses. Abbreviations as in Table 1.

Pseudobrookite-"ferropseudobrookite" join. This join is even less well determined that the other two. Akimoto et al. (1957) synthesized a complete series at 1150°C. Because each end member breaks down to rutile plus a rhombohedral phase (pseudobrookite to hematite and rutile below 585°C, and "ferropseudobrookite" to ilmenite and rutile below 1140°C), it seems clear that intermediate compositions should decompose to rutile plus intermediate rhombohedral phases. Haggerty and Lindsley (1970) presented some data, but noted that their 750°C experiments had failed to reach equilibrium in three years! Certainly a part of the difficulty lies in the slow reaction rates in the hematite-ilmenite system; it is probably no accident that Haggerty and Lindsley had minimal success at temperatures for which the breakdown assemblage was expected to be hematite$_{ss}$ + ilmenite$_{ss}$ + rutile.

1300°C isotherm. The 1300°C isotherm for the system $FeO-Fe_2O_3-TiO_2$ (Fig. 6) illustrates several important aspects of mineral relations in the $FeO-Fe_2O_3-TiO_2$ system. Dominating the diagram are the three complete solid solution series mentioned above, each important in the mineralogy of the Fe-Ti oxides: the orthorhombic series (Fpb-Psb), the rhombohedral series (Ilm-Hem) and the spinel series (Usp-Mt). At 1300°C, the solid solutions can depart from stoichiometry; that gives the single-phase fields a measurable width. It is noteworthy that for all three series, departures from stoichiometry are mainly in the direction of cation deficiency, lying above the binary joins in Figure 6. [Figure 11 of Chapter 6 illustrates that the spinels can also accomodate excess cations at low values of fO_2.] However, even at 1300°C, the spinel field is not sufficiently wide to account for many natural ilmenite-magnetite

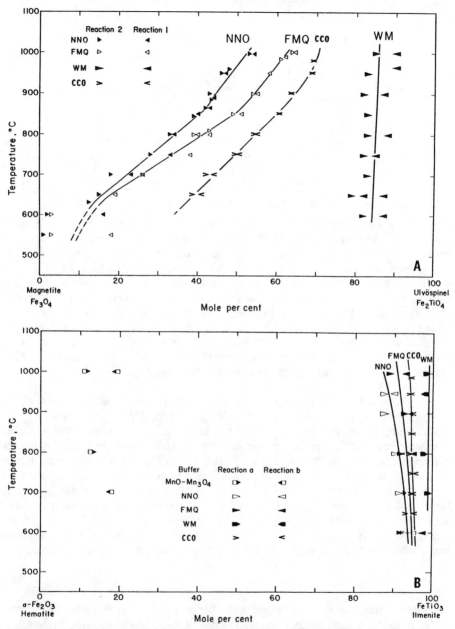

Figure 7A, B. Compositions of coexisting titanomagnetites (7A) and ilmenites (7B) as determined by reversals in buffered hydrothermal experiments at 1 and 2 kbar. Data from Buddington and Lindsley (1964) and Spencer and Lindsley (1981; CCO buffer). To the left of each curve in A, single-phase spinels are stable at the fO_2 of that buffer. Spinel compositions to the right of each curve will oxidize to (Ilm-Hem)$_{ss}$ plus the (Usp-Mt)$_{ss}$ whose composition is given by the appropriate buffer curve at the temperature of the reaction. To the right of each curve in B, single-phase ilmenites are stable at the fO_2 of the corresponding buffer. Ilmenite compositions to the left of each curve will reduce to (Usp-Mt)$_{ss}$ plus that Ilm$_{ss}$ whose composition is given by the appropriate buffer curve at the temperature of reaction. CCO buffer data in A and B from Spencer and Lindsley (1981). NNO and FMQ Ti-Mt compositions at 1000°C from "L" points in C.

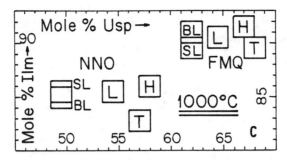

Figure 7C. Data of Hammond et al. (1982) for low pressure (0 to 1 atm.) and 1000°C. H, T, and L indicate experiments of Hammond, Tompkins, and Lindsley, respectively. BL shows the data of Buddington and Lindsley (1964), and SL the compositions calculated from the solution model of Spencer and Lindsley (1981). NNO, FMQ, CCO, and WM designate the, nickel-nickel oxide, fayalite-magnetite-quartz, cobalt-cobalt oxide, and wüstite-magnetite buffers, respectively.

intergrowths by a simple process of exsolution. Magnetite grains containing up to (and sometimes more than) 50% ilmenite lamellae are widespread in igneous and metamorphic rocks and were long interpreted as resulting from exsolution of $FeTiO_3$-Fe_3O_4 solid solutions. Schmahl et al. (1960) reported finding 50% $FeTiO_3$ solid solution in magnetite at 1000°C. Their findings are incompatible with more recent studies, and it appears that their interpretation of their experimental data was strongly influenced by the mineralogical gospel of the time. The Frontispiece of this Volume shows one of the first demonstrations that this long-accepted "exsolution" texture can be formed by oxidation of Fe_2TiO_4 component in titanomagnetite. Leusmann (1979) produced the same texture by oxidizing the Mn_2TiO_4 component of Ti-bearing spinels in the analogous system MnO-Mn_2O_3-TiO_2.

Between the solid-solution series are the two-phase fields wüstite + (Usp-Mt)$_{ss}$, (Usp-Mt)$_{ss}$ + (Ilm-Hem)$_{ss}$, (Ilm-Hem)$_{ss}$ + (Fpb-Psb)$_{ss}$, and (Fpb-Psb)$_{ss}$ + rutile. Compositions of coexisting (Usp-Mt)$_{ss}$ and (Ilm-Hem)$_{ss}$ at lower temperatures are shown in Figure 7. Tie lines connecting the compositions of the coexisting phases are oxygen isobars. As one would expect, the oxygen fugacity increases from left to right in the diagram, that is, as the phases lose FeO and gain Fe_2O_3. Note that in Figure 6 (and also in Figures 8A-G), oxygen lies at plus infinity; therefore any horizontal line is an oxidation-reduction line. The isobars can and generally do curve within the one-phase fields. At 1300°C the tie lines between (Usp-Mt)$_{ss}$ and (Ilm-Hem)$_{ss}$ tend to be parallel (except for very Fe_2O_3-rich compositions). The tie-lines rotate counter-clockwise with decreasing temperature; the exchange equilibrium:

$$Fe_2TiO_4 + Fe_2O_3 = Fe_3O_4 + FeTiO_3 \tag{1}$$

is strongly temperature-sensitive and is the basis of the magnetite-ilmenite thermometer (Chapter 6, Figs. 16a,b; Chapter 7; Andersen and Lindsley, 1988; Andersen et al., 1991). Hammond and Taylor (1982) studied the kinetics of magnetite-ilmenite equilibration over the range 900-1250°C. At 1200°C the rate-determining step was mass transport of the mixed gases used to control fO_2. At 1100°C and below, rates were controlled by diffusion of reactants through a solid product layer. Hammond and Taylor emphasize that re-equilibration can occur within each original phase, so that the initial Fe/Ti within each would remain unchanged. Inasmuch as that ratio is all that is really used to calculate the compositions of the primary phases, in such

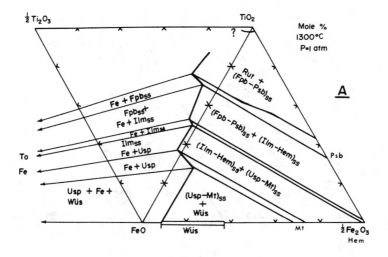

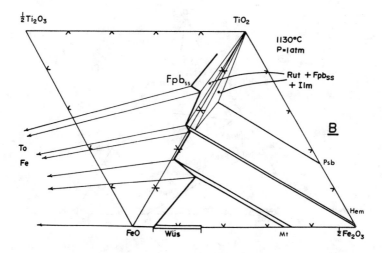

Figure 8. Changes of phase assemblages in the system FeO-Fe$_2$O$_3$-TiO$_2$(-Ti$_2$O$_3$) with decreasing temperature. Most of the phase relations are relatively independent of total pressure; the pressures given are those at which the main experiments were performed. A, 1300°C; B, 1160°C. The heavy black line in A-C is the metallic Fe saturation surface, taken mainly from Simons (1974) and Simons and Woermann (1978). Abbreviations as in Table 1. See following two pages.

84

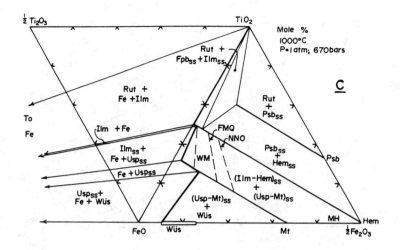

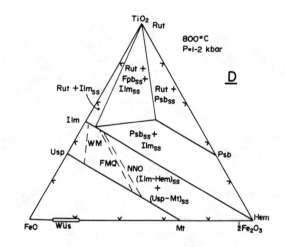

Figure 8, cont. Dashed tie lines in C-E connect the (Usp-Mt)$_{ss}$ and (Ilm-Hem)$_{ss}$ that coexist at the fO$_2$ of the buffer indicated. Note the retreat of the (Fpb-Psb)$_{ss}$ series towards Psb with decreasing temperature. Relations in F are mainly inferred from nature. G shows the possible range of maghemites. Oxidation takes place along lines of constant Fe/Ti, which are horizontal in these diagrams; two examples (with arrows) are shown in G. The diagonal dashed line labelled -Fe in G shows the limits of titanomaghemites that can form by the removal of Fe rather than by addition of oxygen. See following page.

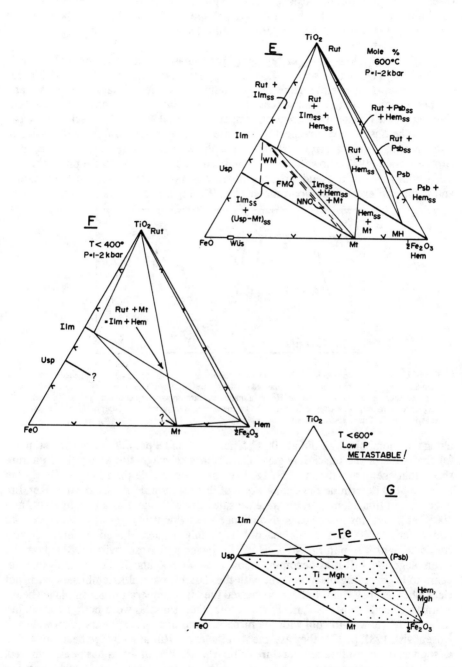

Figure 8, continued.

cases the re-equilibrated oxides retain evidence of their initial temperature of equilibration. On the other hand, if mass transport of Fe and Ti occurs between the primary phases, the original temperature information has been lost.

Phase relations at other temperatures. Phase relations in the system FeO-Fe_2O_3-TiO_2(-Ti_2O_3) at a series of decreasing temperatures are compiled in Figures 8A-G. Noteworthy are the decreasing content of Ti_2O_3 in Fe°-saturated phases and the decreasing width (i.e., nonstoichiometry) of the spinel, rhombohedral, and orthorhombic phase fields with decreasing temperature. By 1130°C "ferropseudobrookite" has become unstable with respect to ilmenite + rutile, and with further decreases in temperature, the most ferrous orthorhombic phase steadily retreats towards Fe_2TiO_5. Brothers et al. (1987) showed that the composition of a rhombohedral phase that coexists with rutile or a pseudobrookite phase is almost independent of temperature and is thus an excellent oxybarometer (Fig. 9). Figure 8E shows a miscibility gap

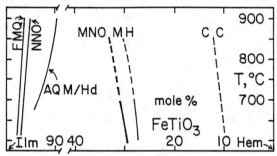

Figure 9. Compositions of ilmenite-hematite$_{ss}$ in equilibrium with rutile (solid portions of curves) or with pseudobrookite$_{ss}$ with or without rutile (dashed portions). Buffered hydrothermal experiments at 1-2 kbar from Brothers et al. (1987). AQM/Hd = almandine-quartz-magnetite-hedenbergite buffer; MH = magnetite-hematite; CC = copper-cuprite; MNO = manganosite-hausmannite. Other buffer abbreviations as in Figure 7. Phase abbreviations as in Table 1.

along the rhombohedral join at 600°C; Chapters 6 and 8 provide improved estimates for the "crest" of the miscibility gap. Coexistence of magnetite and rutile in nature shows that eventually that assemblage becomes more stable than Ilm$_{ss}$ + Hem$_{ss}$. In an attempt to determine the temperature of this reaction, I once held Mt + Rut, Ilm + Hem, and Ilm$_{50}$Hem$_{50}$, in separate capsules and with water as a flux, at 300°C and 400°C at 2 kbar for three years. No change was evident in any of the charges! The upper limit of 400°C for this reaction is mainly a guess, based on other phases coexisting with the natural Mt + Rut. Borowiec and Rosenqvist (1981) show the assemblage Mt + Rut as becoming stable between 800 and 700°C, a result that appears to be in strong disagreement with the data on natural assemblages. It is not clear whether they claim to have converted the high-temperature assemblage (hem$_{ss}$ + ilm$_{ss}$) directly to Mt + Rut. Furthermore, their samples were not quenched, but rather were allowed to cool with the furnace. And finally, it appears (Borowiec and Rosenqvist, 1981, p. 221) they assume that the Mt + Rut assemblage becomes stable as soon as the consolute temperature of the rhombohedral series has been achieved. In fact, there is every reason to believe that ilm$_{ss}$ + hem$_{ss}$ two-phase assemblages are stable over a considerable temperature interval. Nevertheless, their results should not be rejected out of hand. Experiments starting with both Mt + Rut and ilm$_{ss}$ +hem$_{ss}$, and quenched from 700, 600, and 500°C appear called for.

Titanomaghemites. Titanomaghemites are believed to form at relatively low temperatures (generally below 600°C) by non-equilibrium oxidation of $(Usp-Mt)_{ss}$. Figure 8G shows the compositional range possible for these metastable, cation-deficient phases. Further information is provided in Chapter 4, where it is also pointed out that some natural titanomaghemites probably form by the removal of Fe rather than by the addition of O. The removal of Fe would result in increasing Ti/Fe with increasing maghemitization, and thus compositions could lie above the horizontal line extending from Usp to (Psb) in Figure 8G. The limit for titano-maghemites formed strictly by removal of Fe is shown by the dashed line.

Reduction of Fe-Ti oxides. Interest in the reduction of Fe-Ti oxides stems from many sources. Many lunar samples show signs of reduction, and some proposed industrial processes for separating the Fe and Ti involve reduction. Grey and Ward, 1973; McCallister and Taylor, 1973; Grey et al., 1974; Saha and Biggar, 1974; and O'Neill et al., 1988) have studied reduction in the Fe-Ti-O system. Reduction of ilmenite plays an important role in some plans for colonizing the Moon: Hydrogen shipped from Earth would be used to reduce ilmenite, yielding metallic iron and water as a by-product. The iron could be used for construction, the water to support life and to store energy: H_2 and O_2 dissociated from water by solar power during the lunar day could be recombined in a furnace or fuel cell at night (Gibson and Knudsen, 1985; Lewis et al., 1988; Williams, 1983; 1985).

Much as experimentalists and modellers might wish otherwise, titanomagnetite and ilmenite in nature are rarely pure phases. Among the most important other components are MgO, MnO, Al_2O_3, and Cr_2O_3.

Fe-O-MgO-TiO_2 SYSTEM

The addition of MgO to the Fe-Ti oxides is of interest for two main reasons. First, many terrestrial magnetite-ilmenite pairs - especially those from basaltic rocks - contain appreciable amounts of MgO, and successful application of the two-oxide geothermometer-oxybarometer depends on evaluating its effect. Second, oxide minerals from lunar rocks tend to be rich in MgO. For example, the mineral armalcolite is ideally $Fe_{0.5}Mg_{0.5}Ti_2O_5$, although considerable variation from this "endmember" formula is found.

FeO-MgO join

Srecec et al. (1987) made a very careful study of the magnesiowüstite join, and derived its thermodynamic properties. The use of these data to derive properties of other minerals is illustrated in Chapter 7.

FeO-MgO-Fe_2O_3 join

The join FeO-MgO-Fe_2O_3 includes the spinel phase magnesioferrite, which is sometimes found as the earliest oxide mineral in basalts or in highly oxidized rocks (Chapter 12). The join has been studied at high temperatures by Speidel (1967) and by Willshee and White (1967). At 1300°C there is a very large field of magnesiowüstite and a somewhat smaller field of Fe_3O_4-$MgFe_2O_4$ spinel solid solution. Trestman-Matts et al. (1984) determined cation distributions in the spinels.

FeO-MgO-TiO$_2$-(Ti$_2$O$_3$) join

Many armalcolites, ilmenites, and ulvöspinels from the Moon fall near the FeO-MgO-TiO$_2$ join. Johnson et al. (1971) studied this join at liquidus and solidus temperatures with $fO_2 = 10^{-8}$ atm, and at 1300°C in equilibrium with metallic iron (Fig. 10). Superficially, Figure 10 bears a strong resemblance to Figure 6, with three

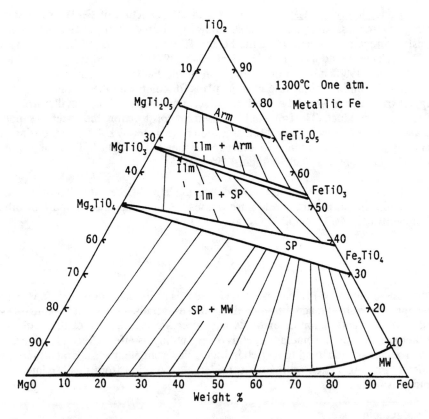

Figure 10. Phase relations in the join FeO-MgO-TiO$_2$ in equilibrium with metallic iron at 1300°C and 1 atm. After Johnson et al. (1971). SP = (Fe,Mg)$_2$TiO$_4$ spinel solid solution; MW = magnesiowüstite; other abbreviations as in Table 1.

binary" joins - spinel, rhombohedral, and orthorhombic - and a field of wüstite. However, to a first approximation, Figure 10 is for nearly constant fO_2.

In some cases, particularly for the armalcolites, the presence of Ti^{3+} requires the component Ti_2O_3 as well. Experimental studies at low pressure include those of Lind and Housley (1972), Hartzmann and Lindsley (1973), and Lindsley et al. (1974);

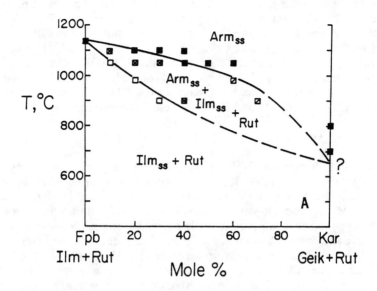

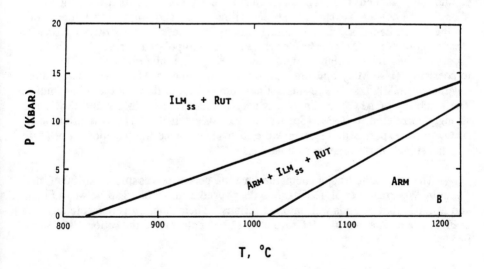

Figure 11. A. Phase relations along the join $FeTi_2O_5$-$MgTi_2O_5$ at low pressures (Lindsley et al., 1974), showing the stability field for armalcolite$_{ss}$. Only the upper curve is binary; the lower curve marks the trace of the trailing Ilm$_{ss}$-Rut edge of the Arm$_{ss}$-Ilm$_{ss}$-Rut three-phase triangles in the system FeO-MgO-TiO_2. B. Effect of pressure on the stability of 50-50 armalcolite (after Friel et al., 1977). Phase boundaries at low pressure have been adjusted slightly so as to be in agreement with Figure 11A. Abbreviations as in Table 1.

they show that ideal armalcolite is stable down to $1010 \pm 20°C$ at 1 atm; below that temperature it breaks down to ilmenite-geikielite solid solution + Rut (Fig. 11A). Kesson and Lindsley (1975) showed that armalcolite is stabilized to still lower temperatures by the presence of Al^{3+}, Cr^{3+}, and Ti^{3+}. Friel et al. (1977) found a dramatic effect of pressure on the stability of armalcolite (Fig. 11B). Stannin and Taylor (1980) studied the reduction of armalcolite, and concluded that the Ti^{3+} content of armalcolite can be used as an oxybarometer: log fO_2 (above that of the IW buffer) = $1.7(Ti^{3+}/Fe^{2+})$, where Ti^{3+} and Fe^{2+} are mole fractions based on 5 oxygens.

$FeO-Fe_2O_3-MgO-TiO_2$ join

The join iron oxide-$MgO-TiO_2$ has been studied at one atm. in air by Woermann et al. (1969), and at 1160-1300°C and a range of oxygen fugacities by Speidel (1970). Two of the Woermann et al. diagrams are reproduced in Figures 8a, b of Chapter 10. Microprobe analysis of coexisting $(Usp-Mt)_{ss}$ and Ilm in Speidel's experiments suggested that Mg preferentially enters the spinel phase, in contrast to most natural pairs, for which Mg tends to be enriched in the ilmenite. Pinckney and Lindsley (1976) studied the join hydrothermally at 1 kbar and 700-1000°C using reversed exchange equilibria and found that Mg preferentially enters the ilmenite (Fig. 12), the preference becoming more pronounced with decreasing temperature. It is not clear whether the discrepancy with Speidel's results is a temperature effect, or whether his experiments - which were not reversed - simply do not all reflect equilibrium. Indeed, the discrepancy may be more apparent than real; Figure 13 plots K_D for FeO-MgO exchange between Ti-magnetite and ilmenite vs. temperature for the experiments of Pinckney and Lindsley. Speidel's data *for oxygen fugacities close to those of the FMQ buffer* plot close to the extrapolated curve. Note also that K_D approaches unity at higher temperatures. Katsura et al. (1976) also added MgO to coexisting $(Usp-Mt)_{ss}$ and Ilm, but in their experiments, MgO was always added together with Al_2O_3, so the results are not comparable to those of the other studies. The solubility of Al_2O_3 in ilmenite is very limited, whereas MgO and Al_2O_3 are strongly correlated in Ti-Mt (Spencer and Lindsley, 1980). Thus the addition of Al_2O_3 to the experimental system will effectively enhance the partitioning of MgO into the spinel phase relative to ilmenite.

The rhombohedral join hematite-ilmenite-geikielite was studied at 1300°C and 1 atm by Woermann et al. (1970), who discovered a miscibility gap between Fe_2O_3 and $MgTiO_3$ that extends well into the ternary. Their diagram is presented as part of Figure 12 in Chapter 10. DeGrave et al. (1975) determined the cation distribution in $Mg_2TiO_4-MgFe_2O_4$ spinels.

$Fe-O-MnO-TiO_2$ SYSTEM

$FeO-Fe_2O_3-MnO$ join

The Ti-free portion of the $Fe-O-MnO-TiO_2$ system was studied at 700°C and 1 bar by Punge-Witteler (1984), who determined tie lines between manganowüstites and Fe-Mn spinels. He does not list duration of his experiments, and one might worry as to whether they reached equilibrium. Because of the Jahn-Teller effect, Mn

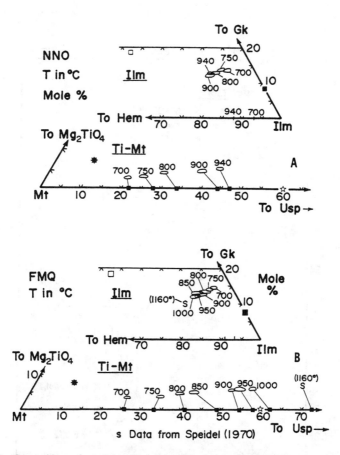

Figure 12. Compositions of coexisting Mg-bearing (Usp-Mt)$_{ss}$ and (Ilm-Hem)$_{ss}$ based on buffered hydrothermal experiments of Pinckney and Lindsley (1976). A. NNO buffer. B. FMQ buffer. The upper part of each diagram shows compositions of ilmenites, the lower part, the compositions of spinels for the indicated temperatures. The solid star shows the composition of the starting spinel; the large solid square, the composition of the starting ilmenite for one pair of starting materials. The open star and open square give the compositions of a second starting pair for reversals. Open "blobs" show the equilibrium compositions allowed by the reversals. Small squares along the Mg-free base show corresponding compositions for Mg-free spinels from Buddington and Lindsley (1964). The range of Mg-free ilmenites is indicated by the vertical lines near 90% Ilm. The 850°C data for FMQ clearly appear aberrant; we have been unable to find any explanation for them; most likely the buffer failed and the charges became too oxidized. Abbreviations as in Table 1.

has a strong preference for the tetrahedral site of the spinel, and a spinel approaching jacobsite composition ($MnFe_2O_4$) coexists with a monoxide having nearly 80% MnO. Muan and Somiya (1962) determined melting relations for the "join" iron oxide-manganese oxide in air (Fig. 14). The diagram is informative on several counts. Note first that the oxides are oxidized at low temperatures (Fe_2O_3, Mn_2O_3), but become more reduced - for the same numeric value of fO_2 (0.21) - with increasing temperature; the rhombohedral phases are replaced by spinel, or by a tetragonal hausmannite phase for Mn-rich compositions. Note also the extreme asymmetry of the rhombohedral miscibility gap. It is unclear to this author whether the near-coincidence of the liquidus and solidus are real, or whether their locations reflect the extreme difficulty distinguishing primary from quench crystals in many oxide melts.

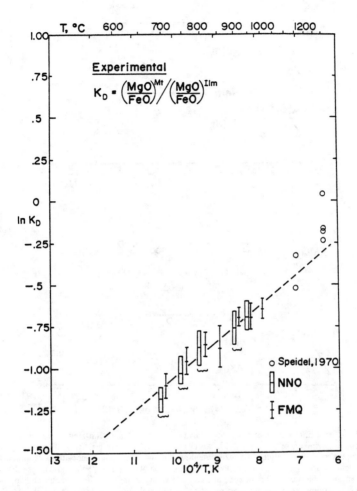

Figure 13. Partitioning of MgO and FeO between titanomagnetite and ilmenite, based on the experiments of Pinckney and Lindsley (1976). Data from Speidel (1970) plot close to the projected curve. Abbreviations as in Table 1.

Mn-O-TiO$_2$ join

Leusmann (1979) studied the join MnO-Mn$_2$O$_3$-TiO$_2$ and compared it to the analogous Fe-Ti oxide system. He found no stability field for either MnTi$_2$O$_5$ or Mn$_2$TiO$_5$, so no rhombohedral series was reported. He found limited solubility between MnTiO$_3$ and α Mn$_2$O$_3$ at 870°C. Leusmann was able to prepare a continuous series of compositions between Mn$_2$TiO$_4$ and Mn$_3$O$_4$ at 1200°C. Inasmuch as Mn$_3$O$_4$ spinel is stable at 1200°C, it is likely that the spinel series is complete at that temperature; however, x-ray studies of the quenched samples show a discontinuous change from cubic for compositions with less than 33% Mn$_3$O$_4$ to tetragonal for those with more than 33%, indicating inversion of more Mn-rich samples upon the quench.

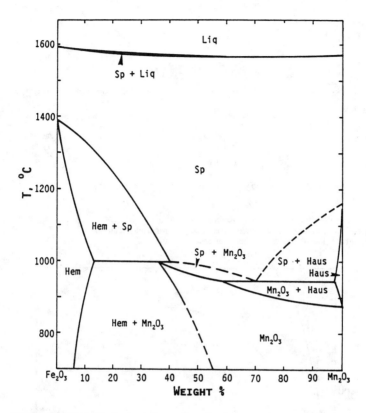

Figure 14. The join Fe_2O_3-Mn_2O_3 in air at one atm, after Muan and Somiya (1962). Abbreviations as in Table 1.

FeO-MnO-TiO$_2$ join

O'Neill et al. (1989) used E.M.F. measurements to determine activity-composition relations between $FeTiO_3$ and $MnTiO_3$ from 1050 to 1300 K. There is a slight positive deviation from ideality. Kress (1986) and Pownceby et al. (1987) had previously obtained data for this ilmenite series by reversing the distribution of Mn and Fe between ilmenite and garnet.

FeO-Fe$_2$O$_3$-MnO-TiO$_2$ join

Magnetite-ilmenite relations in the join FeO-Fe_2O_3-MnO-TiO_2 were studied hydrothermally at 1 kbar and 700-1000°C by Mazzullo et al. (1975). In general the results are similar to those obtained by Pinckney and Lindsley (1976) for the MgO-bearing system: MnO partitions preferentially into ilmenite, the effect becoming stronger with decreasing temperature (Fig. 15). Andersen (1988) developed a thermodynamic solution model for coexisting titanomagnetite and ilmenite in the system Fe-O-MgO-MnO-TiO_2.

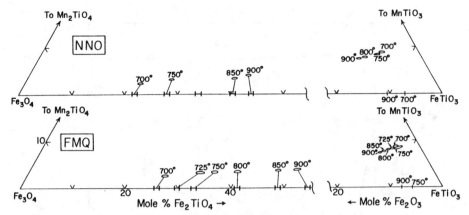

Figure 15. Compositions of coexisting Mn-bearing (Usp-Mt)$_{ss}$ (left portion) and (Ilm-Hem)$_{ss}$ (right portion) based on buffered hydrothermal experiments of Mazzullo et al. (1975) at 1-2 kbar for NNO (upper) and FMQ (lower) buffers. Abbreviations as in Table 1.

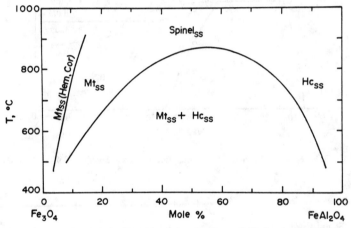

Figure 16. The join Fe_3O_4-$FeAl_2O_4$ (from Turnock and Eugster, 1962) showing the Mt_{ss}-Hc_{ss} miscibility gap. The curve labelled Mt_{ss}(Hem,Cor) shows the composition of Mt_{ss} in equilibrium with hematite plus corundum. Pressures ranged from 1-2 kbar. Abbreviations as in Table 1.

Fe-O-Al_2O_3-TiO_2 AND Fe-O-MgO-Al_2O_3 SYSTEMS

Fe_2O_3-Al_2O_3 join

Phase relations on the hematite-corundum join were determined by Muan and Gee (1956). Below 1318°C, there is a miscibility gap between hematite$_{ss}$ and corundum$_{ss}$. Turnock and Eugster (1962) determined the maximum width of this gap at 500-900°C by dissolution experiments, but did not perform exsolution experiments to reverse it. A curious feature of the join is the stability of the ordered phase $FeAlO_3$ above 1318°C (Muan, 1958). At first glance the formation of that phase with increasing temperature seems incompatible with the generally accepted notion that increasing temperature favors <u>disorder</u>. However, the $FeAlO_3$ phase has a very different structure (orthorhombic rather than rhombohedral) and is presumeably stabilized by a high entropy.

$FeO-Fe_2O_3-Al_2O_3$ join

The $FeO-Fe_2O_3-Al_2O_3$ join has been studied at high temperatures and 1 atm by Roiter (1964) and by Meyers et al. (1980). Turnock and Eugster (1962) used hydrothermal techniques at 1-2 kbar and 600-1000°C, determining the miscibility gap between magnetite and hercynite as well as the composition of Mt_{ss} in equilibrium with hematite and corundum (Fig. 16). The miscibility gap is approximately symmetrical on a weight percent basis, but is asymmetric when plotted in mol percent.

$MgO-Al_2O_3$ join

There is extensive solubility of Al_2O_3 in $MgAl_2O_4$, the extent increasing with increasing temperature (Fig. 17A). Viertel and Seifert (1980) have shown that the solubility - which is a defect solid solution - greatly decreases with increasing pressure (Fig. 17B). Similarly, Kushiro (1960) was able to "squeeze out the holes" by converting maghemite to hematite at 140 bar and 162°C for 21.5 hours. These results have potential importance for other spinels as well. Ghiorso and Sack (Chapter 6) point out that an important direction for future research in the thermodynamics of oxides is the evaluation of the effect of defects. I do not disagree, but the probability that the proportion of defects decreases with increasing pressure may mean that their effect is less critical for oxides formed below Earth's surface.

Wood et al. (1986) studied order-disorder in $MgAl_2O_4$ spinel. They were able to reverse the distribution of Mg and Al on tetrahedral and octahedral sites at 715, 800, and 850°C. The proportion of Al in tetrahedral sites increases from 0.21 at 700°C to 0.39 at 900°C.

$FeO-MgO-Fe_2O_3-Al_2O_3$ system

Lehmann and Roux (1986) used Fe-Mg exchange equilibria with 1M chloride solutions at 800°C, 4 kbar, to determine relations of $(Fe^{2+},Mg)(Al,Fe^{3+})_2O_4$ spinels. They applied an extensive thermodynamic analysis to the system. Mattioli and Wood (1988) determined magnetite activities along the join $MgAl_2O_4-Fe_3O_4$. These results have played important roles in thermodynamic models (Chapters 6, 7).

$FeO-Al_2O_3-TiO_2$ join

The reduced system $FeO-Al_2O_3-TiO_2$ is of interest for lunar rocks and as a base for thermodynamic models of more oxidized systems. Muan et al. (1972) determined phase relations in equilibrium with Fe° at 1300°C and one atm. They also determined the miscibility gaps between $Fe_2TiO_4-FeAl_2O_4$ (Fig. 18A) and $Mg_2TiO_4-MgAl_2O_4$ (Fig. 18B). The reader may wish to compare these with the calculated miscibility gaps for these joins (Fig. 1 of Chapter 6). Lipin and Muan (1974) provided additional data for the $FeO-Al_2O_3-TiO_2$ join. Curiously, Schriefels and Muan (1975) presented a diagram for $Fe_2TiO_4-FeAl_2O_4$ that appears incompatible with Figure 18A (the implied metastable consulute point would lie at a higher temperature for the 1975 diagram) but did not comment on the discrepancy. The 1975 version shows peritectic rather than eutectic melting; its miscibility gap is only dashed, whereas the 1974 version shows data points and thus has been chosen for display here.

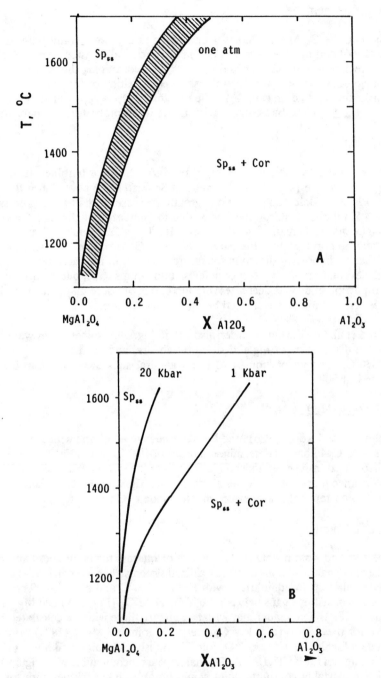

Figure 17. Phase relations along the join $MgAl_2O_4$-Al_2O_3. A. Data at one atm; the shaded band includes a variety of values from the literature. B. Data for 1 and 20 kbar, showing the dramatic effect of pressure in decreasing the equilibrium vacancy content of the spinel. After Viertel and Seifert (1980). Abbreviations as in Table 1.

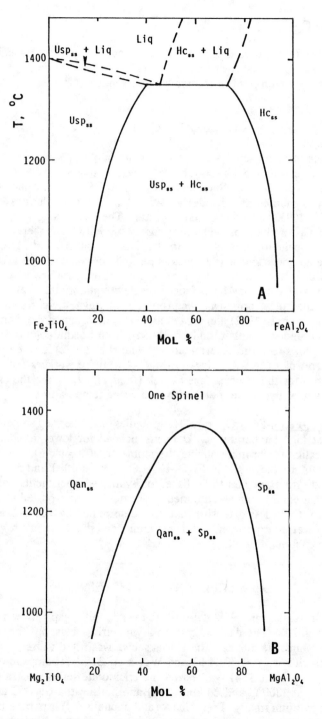

Figure 18. Miscibility gaps in titanate-aluminate spinels. A. Fe_2TiO_4-$FeAl_2O_4$ join. B. Mg_2TiO_4-$MgAl_2O_4$ join. After Muan et al. (1972). Abbreviations as in Table 1.

$FeO-Fe_2O_3-Al_2O_3-TiO_2$ system

Nell et al. (1989) used thermopower and conductivity measurements to determine high-temperature cation distributions in Fe_3O_4-$MgAl_2O_4$-$MgFe_2O_4$-$FeAl_2O_4$ spinels. The data are important in thermodynamic models (Chapter 7).

Cr_2O_3-BEARING SYSTEMS

Both the thermodynamics and phase equilibria of Cr_2O_3-bearing spinels have been reviewed recently by Sack and Ghiorso (1991; Chapters 6 and 9) and thus need not be covered in detail here. Snethlage and Klemm (1975) determined phase relations between magnetite-chromite and hematite-eskolaite solid solutions as a function of fO_2 at 1000, 1095, and 1200°C and one atm. Their data suggest a fairly strong preference of Cr for the rhombohedral phase, especially for higher values of fO_2. These results appear to be at odds with the data of Katsura and Muan (1964) at 1300°C; those authors show nearly constant Fe^{3+}/Cr between spinels and the rhombohedral phase. It is unclear whether there is a strong temperature effect or whether at least one set of data may have failed to reach equilibrium. Petric and Jacob (1982) determined the thermodynamic properties of magnetite-chromite solid solutions. Chatterjee et al. (1982) and Oka et al. (1984) determined the thermodynamic properties of corundum-eskolaite and spinel-picrochromite solid solutions, respectively. Figure 19 shows relations determined by Muan et al. (1972) for the iron-free face of the spinel prism - $MgAl_2O_4$-Mg_2TiO_4-$MgCr_2O_4$; those workers found relatively similar relations for the Mg-free face as well (their Fig. 7). Note the expansion of the miscibility gap (two-spinel field) with decreasing temperature.

$FeCr_2O_4$-Fe_3O_4-$FeAl_2O_4$ join. Many studies on Cr-bearing spinels have been carried out at high temperatures. Data are needed for lower temperatures (for example, to calibrate chromite-olivine thermometers; Chapter 6). Cremer (1965) reported results for the $FeCr_2O_4$-Fe_3O_4-$FeAl_2O_4$ join at 1000 and 500°C. Unfortunately, his diagram appears to be based on synthesis experiments only, and it is unclear whether equilibrium was attained. For example, his reported miscibility gap for Fe_3O_4-$FeAl_2O_4$ is less extensive than that determined by Turnock and Eugster (1962) using reversed experiments. It is recommended that Cremer's results be used with extreme caution, if at all.

MISCELLANEOUS SYSTEMS

Carvalho and Sclar (1988) determined the miscibility gap between gahnite and franklinite (Fig. 20). The data suggest an asymmetric solvus, but they can just be fitted with a symmetric model. The solvus shown was fitted using a slightly asymmetric model with W_{Gah} = 39.44 kJ and W_{Frk} = 42.21 kJ (Thompson convention). Podpora and Lindsley (1984) synthesized the crichtonite-series minerals lindsleyite and mathiasite at 1300°C and 20 kbar (dry) and lindsleyite at 900°C and 22 kbar, both dry and hydrothermally. Preliminary (and unpublished) experiments show that lindsleyite broke down at 1300°C to psuedobrookite- and-rutile bearing assemblages at 1 bar and 10 kbar and to a rutile-bearing assemblage at 35 kbar. [Those are

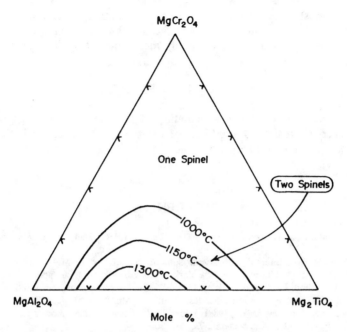

Figure 19. The join $MgAl_2O_4$-Mg_2TiO_4-Mg_2TiO_4 at one atm., showing the ternary miscibility gap at three temperatures (after Muan et al., 1972).

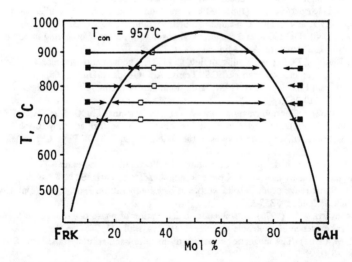

Figure 20. The miscibility gap for the join franklinite ($ZnFe_2O_4$)-gahnite ($ZnAl_2O_4$), after Carvalho and Sclar (1988). Open squares show compositions of single-phase starting materials that exsolved; solid squares show compositions of Frk-rich and Gah-rich phases that were mechanically mixed and allowed to react. Arrowheads show compositions of product phases. Note the much tighter brackets on the Frk-rich side compared to the Gah-rich side. Abbreviations as in Table 1.

doubtless the most inbred sentences in this entire volume.] In view of the potential importance of the crichtonite minerals to the distribution of REE and other trace elements in the upper mantle (Chapter 10), more detailed studies appear needed.

ACKNOWLEDGEMENTS

Ron Frost improved the presentation of this paper by providing a helpful review in a very short time. I also benefitted from the other chapters in this volume as a by-product of the editing process.

REFERENCES

Akimoto, S., Nagata, T., and Katsura, T. (1957) The $TiFe_2O_5$-Ti_2FeO_5 solid solution series. Nature, 179, 37-38.

Andersen, D. J. (1988) Internally Consistent Solution Models for Fe-Mg-Mn-Ti Oxides. Ph.D Dissertation, State University of New York at Stony Brook, xii + 202 pp.

Andersen, D. J., Bishop, F. C., and Lindsley, D. H. (1991) Internally consistent solution models for Fe-Mg-Mn-Ti oxides: Fe-Mg-Ti oxides and olivine. Am. Mineral., 76, 427-444.

Andersen, D. J., and Lindsley, D. H. (1988) Internally consistent solution models for Fe-Mg-Mn-Ti oxides: Fe-Ti oxides. Am. Mineral., 73, 714-726.

Aragon, R., and McCallister, R. H. (1982) Phase and point defect equilibria in the titanomagnetite solid solution. Phys. Chem. Minerals, 8, 112-120.

Beard, W. C., and Foster, W. R. (1967) High-temperature formation of anatase. J. Am. Ceram. Soc., 50, 493.

Bendeliani, N. A., Popova, S. V., and Vereshchagin (1966) New modification of titanium dioxide obtained at high pressures. Geochemistry International, 3, 387-390.

Borowiec, K., and Rosenqvist, T. (1981) Phase relations and oxidation studies in the system Fe-Fe_2O_3-TiO_2 at 700-1100°C. Scandanavian J. Metallurgy, 10, 217-224.

Bowles, J. F. W. (1988) Definition and range of composition of naturally occurring minerals with the pseudobrookite structure. Am. Mineral., 73, 1377-1383.

Brothers, S. C., Lindsley, D. H., and Hadjigeorgiou, C. (1987) The hematite-ilmenite + rutile oxybarometer: Experimental calibration. EOS (Trans., Am. Geophys. Union), 68, 461.

Buddington, A. F., and Lindsley, D. H. (1964) Iron-titanium oxide minerals and synthetic equivalents. J. Petrol., 5, 310-357.

Burton, B. P. (1982) Thermodynamic analysis of the systems $CaCO_3$-$MgCO_3$, αFe_2O_3, and Fe_2O_3-$FeTiO_3$. Ph.D. Dissertation, State University of New York at Stony Brook.

Burton, B. P. (1984) Thermodynamic analysis of the system Fe_2O_3-$FeTiO_3$. Phys. Chem. Minerals, 11, 132-139.

Carel, C. (1974) Discussion de quelques resultats cristallographiques obtenus a l'equilibre thermodynamique sur la wüstite solide. Comte Rendu Acad. Sci., Paris, B278, 417-420.

Carel, C. (1978) Short range order in solid wüstite at high temperature. Proc. Conf. Appl. Crystallog., Kozubnik, Poland, 11, 355-359.

Carel, C., and Gavarri, J. R. (1976) Introduction to description of phase diagram of solid wüstite. I. Structural evidence of allotropic varieties. Mat. Res. Bull., 11, 745-756.

Carmichael, C. M. (1961) The magnetic properties of ilmenite-haematite crystals. Proc. Roy. Soc. A, 263, 508-530.

Carvalho III, A. V., and Sclar, C. B. (1988) Experimental determination of the $ZnFe_2O_4$-$ZnAl_2O_4$ miscibility gap with application to franklinite-gahnite exsolution intergrowths from the Sterling Hill Zinc Deposit, New Jersey. Econ. Geol., 83, 1447-1452.

Chatterjee, N. D., Leistner, H., Terhart, L., Abraham, K., and Klaska, R. (1982) Thermodynamic mixing properties of corundum-eskolaite, α-$(Al,Cr^{3+})_2O_3$, solid solutions at high temperatures and pressures. Am. Mineral., 67, 725-735.

Chou I.-M., and Cygan G. L. (1990) Quantitative redox control and measurement in hydrothermal experiments. in R. J. Spencer and I.-M. Chou, Eds. Fluid-Mineral Interactions: A Tribute to H. P. Eugster, The Geochemical Soc., Special Publication No. 2.

Columbo, U., Fagherazzi, G., Garrarrini, F., Lanzavecchia, G., and Sironi, G. (1964) Mechanisms in the first stage of the oxidation of magnetites. Nature, 202, 175-176.

Columbo, U., Garrarrini, F., Lanzavecchia, G., and Sironi, G. (1965) Magnetite oxidation: a proposed mechanism. Science, 147, 1033.

Cremer, V. (1969) Die Mischkristallbildung im System Chromit-Magnetit-Hercynit zwischen 1000°C und 500°C. Neues Jahrb. Mineral., Abh., 111, 184-205.

Crouch, A. G., Hay, K. A., and Pascoe, R. T. (1971) Magnetite-hematite-liquid equilibrium conditions at oxygen pressures up to 53 bars. Nature, 234, 132-133.

Czanderna, A. W., Rao, C. N. R., and Honig, J. M. (1958) The anatase-rutile transition. Part 1, Kinetics of the transformation of pure anatase. Trans. Faraday Soc., 54, 1069-1073.

Dachille, F., and Roy, R. (1962) A new high-pressure form of titanium dioxide (abstr.). Bull. Am. Ceram. Soc., 41, 225.

Dachille, F., Simons, P. Y., and Roy, R. (1968) Pressure-temperature studies of anatase, brookite, rutile and TiO2-II. Am. Mineral., 53, 1929-1939.

Dachille, F., Simons, P. Y., and Roy, R. (1969) Pressure-temperature studies of anatase, brookite, rutile and TiO2-II: A reply. Am. Mineral., 54, 1481-1482.

Darken, L. S., and Gurry, R. W. (1945) The system iron-oxygen. I. The wüstite field and related equilibria. J. Am. Chem. Soc., 67, 1398-1412.

Darken, L. S., and Gurry, R. W. (1946) The system iron-oxygen. II. Equilibrium and thermodynamics of liquid oxide and other phases. J. Am. Chem. Soc., 68, 798-816.

Deelman, J. C. (1979) Low-temperature synthesis of anatase (TiO_2). N. Jb. Mineral. Monatsh. (6), 253-261.

DeGrave, E., de Setter, J., and Vandenberghe, R. (1975) On the cation distribution in the spinel system yMg_2TiO_4-$(1-y)MgFe_2O_4$. Appl. Phys., 7, 77-84.

Deines, P., Nafziger, R. H., Ulmer, G. C., and Woermann, E. (1974) Temperature-oxygen fugacity tables for selected gas mixtures in the system C-H-O at one atmosphere total pressure. Bulletin of the Earth and Mineral Sciences Experiment Station No. 88 The Pennsylvania State University, University Park, PA, 129 pp.

Dieckmann, R., and Schmalzried, J. (1977a) Defects and cation diffusion in magnetite (II). Ber. Bunsensg. Phys. Chem., 81, 414-419.

Dieckmann, R., and Schmalzried, J. (1977b) Defects and cation diffusion in magnetite (I). Ber. Bunsensg. Phys. Chem., 81, 344-347.

Feitknecht, W., and Gallagher, K. J. (1970) Mechanisms for the oxidation of Fe_3O_4. Nature, 288, 548-549.

Feitknecht, W., and Mannweiler, U. (1967) Der mechanismus der Umwandlung von γ- zu α-Eisensesquioxyd. Helv. Chim. Acta, 50, 570-581.

Friel, J. J., Harker, R. I., and Ulmer, G. C. (1977) Armalcolite stability as a function of pressure and oxygen fugacities. Geochim. Cosmochim. Acta, 41, 403-410.

Gavarri, J. R., Weigel, D., and Carel, C. (1976) Introduction to description of phase diagram of solid wüstite. II. Structural Review. Mat. Res. Bull., 11, 917-926.

Gibson, M. A., and Knudsen, C. W. (1985) Lunar oxygen production from ilmenite. in Mendell, W. W. ed., Lunar Bases and Space Activities of the 21st Century. NASA/Johnson Space Center, Houston, 543-550.

Giddings, R. A., and Gordon, R. S. (1974) Solid-state coulometric titration: critical analysis and application to wüstite. Jour. Electrochem. Soc. Solid State Sci. and Tech., 121, 793.

Glemser, O., and Schwartzmann, E. (1956) Zur Polymorphie des Titandioxyds. Angew. Chem., 68, 791.

Greig, J. W., Posnjak, E., Merwin, H. E., and Sosman, R. B. (1935) Equilibrium relationships of Fe_3O_4, Fe_2O_3, and oxygen. Am. J. Sci., 30 (5th Series), 239-316.

Grey, I. E., Reid, A. F., and Jones, D. G. (1974) Reaction sequences in the reduction of ilmenite: 4 - Interpretation in terms of the Fe-TiO and Fe-Mn-Ti-O phase diagrams. Trans. Inst. Mining Mineral., 83, C105-111.

Grey, I. E., and Ward, J. (1973) An x-ray and Mossbauer study of the $FeTi_2O_5$-Ti_3O_5 system. J. Solid State Chem., 7, 300-307.

Haas, J. L. (1987) Revision of the stability field of wüstite and recommended standard oxygen potentials for the solid oxide buffers. U. S. Geol. Survey Circular, 995, 26-27.

Haggerty, S. E., and Lindsley, D. H. (1970) Stability of the pseudobrookite (Fe_2TiO_5)-ferropseudobrookite ($FeTi_2O_5$) series. Carnegie Inst. Washington Year Book, 68, 247-249.

Hammond, P. A., and Taylor, L. A. (1982) The ilmenite-titanomagnetite assemblage: kinetics of re-equilibration. Earth Planet. Sci. Letters, 61, 143-150.

Hammond, P. L., Tompkins, L. A., Haggerty, S. E., Taylor, L. A., Spencer, K. J., and Lindsley, D. H. (1982) Revised data for coexisting magnetite and ilmenite near 1000°C, NNO and FMQ buffers. Geol. Soc. Am. Abstracts Programs, 14, 506.

Hartzman, M. J., and Lindsley, D. H. (1973) The armalcolite join ($FeTi_2O_5$-$MgTi_2O_5$) with and without excess FeO: Indirect evidence for Ti^{3+} on the moon. Geol. Soc. Am. Abstracts Program, 5, 653-654.

Hayakawa, M., Wagner Jr., J. B., and Cohen, J. B. (1977) Comments on "Phase Diagram of solid wüstite - part I". Materials Research Bulletin, 12, 429-430.

Hazen, R. M., and Jeanloz, R. (1984) Wüstite ($Fe_{1-x}O$): a review of its defect structure and physical properties. Rev. Geophys. Space Phys., 22, 37-46.

Heizmann, J. J., and Baro, R. (1967) Relations topotaxiques entre des cristaux naturels de magnetite Fe_3O_4 et l'hematite Fe_2O_3 qui en est issue par oxydation chimique. Comte Rendu Acad. Sci., Paris, 265, ser. D, 777.

Hentschel, B. (1970) Stoichiometric FeO as metastable intermediate of the decomposition of wüstite at 225°C. Zeit. Naturforschung, 25, 1997-1997.

Holloway, J. R., and Wood, B. J. (1988) Simulating the Earth - Experimental Geochemistry. Unwin Hyman, Inc., Winchester, MA, 196 pp.

Huebner, J. S. (1971) Buffering techniques for hydrostatic systems at elevated pressures. in G. C. Ulmer. ed., Research Techniques for High Pressure and High Temperature. Springer-Verlag, New York, 123-177.

Huebner, J. S. (1973) Experimental control of wüstite activity and mole fraction. Geol. Soc. Am., Abstracts Programs, 5, 676-677.

Huguenin, R. L. (1973a) Photostimulated oxidation of magnetite. 1, Kinetics and alteration phase identification. J. Geophys. Res., 78, 8481-8493.

Huguenin, R. L. (1973b) Photostimulated oxidation of magnetite. 1, Mechanism. J. Geophys. Res., 78, 8495-8506.

Ito, E., and Takahashi, E. (1989) Postspinel transformations in the system Mg_2SiO_4-Fe_2SiO_4 and some geophysical implications. J. Geophys. Res., 94, 10637-10646.

Izumi, F., and Fujiki, Y. (1975) Hydrothermal growth of anatase (TiO_2) crystals. Chem. Lett., 1975, 77-78.

Jamieson, J. C., and Olinger, B. (1968) High-pressure polymorphism of titanium dioxide. Science, 161, 893-895.

Jamieson, J. C., and Olinger, B. (1969) Pressure-temperature studies of anatase, brookite, rutile, and $TiO_2(II)$: a discussion. Am. Mineral., 54, 1477-1481.

Johnson, H. P., and Jensen, S. D. (1974) High temperature oxidation of magnetite to maghemite (abstr.). EOS (Trans., Am. Geophys. Union), 55, 233.

Johnson, H. P., and Merrill, R. T. (1972) Magnetic and mineralogic changes associated with low-temperature oxidation of magnetite. J. Geophys. Res., 77, 334-341.

Johnson, R. E., Woermann, E., and Muan, A. (1971) Equilibrium studies in the system MgO-"FeO"-TiO_2. Am. J. Sci., 271, 278-292.

Katsura, T., Kitayama, K., Aoyagi, R., and Sasajima, S. (1976) High-temperature experiments related to Fe-Ti oxide minerals in volcanic rocks. Kazan (Volcanoes), 1, 31-56 [in Japanese].

Katsura, T., and Muan, T. (1964) Experimental study of equilibria in the system FeO-Fe_2O_3-Cr_2O_3 at 1300°C. Trans. Am. Inst. Mining Metal. Engr., 230, 77-84.

Keesman, I. (1966) Zur hydrothermalen Synthese von Brookit. Zeits. Anorganische Allgemeine Chem., 346, 30-43.

Kesson, S. E., and Lindsley, D. H. (1975) The effects of Al^{3+}, Cr^{3+}, and Ti^{3+} on the stability of armalcolite. Geochim. Cosmochim. Acta, Suppl. 6, 1, 911-920.

Kleman, M. (1965) Proprietes thermodynamiques du protooxyde de fer sous forme solide. Application de resultats experimentaux au trace du diagram d'equilibre. Mem. Sci. Rev. Metal., 62, 457-469.

Knacke, O. (1983) The phase boundaries of wüstite. Ber. Bunsenges. Phys. Chem., 87, 797-800.

Knoll, H. (1961) Zur Bildung von Brookite. Naturwissenschaften, 48, 601.

Knoll, H. (1963) Umwandlung von Anatas im Brookite. Naturwissenschaften, 50, 546.

Knoll, H. (1964) Darstellung von Brookit. Angew. Chem., 76, 592.
Koch, F. B. (1967) Ordering in non-stoichiometric Fe$_x$O. Ph. D. Diss., Northwestern Univ., 266 pp.
Koch, F. B., and Cohen, J. B. (1969) The defect structure of Fe$_{1-x}$O. Acta Cryst., B25, 275-287.
Kress II, V. C. (1986) Iron-manganese exchange in coexisting garnet and ilmenite. M. S. Thesis, State University of New York at Stony Brook, x + 42 pp.
Kushiro, I. (1960) γ-α transition in Fe$_2$O$_3$ with pressure. J. Geomagnet. Geoelectr., 11, 148-151.
Lehmann, J., and Roux, J. (1986) Experimental and theoretical study of (Fe^{2+},Mg)(Al,Fe^{3+})$_2$O$_4$ spinels: activity-composition relationships, miscibility gaps, vacancy contents. Geochim. Cosmochim. Acta, 50, 1765-1783.
Leusmann, D. (1979) Studies in the ternary system MnO-Mn$_2$O$_3$-TiO$_2$. N. Jb. Mineral., Monatsh., 556-569.
Lewis, R. H., Haskin, L. A., and Lindstrom, D. J. (1988) Parameters for electrolysis of molten lunar rocks and soils to produce oxygen and iron. In Symposium on Lunar Bases and Space Activities of the 21st Century, LPI Contribution 652. Houston, 219.
Lind, M. D., and Housley, R. M. (1972) Crystallization studies of lunar igneous rocks: Crystal structures of synthetic armalcolite. Science, 175, 521-523.
Lindh, A. (1972) A hydrothermal investigation of the system FeO, Fe$_2$O$_3$, TiO$_2$. Lithos, 5, 325-343.
Lindsley, D. H. (1962) Investigations in the system FeO-Fe$_2$O$_3$-TiO$_2$. Carnegie Inst. Wash. Year Book, 61, 100-106.
Lindsley, D. H. (1963) Fe-Ti oxides in rocks as thermometers and oxygen barometers. Carnegie Inst. Wash. Year Book, 62, 60-65.
Lindsley, D. H. (1965) Iron-titanium oxides. Carnegie Inst. Wash. Year Book, 64, 144-148.
Lindsley, D. H. (1973) Delimitation of the hematite-ilmenite miscibility gap. Geol. Soc. Am. Bull., 84, 657-661.
Lindsley D. H. (1976) Experimental studies of oxide minerals. In Rumble III, Douglas, ed. Oxide Minerals, Rev. Mineral., v. 3. Mineral. Soc. Am., Washington, D. C., L61-L88.
Lindsley, D. H. (1981) Some experiments pertaining to the magnetite-ulvöspinel miscibility gap. Am. Mineral., 66, 759-762.
Lindsley, D. H., Kesson, S. E., Hartzman, M. J., and Cushman, M. K. (1974) The stability of armalcolite: Experimental studies in the system MgO-Fe-Ti-O. Geochim. Cosmochim. Acta, Suppl. 5, 1, 521-534.
Lindsley, D. H., and Lindh, A. (1974) A hydrothermal investigation of the system FeO, Fe$_2$O$_3$, TiO$_2$: A discussion with new data. Lithos, 7, 65-68.
Lipin, B. R. and Muan, A. (1974) Equilibria bearing on the behavior of titanate phases during crystallization of iron silicate melts under strongly reducing conditions. Proceedings of the Lunar Sci. Conference, 5th, 1, 535-548.
Liu, L.-G., and Bassett, W. A. (1986) Elements, Oxides and Silicates: High-pressure phases with implications for the earth's interior. Oxford University Press, New York.
MacChesney, J. B., and Muan, A. (1959) Studies in the system iron oxide-titanium oxide. Am. Mineral., 44, 926-945.
MacChesney, J. B., and Muan, A. (1960) The system iron oxide-TiO$_2$-SiO$_2$ in air. J. Am. Ceram. Soc., 43, 586-591.
Manenc, J. (1968) Structure du protoxyd de fer, resultats recents. Bull. Soc. Francais Mineral. Cristallogr., 91, 594-599.
Mao, H.-K., Takahashi, T., Bassett, W. A., Kinsland, G. A., and Merrill, L. (1974) Isothermal compression of magnetite to 320 kbar and pressure-induced phase-transition. J. Geophys. Res., 79, 1165-1170.
Matsuoka, K. (1971) Syntheses of iron-titanium oxides under hydrothermal conditions. Bull. Chem. Soc. Japan., 44, 719-722.
Matthews, A. (1976) The crystallization of anatase and rutile from amorphous titanium dioxide under hydrothermal conditions. Am. Mineral., 61, 419-422.
Mattioli, G. S., and Wood, B. J. (1988) Magnetite activities across the MgAl$_2$O$_4$-Fe$_3$O$_4$ join, with application to thermobarometric estimates of upper mantle oxygen fugacity. Contrib. Mineral. Petrol., 98, 148-162.
Mazzullo, L. J., Dixon, S. A., and Lindsley, D. H. (1975) T-fO$_2$ relationships in Mn-bearing Fe-Ti oxides. Geol. Soc. Am. Abstracts Program, 7, 1192.
McCallister, R. H., and Taylor, L. A. (1973) The kinetics of ulvöspinel reduction: synthetic study and applications to lunar rocks. Earth Planet. Sci. Letters, 17, 357-364.

McCammon, C. A., and Liu, L.-G. (1984) The effects of pressure and temperature on nonstoichiometric wüstite, Fe_xO: The iron-rich phase boundary. Phys. Chem. Minerals, 10, 106-113.

McQueen, R. G., Jamieson, J. C., and Marsh, S. P. (1967) Shock-wave compression and x-ray studies of titanium dioxide. Science, 155, 1401-1404.

Merrill, R. B., and Wyllie, P. J. (1973) Absorption of iron by platinum capsules in high pressure rock melting experiments. Am. Mineral., 58, 16-20.

Meyers, C. E., Mason, T. O., Petuskey, W. T., Halloran, J. W., and Bowen, H. K. (1980) Phase equilibria in the system Fe-Al-O. J. Am. Ceram. Soc., 63, 659-663.

Muan, A. (1958) On the stability of the phase $Fe_2O_3 \cdot Al_2O_3$. Am. J. Sci., 256, 413-422.

Muan, A. (1963) Silver-palladium alloys as crucible materials in studies of low-melting iron silicates. Bull. Ceram. Soc. Am., 42, 344-347.

Muan, A., and Gee, C. L. (1956) Phase equilibrium studies in the system iron oxide-Al_2O_3 in air and at 1 atm O_2 pressure. J. Am. Ceram. Soc., 39, 207-214.

Muan, A., Hauck, J., and Lofall, T. (1972) Equilibrium studies with a bearing on lunar rocks. Proc. Lunar Sci. Conference 3rd, 1, 185-196.

Muan, A., and Osborn, E. F. (1965) Phase equilibria among oxides in steelmaking. Addison-Wesley Publishing Co., xx + 236 p.

Muan, A., and Somiya, S. (1962) The system iron oxide-manganese oxide in air. Am. J. Sci., 260, 230-240.

Nafziger, R. H., Ulmer, G. C., and Woermann, E. (1971) Gaseous buffering for the control of oxygen fugacity at one atmosphere. in G. C. Ulmer. Ed., Research Techniques for High Pressure and High Temperature. Springer-Verlag, New York, p. 9-41.

Nakagiri, N., Manghnani, M. H., Ming, L. C., and Kimura, S. (1986) Crystal structure of magnetite under pressure. Phys. Chem. Minerals, 13, 238-244.

Nell, J., Wood, B. J., and Mason, T. O. (1989) High-temperature cation distributions in Fe_3O_4-$MgAl_2O_4$-$MgFe_2O_4$-$FeAl_2O_4$ spinels from thermopower and conductivity measurements. Am. Mineral., 74, 339-351.

Nord Jr., G. L., and Lawson, C. A. (1989) Order-disorder transition-induced twin domains and magnetic properties in ilmenite-hematite. Am. Mineral., 74, 160-176.

O'Neill, H. S. C., Pownceby, M. I., and Wall, V. J. (1988) Ilmenite-rutile-iron and ulvöspinel-ilmenite-iron equilibria and the thermochemistry of ilmenite ($FeTiO_3$) and ulvöspinel (Fe_2TiO_4). Geochim. Cosmochim. Acta, 52, 2065-2072.

O'Neill, H. S. C., Pownceby, M. I., and Wall, V. J. (1989) Activity-composition relations in $FeTiO_3$-$MnTiO_3$ ilmenite solid solutions from EMF measurements at 1050-1300K. Contrib. Mineral. Petrol., 103, 216-222.

Ohtani, E., and Sawamoto, H. (1976) Phase relations in the systems Fe_2SiO_4-$FeAl_2O_4$ and Co_2SiO_4-$CoAl_2O_4$ at high pressure and high temperature. Mineral. J., 8, 226-233.

Oka, Y., Steinke, P., and Chatterjee, N. D. (1984) Thermodynamic mixing properties of $Mg(Al,Cr)_2O_4$ spinel crystalline solid solution at high temperatures and pressures. Contrib. Mineral. Petrol., 87, 197-204.

Osborn, E. F. (1953) Subsolidus relationships in oxide systems in presence of water at high pressures. J. Am. Ceram. Soc., 36, 147-151.

Ozdemir, O., and Banerjee, S. K. (1984) High temperature stability of maghemite (gamma-Fe2O3). Geophys. Res. Letters, 11, 161-164.

Petric, A., and Jacob, K. T. (1982) Thermodynamic properties of Fe_3O_4-FeV_2O_4 and Fe_3O_4-$FeCr_2O_4$ spinel solid solutions. J. Am. Ceram. Soc., 65, 117-123.

Phillips, B., and Muan, A. (1960) Stability relations of iron oxides: phase equilibria in the system Fe_3O_4-Fe_2O_3 at oxygen pressures up to 45 atmospheres. J. Physical Chem., 64, 1451-1453.

Pinckney, L. R., and Lindsley, D. H. (1976) Effects of magnesium on iron-titanium oxides. Geol. Soc. Am. Abstracts Programs, 8, 1051.

Podpora, C., and Lindsley, D. H. (1984) Lindsleyite and Mathiasite: Synthesis of chromium-titanates in the crichtonite ($A_1M_{21}O_{38}$) series. EOS (Trans., Am. Geophys. Union), 65, 293.

Post, J. E., and Burnham, C. W. (1986) Ionic modelling of mineral structures and energies in the electron gas approximation: TiO_2 polymorphs, quartz, forsterite, diopside. Am. Mineral., 71, 142-150.

Pownceby, M. I., Wall, V. J., and O'Neill, H. S. C. (1987) Fe-Mn partitioning between garnet and ilmenite: experimental calibration and applications. Contrib. Mineral. Petrol., 97, 116-126.

Price, G. D. (1981) Subsolidus phase relations in the titanomagnetite solid solution series. Am. Mineral., 66, 751-758.
Punge-Witteler, B. (1984) Phasengleichgewichte zwischen Spinell und Wuestit im System Fe-Mn-O bei 700 degrees C, 1 bar. Zeit. Physikalische Chemie (Neue Folge), 142 (Part II), 239-248.
Rao, C. N. R. (1961) Kinetics and thermodynamics of the crystal structure transformation of spectroscopically pure anatase to rutile. Can. J. Chem., 39, 498-500.
Rao, C. N. R., Yoganarasimhan, S. R., and Faeth, P. A. (1961) Studies on the brookite-rutile transformation. Trans. Faraday Soc., 57, 504-510.
Ribaud, P. V., and Muan, A. (1962) Phase equilibria in a part of the system "FeO"-MnO-SiO_2. Trans. Am. Inst. Mining Metal. Engr., 224, 27-33.
Roiter, B. D. (1964) Phase equilibria in the spinel region of the system FeO-Fe_2O_3-Al_2O_3. J. Am. Ceram. Soc., 47, 509-511.
Sack, R. O., and Ghiorso, M. S. (1991) Chromian spinels as petrogenetic indicators: Thermodynamics and petrological applications. Am. Mineral., 76, in press.
Saha, P., and Biggar, G. M. (1974) Subsolidus reduction equilibria in the system Fe-Ti-O. Indian J. of Earth Sci., 1, 43-59.
Sato, H., Endo, S., Sugiyama, M., Kikegawa, T., Shimomura, O., and Kusaba, K. (1991) Baddeleyite-type high pressure phase of TiO_2. Science, 251, 786-788.
Sato M. (1972) Intrinsic oxygen fugacities of iron-bearing oxide and silicate minerals under low total pressure. in Studies in Mineralogy and Precambrian Geology, Geol. Soc. Am. Memoir 135
Sato, M., Hickling, N. L., and McLane, J. E. (1973) Oxygen fugacity values of Apollo 12, 14, and 15 lunar samples and reduced state of lunar magmas. Proc. Lunar Sci. Conference 4th, 1, 1061-1079.
Sato, M., and Wright, T. L. (1966) Oxygen fugacities directly measured in magmatic gases. Sci., 153, 1103-1105.
Schmahl, N. G., Frisch, B., and Hargartner, E. (1960) Zur Kenntnis der Phasenverhaltnisse im System Fe-Ti-O bei 1000°C. Zeit. Anorgan. Allgem. Chemie, 305, 40-54.
Schriefels, W. A., and Muan, A. (1975) Liquid-solid equilibria involving spinel, ilmenite and ferropseudobrookite in the system "FeO"-Al_2O_3-TiO_2 in contact with metallic iron. Geochim. Cosmochim. Acta Suppl. 6, 1, 973-985.
Shannon, R. D., and Pask, J. A. (1965) Kinetics of the anatase-rutile transformation. J. Am. Ceram. Soc., 48, 391-398.
Shen, P., Bassett, W. A., and Liu, L.-G. (1983) Experimental determination of the effects of pressure and temperature on the stoichiometry and phase relations of wüstite. Geochim. Cosmochim. Acta, 47, 773-778.
Simons, B. (1974) Zusammensetzung und Phasenbreiten der Fe-Ti-Oxide im Gleichgewicht mit metallischem Eisen. Diplomarbeit, Institut fur Kristallographie, Technische Hochschule Aachen
Simons, B., and Woermann, E. (1978) Iron titanium oxides in equilibrium with metallic iron. Contrib. Mineral. Petrol., 66, 81-89.
Snetlage, R., and Klemm, D. D. (1975) Das System Fe-Cr-O bei 1000, 1095, und 1200°C. N. Jb. Mineral., Abh., 125, 227-242.
Sosman, R. B., and Hostetter, J. C. (1916) The oxides of Iron: I. Solid solution in the system Fe_2O_3-Fe_3O_4. J. Am. Chem. Soc., 38, 807-833.
Speidel, D. H. (1967) Phase equilibria in the system MgO-FeO-Fe_2O_3: the 1300°C isothermal section and extrapolations to other temperatures. J. Am. Ceram. Soc., 50, 243-248.
Speidel, D. H. (1970) Effect of magnesium on the iron-titanium oxides. Am. J. of Sci., 268, 341-353.
Spencer, K. J., and Lindsley, D. H. (1980) Statistical study of minor elements in coexisting Fe-Ti Oxides. Geol. Soc. Am., Abstracts Programs, 1980, 527.
Spencer, K. J., and Lindsley, D. H. (1981) A solution model for coexisting iron-titanium oxides. Am. Mineral., 66, 1189-1201.
Srecec, I., Ender, A., Woermann, E., Gans, W., Jacobsson, E., Eriksson, G., and Rosen, E. (1987) Activity-composition relations of the magnesiowüstite solid solution series in equilibrium with metallic iron in the temperature range 1050-1400K. Phys. Chem. Minerals, 14, 492-498.
Stanin, F. T., and Taylor, L. A. (1980) Armalcolite: an oxygen fugacity indicator. Geochim. Cosmochim. Acta, Suppl. 11, 1, 117-124.
Swaroof, B., and Wagner, J. B. (1967) On the vacancy concentrations of wüstite (FeOx) near the p to n transition. Trans, Metallurg. Soc. AIME, 239, 1215-1218.

Taylor, L. A., Williams, R. J., and McCallister, R. H. (1972) Stability relations of ilmenite and ulvöspinel in the Fe-Ti-O system and application of these data to lunar mineral assemblages. Earth Planet. Sci. Letters, 16, 282-288.

Taylor, R. W. (1963) Liquidus temperatures in the system $FeO-Fe_2O_3-TiO_2$. J. Am. Ceram. Soc., 46, 276-279.

Taylor, R. W. (1964) Phase equilibria in the system $FeO-Fe_2O_3-TiO_2$ at 1300°C. Am. Mineral., 49, 1016-1030.

Tompkins, L. A. (1981) Experimental studies in the system $FeO-Fe_2O_3-TiO_2$ with a bearing on the Fe-Ti oxide geothermometer-oxygen geobarometer. Senior Honors Thesis, University of Massachusetts, 79 p.

Trestman-Matts, A., Dorris, S. E., Kumarakrishnam, S., and Mason, T. O. (1983) Thermoelectric determination of cation distributions in $Fe_3O_4-Fe_2TiO_4$. J. Am. Ceram. Soc., 66, 829-834.

Trestman-Matts, A., Dorris, S. E., and Mason, T. O. (1984) Thermoelectric determination of cation distributions in $Fe_3O_4-MgFe_2O_4$. J. Am. Ceram. Soc., 67, 69-74.

Turnock, A. C., and Eugster, H. P. (1962) Fe-Al oxides: Phase relationships below 1000°C. J. Petrol., 3, 533-565.

Ulmer, G. C., Ed. (1971) Research Techniques for High Pressure and High Temperature. Springer-Verlag, New York, 367 pp.

Vallet, P., and Raccah, P. (1964) Sur les limites du domaine de la wüstite solide et le diagramme general qui en resulte. Comte Rendu Acad. Sci., Paris, 258, 3679-3682.

Vallet, P., and Raccah, P. (1965) Contribution a l'etude des proprietes thermodynamiques du protoxyde de fer solide. Mem. Sci. Rev. Metal, 62, 1-29.

Viertel, H. U., and Seifert, F. (1980) Thermal stability of defect spinels in the system $MgAl_2O_4-Al_2O_3$. N. Jb. Mineral., Abh., 140, 89-101.

Vincent, E. A., Wright, J. B., Chevallier, R., and Mathieu, S. (1957) Heating experiments on some natural titaniferous magnetites. Mineral. Mag., 31, 624-655.

Voigt, R., and Will, G. (1981) Das System $Fe_2O_3-H_2O$ unter hohen Drucken. N. Jb. Mineral., Monatsh., 89-96.

Webster, A. H., and Bright, N. F. H. (1961) The system iron-titanium-oxygen at 1200°C and oxygen partial pressures between 1 atmosphere and 2×10^{-14} atmospheres. J. Am. Ceram. Soc., 44, 110-116.

Wechsler, B. A., Lindsley, D. H., and Prewitt, C. T. (1984) Crystal structure and cation distribution in titanomagnetites [$Fe_{3-x}Ti_xO_4$]. Am. Mineral., 69, 754-770.

Weidner, J. R. (1982) Fe oxide magmas in the system Fe-C-O. Canadian Mineral., 20, 555-566.

Williams, R. J. (1983) Enhanced production of water from ilmenite: an experimental test of a concept for producing lunar oxygen. Fourteenth Lunar and Planetary Sci. Conference, Special Session Abstracts. LPI Contribution, 500, 35-35.

Williams, R. J. (1985) Oxygen extraction from lunar materials; an experimental test of an ilmenite reduction process. in Mendell, W. W. ed., Lunar Bases and Space Activities of the 21st Century. NASA/Johnson Space Center, Houston, TX, 551-558.

Willshee, J. C., and White, J. (1967) An investigation of equilibrium relationships in the system $MgO-FeO-Fe_2O_3$ up to 1750°C in air. Trans. Brit. Ceram. Soc, 66, 541-555.

Woermann, E., Brezny, B., and Muan, A. (1969) Phase equilibria in the system MgO-iron oxide-TiO_2 in air. Am. J. Sci., 267A (Schairer Vol.), 463-479.

Woermann, E., Hirschberg, A., and Lamprecht, A. (1970) Das System Hematit-Ilmenit-geikielith unter hohen Temperatures und und hohen Drucken (Abs.). Fortschritte der Mineralogie, 47, 79-80.

Wood, B. J., Kirkpatrick, R. J., and Montez, B. (1986) Order-disorder phenomena in $MgAl_2O_4$ spinel. Am. Mineral., 71, 999-1006.

Wu, C. C., and Mason, T. O. (1981) Thermopower measurement of cation distribution in magnetite. J. Am. Ceram. Soc., 64, 520-522.

Chapter 4   Subir K. Banerjee

MAGNETIC PROPERTIES OF FE-TI OXIDES

INTRODUCTION

O'Reilly's (1984) comprehensive monograph on Rock Magnetism deals with a variety of minerals including Fe-Ti oxides. Dunlop's (1990) excellent review paper is a more recent source but it is briefer than O'Reilly's, and the chief strength of Dunlop's review is not so much the mineralogic-petrologic control of the observed magnetism as his emphasis on a unifying magnetic domain theory model to explain grainsize variation of the magnetic properties of a given mineral, magnetite. This review aims to briefly place in context our up-to-date knowledge of the magnetic properties of the Fe-Ti oxide group of minerals. In this article the control exercised by chemical composition and crystallography is emphasized at the cost of the more extrinsic variable: grainsize. Also, this chapter does not deal with the various natural remanent magnetizations of Fe-Ti oxide-containing igneous, metamorphic and sedimentary rocks.

It cannot be said that new research on all the magnetic properties of pure Fe-Ti oxide minerals has continued till recent times with equal vigor, hence many references will still be from older times. What is clear though is that with an increasing knowledge of magnetism of solids in general it is now possible to place much of the older data in a common perspective, which we hope will aid the newcomer to the field in grasping many apparently scattered and disparate records of individual mineral behavior.

The older texts that may be of interest to the reader are Nagata's (1961) *Rock Magnetism* (2nd edition) which, in reality, is a multi-authored volume, and the *Physical Theory of Rock Magnetism* by Stacey and Banerjee (1974) which covers succinctly both the mineral magnetism and remanence properties of (mainly) the Fe-Ti oxide group of minerals.

CRYSTAL STRUCTURE AND COMPOSITION

Before we discuss the various Fe-Ti oxides based primarily on their crystal structures and secondarily on their specific compositions (i.e., chemical formula unit), it may help the reader to refer to Figure 1, taken from O'Reilly (1984). This ternary diagram shows conveniently the relationship between compositions and geologic processes for the various Fe-Ti oxide minerals found in the earth's crust. While the FeO and TiO_2 corners represent wüstite and rutile, the corner labeled $(1/2)Fe_2O_3$ represents either hematite (rhombohedral) or maghemite (cation-deficient spinel). Combinations of appropriate fractions of these end member compositions lead to the locations of ulvöspinel, pseudobrookite and magnetite on this ternary composition diagram. Additionally, O'Reilly has indicated the various geological processes such as low temperature oxidation ("maghemitization"), high temperature or deuteric oxidation, hydration, kinetically controlled inversion of structures from metastable to stable forms, etc., to indicate how the magnetic mineralogy will respond to these processes. The thick, cross-hatched join between ulvöspinel and magnetite represent the spinel solid solution, known as titanomagnetite. The cross-hatching is intended to remind the reader that some amount of high temperature non-stoichiometry is expected when titanomagnetites are synthesized, or formed naturally, at elevated temperature (>900°C) and then quenched to room temperature.

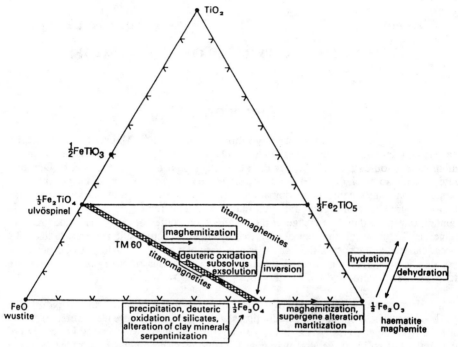

Figure 1. Ternary composition diagram of the FeO - Fe$_2$O$_3$ - TiO$_2$ system. High temperature non-stoichiometry is indicated by the cross-hatching on the Fe$_2$TiO$_4$ - Fe$_3$O$_4$ join. Low temperature non-stoichiometry due to maghemitization is present for titanomaghemite compositions represented by locations within the quadrilateral Fe$_2$O$_3$ - FeO - FeTiO$_3$ - Fe$_2$TiO$_5$. The FeTiO$_3$ - Fe$_2$O$_3$ join is deleted for clarity. Modified from O'Reilly (1984).

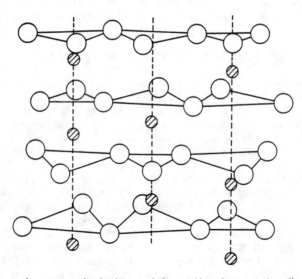

Figure 2. The corundum structure showing hexagonal close-packing of oxygen anions (larger open circles) within which the iron cations (smaller hatched circles) reside. The [0001] axis is indicated by dashed lines. From O'Reilly (1984).

For the sake of clarity, a similar cross-hatched join between ilmenite and hematite has been omitted from the diagram, but this join represents the rhombohedral solid solution discussed below.

Rhombohedral oxides

Hematite (α-Fe_2O_3) crystallizes in the corundum structure (space group $\bar{R}3c$) with the divalent oxygen anions arranged in a hexagonal close-packed lattice. The hexagonal unit cell parameters are a = 5.034 Å and c = 13.749 Å. The cations, Fe^{3+}, occupy sites coordinated octahedrally by the anions. However, the (0001) plane cations normal to the c-axis constitute slightly "puckered" layers with some cations shifted vertically with respect to one another (Fig. 2). Mössbauer absorption spectra of hematite reveal only one type of local symmetry (octahedral) with a hyperfine field at the nucleus of 518 kOe at room temperature (Warner et al., 1972).

Upon cooling to -10°C (263 K) hematite undergoes a first-order transition (the Morin transition) and a shortening of the c-axis. The Morin transition can be shifted by pressure, magnetic field or atomic substitutions (Liebermann and Banerjee, 1971; Knittle and Jeanloz, 1986). It is also suppressed in fine grains less than 200 Å because of internal dilatational strain (Nininger and Schrooer, 1978). The important magnetic effects at the Morin transition will be dealt with below. Although natural hematite often displays impurities of less than 0.1 per molecule of Al^{3+} and Mn^{3+}, Ti^{4+} is the most important replacement ion for Fe^{3+}, resulting in a complete solid solution, hematite - ilmenite above ~700°C and quenchable to room temperature. Substitution of Ti^{4+} for Fe^{3+} causes reduction of Fe^{3+} ions as explained below.

The opposite end-member of the hematite-ilmenite solid solution series (also called titanohematite series) is, of course, ilmenite ($FeTiO_3$). It also crystallizes in the hexagonal system ($R\bar{3}$) with unit cell parameters a = 5.088 Å and c = 14.080 Å. Ti^{4+} ions are accommodated in alternate ($0\bar{1}11$) planes of cations leading to two types of layers that either contain only Fe^{2+} or Ti^{4+} ions. Thus, Fe^{3+} ions of hematite are replaced by a combination of Fe^{2+} and Ti^{4+} ions to conserve charge balance:

$$2\ Fe^{3+} = Fe^{2+} + Ti^{4+}\ .$$

Vegard's "law" of a linear variation of the unit cell parameter is not obeyed strictly on going from hematite to ilmenite because of curvatures at either end. Mössbauer data at 5 K show that there is only one value of hyperfine field (47 kOe) at the Fe^{2+} nucleus corresponding to the pure iron containing plane normal to the c-axis.

An important corollary of the mixed Fe^{2+} - Ti^{4+} occupation in ilmenite is the vanishing of a center of symmetry in ilmenite (space group = $R\bar{3}$). Hematite (space group = $R\bar{3}c$), of course, has a center of symmetry. This difference is of immense consequence to the magnetic properties of the titanohematite solid solution series, and will be dealt with later. Following the convention among rock magnetists and physicists, "titanohematite" includes the complete hematite-ilmenite solid solution, parts of which are known as "hemo-ilmenites" and "ilmeno-hematites". At room temperature around the midpoint of the solid solution, $Fe_{1.5}Ti_{0.5}O_3$, there is a transition from $R\bar{3}c$ symmetry to $R\bar{3}$ for perfectly ordered (slowly cooled) samples. For quenched samples in the laboratory or in nature (volcanic tuff), there is imperfect order which is subject to change as a function of annealing temperature and time. This has important ramifications for magnetic properties (Ishikawa and Syono, 1963; Nord an Lawson, 1989), including the phenomenon of self-reversal of thermoremanent magnetization (TRM) and is a subject of discussion in the last section of this chapter. With the progress of our knowledge in condensed matter physics, it is now realized that quenched titanohematites, rich in titanium, constitute natural examples of an advanced industrial magnetic material: "spin glass", so named because of the lack of long range order seen in its magnetic spin moments. Mössbauer studies of these intermediate compositions show a number of hyperfine fields, reflecting fluctuating local site symmetries and consequently broadened non-Lorentzian absorption lines (Warner et al., 1972).

Spinel oxides

The two end members of the titanomagnetite solid solution series are magnetite (Fe_3O_4) and ulvöspinel (Fe_2TiO_4). The complete solid solution can be quenched from above 600°C given appropriate oxygen fugacity (Taylor, 1964; Lindsley, Chapter 3). We shall first discuss the structure of magnetite and its transitions at low temperature.

Known to humankind from medieval times as naturally occurring lodestone, magnetite (space group Fd3m) has an inverse spinel structure (unit cell = 8.395Å) at room temperature, so that Fe^{3+} ions occupy the tetrahedrally coordinated or "A" sites, and both Fe^{3+} and Fe^{2+} ions occupy randomly the octahedrally coordinated or "B" sites (Fig. 3). Partial covalency due to hybridized orbitals of tetrahedral Fe^{3+} ions is said to be responsible for an A site preference of Fe^{3+} (Blasse, 1964). This arrangement has been confirmed by neutron diffraction as well as Mössbauer effect studies (Stacey and Banerjee, 1974). The random distribution of B site Fe^{2+} and Fe^{3+} ions leads to "electron-hopping" or polaron conduction in magnetite at room temperature. Activation energy for this process is low, $\sim 10^{-2}$ eV, and therefore, although an insulator below -155°C, magnetite has a moderately high electronic conductivity (1.43×10^4 mho/m) at room temperature (Verwey et al., 1947). The electron hopping from Fe^{2+} ($3d^6$) to Fe^{3+} ($3d^5$) ions at room temperature is faster than 10^{-8} sec, the time constant for the nuclear transition processes

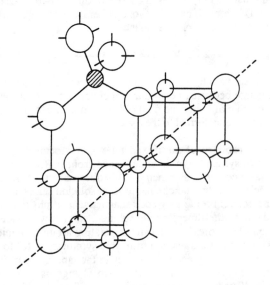

Figure 3. A part of the spinel structure to display octahedral B site iron ions (small open circles) and tetrahedral A site iron ions (small hatched circles). The larger open circles represent oxygen anions. The dashed line is the body diagonal [111] axis, the easy axis of magnetization in magnetite. From O'Reilly (1984).

responsible for Mössbauer absorption. The result is that the Mössbauer spectra for magnetite at room temperature can be resolved in terms of only two nuclear hyperfine fields, 490 kOe for A site Fe^{3+} and 450 kOe for B site $Fe^{2.5+}$ (effectively) (Jensen and Shive, 1973).

A most interesting, and as yet not completely understood, process occurs in magnetite structure when it undergoes a first order transition upon cooling through -155°C (118 K), known as the Verwey transition (Verwey et al., 1947). In contrast to the pathbreaking study of Verwey, it is now believed that the low-temperature structure is monoclinic or even triclinic, but not orthorhombic, as it was earlier believed (Miyamoto and Chikazumi, 1988). One of the room temperature cube edges becomes the monoclinic c axis, and twinning results in the low temperature phase of a single crystal, which can be removed by applying stress or a large applied magnetic field along a chosen cube edge. Verwey's pioneering contribution was to study the transition using electrical conductivity, which shows a phenomenal drop upon cooling. At present, there is a rich literature on this topic but as Verwey discovered, this first order transition also produces an ordered arrangement of Fe^{3+} and Fe^{2+} ions in B sites resulting in a large increase in activation energy for polaron conduction. Recent research involving nuclear magnetic resonance (NMR), magnetoelectric and dielectric studies, and even muon-spin resonance, have shown that ionic ordering, vacancy ordering, antiphase domain creation, etc. result in multiple local site symmetries. For our purpose here, it is sufficient to take note that the transition is also accompanied by magnetoelectricity and a sharp change in magnetocrystalline anisotropy, the latter will be dealt with later.

Ulvöspinel (Fe_2TiO_4) is the other end-member of the titanomagnetite solid solution series. It is also an inverse spinel (a = 8.536 Å) with Fe^{2+} ions in both the tetrahedral A and octahedral B sites. This mineral does not have a specific low temperature transition like that in magnetite, but on cooling to temperatures below liquid nitrogen temperature (77 K), lattice distortions set in which have been variously identified as Jahn-Teller distortion of tetrahedral Fe^{2+} (Ono et al., 1968) and as the effect of unquenched orbital angular momentum of octahedral Fe^{2+} (Readman et al., 1967). The chief effect of this distortion is a phenomenally large magnetic anisotropy.

The intermediate members of the solid solution display a smooth variation of unit cell parameter with small departures at the magnetite and ulvöspinel ends (Lindsley, 1976; Wechsler et al, 1984). The distribution of the cations, Fe^{2+}, Fe^{3+} and Ti^{4+}, among the tetrahedral and octahedral sites has been a topic of much research. Since the cation distribution affects directly the saturation magnetization per molecule, we will defer this discussion to Section IV. Electrical conductivity is again controlled by occupation of mixed valences (Fe^{2+} and Fe^{3+}) in the octahedral sites, and with increasing Ti^{4+} substitution conductivity decreases and activation energy for polaron conduction increases. The low temperature transition in magnetite is suppressed when Ti^{4+} content increases to more than 0.1 atom per molecule. In nature, the most important occurrence of titanomagnetites is in the center of the median valley of ocean-spreading centers while cation-deficient titanomaghemites occur in the altered pillow basalts on either side of the median valley central axes.

Cation-deficient spinel

The one Fe bearing cation-deficient spinel that has been the most thoroughly studied is maghemite (γ-Fe_2O_3), a product of oxidation of finegrained (~10^3 Å) magnetite at temperatures below ~250°C. Although its chemical composition, i.e., Fe^{3+}/O^{2-} ratio, is identical to that of hematite (α-Fe_2O_3), maghemite crystallizes in the inverse spinel structure. In order to emphasize this, the formula is sometimes written as $Fe_{8/3} \square_{1/3} O_4$, where $\square$ denotes vacant cation sites and the four oxygen ions per formula unit reminds the reader of maghemite's close similarity in crystal structure and magnetic properties to magnetite (Fe_3O_4).

Maghemite is metastable at room temperature and pressure, and converts quickly to the thermodynamically stable form, hematite, upon heating to ~300°C and/or compression (Kushiro, 1960). Its presence in nature suggest diagenesis or authigenesis for sediments and sedimentary rocks, and low temperature alteration ("brownstone" facies) for igneous and metamorphic rocks.

It should be noted that it is not yet clear whether dry (heated in air) versus wet (heated in a solution), oxidation of finegrained magnetite produces maghemite with the same physical properties (O'Reilly, 1983, also see section below on *Theories of Magnetic Superexchange*). There are reports also of forming thin film (~10^3Å) single crystals of mm size maghemite by epitaxial growth on single crystal substrates at temperatures that are much higher than the conventionally accepted temperature limit (~250°C) for stability of maghemite at one atmosphere pressure (Chen et al., 1988). We should note further the successful synthesis by Hauptman (1974) of slightly oxidized titanomagnetite (or titanomaghemite) at 1275°C by equilibriating $Fe_{2.4}Ti_{0.6}O_4$ under controlled oxygen fugacity. Hauptman compared the rate of change of Curie point (T_C) with vacancy concentration for his samples with those produced by low temperature oxidation and found that his high temperature samples showed a nine times steeper increase of T_C with vacancy concentration than low temperature maghemite. This phenomenon is also not properly understood at the moment (O'Reilly, 1984; Moskowitz, 1987).

The cubic unit cell parameter for maghemite is 8.32 Å and one in three of the Fe^{3+} ion sites are occupied by vacancies, as required by change balance: $3 Fe^{2+} = 2 Fe^{3+} + \square$. The compositions between pure magnetite and pure maghemite can be considered a solid solution. Its magnetite end members have been studied thoroughly for their electrical conductivity and magnetic properties, including the suppression of the Verwey transition with increasing vacancy concentration. For a recent review to this literature the reader is referred to Rasmussen and Honig (1988) who found that above a vacancy concentration of 0.005 per molecule the transition is second order.

Transition ion impurities such as Al^{3+} are known to suppress completely the tendency of maghemite to invert to the stable form, hematite (Stacey and Banerjee, 1974). However, the commonest impurity in nature Ti^{4+}, is not known to inhibit this irreversible crystallographic transition (Readman and O'Reilly, 1970). A number of authors have synthesized titanomagnetites of different titanium content and then studied the changes in their physical properties with variable degrees of low temperature oxidation (in air) and the consequent increase in vacancy concentration. The impetus for such studies came from the desire to simulate the natural titanomaghemites found in the altered pillow lavas of the ocean crust. O'Reilly (1984) should be consulted for a comparison of the mechanics responsible for such oxidation (in air) of sintered titanomagnetites with the submarine oxidation of oceanic pillow basalts in low pH solutions.

This large variety of methods for preparing titanomaghemites, and the difficulty in determining their Fe^{2+}/Fe^{3+} ratios by titration have led to mutually incompatible data for cubic cell edge variations with degrees of oxidation. However, the most frequently used data are those by Readman and O'Reilly (1972) and Nishitani and Kono (1983). The degree of oxidation (by oxygen addition) is usually measured by the oxidation parameter labeled 'z' by O'Reilly and Banerjee (1967). It refers to the fraction of Fe^{2+} ions per unit titanomagnetite molecule that has undergone low temperature oxidation to Fe^{3+} produce a cation-deficient spinel. Hence, for change balance:

$$Fe^{2+} + (1/2)z\ O = z\ Fe^{3+} + (1-z)\ Fe^{2+} + (1/2)z\ O^{2-}$$

For a titanomagnetite with xTi^{4+} per molecule, the oxidation process can be written as:

$$Fe_{3-x}Ti_xO_4 + (z/2)(1+x)\ O = (1/R)\ Fe_{(3-x)R}\ Ti_x\ R[\ ]_{3(1-R)}O_4$$

$$\text{where } \frac{1}{R} = \frac{4 + (z/2)\ (1+x)}{4}$$

For measuring aqueous oxidation degree, O'Reilly (1984) has proposed a similar parameter z^1, based on a proposed cation removal mechanism. The quantitative experiments to ascertain the exact mechanism of cation removal by aqueous oxidation have yet to be carried out, although Worm and Banerjee (1984) have suggested an equation based on their preliminary studies.

INTRODUCTION TO THEORIES OF MAGNETIC SUPEREXCHANGE

Magnetic Fe-Ti oxides are ionic compounds in which the iron ions (Fe^{3+}, Fe^{2+}) carry net magnetizations ("spin moments") proportional to the number of unpaired electron spins on each iron. The magnitude of unit spin moment is called a Bohr magneton (μ_B), where 1 μ_B = he/4πm (in cgs electromagnetic units) and h = Planck's constant, e = electronic charge and m = electronic mass. Transition metal elements of the 3d series (Fe, Mn, Ni, Co etc.) are the main magnetic carriers in Fe-Ti oxides but iron (Fe) by far is the most important source of magnetism. Simultaneous application of Pauli exclusion principle and Hund's rule of maximum multiplicity lead to the derivation of spin magnetic moments for transition metal ions (O'Reilly, 1984; Stacey and Banerjee, 1974). Thus, for the most common ions in Fe-Ti oxides the following spin moments can be calculated:

Ion	Spin Moment (μ_B per molecule)
Fe^{3+}	5
Fe^{2+}	4
Mn^{3+}	5
Mn^{2+}	4
Co^{2+}	3
Ni^{2+}	2

Actual measurements confirm most of the above values with two important exceptions. In spinel lattices, octahedrally located Fe^{2+} carries a spin moment of 4.2 μ_B while octahedral Co^{2+} has 3.7μ_B. Both are due to unquenched orbital angular momentum which is ignored in first order calculations. Orbital magnetic moment is highly anisotropic due to spin-orbit coupling (for a clear exposition see O'Reilly, 1984), and therefore we shall note in the last of this paper that large magnetocrystalline anisotropy in spinels is invariably associated with octahedral Fe^{2+} and Co^{2+}. It has also been suggested that tetrahedral Fe^{2+}, because of the Jahn-Teller effect, can contribute additional lattice-sensitive anisotropy (Schmidbauer and Readman, 1982). We shall now discuss the different types of long range coupling between ions carrying spin magnetic moments, resulting in magnetically ordered solids.

Antiferromagnetism

Figure 4 shows the temperature variation of the reciprocal magnetic susceptibility, χ = (induced magnetization) / (applied field parallel and perpendicular to the spin alignment axis of a single crystal antiferromagnet. T_N is the critical temperature for second order transition, called the Néel point in honor of L. Néel who is the originator of the theory of antiferromagnetism. Above T_N, both $\chi_\parallel$ and $\chi_\perp$ have the same value. For further details consult either O'Reilly (1984) or Stacey and Banerjee (1974).

Figure 5 shows the antiferromagnetically ordered manganous oxide (MnO). It crystallizes in the face-centered cubic structure and is thus analogous to the structure of the inverse spinel magnetite or ulvöspinel. In MnO, the cation is Mn^{2+}, and being a $3d^5$ ion, it carries a spin moment of 5 μ_B. Note that each (111) plane contains parallel (or ferromagnetically oriented) magnetic moments, the antiferromagnetism refers to the antiparallel orientation of successive (111) cation layers. Figure 6 illustrates how the d-shell electron of a Mn^{2+} ion with one type of spin (up) orientation is coupled through the 2p electron orbit of a O^{2-} ion to the opposite type (down) of spin orientation of another Mn^{2+} ion. This phenomenon is called magnetic superexchange because the coupling takes place through an intermediary (O^{2-}). Either O'Reilly (1984) or Stacey and Banerjee (1974) should be consulted for a more detailed explanation.

Among Fe-Ti oxides, ulvöspinel (Fe_2TiO_4) and ilmenite ($FeTiO_3$) are examples of antiferromagnetic order below their Néel points of 120 K and 80 K, respectively. Ulvöspinel, however, has a weak magnetic moment of about 0.2 μ_B per molecule which could arise from one of two reasons. Either Fe^{2+} in tetrahedral and octahedral sites does not carry the same ionic magnetic moment, or the oppositely coupled spins are non-collinear. More about this will be mentioned in a later section. Hematite (α-Fe_2O_3) is not a perfect antiferromagnet either. Its Fe^{3+} spin moments on antiparallel (0001) basal planes are slightly canted with regard to each other at room temperature, leading to a weak magnetic moment. Below the Morin transition (T_M), the spins align along the [0001] axis, the canting disappears and perfect antiferromagnetism returns. This will be discussed later.

Ferrimagnetism

This name was coined to describe antiferromagnetic ordering of non-equal spin moment magnitudes, as in magnetite. It is different from true ferromagnetism because the ionic spin moments are antiparallel. However, there is a substantial magnetic moment per molecule (as in ferromagnets) because of non-cancellation of antiparallel sublattice magnetizations. Figure 4 shows how ferrimagnetism can be distinguished from both antiferromagnetism and ferromagnetism through temperature variation of reciprocal susceptibility.

In the case of magnetite, the arrangement of molecular magnetic moments of each sublattice (tetrahedral A and octahedral B, the latter shown within square brackets) can be indicated as follows:

$\rightarrow$	$\leftarrow$	$\leftarrow$
Fe^{3+}	$[Fe^{3+}$	$Fe^{2+}]$
–5 μ_B	5 μ_B	4 μ_B

The net magnetic moment (or saturation magnetization) is therefore 4 μ_B per formula unit, and not $5 + 5 + 4 = 14$ μ_B, as would have been the case for ferromagnetism.

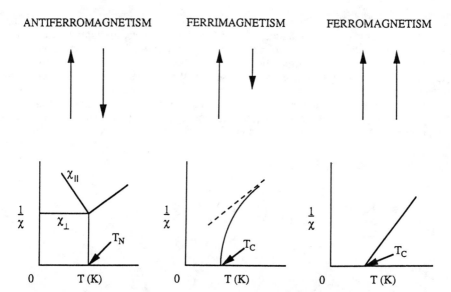

Figure 4. Antiferromagnetic, ferrimagnetic and ferromagnetic spin alignments with paired arrows. Characteristic variation of reciprocal susceptibility ($1/\chi$) with temperature plotted below each type of spin alignment. For the antiferromagnet, parallel and perpendicular susceptibilities are the same above the Néel point, T_N. T_C is the Curie point.

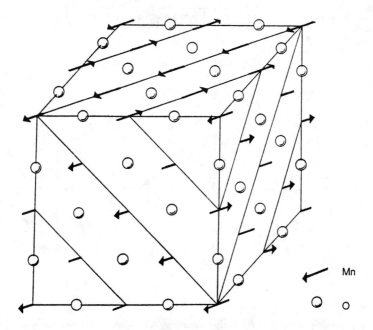

Figure 5. Antiferromagnetic spin alignment between (111) planes in manganous oxide, MnO. The arrows represent spin moments, the circles oxygen anions. Modified from Martin (1967).

Although our discussion of superexchange has centered on the antiparallel coupling *between* iron atoms on two sublattices (A and B in ferrimagnets, for example), there are also weaker magnetic couplings *within* a sublattice (AA and BB interactions). Unlike the predominantly negative or antiparallel interaction between sublattices, intrasublattice interactions can be positive or negative. Blasse (1964) has discussed these interactions in spinels most thoroughly and has reemphasized how in the presence of such additional interactions, the M_S vs. T curves for some ferrimagnets can show anomalies (Fig. 7) as predicted by Néel. A conventional thermal dependence was called Q-type by Néel, while the presence of a 'hump' was predicted for P-types and a reversal of sublattice

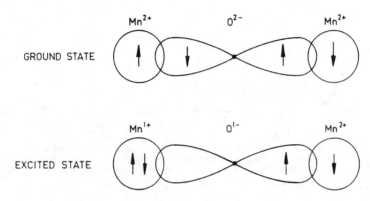

Figure 6. Magnetic superexchange between manganese ions on neighboring (111) planes. In the excited state, an electron from the p-orbital of oxygen anions occupy the d-orbital of the left manganese ion. Because both Hund's nile and Pauli exclusion principle must be obeyed simultaneously, the orientations of the spin moments of the left and right manganese ions are antiparallel, brought on by the two antiparallel spins of the oxygen ion momentarily occupying the iron ion orbitals. From Stacey and Banerjee (1974).

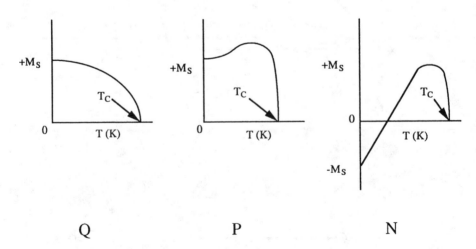

Figure 7. Saturation magnetization (M_S) versus temperature (T, in Kelvins) plots to illustrate schematically Néel's Q, P and N-type curves of ferrimagnetic temperature dependence. The same M_S can be obtained at more than one temperature value for the P-type. The N-types can display self-reversal of thermoremanent magnetization.

preponderance was predicted for N-types. Blasse was able to synthesize ferrimagnetic oxides that confirmed each of these predictions; and P-type curves were observed by Schult (1968) for natural basalts and by Readman and O'Reilly (1972) for a highly oxidized $Fe_{2.3}Ti_{0.7}O_4$ synthetic sample. For Readman and O'Reilly's sample, the easiest explanation would be a negative (antiparallel) BB interaction among octahedral iron ions, hastening the thermal decrease of B sublattice moments. Upon further oxidation, A sublattice moments become the dominant source of magnetization at low temperature and N-type curves were shown to result. Néel had suggested that rocks carrying such ferrimagnets could cause self-reversal of natural remanent magnetization but no such N-type has yet been found which reverses its magnetization above room temperature to be of relevance to paleomagnetism.

Non-collinear structures

In the discussions above it has been tacitly assumed that the equilibrium orientation for antiferromagnetism is a collinear spin arrangement. However, under circumstances dominated by competing intra-sublattice superexchange (e.g., negative BB interaction) or by crystal field effects and substantial presence of spin-orbit coupling, it may be incumbent upon the spins to take up canted, triangular or even spiral configurations to minimize the energy. The first example of this is the canted antiferromagnet hematite, also called a weak ferromagnet before the origin of its net magnetic moment was understood.

In a hematite molecule at room temperature, Fe^{3+} ions in alternate (0001) layers should be antiferromagnetically aligned, while in a given layer they are aligned parallel to one another. Polarized neutron diffraction (Nathans et al, 1968) shows, however, that alternate cation layers are slightly canted producing a net weak magnetic moment at right angles to the average spin alignment axis. Dzyaloshinski (1958) and Moriya (1960), on theoretical grounds, had predicted such a canted moment to explain the observed weak magnetism in hematite.

Another example of non-collinearity *may* be the case of weak magnetic moment observed in ulvöspinel (Fe_2TiO_4) which has Fe^{2+} ions in equal numbers on tetrahedral and octahedral sub-lattices of a spinel structure. As we will discuss in the section on *Magnetocrystalline Anisotropy and Magnetostriction*, at low temperature ulvöspinel shows a highly anisotropic magnetization. If this resulted from octahedral Fe^{2+} ions, the preferred (or "easy") alignment of individual Fe^{2+} spins would be along the cubic [111] axis while the tetrahedral Fe^{2+} spins would point along [100] axis, the easy axis for the single crystal as a whole. Furthermore, since a negative intra-sublattice superexchange interaction is present in the B sites of spinel (Blasse, 1964), one can postulate a triangular arrangement in a magnetic cell in which two tetrahedral Fe^{2+} spins will point along [100] while the opposed two octahedral Fe^{2+} will form a slightly misaligned v-shaped pair. This arrange-ment will result in a net weak magnetization along [100] the axis of the crystal. Since only unpolarized neutron diffraction of powdered and single crystal samples of ulvöspinel has been attempted to date (Boysen and Schmidbauer, 1984), a triangular spin arrangement as mentioned above has not yet been examined or sought for.

SATURATION MAGNETIZATION AND CURIE (AND NEEL) POINTS

Having been armed with some basic information about magnetic superexchange, we can now investigate the by-products of superexchange: saturation magnetization (M_s) and Curie point (T_C), the latter being the temperature at which thermal energy becomes equal to superexchange energy and magnetic order breaks down. Both ferromagnets (Fe, Ni etc.) and ferrimagnets (titanomagnetites) lose their saturation magnetization and long range magnetic order at T_C. Under a large applied magnetic field, say 10 kOe (1 Tesla), ferromagnets and ferrimagnets display a weak magnetization even above T_C. This results from the persistence of short range magnetic order of a few tens of Ångströms. M_S vs. T curves should therefore be carefully evaluated at or close to T_C in order to obtain the true value of T_C (Moskowitz, 1981). True antiferromagnets (pure ilmenite, $FeTiO_3$) lose their magnetic order at a similar disordering temperature called the Néel point (T_N). Induced magnetization at a low (~10 Oe) field, i.e., low field susceptibility, shows a characteristic peak at T_N.

Since we deal mostly with ferrimagnets in Fe-Ti oxides, saturation magnetization measured at a high field (~ 20 kOe or 2 Tesla) and at low temperatures (~4.5 K) provides a measure of the net magnetization in Bohr magnetons, μ_B per formula unit, which can be compared with the predicted molecular magnetization for model cation distributions between tetrahedral and octahedral sites. Alternative distributions producing the same M_S can be distinguished by additional physical property measurements, such as Mössbauer effect, polarized neutron diffraction, electrical conductivity, thermoelectric power etc.

Measurements of T_C and T_N, like M_S, are intrinsic properties depending only on chemical composition and crystal structure. Thus, T_C and T_N values help paleomagnetists and mineral physicists to have an idea of the composition of a magnetic mineral from a rapid thermomagnetic measurement. Local (non-equivalent) site occupation by magnetic ions can also be inferred from T_C and its shift with impurities. But while M_S is very sensitive to cation distribution because net magnetization is the *difference* between A and B sublattice magnetic moments, T_C is determined by the *product* of the sublattice spin moments. Hence, T_C is relatively less sensitive to cation distribution variations than M_S.

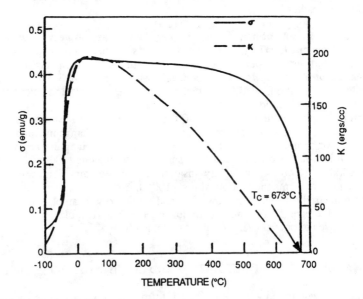

Figure 8. Saturation magnetization (σ) and basal plane anisotropy constant (K) versus temperature (in degrees Celsius) for a natural single crystal of hematite from Ascension. Note the unusual decay in magnetization and anisotropy upon cooling below -10°C, the Morin transition. Modified from Flanders and Schuele (1964).

Rhombohedral oxides

The titanohematites ($Fe_{2-x}Ti_xO_3$) have a complex variation of room temperature saturation magnetization with composition and hence M_S does not lend itself easily to interpretations regarding cation distribution. Hematite (α-Fe_2O_3), one end member, has an interesting variation of M_S vs. T as shown in Figure 8, taken from Flanders and Schuele (1964). Above the Morin transition, T_M (-10°C), M_S vs. T has a conventional look, the weak ferromagnetism vanishing at a T_C = 675°C which is also the Néel point T_N, or the disordering temperature of the antiferromagnetic ordering. Thus the canting of the sublattices and the ordering within the sublattice vanish simultaneously. Occasional suspicions to the contrary (Smith and Fuller, 1967), T_N is not different from T_C of the weak ferromagnetic phase.

Below T_M, however, canting vanishes, and M_S drops rapidly on cooling but does not go to zero because of a very weak magnetism due possibly to a slight non-equivalence in the sublattice magnetic moments. Whether the effect is due to vacancies or impurity substitutions, they have to be ordered on a preferred sublattice to give rise to a net M_S, and it is not yet known what the details of such an ordering are. The value of the low temperature M_S, however, is clearly variable from crystal to crystal, and exists even in pure synthetic crystals.

The other end member, ilmenite ($FeTiO_3$) is a perfect antiferromagnet with zero net magnetization at absolute zero and a Néel point of -183°C (80 K). There has been much confusion in the literature about the presence of natural remanent magnetization in ilmenite and about the utility of magnetic separation of "ilmenite". In the former case, one usually is dealing with solid solution members $FeTiO_3$ - $Fe_{1.5}Ti_{0.5}O_3$, while in the latter, the explanation may be the same; sometimes it is ilmenite with fine-scale exsolved ferrimagnetic magnetite.

Much has been written about the unusual magnetic properties of that part of the solid solution which is magnetic at room temperature: $Fe_{1.5}Ti_{0.5}$ - $Fe_{1.27}Ti_{0.73}O_3$ (Ishikawa and Syono, 1963; Nord and Lawson, 1989). These are ferrimagnetically and crystallographically ordered compounds, with one cation layer containing iron ions only, while the oppositely coupled layer contains both iron and titanium ions, leading to a net magnetic moment. The predicted variation in M_S with titanium content is indeed observed for slowly cooled, ionically ordered samples. When measured at 4.5 K, M_S vs. composition shows a smooth variation with a maximum near the ilmenite-rich end. The Néel points vary smoothly between -183°C and 675°C and are anchored between these two points by $FeTiO_3$ and α-Fe_2O_3 respectively.

The most fascinating aspect of this solid solution, however, is the appearance of reversed thermo-remanent magnetization (RTRM) in the members $Fe_{1.5}Ti_{0.5}O_3$ - $Fe_{1.27}Ti_{0.73}O_3$ when they are quenched and heat-treated. Research has gone on in this area for more than 30 years because these minerals are the best candidates for recording a reverse polarity when cooled in a normal geomagnetic field. Since the confirmation of global geomagnetic field reversals from isotopically dated rocks worldwide, these minerals have lost some of their charm and notoriety. But they still constitute a classic example of order-disorder phenomenon in the solid state leading to superexchange coupling of ionic spins across a crystallographic domain boundary. Although some critics may question my confidence, the article by Nord and Lawson (1989) on this topic provides a thoroughly satisfactory explanation for most of these unusual properties. When quenched from above a composition-dependent critical temperature (usually 700°-1300°C), the structure is disordered (R$\bar{3}$c) and net magnetization is weak at room temperature. After annealing for minutes to hours at elevated temperatures (700°-900°C), however, an ordered (R$\bar{3}$) structure emerges which has strong ferrimagnetic properties. The boundaries between the ordered and disordered domains are highly anisotropic and weakly ferrimagnetic. If there is a sufficient amount of these domain boundaries, the iron ions within the boundaries magnetize first on cooling (because of a higher T_C) and then reversely couple the spins of the strongly magnetic (ferrimagnetically) ordered phases, producing a net RTRM. Future studies with a transmission electron microscope may be able to test this model by simultaneously observing both the magnetic domains and the antiphase domain boundaries, thus testing the magnetic coupling directly.

Spinel oxides

Just as the order-disorder transition in titanohematites has been a subject of intense interest so has been the cation distribution among tetrahedral and octahedral sites of stoichiometric titanomagnetite spinels. The reason for the interest lies in the control of saturation magnetization in these ferrimagnets by the antiparallel sublattice spin moments, and for the fact that titanomagnetites and their cation-deficient non-stoichiometric products (titanomaghemites) constitute most of the strongly magnetic carriers in nature.

It is, therefore, important to recapitulate what light saturation magnetization per molecule can truly throw upon cation distribution in a ferrimagnet. As O'Reilly (1984) has remarked, only when there are just two species of cations (and this includes vacancies), can M_S vs. composition provide unique information about cation distribution. In the case of titanomagnetites, the species are four: Fe^{3+}(5 μ_B), Fe^{2+}(4 μ_B), Ti^{4+}, and vacancies. Variation in M_S, therefore, provides valuable but non-unique information about site occupancy. Supporting, but still indirect and non-unique, information comes from thermoelectric e.m.f., electrical conductivity, conventional neutron diffraction and Mössbauer absorption data. However, only low temperature polarized neutron diffraction and low temperature Mössbauer absorption spectra in the presence of a large magnetic field can provide sublattice magnetic moments directly.

Since Akimoto's (1954) pioneering attempt at determining the cation distribution of the titanomagnetite solid solution series from measurements of M_S, there has been a number of studies whose details can be found in O'Reilly (1984) and Wechsler et al. (1984) Broadly, the problem consists of finding the cation distribution pathway(s) by which (a) the octahedral B sites, on going from magnetite to ulvöspinel, end up having only Fe^{2+} ions, instead of an equal Fe^{2+}/Fe^{3+} ratio, and (b) the tetrahedral A sites start with one Fe^{3+} ion per molecule in magnetite, but end up with only one Fe^{2+} ion instead. It is suspected (O'Reilly and Banerjee, 1965) that the intermediate titanomagnetites may not display a linear variation of M_S (at low temperatures and high magnetic fields), which would be the sign of mixed Fe^{2+} - Fe^{3+} occupation of A *and* B sites throughout, but the M_S data to date have remained murky. For the most useful results, M_S should be measured for single crystals at 4.5 K in magnetic field ~50 kOe; this has yet to be done.

Neither Mössbauer nor neutron diffraction data have helped distinguish between models because increasing titanium substitution in magnetite produces varying local site symmetry for Fe ions, causing large values of diffuse scatter or line broadening (Wechsler et al., 1984; Boysen and Schmidbauer, 1984; Banerjee et al., 1967; Jensen and Shive, 1973). Even for the same technique, when neutron diffraction is applied to polycrystals, of a critical composition, $Fe_{2.4}Ti_{0.6}O_4$, the ratio of magnetic moment per iron ion (μ_{Fe}) as seen in the two sublattices turns out to be different for different authors (Wechsler et al.,

1984; Boysen and Schmidbauer, 1984). A resolution can perhaps come from polarized neutron diffraction and Mössbauer absorption in high fields and low temperature. As stated before, only such experiments provide individual sublattice magnetic moments, and not merely their difference as given by saturation magnetization measurements.

<u>Cation-deficient spinels</u>

As shown in Figure 1 when titanomagnetites in nature are oxidized at low temperature (<300°-400°C), especially in the presence of hydrothermal solutions, cation-deficient spinels (titanomaghemites) are formed in which some vacancies appear in the iron cation sites. They constitute important carriers of natural remanent magnetization (NRM) of submarine pillow basalts, responsible for the larger part of linear marine magnetic anomalies. Among sub-aerial basalts, occasionally reversed polarity of NRM has been observed (Domen, 1969), resulting from the presence of titanomaghemites. Starting with a given cation distribution for titanomagnetites, it is possible to follow the changes in their net ferrimagnetism (M_S per molecule) with oxidation, and predict when such a process will cause a self-reversal of magnetization in a basalt after prolonged titanomag-hemitization (Verhoogen, 1962; O'Reilly and Banerjee, 1967). Because these minerals are metastable and invert rapidly on heating to a magnetite-ilmenite intergrowth, such self-reversals are difficult to reproduce in the laboratory. In the example given by Domen (1969) and other cases involving submarine basalts, circumstantial evidence is strong and it behooves us to arrive at a definitive model for cation distribution in titanomaghemites from saturation magnetization and other studies.

Readman and O'Reilly (1972) and Nishitani and Kono (1983) have carefully oxidized finegrained (less than 7000 Å) titanomagnetite in air at moderate temperatures (~300°C) to produce single phase titanomaghemites whose compositions lie inside the quadrilateral $Fe_2O_3 - Fe_3O_4 - Fe_2TiO_4 - Fe_2TiO_5$ in Figure 1. Nishitani and Kono (1982) explained why attempts other than theirs and Readman and O'Reilly's to produce truly single phase metastable titanomaghemite had failed. They point out that if the initial grain size of the starting titanomagnetite is not less than 7000 Å, the diffusion rate at, say, 300°C is too slow compared to the rate of oxygen addition at the surface, resulting in the production of a magnetite-ilmenite exsolution phase.

Although Readman and O'Reilly (1972) and Nishitani and Kono (1983) both measured M_S vs. z (oxidation parameter), they did not provide standard tables of M_S values vs. z of titanomaghemites. Part of the reason could lie in the fact that M_S vs. T of highly oxidized titanomaghemites display Néel's P-type thermal dependence (Fig. 7), resulting in non-unique values of M_S as a function of temperature. Therefore, an oxidized sample may have a higher M_S at room temperature compared to its unoxidized parent titanomagnetite and yet, the situation could be reversed if M_S was measured at 4.5 K. Thus room temperature M_S values, the ones more often measured in various laboratories, would not provide definitive measures of degree of oxidation.

In contrast to M_S, T_C was found to increase predictably with oxidation. It increases because Fe^{2+} ion, when converted to Fe^{3+}, has a larger number of spin moments (5 instead of $4\mu_B$) per ion than before. Even here, observations by different groups are not uniform, and O'Reilly (1984) suggested that the reason lay in a variable initial non-stoichiometry caused when titanomagnetite sintering took place at different elevated temperatures (>1200°C). A very thorough study by Moskowitz (1987), however, has shown that this is not the case. According to him the most likely causes are non-uniform techniques for determining T_C and z parameter (oxidation degree) of titanomaghemites in different laboratories. Be that as it may, when appropriate error bars and replicate measurements of

Moskowitz are compared with those of prior work, the data of Nishitani and Kono (1983) appear to be confirmed as accurate. They have provided the data for variations of T_C and cell parameter with oxidation. These plots can therefore be used to determine oxidation state from magnetic parameters alone, for compounds that are difficult to dissolve (for wet chemical analysis) without compromising the initial Fe^{2+}/Fe^{3+} ratio.

The newest twist in the study of magnetic properties of titanomaghemites is the confirmation of an early suspicion that natural cation-deficient spinels, especially in the submarine pillow basalts, are produced not by addition of O^{2-} to a constant initial Fe/Ti ratio, but by leaching and removal of Fe^{2+} ions by hydrothermal fluids with the Ti/O ratio remaining constant. Many authors have contributed to this discovery but a recent and thoroughly quantitative study is that by Furuta et al. (1985) who used a JEOL JXA-733 microprobe to obtain both backscattered electron images (BEI) and quantitative elemental data including oxygen, the latter measured with a lead stearate detector. Furuta et al. have shown decisively that titanomaghemites from a subaerial andesite and a submarine basalt both have cation-deficiency caused by iron loss, not oxygen-addition as has been done to date in the laboratory simulations. Some preliminary attempts have been made to measure the magnetic properties of synthetic titanomaghemites produced by aqueous oxidation (Worm and Banerjee, 1984; Suk, 1985) but much work yet remains to be done.

MAGNETOCRYSTALLINE ANISOTROPY AND MAGNETOSTRICTION

Basic theory

The magnetic superexchange theory provides a good explanation for long range spin alignment of magnetic moments in an ionic solid but it cannot explain, a priori, why certain crystallographic directions are preferred by the spins to point along ("easy axes"), and certain others are avoided ("hard axes"). Phenomenologically this is expressed as a magnetocrystalline energy term (E_K) expressed as a function of two constants and the direction cosines for a cubic lattice (Stacey and Banerjee, 1974):

$$E_K = K_1(\alpha_i^2 \alpha_j^2 + ... + ...) + K_2(\alpha_i^2 \alpha_j^2 \alpha_K^2 + ... + ...)$$

For a hexagonal crystal E_K can be expressed in terms of one uniaxial constant that determines the anisotropy between the c-axis and the (0001) plane, while the other triaxial constant provides the magnitude of the 6-fold in-plane anisotropy perpendicular to the c-axis.

Magnetostriction is the phenomenon of an increase or decrease in length of a magnetic material when it is magnetized parallel to that chosen direction. Magnetostriction can also be understood as the strain dependence or strain-induced modification of the magnetocrystalline energy of a single crystal. The average magnetostriction constant λ_s of a cubic crystal assemblage is expressed in terms of two constants and direction cosines:

$$\lambda_s = \lambda_{100} + 3(\lambda_{111} - \lambda_{100})(\alpha_i^2 \alpha_j^2 + ... + ...)$$

Why is it important to know about magnetostriction? The first reason is the origin and stability of any type of magnetization, including natural remanent magnetization (NRM). For a strongly magnetic material such as titanomagnetite, the direction of magnetization is controlled primarily by shape anisotropy and next, by magnetocrystalline anisotropy and magnetostriction. For a sphere, however, shape anisotropy is zero. And for a weakly magnetic material like hematite, shape anisotropy is negligible and it must be either the (0001) plane anisotropy, or internal strain and magnetostriction that control the stability of magnetization.

The next reason for a knowledge of magnetocrystalline anisotropy, especially of its thermal dependence, is for the discovery of isotropic points at which the characteristic easy directions change their orientation. For magnetite, there is an isotropic point T_K at -143°C (130K), which in bulk magnetite happens to be about 12° higher than the Verwey transition discussed earlier. The Morin transition (T_M) in hematite is an isotropic point, too.

Finally, a knowledge of magnetocrystalline anisotropy constant K and its temperature dependence are critically important for determining the wall width and wall energy of magnetic domain walls that are present in all magnetic grains larger than the critical size for uniform single domain behavior. Although in this chapter we have not dealt with grainsize dependence of magnetic properties they have been dealt with recently by Dunlop (1990). Suffice it to say here that as magnetic grains assume sizes larger than the single domain size, differently oriented magnetic domains (>10,000 Å) are created to lower the net energy due to self-demagnetization. The width (~1000 Å) of the domain wall is determined by the balance between superexchange constant (A) and anisotropy constant (K). Thus, the wall width (δ_w) is given by:

$$\delta_w = \text{constant } (A/K)^{1/2}$$

For ferrimagnets and antiferromagnets the magnitude of K, and its variation with temperature and impurity substitution are explained in terms of the so-called "one ion model", described especially well in O'Reilly (1984). Just as the net magnetization is determined by the algebraic addition of oppositely directed ionic spin moments per molecule, the net magnetocrystalline anisotropy is also determined by the algebraic addition of the anisotropy contributions of the ionic moments situated in the two sublattices. Each sublattice anistropy, however, has its characteristic temperature dependence. The net anisotropy vs. temperature, therefore, can have striking behaviors, including going through zero, i.e. an isotropic point.

The basic reason for an ion to have anisotropy, i.e. to become lattice-sensitive, comes from spin-orbit coupling. For most magnetic ions in Fe-Ti oxides this coupling constant is small. Examples are the spherically symmetric $3d^5$ ions like Fe^{3+} and Mn^{2+}. On the other hand, octahedral Fe^{2+} ($3d^6$) and Co^{2+} ($3d^7$) have large spin-orbit coupling and their presence produces significant contributions to magnetocrystalline anisotropy (O'Reilly, 1984).

Isotropic points

Figure 9 (left hand side) shows the calculated variations of two terms, H_{SI} and H_{MD} (with opposite signs), responsible for the net anisotropy of hematite. They cross over at T_M, the Morin transition (= 0.281 T_N where $T_N = T_C$, as described here). The spins turn at this temperature from the (0001) plane to the c-axis upon cooling. Figure 9 (right hand side) shows how dramatically T_M can change with the substitution of the platinum-group element Ir^{4+} which in itself is a strongly anisotropic ion. Notice how little the T_C (called T_N in the figure) has been affected over the substitution range of only 0.02 atoms per molecule.

In the case of magnetite, data for K_1 (the first order cubic anisotropy constant) and λ_s (averaged for polycrystals) have been presented in Figure 10. In the absence of shape

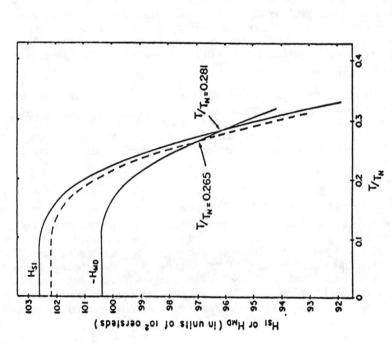
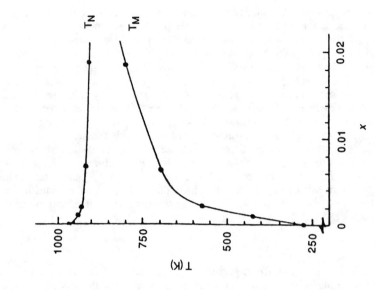

Figure 9. The left-hand figure shows how two opposing crystalline anisotropy fields (H_{SI} and H_{MD}) cancel each other at the Morin transition, indicated by the point $T/T_N = 0.281$, for pure hematite. Upon substitution of 1.28 mole percent aluminum, the Morin transition is depressed to the lower value, $T/T_N = 0.265$. From Besser et al. (1967). The right-hand figure shows the opposite effect. Substitution of 1.8 mole percent iridium causes the Morin transition to rise to nearly 800 K or 527°C. From Liu (1986).

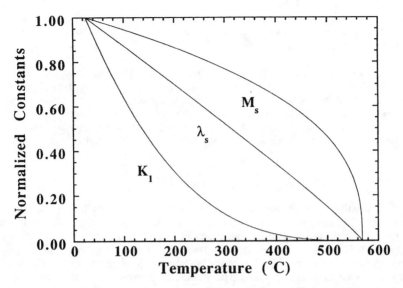

Figure 10. Saturation magnetization (M_s), average magnetostriction constant (λ_s) and cubic anisotropy constant (K_1) of magnetite versus temperature in °C. Data measured and collated by B.M. Moskowitz.

anisotropy, therefore, it is magnetostriction that exercises significant control over direction and stability of magnetization in magnetite, not the cubic anisotropy. This means that internal non-uniform strain and strain history will determine the stability of magnetization in coarse-grained ("multidomain", >1 μm) magnetite. We have already discussed the low temperature electrical conductivity transition or Verwey transition in the section on *Crystal Structure and Composition*. Approximately 10°C above the Verwey transition (variously quoted as -155°C to -148°C), the easy magnetic axis rotates from the (room temperature) cubic [111] to one of the cubic [100] axes. We use the cubic orientation designation here for clarity of understanding although the low temperature structure is now regarded as monoclinic or triclinic (Miyamoto and Chikazumi, 1988). Since any one of the three [100] axes may be chosen as the new easy axis, twinning usually results which can be removed by applying a biasing magnetic field or a linear stress along a chosen [100] axis.

Rock magnetists have taken advantage of the above phenomenon to cool NRM-carrying magnetite grains to liquid nitrogen temperature in the absence of a biasing field. This results in the demagnetization of the remanent magnetization of all grains large enough carry multiple magnetic domains, which are now equally divided, parallel and antiparallel to all three [100] axes. This leaves only the shape anisotropic uniformly magnetized single domain grains with a memory of their room temperature magnetization. Upon reheating to room temperature, also in zero applied field, this memory magnetization reappears and is the preferred stable component of NRM, likely to be the most stable over geological times (Kobayashi and Fuller, 1968; Merrill, 1970; Dunlop and Argyle, 1991).

The physical origin of the isotropic point in magnetite is similar to that of the Morin transition in hematite. At low temperatures, electron hopping stops and Fe^{2+} ions display a very large positive anisotropy along the [100] axis. At room temperature, electron hopping causes a pure Fe^{2+} contribution to disappear and tetrahedral Fe^{3+} ions have the dominating negative anisotropy, displayed as an easy [111] axis. Upon cooling, the pure Fe^{2+} contribution becomes strong and at the isotropic point (T_K) overwhelms the Fe^{3+}

contribution. Recent studies of this transition have involved nuclear magnetic resonance (NMR) and magnetoelectric studies, and the reader is referred to Mizoguchi (1985) and Miyamoto and Chikazumi (1988) for details.

Magnetostriction constants have been measured for magnetite and hematite (O'Reilly, 1984; Stacey and Banerjee, 1974). The low temperature increases of magnetostriction constants of magnetite are extremely large, while hematite at room temperature is found to have appreciable magnetostriction even though it is only a weak ferromagnet. Future studies of magnetostriction constants, and especially their temperature dependence, may throw important light on the acquisition and stability of magnetization in magnetite and hematite.

OVERVIEW AND CONCLUSION

Two constituencies of research scientists are served directly by advances in our understanding of the magnetic properties of Fe-Ti oxides. They are the paleomagnetists and the rock magnetists. Let us deal with the paleomagnetists first. The questions about stability of natural remanent magnetization (NRM) are foremost in their minds, and as mentioned in the *Introduction*, NRM depends both on the intrinsic properties discussed in this chapter as well as their variations with grainsize. It would be wrong to conclude, however, that a study of the intrinsic properties is not relevant to paleomagnetism. A rapidly burgeoning area of research in paleomagnetism is related to orogenic belts, deformed rocks and ancient rocks, all of whom have a complex stress history. In order to separate NRM components of such rocks, it is essential to know single crystal properties such as magnetocrystalline anisotropy and magnetostriction, and how they vary with composition and temperature. These parameters have been measured only once (Syono, 1963), and for a limited number of titanomagnetite single crystals. Titanohematite single crystals of wide ranging compositions have not been synthesized or studied. A knowledge of temperature dependence of these constants above room temperature may provide paleomagnetists with useful guidance for thermal demagnetization and separation of multiple components of NRM in rocks with a complex stress history.

Of course, for rock magnetists, future magnetic studies of Fe-Ti oxides have many important uses. To begin with, the same temperature dependencies of magnetocrystalline anisotropy and magnetostriction constants mentioned above are critical for new advances in our understanding of the origin of a primary magnetization such as thermoremanent magnetization (TRM) and secondarily acquired magnetizations such as chemical remanent magnetization (CRM) and viscous remanent magnetization (VRM).

Values of saturation magnetization, magnetocrystalline anisotropy, magnetostriction and exchange constants are all extremely important to experimental rock magnetists for interpreting observations on natural and synthetic samples, and for theoretical rock magnetists to model the behavior of ultrafine (<1 µm) grains that cannot be synthesized in controlled shapes or as truly non-interacting grains. For the prediction of seismomagnetic effects or for the interpretation of long wavelength magnetic anomalies, rock magnetists again need to know the stress and temperature dependence of the intrinsic magnetic constants. For understanding submarine alteration (maghemitization) of the igneous oceanic crust and its effect on the marine magnetic anomalies, we need to obtain values of saturation magnetization and anisotropy constant of titanomaghemites of well-characterized composition. Such data are sorely lacking.

In conclusion, we can take pride in the progress over the last two decades in our theoretical understanding and experimental evaluation of the basic magnetic properties of pure magnetite, hematite and maghemite. What we still need to do is to extend these studies to the titanium-bearing equivalent of these minerals, in single crystal form, and as a function of stress and temperature.

ACKNOWLEDGMENTS

Professors D.H. Lindsley and B. Moskowitz helped with the review of this chapter. I am most grateful to them for their comments and criticism. Mr. Christopher Hunt was immensely helpful with the drafting of diagrams, while Ms. Laura Holmstrand and Ms. Sue Linehan provided invaluable assistance with the preparation of the manuscript. I thank them all. I acknowledge gratefully the support of the National Science Foundation through grant number EAR-8804853.

REFERENCES

Akimoto, S. (1954) Thermomagnetic study of ferromagnetic minerals contained in igneous rocks. J. Geomagn. Geoelectr. 6, 1-14.

Banerjee, S.K., O'Reilly, W., Gibb, T.C. and Greenwood, N. (1967) The behavior of ferrous ions in iron-titanium spinels. J. Phys. Chem. Solids 28, 1323-1335.

Besser, P.J. Morrish, A.H. and Searle, C.W. (1967) Magnetocrystalline anisotropy of pure and doped hematite. Phys. Rev. 153, 632-640.

Blasse, G. (1964) Crystal chemistry and some magnetic properties of mixed metal oxides with spinel structure. Philips Research Rept. Suppl. 3, 1-139.

Boysen, H. and Schmidbauer, E. (1984) Neutron diffraction study of titanomagnetites. Geophys. Res. Lett. 11, 165-168.

Chen, M.M., Lin, J., Wu, T.W. and Castillo, G. (1988) Wear resistance of iron oxide thin films. J. Applied Phys. 63, 3275-3277.

Domen, H. (1969) An experimental study of the unstable natural remanent magnetization of rocks as a paleogeomagnetic fossil. Bull. Fac. Educ. Yamaguchi Univ. 18, (2):1. Quoted in detail in Stacey and Banerjee (1974)

Dunlop, D.J. (1990) Developments in rock magnetism. Rep. Prog. Phys. 53, 707-792.

Dunlop, D.J. and Argyle, K.S. (1991) Separating multi domain and single domain like remanences in pseudo single domain magnetites (215-540nm) by low temperature demagnetization. J. Geophys. Res. 96, 2007-2018.

Dzyaloshinski, I. (1958) A thermo dynamic theory of "weak" ferromagnetism of antiferromagnetics. J. Phys. Chem. Solids 4, 241-255.

Flanders, P.J. and Scheuele, W.J. (1964) Temperature-dependent magnetic properties of hematite single crystals. Proc. Int'l Cof. Magnetism, Nottingham, U.K., 594-596.

Furuta, T. and Otsuki, M. (1985) Quantitative Electron Probe Microanalysis of Oxygen in titanomagnetites with implications for oxidation processes. J. Geophys. Res. 90, 3145-3150.

Hauptman, Z. (1974) High temperature oxidation, range of non-stoichiometry and curie point variation of cation deficient titanomagnetite $Fe_{2.4}Ti_{0.6}O_{4.8}$. Geophys. J. Roy. Astron. Soc. 38, 29-47.

Ishikawa, Y. and Syono, Y. (1963) Order-disorder transformation and reverse thermo-remanent magnetism in the $FeTiO_3$ - Fe_2O_3 system. J. Phys. Chem. Solids 24, 517-528.

Jensen, S.D. and Shive, P.N. (1973) Cation distribution in sintered titanomagnetites. J. Geophys. Res. 78, 8474-8480.

Knittle, E., Jenaloz, R. (1986) High-pressure electrical resistivity measurements of Fe_2O_3: comparison of static-compression and shock-wave experiments to 61 GPA. Solid State Comm. 58, 129-131.

Kobayashi, K. and Fuller, M.D. (1968) Stable remanence and memory of multidomain materials with special reference to magnetite. Phil. Mag. 18, 601.

Kushiro, I. (1960) $\gamma \rightarrow \alpha$ transition in Fe_2O_3 with pressure. J. Geomagn. Geoelectr. 11, 148-151.

Liebermann, R.C. and Banerjee, S.K. (1971) magnetoelastic interactions in hematite: Implications for geophysics. J. Geophys. Res. 76, 2735-2756.

Lindsley, D.H. (1976) The crystal chemistry and structure of oxide minerals. In D. Rumble (ed.), Oxide Minerals, Rev. Mineral. 3, L1-L60.

Liu, J.Z. (1986) Morin transition in hematite doped with iridium ions. J. Magnetism Magnetic Materials 54-57, 901-902.

Martin, A.H. (1967) Magnetism in Solids. London Iliffe Books Ltd., London, U.K.

Merrill, R.T. (1970) Low temperature treatments of magnetite and magnetite-bearing rocks. J. Geophys. Res. 75, 3343-3349.

Miyamoto, Y. and Chikazumi, S. (1988) Crystal symmetry of magnetite in low temperature phase deduced from magneto electric measurements. J. Phys. Soc. Japan 57, 2040-2050.

Mizoguchi, M. (1985) Abrupt change of NMR line shape in the low temperature phase of Fe_3O_4. J. Phys. Soc. Japan 54, 4295-4299.

Moriya, T. (1960) Anisotropic exchange interaction and weak ferromagnetism. Phys. Rev. 120, 91.

Moskowitz, B.M. (1981) Methods for estimating Curie temperatures of titanomaghemites from experimental Js-T data. Earth Planet. Sci. Lett. 53, 84-88.

Moskowitz, B.M. (1987) Towards resolving the inconsistencies in characteristic physical properties of synthetic titanomaghemites. Phys. Earth Planet. Interiors 46, 173-183.

Nagata, T. (1961) Rock Magnetism (2nd edition). Tokyo: Maruzen.

Nathans, R., Pickart, S.J., Alperin, H.J. and Brown, P.J. (1964) Phys. Rev. 136A, 1641.

Nininger, R.C. Jr. and Schroeer, D. (1978) Mössbauer studies of the morin transition in bulk and microcrystalline α-Fe_2O_3. J. Phys. Chem. Solids 39, 137-144.

Nishitani, T. and Kono, M. (1982) Grainsize effect on the low-temperature oxidation of titanomagnetite. J. Geophys. 50, 137-142.

Nishitani, T. and Kono, M. (1983) Curie temperature and lattice constant of oxidized titanomagnetite. Geophys. J. Royal Astron. Soc. 74, 585-600.

Nord, G.L. and Lawson, C.A. (1989) Order-disorder transition-induced twin domains and magnetic properties in ilmenite-hematite. Am. Mineral. 74, 160-176.

O'Reilly, W. (1983) The identification of titanomaghemites: Model mechanisms for the model maghemitization and inversion processes and their magnetic consequences. Phys. Earth Planet. Int. 31, 65-76.

O'Reilly, W. (1984) Rock and Mineral Magnetism. Glasgow: Blackie.

O'Reilly, W. and Banerjee, S.K. (1965) Cation distribution in titanomagnetites (1-x)Fe_3O_4-xFe_2TiO_4. Phys. Lett. 17, 237-238.

O'Reilly, W. and Banerjee, S.K. (1967) Mechanism of oxidation in titanomagnetites, a magnetic study. Mineral. Mag. 36, 29-37.

Ono, K., Chandler, L. and Ito, A. (1968) Mössbauer study of the ulvöspinel, Fe_2TiO_4. J. Phys. Soc. Japan 25, 174-176.

Rasmussen, R.J. and Honig, J.M. (1988) AC electrical transport properties of magnetite. J. Applied Physics 64, 5666.

Readman, P.W. and O'Reilly, W. (1970) The synthesis and inversion of non-stoichiometric titanomagnetites. Phys. Earth Planet. Interiors 4, 121-128.

Readman, P.W. and O'Reilly, W. (1972) Magnetic properties of oxidized (cation-deficient) titanomagnetites $(Fe,Ti)_3O_4$. J. Geomagn. Geoelectr. 24, 69-90.

Readman, P.W., O'Reilly, W. and Banerjee, S.K. (1967) An explanation of the magnetic properties of Fe_2TiO_4. Phys. Lett. 25A, 446-447.

Schmidbauer, E. and Readman, P.W. (1982) Low temperature magnetic properties of Ti-rich Fe-Ti spinels. J. Magnetism Magnetic Mater. 27, 114-118.

Schult, A. (1968) Self-reversal of magnetization and chemical composition of titanomagnetites in basalts. Earth Planet. Sci. Lett. 4.

Smith, R.W. and Fuller, M. (1967) Alpha-hematite: stable remanence and memory. Science 156, 1130-33.

Stacey, F.D. and Banerjee, S.K. (1974) The Physical Principles of Rock Magnetism. Amsterdam: Elsevier.

Suk, D. (1985) Low temperature aqueous oxidation of titanomagnetite. M.S. thesis, University of Minnesota, Minneapolis, MN.

Syono, Y. (1965) magnetocrystalline anisotropy and magnetostriction of Fe_3O_4 - Fe_2TiO_4 series - with special application to rock magnetism, Japan. J. Geophys. 4, 71-143.

Taylor, R.W. (1964) Phase equilibria in system FeO-Fe_2O_3-TiO_2 at 1300°C. Am. Mineral. 49, 1016-30.

Verhoogen, J. (1962) Oxidation of iron-titanium oxides in igneous rocks. J. Geol. 70, 168-181.

Verwey, E.J., Haayman, P.W. and Romeijn, F.C. (1947) Physical properties and cation arrangements of oxides with spinel structure. J. Chem. Phys. 15, 181-187.

Warner, B.N., Shive, P.N., Allen, J.L., Terry, C. (1972) A study of the hematite-ilmenite series by the Mössbauer effect. J. Geomagn. Geoelectr. 24, 353-367.

Wechsler, B.A., Lindsley, D.H. and Prewitt, C.T. (1984) Crystal structure and cation distribution in titanomagnetites ($Fe_{3-x}Ti_xO_4$). Am. Mineral. 69, 754-770.

Worm, H.U. and Banerjee, S.K. (1984) Aqueous low temperature oxidation of titanomagnetite. Geophys. Res. Lett. 11, 169-172.

Chapter 5

Stephen E. Haggerty

OXIDE TEXTURES – A MINI-ATLAS

"My books on ore microscopy have obviously inspired many interested scientists to investigate and to clear many questions. Certainly it turned out that not everything, which I thought to have obtained by my work and which I explained, was in all details correct. But that is a natural consequence of the progress in science."

> Paul Ramdohr
> *The Ore Minerals and Their Intergrowths*
> 2nd Edition, Volume 1, 1980.

INTRODUCTION

Ramdohr made a major contribution over a period of more than 70 years to the metallography of ore deposits. In turning his light source to perfectly polished specimens of igneous, metamorphic and sedimentary rocks, he turned the tables on the classical petrographer by readily identifying, but relegating to mere gangue, the lowly silicate minerals. Unfortunately the ore and gangue camps remain divided. In fact the chasm between transmitted and reflected light microscopists appears to be widening rather than converging. But all of this is of little consequence in this modern age of glamorous analytical black boxes and the virtual abandonment of the optical microscope; some of this is implicit in the quote from Ramdohr. The explicit statements, however, throughout this volume on the sensitivity to P-T and redox changes, and the widespread application of the oxides to magnetic and petrogenetic studies, makes ore microscopy essential if sensible interpretation is to follow. Many geochemical and isotopic studies could well have benefited from an early ore microscopic view of uncrushed samples which would have cut short the later realization that diagenesis, deuteric modification, or metasomatism had been operative.

The objective in this Chapter is to provide a broad survey of the oxide textures and assemblages most commonly encountered in terrestrial igneous, metamorphic, and sedimentary rocks; two comparisons are drawn from lunar basalts, but none from meteorites (see *Reviews in Mineralogy*, 1976). Ramdohr (1980 and earlier editions) is the standard text, but other noteworthy publications include Edwards (1965), Galopin and Henry (1972), and Uytenbogaaardt and Burke (1972). The classic paper on oxidation exsolution and reduction exsolution is by Buddington and Lindsley (1964). It was this and other papers on the relationship between high temperature oxidation and exsolution - like textures, that led Ramdohr to his spirited quote on "...the progress in science." Important new work, using TEM and HRTEM, is resolving the fine structural detail of intergrowths and defects in the oxide minerals (e.g. Bursill et al., 1984; Bursill and Smith, 1984; Price, 1980; Putnis and Price, 1979; Nord and Lawson, 1989), and in the recognition of new phases (e.g., Banfield et al., 1991). Scanning tunneling microscopy and atomic force microscopy are tools of the future (e.g., Hansma et al., 1988).

130

Mini-Atlas

All photomicrographs are in reflected light and under oil-immersion objectives, unless otherwise stated. All optical identifications have been confirmed by either electron microbeam analysis or x-ray crystallography or both; where uncertainties exist, these are stated. Some of the photomicrographic plates are reproduced from *Reviews in Mineralogy* (1976; Chapters 4 and 8), but most of the text, bibliography, mineral analyses, and details of textural interpretation are not repeated here.

The major rock-forming mineral oxides are spinel, ilmenite, pseudobrookite-armalcolite, perovskite, and rutile. High temperature oxidation minerals are ilmenite, hematite, rutile, and pseudobrookite. Lower temperature replacement minerals include maghemite, titanite, and hematite. A class of minerals of growing interest to petrogenetic models include lindsleyite, mathiasite, armalcolite, hawthorneite, and Nb-Cr rutile. The textures of these high pressure, upper mantle assemblages are given in this Mini-Atlas with additional discussion in Chapter 10. Other minerals of interest include perovskite, baddelyite, and zirconolite, which are large-ion-lithophile in character. The Mini-Atlas gives greatest coverage to oxides in igneous rocks (including upper mantle metasomites), and the least coverage to the economically important ore deposits of Fe oxides and Fe-Ti oxides. This is in part a reflection of active research, but it also reflects the greater diversity of phases and processes and hence textures in oxides of igneous derivation, relative to metamorphic and sedimentary environments. Figure 39 is dedicated to the wide array of oxides possibly contributing to ocean floor magnetism (e.g. Wooldridge et al., 1990). The very foundation of the basic definition of a mineral (i.e. formed by inorganic processes) is challenged by the growing interest and recognition that biogenetically precipitated oxides are potentially important contributors to the magnetism of sediments. Examples of biogenic magnetite are offered in Figure 40 (Stolz et al., 1990), perhaps as a fitting conclusion to future studies in textures of the oxide minerals.

While it is true that the photomicrographic coverage in the Mini-Atlas is a product of high-grading, most examples are typical and are of widespread distribution; and few are either monomineralic or in binary association. The message is that if models are to be applied to natural systems, the groundstate and suitability of the oxide system must be established.

OXIDATION OF Fe-Ti SPINELS AND ILMENITES

While this Mini-Atlas concentrates on the textures rather than their description, it has been deemed useful to reproduce a part of the descriptions of oxidation textures from *Reviews in Mineralogy* (1976, Chapter 4). The oxide classification, C1 to C7 for titanomagnetite, and R1 to R7 for ilmenite, that is applied to studies in rock magnetism is covered in some detail in this Mini-Atlas.

Oxidation of titanomagnetite

Cubic titanomagnetite $(Usp-Mt)_{ss}$ can be oxidized by two mechanisms: (a) Oxidation at low pressure and below 600°C to yield cation-deficient spinels of the metastable titanomaghemite series (Usp-Mt-γ Fe$_2$O$_3$), which in some cases may then subsequently convert to members of the Hem-Ilm$_{ss}$ series. (b) Oxidation at low to moderate pressures and above 600°C with the direct formation of Ilm-Hem$_{ss}$. Most of the discussion centers on the latter type. Ilmenite$_{ss}$ intergrowths in titanomagnetite are divided into the following forms (Buddington and Lindsley, 1964):

(a) Trellis types: ilmenite along the {111} planes of the host (Fig. 6).
(b) Sandwich types: thick ilmenite lamellae restricted to one set of {111} planes (Fig. 7).
(c) Composite types (Fig. 8).

<u>Trellis types</u>. The {111} planes of spinel and the (0001) planes of ilmenite are closely similar in structure (Chapter 2), which accounts for the common orientation of ilmenite lamellae along {111} of the titanomagnetite host. While Buddington and Lindsley considered both (a) and (b) types to have formed by "oxidation-exsolution" or "oxyexsolution", the terms are applied here only to the trellis type because of possible ambiguities in the origin of the sandwich type.

A complete transition from fine (<1-10 μm) spindles of ilmenite along one set of {111} planes to crowded lamellae along all sets of the octahedral planes are commonly observed (Fig. 6). This intergrowth is identical in appearance to the well-known Widmanstatten texture in iron meteorites. Ilmenite lamellae that are parallel to any one set of octahedral planes are in optical continuity, show identical degrees of anisotropy, and similar pleochroic schemes. Oxyexsolved lamellae show sharp, well defined contacts with their titanomagnetite hosts and although the lamellae are smooth in outline they are rarely parallel. Tapered terminations develop at the intersection of two or more lamellae, which is indicative and typical of diffusion, but it is important to note that tapering also occurs in ilmenite lamellae that extend to the limits of the titanomagnetite grain boundaries, a feature that is in marked contrast to the sandwich laths described below. The width of lamellae in any one grain tends to be fairly uniform (Fig. 6) but super trellis-works of larger laths (10-20 μm) infilled by several generations of finer lamellae (1-10 μm) are also common (Fig. 6).

Ilmenite lamellae are typically concentrated along cracks, around silicate inclusions, and, most significantly of all, along titanomagnetite grain boundaries (Fig. 6a). These textural features strongly support the experimental evidence that oxidation and not true exsolution is responsible for the formation of ilmenite lamellae from primary Usp-Mt$_{ss}$. Ilmenite in these oxidized lamellar zones increases in size and abundance towards the grain boundaries (Fig. 6a). Lamellae that project into the titanomagnetite are sharply tapered and gradually disappear, on a microscopic scale, as the unoxidized areas are approached. There are significant changes in color, reflectivity and in the composition of the oxidized titanomagnetite: the mineral becomes whiter with an increase in reflectivity as titanium diffuses from the host and the iron is oxidized; the ratios of Fe:Ti and Fe^{3+}:Fe^{2+} increase. If a basic lamellar framework is established at the edge of a grain, microcapillary access to the center becomes possible and continued oxidation of the crystal interior may result largely by propagation along these lattice discontinuities. From the foregoing discussion the basic premise being made is that oxidation of Mt-Usp$_{ss}$ results in "exsolved" lamellae of ilmenite, as confirmed experimentally (Frontispiece).

Distinct textural stages of oxidation are recognized and have been classified as follows (the prefix C distinguished primary cubic phases from primary ilmenite - rhombohedral or R, described later):

C1 stage. Optically homogeneous Usp-rich magnetite solid solutions.
C2 stage. Magnetite-enriched solid solutions with a samll number of "exsolved" ilmenite lamellae parallel to {111}.
C3 stage. Ti-poor magnetite$_{ss}$ with densely crowded "exsolved" ilmenite lamellae parallel to {111} of the host.

Typical reactions that apply to the C2 and C3 assemblages, with the partial and more complete oxidation of ulvospinel are as follows:

$$C2: \quad 6\ Fe_2TiO_4 + O_2 = 6\ FeTiO_3 + 2\ Fe_3O_4$$
$$C3: \quad 4\ Fe_2TiO_4 + O_2 = 4\ FeTiO_3 + 2\ Fe_2O_3$$

Sandwich types. Thick (25-50 μm) sandwich laths of ilmenite along one set of the octahedral planes are common (Fig. 7). These laths generally occur in small numbers, and a single lath is the most common form. These laths have sharply defined contacts with the titanomagnetite, in common with composite ilmenite (see below), and rarely have parallel sides or tapered terminations typical of trellis ilmenite. Sandwich laths that coexist with trellis lamellae predate these lamellae, but sandwich laths that coexist with composite inclusions are rarely in contact, and their relative paragenesis is therefore indeterminate.

Composite types. Euhedral to anhedral inclusions of ilmenite are commonly present in titanomagnetite (Fig. 8). These inclusions may show sharp contacts with their titanomagnetite hosts and when compared with trellis lamellae are rarely oriented along either {111} or {100} planes. These inclusions are termed internal or external, depending on whether the ilmenite is totally or partially included in the titanomagnetite host. The term composite, rather than "granule-exsolution" as used by Buddington and Lindsley (1964) is used here because the inclusions may be either primary grains or "oxidation-exsolution" products of Mt-Usp$_{ss}$.

In summary, trellis lamellae clearly result from "oxidation-exsolution", whereas sandwich and composite ilmenites can be products of either oxidation or primary crystallization.

Oxidation of titanomagnetite-ilmenite intergrowths. In some rocks the oxidation of titanomagnetite proceeds beyond the C3 stage described above. For some the oxidation may be continuous; for others it may occur in more than one step. The more advanced oxidation typically results in the pseudomorphing of early spinel and ilmenite by phases such as hematite, pseudobrookite, and rutile.

Ilmenite that is produced by oxidation exsolution is structurally controlled within the titanomagnetite host. Phases that subsequently develop from this ilmenite by more intense oxidation, continue to reflect this structural control, but through composition. Thus rutile tends to be concentrated in areas of former ilmenite lamellae, and hematite dominates in the former host magnetite. Thus former {111} planes can typically still be recognized at the most advances stages of oxidation.

In detail the following oxide assemblages develop upon more intense oxidation of either the C2 or C3 stages described above):

C4 stage. The first sign of additional oxidation that is observed optically is an indistinct mottling of the ilmenomagnetite intergrowth (Fig. 10a). This mottling results from (a) the fine serrations that develop at the "exsolution" interface, (b) the development of minute exsolved transparent spinels in the titanomagnetite, and (c) the development of ferri-rutile (metailmenite) in the ilmenite.

With increasing oxidation, the "exsolved" metailmenite becomes lighter in color (Hem_{ss}), and the titanomagnetite changes from tan to dark brown (Mt_{ss}). There is a considerable increase in reflectivity in metailmenite, internal reflections are more apparent, and the lamellae show variable degrees of anisotropy caused by finely disseminated ferri-rutile. Lenses of ferri-rutile are oriented parallel to the length of metailmenite lamellae (0001) and are also present at an acute angle to the lamellae (probably $\{01\bar{1}1\}$). The lamellar assemblage corresponds to the R2 and R3 stages of discrete ilmenite that are defined below. Although the entire grain is subjected to the oxidation process, the most obvious changes are those within the metailmenite. Coarse and fine {111} lamellae are equally affected, although grains near the edges of titanomagnetite grains are always more intensely oxidized than those towards the center. The spinel rods, the internal reflections, and the mottled appearance are the characteristic visible properties of this oxidation stage.

C5 Stage. This stage of oxidation is characterized by the developemnt of rutile + titanohematite. Ferri-rutile may persist in transitionally oxidized grains but is absent in the more advanced stages. Rutile and titanohematite develop extensively within the "exsolved" metailmenite lamellae; complete replacement of these lamellae may occur. This is equivalent to the R5 stage of discrete ilmenite; the optical, textural form, and orientation properties are similar, but the ratio of titanohematite:rutile is slightly greater. The serrated contacts between the "exsolved" lamellar planes and the host titanomagnetite are more pronounced than at the C4 stage. With more intense oxidation, the contacts become irregular and the lamellar rutile-titanohematite assemblage extends into and begins to develop within the titanomagnetite. The contacts have the appearance of replacement-fronts, although at least part of the rutile probably originates from the decomposition of titanomagnetite. The progressive enlargement of the lamellar assemblage and the complete breakdown of the titanomagnetite are illustrated in Figure 10e. Although replacement at this stage is complete, the "exsolved" {111} texture may still be recognized.

The relic {111} fabric is optically prominent, particularly under crossed-nicols, and is dominantly controlled by the ratio of rutile:titanohematite, which may be several orders of magnitude greater in the regions of former ilmenite lamellae than in the former host. Other contributory factors are the darker color of the lamellar titanohematite and the optical continuity displayed by relic {111} lamellar sets (Fig. 10f). It appears that there is little or no diffusion across these former {111} planes and that the primary establishmentof the Ti-rich zones is maintained.

C6 stage. This stage, in common with the R6 stage of discrete ilmenite, is defined by the incipient formation of pseudobrookite$_{ss}$ (Psb$_{ss}$) from rutile + titanohematite. Psb$_{ss}$ develops almost exclusively along relic {111} planes (Figs. 10g-h), a reflection of the initially higher Ti content in them. Psb$_{ss}$ is typically heterogeneous in color, reflectivity, and degree of optical anisotropy. Accompanying rutile lenses are much finer that those observed at the R5 stage of discrete ilmenite, and although selective replacement of rutile by Psb$_{ss}$ (refer to R5) may occur, it is much more difficult to observe or follow. Electron microprobe analyses of these color-contrasted pseudobrookites are inconclusive although there is the suggestion of Fe^{2+}-Fe^{3+} variability. The lamellar Psb$_{ss}$ is typically jagged with finely textured cuspate contacts against the intralamellar areas. Total replacement of the host titanomagnetite by rutile + titanohematite is not a pre-requisite for the formation of Psb$_{ss}$; some areas of relic titanomagnetite, especially near the center of grains, may persist.

The development of Psb$_{ss}$ is indicative of more intense oxidation. The three-phase assemblage Psb$_{ss}$ + rutile + titanohematite, with or without relic titanomagnetite, defines this stage of oxidation.

C7 stage. C7 is the most advanced stage of oxidation of original spinels and is characterized by the assemblage Psb$_{ss}$ + hematite$_{ss}$. Psb$_{ss}$ typically has two distinct textural forms: as pseudomorphic {111} lamellae or as graphic intergrowths with titanohematite (Figs. 11a-c). These textural forms have slightly different origins, although the transition of lamellar to graphic fabrics can be traced and is the result of progressive redistribution of Psb$_{ss}$ within titanohematite. Lamellar pseudobrookite appears to develop most commonly in grains that had previously indergone extensive "oxidation exsolution"; such grains have high Ti lamellae and high Fe hosts with the result that Psb$_{ss}$ develops selectively in those areas that are compositionally favorable. In cases where initial "oxidation exsolution" was limited, or in titanomagnetite grains containing sandwich or composite inclusions, the spinel host continues to maintain large concentrations of Ti which may result in titanohematite + rutile (C6) or in pseudobrookite. Such Psb$_{ss}$ intergrowths are graphic, but these are always accompanied by Psb$_{ss}$ which is also lamellar in form, reflecting an original {111} fabric.

Inclusions in lamellar Psb$_{ss}$ are dominantly rutile, whereas inclusions in intralamellar Psb$_{ss}$ are dominantly titanohematite. These inclusions may result, in part, from decomposition of Psb$_{ss}$ upon slow cooling, but more likely are the result of incomplete reaction, respectively, of the original ilmenite and titanomagnetite. Both forms of Psb$_{ss}$ are optically similar, both show variations in color, and both types are polycrystalline; the variations in color and degree of anisotropy are not due to random crystal orientation, because the contacts between light bluish-gray and dark gray areas are typically gradational.

C7 Titanohematite is considerably whiter than that observed in either C5 or C6; it is more intensely anisotropic, and red internal reflections are apparent in thin plates. Titanohematite associated with graphic Psb$_{ss}$ is typically polycrystalline, whereas that in association with lamellar Psb$_{ss}$ may be either polycrystalline or in laminated optical continuity.

The C7 oxidation stage of titanomagnetite is analogous to the R7 stage of discrete ilmenite, and both are present in close association. High concentrations of titanohematite in C7 contrast with trace amounts in R7. These variations are distinctive, and allow the origins of completely pseudomorphed crystals to be distinguished. In addition to these modal variations, the persistence of relic {111} planes and crystal morphology are also useful distinguishing features.

Oxidation of discrete primary ilmenite

Seven clearly defined high-temperature oxidation stages have been defined for primary ilmenites on the basis of the same samples used for the spinels. They are:

1) homogeneous ilmenite
2) ferrian ilmenite + ferrian rutile
3) ferrian rutile + (ferrian ilmenite)
4) rutile + titanohematite + ferrian rutile + ferrian ilmenite
5) rutile + titanohematite
6) rutile + titanohematite + (pseudobrookite)
7) pseudobrookite = (rutile + titanohematite)

Although the sequence given above is the most commonly observed one, in some cases pseudobrookite forms directly, bypassing intermediate assemblages. Probably the bypassing is favored by higher temperatures and values of fO_2.

Details of the seven stages of ilmenite oxidation, as observed optically, are as follows (each stage is preceded by R to distinguish it from the cubic C series):

R1 stage. Homogeneous, unoxidized primary ilmenite (Fig. 17a).

R2 stage. The first optically observable signs of oxidation are an increase in reflectivity and a change in the color of the ilmenite from reddish-brown to light-tan. Fine (1-5 μm) wisp-like sigmoidal lenses of ferrian rutile develop along (0001) and {01$\bar{1}$1} planes of the ilmenite. These lenses generally develop uniformly throughout the ilmenite crystal, although concentrations towards grain boundaries and along fractures may be present. There is a tendency for lamellae along (0001) to be coarser than those along the rhombohedral planes (Fig. 17b).

R3 stage. With more advanced diffusion the lenses become thicker and more abundant. There are slight changes in color from pale gray-white to white as the lenses thicken, with marked increases in reflectivity, the appearance of white internal reflections, and a pronounced increase in anisotropy. Lenses within any one crystallographic plane are in optical continuity, are doubly terminated, and are relatively well defined. Ferrian rutile lenses are mantled by haloes of bleached ferrian ilmenite that are gradationally lighter in color as the ferrian ilmenite host is approached. This variation in color is due in part to differential concentrations of Fe^{2+} and Fe^{3+} in the host, and in part to the concomitant diffusion of Ti to the ferrian rutile lamellae. The (0001) and {01$\bar{1}$1} lenses rarely intersect or continue along precisely the same planes for any distance; tapering occurs on contact, and the sigmoidal form of the lenses gives rise to classic syneusis textures (Figs. 17c-d).

R4 stage. The R4 stage is considerably more complex than either R3 or R5. It is marked by a four-phase assemblage comprising ferrian ilmenite, titanohematite, ferrian rutile, and rutile (Figs 17e-g). The rhombohedral ferrian ilmenite and titanohematite are present in approximately equal concentrations and constitute the host; ferrian rutile and rutile occur as sigmoidal lenses or as finely disseminated lamellae, and these phases are oriented along presumed (0001) and $\{01\bar{1}1\}$ planes of the original ilmenite. Optical continuity of phases along these planes is maintained, but there are marked reflectivity increases in both the host and the lamellar components compared to the R3 stage. The compositions of the phases are highly variable.

R5 stage. Rutile and titanohematite develop extensively at the R5 stage of oxidation (Fig. 18a). Ferrian rutile and ferrian ilmenite may appear as finely disseminated unreacted phases but these gradually disappear with increased diffusion and oxidation. This stage is represented by a simple two-phase assemblage: the titanohematite host is, in general, optically homogeneous and each grain has the properties of a single crystal. Titanohematite is considerably whiter, has a higher reflectivity, and is more strongly anisotropic than its incipient counterpart in R4. These properties result from an increase in Fe^{3+} and a loss of TiO_2. The ferrian rutile lenses are sharply defined and crystallographically controlled along (0001) and $\{01\bar{1}1\}$ planes of the host. Within each orientation, the lenses are in optical continuity; they are strongly anisotropic and show distinct yellow or yellow-red internal reflections. From a textural standpoint, stages R3, R4, and R5 are identical, insofar as each assemblage contains a member of the ilmenite-hematite series and each is characterized by lenses of a Ti-rich oxide that are crystallographically controlled within the host.

R6 stage. The R6 stage is characterized by the development of Psb_{ss} from the R5 assemblage. In the incipient stages, pseudobrookite is concentrated along cracks and grain boundaries and gradually replaces the central-most regions of the grain by preferential migration along rutile lenses. Both titanohematite and rutile participate in the reactions, but the development of Psb_{ss} is most clearly controlled by the distribution and orientation of rutile (Figs. 18b-e). The pseudobrookite is polycrystalline and varies from pale to dark gray in color, being slightly darker adjacent to rutile, and lighter near titanohematite. These optical differences reflect the higher TiO_2 (and thus higher $FeTi_2O_5$) contents of Psb_{ss} adjacent to rutile. Reflectivity and the degree of anisotropy are almost identical, as are the reddish internal reflections.

The well-defined lensoidal texture of rutile in R3-R5 is totally disrupted as Psb_{ss} forms. Replacement is rarely complete, and even in the most advanced stages, unreacted subgraphic inclusions of rutile and titanohematite tend to persist within the pseudobrookite host. These inclusions are neither ordered in distribution nor crystallographically controlled, which helps distinguish pseudobrookite formed by high-temperature oxidation from primary pseudobrookite that has undergone partial decomposition to the low-temperature assemblage rutile + $hematite_{ss}$.

R7 stage. The most advanced stage is R7, which is represented by the predominance of pseudobrookite (Figs. 18f-h). Intermediate stages between R6 and R7 are common. Somewhat similar optical and textural variations are observed in both stages. R7 pseudobrookites are heterogeneous, and faint ghost-like traces of ferrian rutile are sometimes apparent; these relic lenses are particularly evident in crossed-nicols, and two distinct sets of extinction are observed upon rotation of the stage. Psb_{ss} generally constitutes 80-90% of the pseudomorph, ferrian rutile is next in abundance, and titanohematite is minor or totally absent. Ferrian rutile and titanohematite occur as drop-like inclusions or as subgraphic intergrowths and show no preferred orientation in the Psb_{ss}.

REFERENCES

Banfield, J. F., Veblen, D. R. and Smith, D. J. (1991) The identification of a new mineral as TiO2(B) by structure determination using HRTEM, image simulation, and DLS refinement. Am. Mineral. in press.

Buddington, A. F. and Lindsley, D. H., (1964) Iron-titanium oxide minerals and synthetic equivalents. J. Petrol. 5, 310-357.

Bursill, L. A., Banchin, M. G. and Smith, D. J. (1984) Precipitation phenomena in non-stoichiometric oxides II. {100} platelet defects in reduced rutiles. Proc. Roy. Soc. Lond. A391, 373-391.

Bursill, L. A. and Smith, D. J. (1984) Interaction of small and extended defects in non-stoichiometric oxides. Nature 309, 319-321.

Edwards, A. B. (1965) Textures of the Ore Minerals. Australian Inst. Min. and Mett. Melbourne.

Galopin, R. and Henry, N. F. M. (1972) Microscopic Study of Opaque Minerals. Heffer and Sons, Ltd. Cambridge.

Hansma, P. K., Elings, V. B., Marti, O. and Bracker, C. E. (1988) Scanning tunneling microscopy and atomic force microscopy: Applications to biology and technology. Science 242, 209-216.

Nord, A. L. and Lawson, C. A. (1989) Order-disorder transition induced domains and magnetic properties in ilmenite-hematite. Am. Mineral. 74, 160-176.

Price, G. D. (1980) Exsolution microstructures in titanomagnetites and their magnetic significance. Phys. Earth Planet. Int. 23, 2-12.

Putnis, A. and Price, G. D. (1979) The nature and significance of exsolved phases in some chrome spinels from the Rhum layered intrusion. Mineral. Mag. 43, 519-526.

Ramdohr, P. (1980) The Ore Minerals and their Intergrowths. Int. Ser. in Earth Sci. vol. 35, Pergamon Press, Frankfurt.

Stolz, J. F., Lovley, D. R. and Haggerty, S. E. (1990) Biogenic magnetite and the magnetization of sediments. J. Geophys. Res. 95, 4355-4361.

Uytenbogaardt, W. and Burke, E. A. J. (1971) Tables for the Microscopic Identification of Ore Minerals. Elsevier Pub. Co., Amsterdam.

Wooldridge, A. L., Haggerty, S. E., Rona, P. A. and Harrison, C. G. A. (1990) Magnetic properties and opaque mineralogy of rocks from selected seafloor hydrothermal sites at oceanic ridges. J. Geophys. Res. 95, 12,351-12,374.

Figure 1

Crystal morphologies of oxide minerals in igneous rocks

(a) Dark gray chromian-spinel core mantled by titanomagnetite which is partially oxidized along the peripheral margins to titanomaghemite (white). 0.09 mm.*

(b) Euhedral primary armalcolite crystal containing a cylindrical silicate glass inclusion with an attached crystal of chromian-titanomagnetite. 0.09 mm.

(c) Skeletal ilmenite with glass and finely crystalline inclusions. 0.09 mm.

(d) An ilmenite crystal with glass inclusions and with renewed second generation growth of T-shape crystallites along the basal plane. 0.09 mm.

(e) - (h) Skeletal growth morphologies of titanomagnetite illustrating a progressive trend towards euhedral morphology. This trend is classified as the cruciform type. 150 um.

(i) This titanomagnetite crystal is also of the cruciform type with the distinction that growth, parallel to {111} spinel planes, takes place along the entire length of the primary cross-arms (c.f. with e-h). 0.11 mm.
(j) Titanomagnetite crystal of the multiple cross-arm type. 80 um.

(k) - (m) Titanomagnetite crystals of the complex type contain orthogonal and non-orthogonal multiple cross-arms. Growth patterns are evenly or haphazardly initiated at terminations, and along primary, secondary or tertiary cross-arms. k = 0.13 mm.; l = 0.12 mm.; m = 750 um.

*Scale = width of photomicrograph.

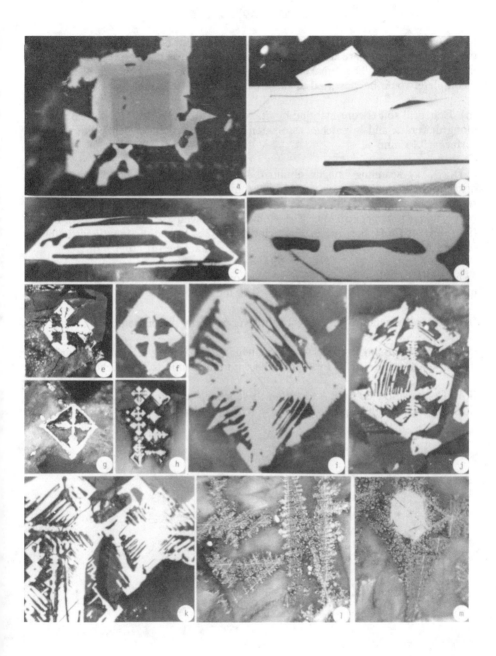

Figure 2

Chromian spinel solid solutions

(a) Asymmetrically mantled, chromian-spinel cluster in basalt partially enclosed in olivine. The mantles are titanomagnetite and the intermediate zones are chromian titanomagnetites. The adjacent discrete crystal of titanomagnetite, although incomplete would be classified as the cruciform type. 0.16 mm.

(b) Euhedral spinel core mantled by titanomagnetite in basalt. Mt_{ss} is also present along the crack and in patches associated with inferred cracks below the polished surface. 0.16 mm.

(c-f) X-ray scanning images obtained by electron microprobe illustrating the distribution of major elements for a chrome spinel-titanomagnetite core-mantle relationship from a basalt. The intensity of the white spots is approximately proportional to concentration. The elements displayed are Fe (in c), Ti (in d), Cr (in e) and Al (in f); the core is thus virtually Ti-free (d) and the mantle virtually Cr (e) and Al-free; both the mantle and the core contain Fe (c). 0.11 mm.

(g-h) Multiple zoning in chromian-spinels from kimberlites with chromite cores and magnetite mantles. g = 0.15 mm.; h = 0.08 mm.;

(i) Euhedral magnetite core, epitaxially overgrown by picroilmenite. 0.09 mm.

(j) Oscillatory zoning in spinel where the white areas are Mg, Al and Fe^{3+}-enriched, the darker areas are Cr + Fe^{2+}-enriched, and the mantle is Ti-enriched.

(k-l) Compositionally similar to (j) but these spinels are present in garnet kelyphite rims associated with Ti-phlogopite. k = 0.09 mm.; l = 0.07 mm.

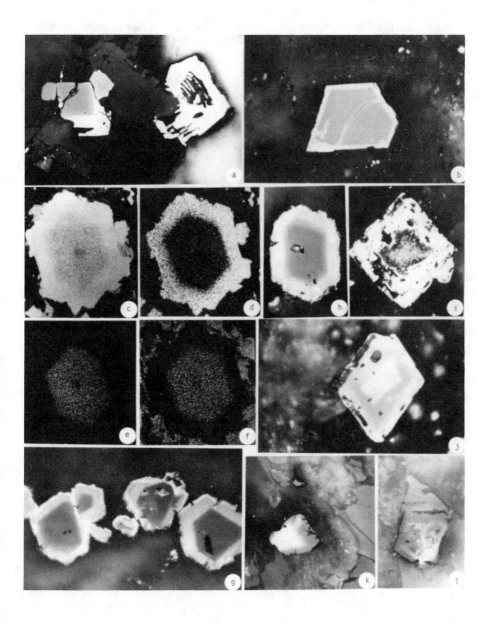

Figure 3

Xenocrystic chromian spinel reactions in basalts

(a) Euhedral spinel core mantled in a sharp contact by chromian titanomagnetite. The rounded crystal to the left is only partially zoned because it is enclosed in olivine. 0.3 mm.

(b) An irregularly shaped chromian spinel crystal mantled by chromian titanomagnetite, and at the outer margin by titanomagnetite. Intergrown silicates at the edge of the crystal are plagioclase and pyroxene. 0.3 mm.

(c) Several adjoining spinels with diffuse reaction contacts against chromian titanomagnetite mantles. 0.3 mm.

(d) A zoned chromian spinel core mantled by the assemblage titanomagnetite + trellis ilmenite. Both mantle phases show partial oxidation as evidenced by the irregular mottled appearance. The white stringers are late-stage hematite. 0.3 mm.

(e) An oxidized chromian spinel inclusion (black) partially enclosed in olivine mantled by an original titanomagnetite crystal in the groundmass. The spinel core has decomposed to Mt_{ss} + $pleonaste_{ss}$; associated trellis and external composite assemblages, after ilmenite, are now Rut + Hem_{ss}. The chromian spinel has hematite lamellae along {111} spinel planes which are parallel to the {111} spinel planes of the mantle assemblage. 0.3 mm.

(f) A discrete chromian spinel with oxidation hematite lamellae along {111} spinel and a diffusion rim of hematite along the spinel grain boundary. The spinel is enclosed in a highly oxidized olivine crystal and the bright white flecks are magnetite + hematite. 0.2 mm.

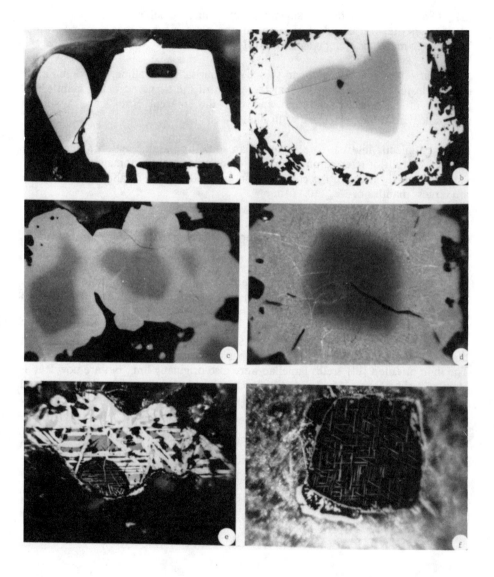

Figure 4

Chromian spinel reactions

(a) Glass inclusions in a phenocrystic chromian spinel in basalt mantled by titanomagnetite; an island of chromite in the glass suggests magmatic corrosion. The lighter mantle is titanomagnetite. 0.14 mm.

(b) Coarse web-shaped chromian spinel with partially crystalline, and dominantly glass inclusions in basalt. Note that the extent of the titanomagnetite mantle is a function of the size of chromite and that the internal cores have a similar morphology to the outer edges. 0.16 mm.

(c) A symplectic internal core of chromian spinel in basaltic andesite mantled by successive atolls of later chromite and an outer margin of titanomagnetite; most of the cuniform segments in the core are also mantled and the smallest areas contain the largest mantles of Mt_{ss}. 0.15 mm.

(d) Euhedral chromian spinel core mantled by subgraphic chromite + glass + olivine in a basalt bomb. The overall outer-morphology of the symplectite is broadly similar to that of the core. 0.16 mm.

(e) Irregular glassy inclusions in chromite in basalt mantled by titanomagnetite which has undergone oxidation exsolution and subsequent partial decomposition to Rut + Hem_{ss}. 0.16 mm.

(f) The dark central core is chromian spinel, the outer assemblage is oxidized titanomagnetite and the white attached areas were Ilm_{ss}, but are now Rut + Hem_{ss}. The light oriented {111} trellis lamellae were also originally Ilm_{ss} but are now Rut + Hem_{ss}; residual host Mt_{ss} are still apparent and these contain dark, oriented pleonaste rods. The large subhedral and mostly white crystals were ilmenite now decomposed to Rut + Psb. 0.15 mm.

(g-h) These two examples illustrate the features typical of "ferritchromit" which results during the serpentinization of chromite (gray cores) in ultramafic rocks. The outer mantles are Fe^{3+}-rich chromian magnetites; the exchange chemistry is complex and varied and typically involves the formation of chlorite and other hydrous silicates. 0.15 mm.

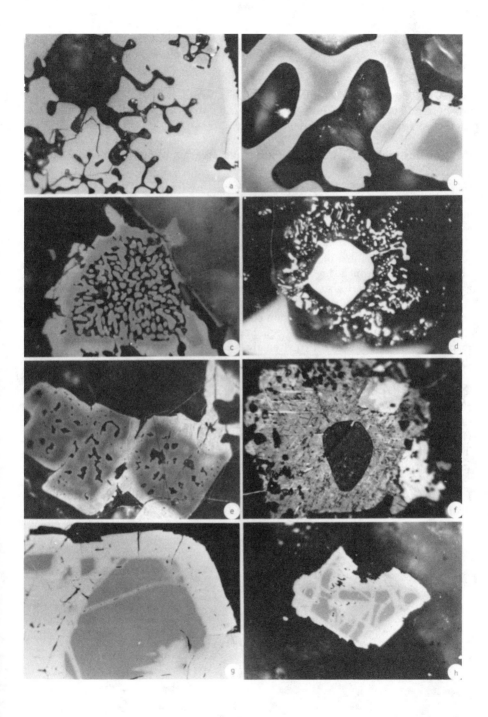

Figure 5

Equilibration of ulvospinel-magnetite solid solutions in mafic intrusions

(a) A parquet-textured intergrowth of exsolved Mt_{ss} (white) and Usp_{ss} (gray but brown in reflected light oil immersion) from an initial composition which must have been close to $Usp_{50}Mt_{50}$. 0.15 mm.

(b) Blitz-textured Usp_{ss} in Mt_{ss} with an attached grain of pyrrhotite. 0.15 mm.

(c) A transitional series from lit-par-lit Usp_{ss} to lamellar Usp_{ss}. The planes of exsolution are parallel to {100} spinel planes. The assemblage is weakly anisotropic, and the presence of Ilm_{ss} is inferred, although none can be identified as discrete oxidation products. Pyrrhotite is present along the crack. 0.15 mm.

(d) The black rods are pleonaste and the host is exsolved Usp_{ss} from Mt_{ss}; the pleonaste results from a higher temperature solvi intersection than the $Usp-Mt_{ss}$ assemblage. Note that Usp_{ss} is absent in zones adjacent to pleonaste. 0.15 mm.

(e) The area to the right of this grain contains abundant lamellae of Ilm_{ss} whereas the area to the left contains almost equal proportions of $Usp_{ss} + Mt_{ss}$ in a cloth-like texture. This is indicative of extremely steep oxidation gradients and although ilmenite is not observed within the Usp_{ss}-rich portion, weak optical anisotropy is apparent and incipient Ilm_{ss} is inferred. 0.15 mm.

(f-g) Blocky Mt_{ss} (white) with finely dispersed original Usp_{ss}. The dark gray areas display distinct optical anisotropy and Ilm_{ss} is inferred. The term proto-ilmenite (Ramdohr, 1980) has been employed to describe these optically unresolvable areas. These grains show a distinct marginal texture which is apparent also in (a) and in (e). In addition to proto-ilmenite, grain (g) also has a sandwich lath of Ilm_{ss}. 0.15 mm.

(h) A thick sandwich lath of Ilm_{ss} divides this grain into: a relatively unoxidized and proto-ilmenite-rich zone (left); and a highly oxidized zone with Ilm_{ss} trellis lamellae and minor Usp_{ss} exsolution. 0.15 mm.

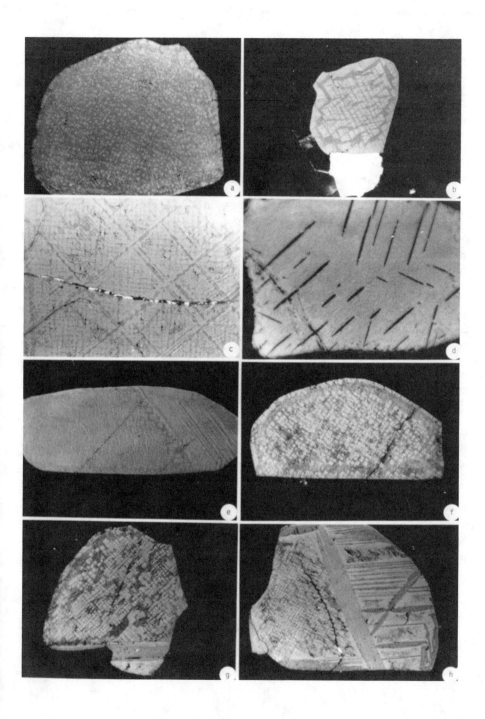

Figure 6

Oxidation exsolution in Usp-Mt solid solutions: Trellis types in basalts

(a) Subhedral grain of titanomagnetite showing a preferred concentration of ilmenite trellis lamellae along the grain boundary. Note the differences in color between the Usp_{ss}-poor area with ilmenite and the Usp_{ss}-rich central core which is ilmenite free. 0.2 mm.

(b) Fine trellis lamellae of ilmenite radiating from cracks in a titanomagnetite crystal. The largest concentration of ilmenite is adjacent to the cracks and lamellae gradually thin out towards the grain boundary. 0.15 mm.

(c) Two dominant sets of ilmenite lamellae along {111} spinel planes and a third minor set which has developed in areas of maximum ilmenite concentration. 0.15 mm.

(d) Short discontinuous lamellae of ilmenite along two sets of {111} spinel planes and a third set (along a direction NW-SE) which appears to be earlier. The adjacent titanomagnetite crystal (lower left) shows ilmenite in a rectangular pattern; these are not {100} spinel planes but are due in differences in crystal orientation. 0.15 mm.

(e) Well developed early sets of coarse ilmenite lamellae infilled by several generations of finer lamellae. Note the lensoidal form of the coarse lamellae and the constricted diffusion contacts with other lamellae. 0.15 mm.

(f) Very coarse ilmenite lamellae illustrating offsets of other lamellae, lamellae-interference, and the development of a fine lamellae set along one dominant {111} spinel plane. 0.15 mm.

Grains (a) and (b) are a clear demonstration of "oxidation exsolution" and are classified as C2; the remaining grains are classified as C3.

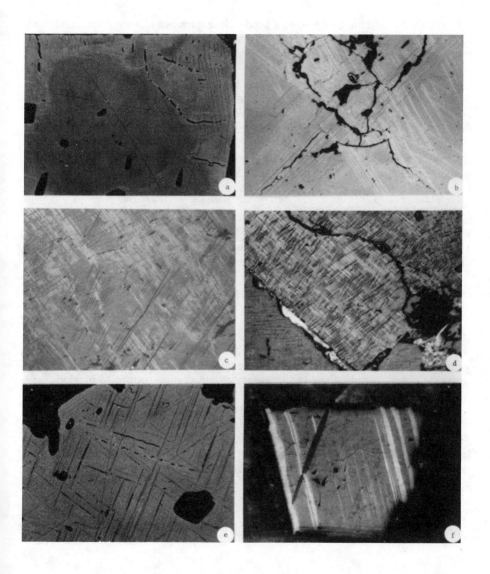

Figure 7

Oxidation exsolution in Usp-Mt solid solutions:
Sandwich types in basalts

(a) This is an extreme example of a sandwich texture but is a clear demonstration of an early crystal of ilmenite (darker) on which titanomagnetite has nucleated. 0.3 mm.

(b) Two thick laths of ilmenite which appear to have a crude {111} spinel orientation. Note that the laths are approximately parallel-sided and that the laths terminate in edges coincident with those of the titanomagnetite grain boundaries. 0.3 mm.

(c) An internally truncated lath and a second lath that extends to the edges of the titanomagnetite crystal and shows increased width at the spinel grain boundary. 0.2 mm.

(d) A thick ilmenite sandwich lath with irregular sides and with second and tertiary trellis sets along {111} spinel planes. The sandwich lath conforms to one of the {111} spinel planes. Note the diffusional narrowing of the trellis ilmenite lamellae at the sandwich ilmenite contact. 0.2 mm.

(e) Well developed sandwich laths of ilmenite extending into the silicate matrix as euhedral crystals. 0.3 mm.

(f) A lensoidal sandwich lath of ilmenite which is terminated at one edge of the titanomagnetite crystal but continues into the silicate matrix at the other edge in a graphic texture. The graphic ilmenite contains olivine + plagioclase and the apparent crystal terminations result from a coarse ophitic texture of plagioclase + pyroxene. 0.3 mm.

Grains (a) and (e) contain primary ilmenite whereas the ilmenite in grains (b), (c), and (d) may be either primary or may have resulted by oxidation exsolution. The lath in grain (f) may have exsolved, but the graphic extension to this grain is clearly due to primary crystallization and not oxidation exsolution.

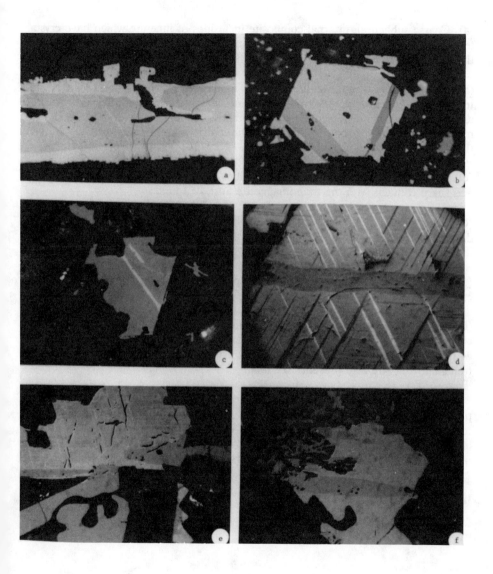

Figure 8

Oxidation exsolution in Usp-Mt solid solutions: composite types in basalts

(a) Coarse subhedral, internal composite inclusions of ilmenite which have crystal terminations that closely parallel those of the octahedral planes of the host titanomagnetite. Finer sets of ilmenite trellis lamellae are also present. 0.15 mm.

(b) An external composite inclusion of ilmenite that has at least two crystal terminations precisely parallel to {111} spinel as shown from parallel directions of associated trellis lamellae. Note that the composite ilmenite extends into the silicate matrix and that the cuspate contact with the matrix is continuous with the contact that is shared by titanomagnetite. 0.2 mm.

(c) Internal and external composite inclusions of ilmenite that have both irregular and sharp contacts with the titanomagnetite host. 0.2 mm.

(d) An external composite inclusion of ilmenite in titanomagnetite with a well developed euhedral arm extending into the silicate matrix. The dark blebs within the ilmenite are inclusions of glass. 0.2 mm.

(e) External composite ilmenite occupying approximately 30% of the titanomagnetite crystal with abundant trellis lamellae along {111} spinel planes. 0.2 mm.

(f) An external composite crystal of ilmenite extending into the silicate matrix in a coarse graphic texture. A second external composite grain (upper right) is contained within the titanomagnetite but shows an irregular contact with the silicates. Fine trellis lamellae are also present. 0.2 mm.

The conformity of ilmenite crystal faces and titanomagnetite crystallographic planes in grains (a) and (b) could argue for either oxidation exsolution or for topotactic growth of titanomagnetite on ilmenite. The case for grain (c) is ambiguous, whereas (d), (e), and (f) are almost certainly simultaneous crystallization of spinel and ilmenite, and not oxidation exsolution.

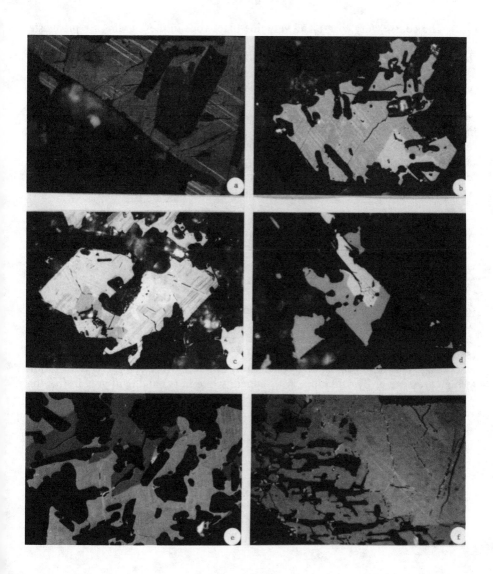

Figure 9

Titanomagnetite-spinel solid solutions in mafic intrusives

(a-d) The coarse, very coarse and very fine black rods in these grains are pleonaste spinels along the join $MgAl_2O_4$-$FeAl_2O_4$. The plane of exsolution, in common with that of ulvospinel, is along {100} spinel planes. Pleonaste exsolution predates that of Usp exsolution (a) and in some cases several generations of exsolution may be present (b and d). In relation to coexisting oxidation - exsolved Ilmss, pleonaste exsolution is either earlier than the oxidation of titanomagnetite (c), or contemporaneous with oxidation (d). Ilm_{ss} lamellae are either lighter or darker gray in the photomicrographs depending on the orientation. Some pleonaste rods are offset by Ilm_{ss} (d), whereas in other regions (d also) it appears that oxidation and nucleation of Ilm_{ss} has occurred along the discontinuities created by high temperature exsolution. In these cases the exsolved spinels are fragmented and are totally enclosed in ilmenite (thick sandwich lath in grain d). 0.15 mm.

(e-f) These are two examples of coexisting Cr-Mg hercynite ($FeAl_2O_4$, black) and Al-Cr magnetite. The extremely fine cloth-like fabric in the lighter Mt_{ss} areas, although similar in appearance to Usp-Mt_{ss} exsolution, has not been identified and is not Usp because of analytically determined low Ti-contents. An Ilm_{ss} lath is present in (e) and the bright fleck in (f) is pyrrhotite. 0.15 mm.

(g-h) The relationships of chromite and of magnetite differ only in the degree of phase separation. The series is complete at high temperatures and coexistence is dependent on fO_2. These assemblages are present in highly reduced rocks (serpentinites) which probably formed at T < 500°C. Exsolution is inferred but the assemblage may have originated by an entirely different process, e.g. recrystallization at high pressure or by oxidation. Mt_{ss} appear white in (g) and light gray in (h); the white areas in (h) are metallic awaruite (Ni3Fe). 0.15 mm.

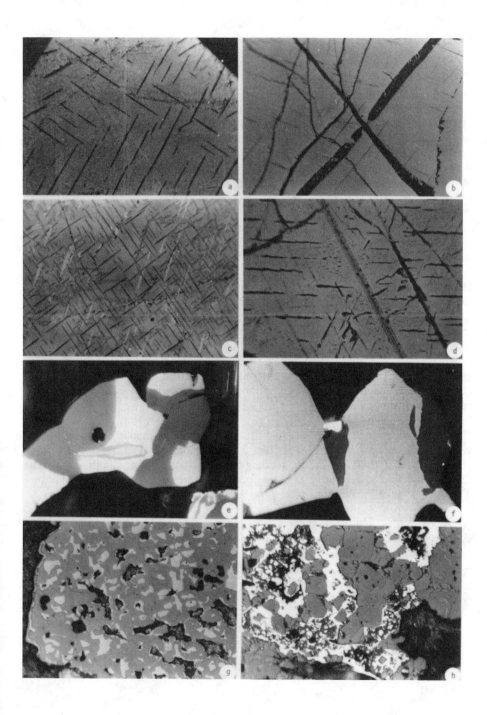

Figure 10

Oxidation of titanomagnetite-ilmenite intergrowths: Stages C4, C5, and C6 in basalts.

(a-d) This series illustrates the progressive decomposition of fine and coarse trellis ilmenite lamellae to ferrian rutile + ferrian ilmenite and to rutile + Hem_{ss}. The lamellae are mottled in (a) and (b) but well developed rutile lenses can be discerned in (c) and (d). Grain (d) also contained an original composite inclusion, the large central area which is now Rut + Hem_{ss}. The host magnetite dark gray in the photomicrographs but brown in reflected light oil immersion) contains abundant spinel lamellae (pleonaste-magnesioferrite$_{ss}$) which are black and are oriented along {100} spinel planes. This series is typical of stage C4. 0.05 mm.

(e) Coarse lenses of rutile + Hem_{ss} replacing original ilmenite lamellae but extending also to replace magnetite. Spinel lamellae in magnetite are black and the white lamellae are hematite. 0.05 mm.

(f) A completely pseudomorphed grain of original titanomagnetite + ilmenite. The assemblage is Rut + Hem_{ss} with associated disseminated blebs of black pleonaste$_{ss}$. The original {111} spinel fabric is still maintained with larger concentrations of rutile in areas originally occupied by ilmenite and higher concentrations of Hem_{ss} in areas that were originally titanomagetite. This is a perfect example of stage C5. 0.06 mm.

(g) This grain would be classified as C6 because of the first appearance of Psb_{ss} (irregular and medium gray, lower left). Residual magnetite is present and the remainder of the grain consists of Rut + Hem_{ss}. 0.08 mm.

(h) The dark gray irregular lamellae are Psb_{ss}, the fine medium-gray lenses are rutile and the gray to white host is Hem_{ss}. An original {111} fabric is still evident with Psb_{ss} and with Rut + Hem_{ss} tending to concentrate along relic ilmenite. The fine mottled black specks are pleonaste$_{ss}$ and the larger black inclusions are silicates. This is a C6 grain. 0.08 mm.

(i) The entire sequence from C4 to C7 may be followed in this original titanomagnetite + ilmenite grain starting in the upper right (with magnetite + metailmenite) and proceeding anticlockwise to Rut + Hem_{ss} (C5) to Psb_{ss} + Hem_{ss} + Rut (C6), and to Psb_{ss} + Hem_{ss} (C7). 0.15 mm.

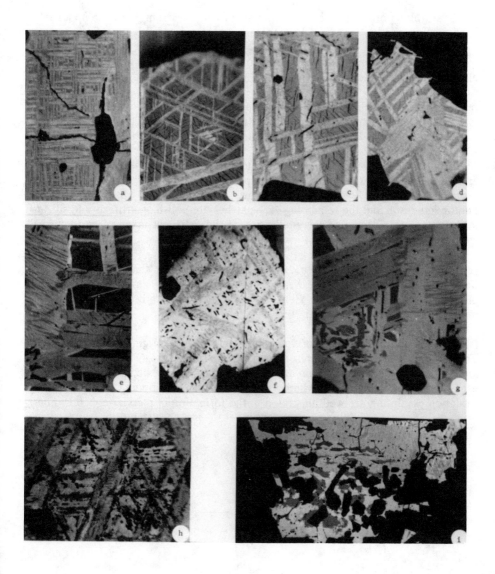

Figure 11

Oxidation of titanomagnetite-ilmenite intergrowths: Stage C7 in basalts

(a) The trellis portion of this grain consists of Psb_{ss} + Hem_{ss} and the upper left-hand portion consists largely of Rut + Hem_{ss} + associated Psb_{ss}. 0.15 mm.

(b) A completely pseudomorphed grain consisting of Psb_{ss} along {111} spinel relic planes in a host of Hem_{ss}. 0.15 mm.

(c) Thick original trellis lamellae and composite inclusions of ilmenite now oxidized to Psb_{ss}. The white areas are Hem_{ss} and the medium-gray inclusions in Psb_{ss} are largely Rut although Hem_{ss} is also present. 0.15 mm.

(d) An original sandwich ilmenite lath pseudomorphed by Psb_{ss} + Hem_{ss}. The Psb_{ss} area to the lower right was probably a composite inclusion, the inclusions are Rut + associated Hem_{ss}. $Pleonaste_{ss}$ blebs are present in the upper left of the photomicrograph and the remaining dark patches are polishing artifacts and silicate inclusions. Some degree of recrystalization of the Psb_{ss} has taken place (compare with grain (b) for example). 0.15 mm.

(e) Although a crude linear fabric of the Psb_{ss} is present in this grain, the original titanomagnetite probably underwent very rapid oxidation which resulted in the breakdown directly to Psb_{ss} + Hem_{ss} + associated Rut, rather than through the intermediate stages of C3 to C6. 0.15 mm.

(f) This is another example of very rapid oxidation of original titanomagnetite. There are no residual spinels present and rutile is sparse. MgO and Al_2O_3 are now largely in the Psb_{ss} and this grain as well as grain (e) should be contrasted with the examples shown in Figure 12 where the effects of the partitioning of Mg + Al by oxidation exsolution are well illustrated in the formation of $pleonaste_{ss}$. 0.15 mm.

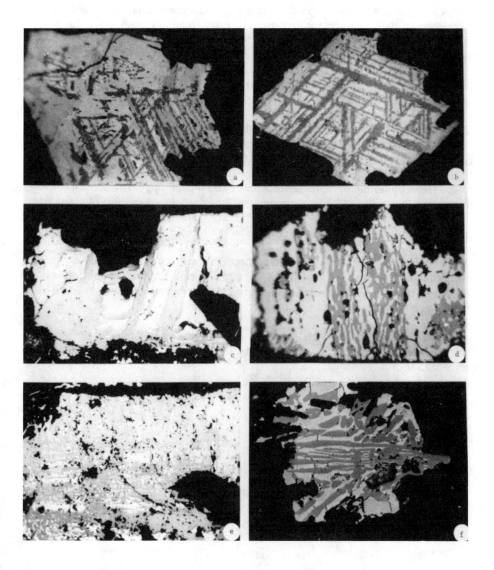

Figure 12

Oxidation of Mg-Al titanomagnetites in high-Al basalts

(a) A subhedral grain of magnetite containing metailmenite lamellae along {111} spinel planes; there is incipient formation of Psb_{ss} (medium gray) along the largest of the thick lamellae. The magnetite contains coarse lamellae of pleonaste-magnesioferrite$_{ss}$ oriented along {100} planes of the spinel host. 0.2 mm.

(b) A more advanced oxidation stage from the assemblage in grain (a) with both trellis and composite ilmenite oxidized to Psb_{ss} + Hem_{ss}. 0.2 mm.

(c) Oxidation (from b) is extended to include the magnetite, which is replaced by hematite; the previously exsolved spinel (black) lamellae are scattered throughout the hematite as finely disaggregated blebs. 0.05 mm.

(d-e) These photomicrographs are at a lower and higher magnification respectively, illustrating the oxidation of original ilmenite to Psb_{ss}, the exsolution of pleonaste$_{ss}$ in magnetite, the replacement of Mt_{ss} by Hem_{ss}, and the resistance of unresorbed spinels that are dispersed within Hem_{ss}. d = 0.2 mm.; e = 0.15 mm.

(f) A subhedral grain of original titanomagnetite with satellite skeletal extensions into groundmass consisting of Psb_{ss} + Hem_{ss}; the two large rectangular areas are magnesioferrite$_{ss}$ (black, lower left). None of the original magnetite (apart from crystal morphology) is any longer present. 0.2 mm.

(g) A higher magnification photomicrograph of a grain similar to that of grain (f) but with the magnesioferrite$_{ss}$ being partially replaced along {lll} spinel planes by hematite. 0.15 mm.

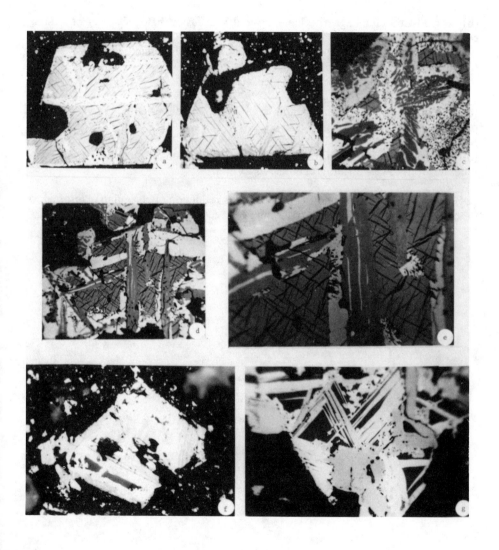

Figure 13

Subsolidus reduction of ilmenite

(a) Well oriented but discontinuous lenses of Mg-Al titanomagnetite in kimberlitic picroilmenite. 0.15 mm.

(b) Mg-Al-Cr titanomagnetite lamellae and a diffusion halo of spinel of similar composition in kimberlitic
picroilmenite. The bright area to the left of the photomicrograph is a sulfide (pyrite, pyrrhotite,
chalcopyrite) and magnetite complex. 0.15 mm.

(c) Titanomagnetite (white) and Mg-Cr-Al spinel oriented in kimberlitic picroilmenite. 0.15 mm.

(d-f) Ulvospinel lamellae (gray to dark gray) and oriented metallic Fe lenses in ilmenite from lower crustal granulites. 0.15 mm.

(g-j) Lunar examples (Apollo 14) of subsolidus reduction of chromian ulvospinel to ilmenite + metallic Fe (white). 0.05 mm.

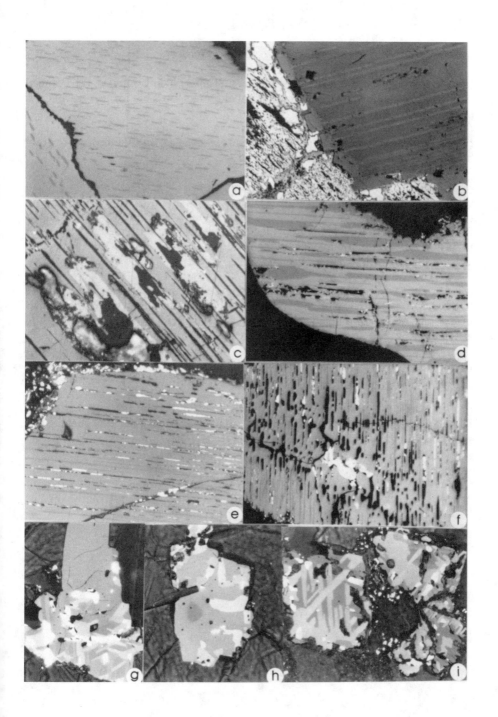

Figure 14

Titanite-oxide reactions

(a-b) Primary phenocrystic titanite (sphene, $CaTiSiO_5$) in ankaramites. (a) Twinned titanite crystal mantled by ilmenite in parallel growth, possible formed by reaction of titanite + liquid. 0.25 mm.

(b) Euhedral crystals of titanite (dark gray and wedge-shaped) in interference grain boundary contact with magnetite. 0.35 mm.

(c-d) Well preserved trellis networks of titanite after {111} ilmenite in finer grained titanite after earlier titanomagnetite. Titanite compositions may be aluminous but typically have very low concentrations of FeO. Metasomatic introduction of CaO + SiO_2 from an external source or from plagioclase decomposition, and removal of FeO to form chlorite or amphibole, or sulfide (white crystals in d) is generally invoked to account for secondary titanite formation. 0.15 mm.

(e) The close association of pyrite (white) with {111} controlled lamellae of titanite in titanite from a previous titanomagnetite and ilmenite intergrowth suggests that S may also be an important element in the metasomatic fluid; the Fe to produce pyrite is derived from the decomposition of the oxides. 0.15 mm.

(f) Advanced replacement of titanomagnetite and trellis ilmenite lamellae, relative to the marginal replacement of ilmenite by titanite in both cases. 0.15 mm.

(g-h) Curvilinear replacement of titanomagnetite by titanite (g), possibly closely related to the low temperature ($<350°C$) oxidation of magnetite to maghemite (h), shown more clearly in Figure 15. 0.15 mm.

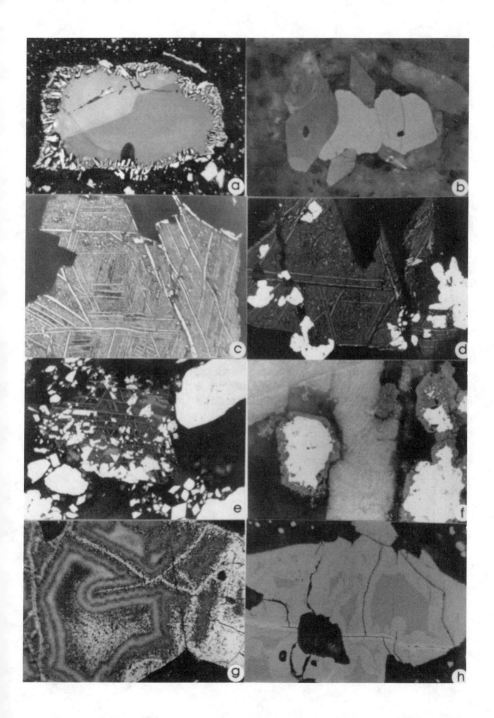

Figure 15

Titanomaghemite and titanite relations

(a-b) Titanomaghemite (white) replacement along curvilinear fractures of titanomagnetite. Note that ilmenite is optically unaffected by this form of low-temperature oxidation (b). 0.25 mm.

(c-d) Titanite (gray) replacement of titanomagnetite and titanomaghemite. 0.25 mm.

(e) High temperature oxidation exsolution of ilmenite from titanomagnetite (note that the core of the spinel is unoxidized), with superimposed replacement of magnetite by maghemite at and along the spinel-ilmenite interfaces (note that low T oxidation has penetrated just about as far as the higher temperature ilmenite lamellae have allowed). 0.15 mm.

(f) Variable compositions and a spectrum of intermediate titanomaghemites are apparent in this partially oxidized titanomagnetite grain. The lamellae are Ilm_{ss} and the variations in color among segments, which are bounded by these lamellae, are representative of stages in the transition from titanomagnetite to titanomaghemite. Note that some areas of the crystal interior are more oxidized than the crystal margins. 0.15 mm.

(g-h) High temperature (oxidation exsolution) and lower temperature (maghemitization) oxidation and superimposed titanite metasomatism in titanomagnetite. Note that (g) has both sandwich and external composite (graphic) ilmenite, and that (h) has trellis ilmenite and advanced titanite formation concentrated at one end of the crystal. 0.2 mm.

Optically, titanomaghemite is either white or blue and isotropic; hematite is also white but has a higher reflectivity, red internal reflections and is anisotropic. The curved cracks so typical of maghemitization
are due to a volume change in the transformation from stoichiometric spinel to cationic-deficient spinel.

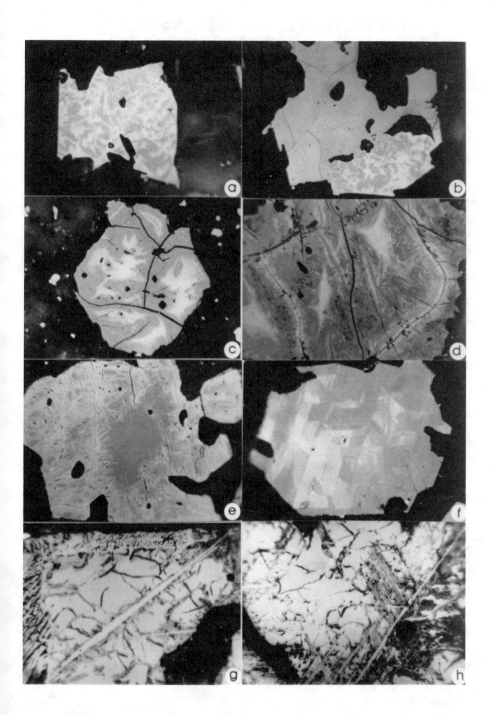

Figure 16

Equilibration of ilmenite-hematite solid solutions in intrusive rocks

(a-b) The light-gray hosts are titanohematite and the oriented darker-gray lenses are ferrian ilmenite. In (a) individual grains show sharp terminations and large concentrations of ilmenite at the grain boundaries. In (b) the grains are rounded and these preferred concentrations are not as evident. Each of the larger ilmenite lenses is surrounded by a depletion zone, and the regions between the large lenses are occupied by finer lenses which are assumed to have formed at lower T than those of the larger bodies. This distribution is the synneusis texture. 0.15 mm.

(c-d) The cores of these Hem_{ss} grains contain similar distributions in Ilm_{ss} lenses to those shown in (a-b), but here the development of thick mantles of ilmenite has resulted from very extensive migration during exsolution. These mantles contain a second generation of exsolved Hem_{ss} and for all sets, which are in optical continuity, the plane of exsolution is (0001). 0.15 mm.

(e-f) Low and high magnifications respectively of Hem_{ss} cores with exsolved lenses of Ilm_{ss} and with mantles of Ilm_{ss}. In addition to these lenses are coarse lamellae of pleonaste which share the same plane of exsolution as those of the ilmenite. The outer margins result in a symplectic intergrowth with Mt_{ss}. At high magnification it is evident that each of the pleonaste lamellae is surrounded by Ilm_{ss} suggesting that the spinel predated Ilm_{ss} exsolution. Whether these spinels are the result of exsolution or an exsolution-like process is not clearly understood. e = 0.15 mm.; f = 400 um.

(g-h) The reverse relationships are illustrated here for ilmenite hosts and Hem_{ss} exsolution. The plane of exsolution is still (0001) but the notable difference here is that the lenses are extremely fine grained and that depletion zones, or zones of non-exsolution are concentrated along the grain boundaries. 0.15 mm.

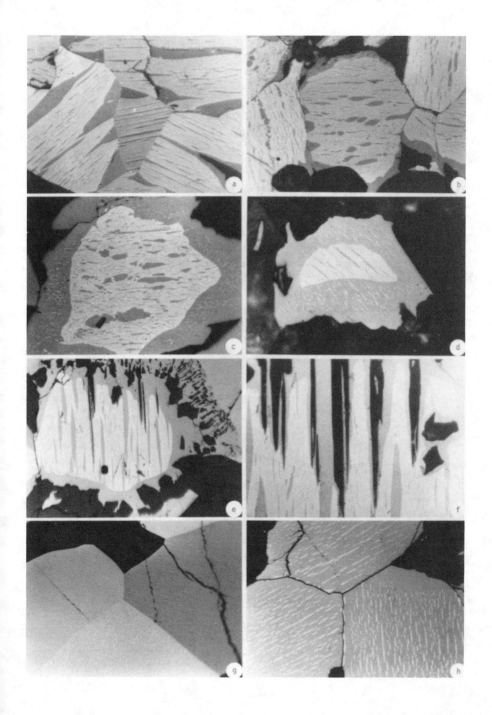

Figure 17

High temperature deuteric oxidation of primary ilmenite: Stages R1 to R4 in basalts

(a) Homogeneous discrete crystal of primary ilmenite. Stage R1. 0.1 mm.

(b) Subhedral crystal of ilmenite showing one coarse and many fine lamellae of ferrian rutile. Stage R2. 0.15 mm.

(c) Fine sigmoidal lenses of ferrian rutile in ferrian ilmenite. Stage R3. 0.15 mm.

(d) Similar to grain (c) but with one particularly well developed set of ferrian rutile lamellae along (0001) ilmenite parting planes. The subsidiary planes are $\{01\bar{1}1\}$ or $\{01\bar{1}2\}$. Stage R3. 0.15 mm.

(e-g) Three metailmenite crystals illustrating the progressive transformation of ferrian rutile to rutile, and of ferrian ilmenite to titanohematite. All grains contain the four-phase metastable assemblage typical of Stage R4. The light gray sigmoidal lamellae are rutile, the lighter gray lamellae (c.f. grains (b), (c), (d)) are ferrian rutile and the host phases are ferrian ilmenite (medium gray) and titanohematite (gray to white). The oxide complex to the upper right of grain (g) is an oxidized olivine crystal. 0.1 mm.

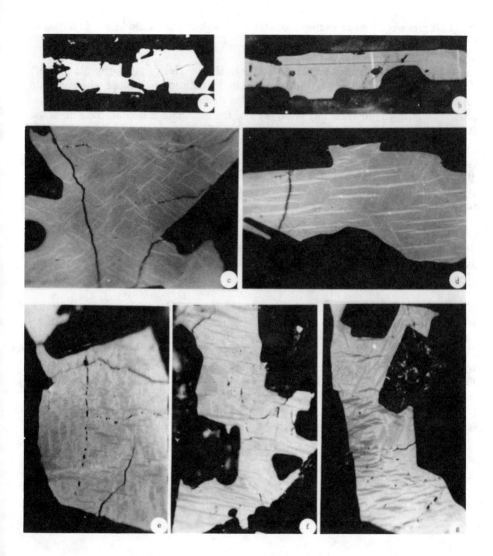

Figure 18

High temperature deuteric oxidation of primary ilmenite: Stages R5-R7 in basalts

(a) This grain is a classic example of Stage R5 with coarse rutile lenses in a host of titanohematite. 0.1 mm.

(b-c) Psb_{ss} (gray) mantles on the assemblage Rut + Hem_{ss} after original primary ilmenite. Grain (b) contains abundant rutile in the mantle assemblage whereas grain (c) is mostly pseudobrookite. These mantles develop progressively from the two-phase assemblage illustrated by grain (a) and the assemblage is characteristic of the R6 stage of oxidation. 0.1 mm.

(d) In contrast to grains (b) and (c) Psb_{ss} has developed along cracks in an ilmenite grain which is otherwise at the R5 stage (R + Hem_{ss}). Note that the protruding crystals of Psb, which are normal to the cracks, are parallel to the directions of the rutile lamellae. Stage R6. 0.1 mm.

(e) The assemblage in this grain is similar to those in grains (b), (c), and (d) and illustrates the disruption of rutile lenses by the encroaching Psb_{ss} which replaces the assemblage Rut + Hem_{ss}. Cuniform inclusions in Psb_{ss} are rutile (light gray) and Hem_{ss} (white). Stage R6. 0.1 mm.

(f) The core of this oxidized crystal consists of Rut + Psb_{ss} with associated Hem_{ss}, whereas the outermost part of the crystal consists of a Hem_{ss} matrix with Psb_{ss} lamellae along a crude {111} relic spinel fabric. The relative ratios of R:Psb_{ss} and Hem_{ss}: Psb_{ss} suggest that the core region was originally Ilm_{ss} and that the mantle was originally Usp-Mt_{ss}, an example comparable to the relationship illustrated in Figure 8a. Stages C7 and R7. 0.1 mm.

(g) Completely pseudomorphed crystals after original primary ilmenite. Dark gray is Psb_{ss}, medium gray is rutile, and the whitish areas are Hem_{ss}. Stage R7. 0.1 mm.

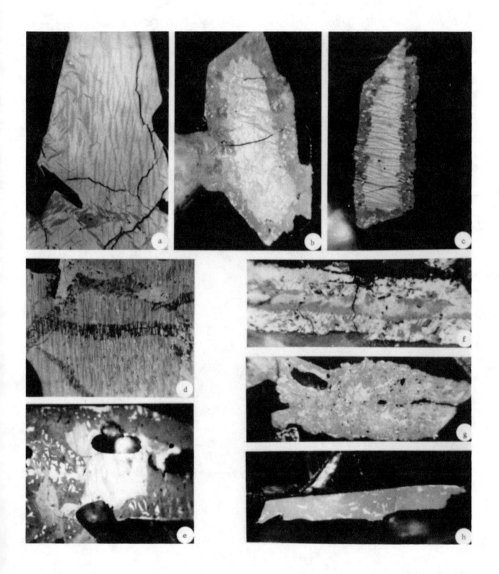

Figure 19

Textural genealogy for oxide minerals at high temperatures in basalts

Phase compatibility diagram for the system $FeO-Fe_2O_3-TiO_2$ at about 700 degrees C. Solid solution joins are continuous heavy lines, and miscibility gaps are dashed lines. The deuteric oxidation sequence may be followed by a comparison of tie lines (lighter continuous lines) with the oxide assemblages illustrated along the perimeter of the ternary. The cubic (C1 to C7) titanomagnetite sequence is anticlockwise, and the rhombohedral ilmenite sequence (R1 to R7) is clockwise.

The trends of oxidation are from the $FeO-TiO_2$ sideline towards the $Fe_2O_3-TiO_2$ join; for example, the terminal oxidation of titanomagnetite is Psb_{ss} + Hem_{ss} (C7), whereas the final oxidation of ilmenite is Psb_{ss} + Rut (R7). Other examples of the oxidation assemblages are given in Figures 6, 7, 8, 10, and 11 for titanomagnetite, and in Figures 17 and 18 for ilmenite.

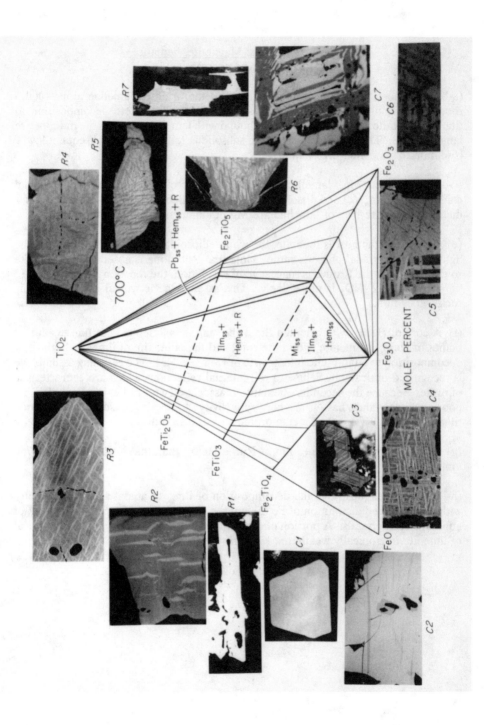

Figure 20

Ilmenite reactions: Magnetite, hematite, rutile and anatase

(a-b) The hosts in these grains are Ilm_{ss} and the oriented lamellae along (0001) rhombohedral planes are Mt_{ss} which show partial to complete decomposition to titanomaghemite. Note that the cores abound with lamellae whereas the margins are lamellae-free. The lamellae result from subsolidus reduction and subsequent low T oxidation in andesite. 0.15 mm.

(c-d) Deformed Mt_{ss} lamellae in kimberlitic picroilmenite which show partial oxidation to Hem_{ss} (c) and dissolution of these lamellae as shown in (d). The darker mantles which are free of lamellae are Mg titanomagnetites. 0.15 mm.

(e) The dark gray host is ilmenite, the white ellipsoidal bodies are Hem_{ss}, and the lighter gray sigmoidal phase is rutile. Note that Rut alone is absent in the ilmenite so that although the texture is suggestive of exsolution, the reaction $2\ FeTiO_3 + 1/2\ O_2 = Fe_2O_3 + 2\ TiO_2$ is more likely. This assemblage is typical of metamorphic rocks. 0.09 mm.

(f) A core of Hem_{ss} with oriented Ilm_{ss}; the mantle is ilmenite with fine exsolution bodies of Hem_{ss}. Ilmenite within the core, and the outer rim of ilmenite both show decomposition to an intimate intergrowth of Hem_{ss} + Rut. The resulting assemblage is recognized optically by yellow to red internal reflections which are characteristic of rutile. Because the secondary hematite is associated with rutile, it is Fe^{3+}-rich, Ti-poor, whiter, and has a higher reflectivity than the exsolution-associated titanohematites. Typical of metamorphic rocks. 0.13 mm.

(g) A more advanced stage of Ilm_{ss} - Rut + Hem_{ss} than that illustrated in (h). 750 um.

(h-j) This series illustrates the decomposition of Ilm_{ss} to anatase (TiO_2) + Hem_{ss} with the preferred dissolution of Fe_2O_3 and the development of a porous network of euhedral TiO_2 crystals. A portion of the unaltered ilmenite is shown in (h). Typical of ilmenite in tropically weathered kimberlites. 0.1 mm.

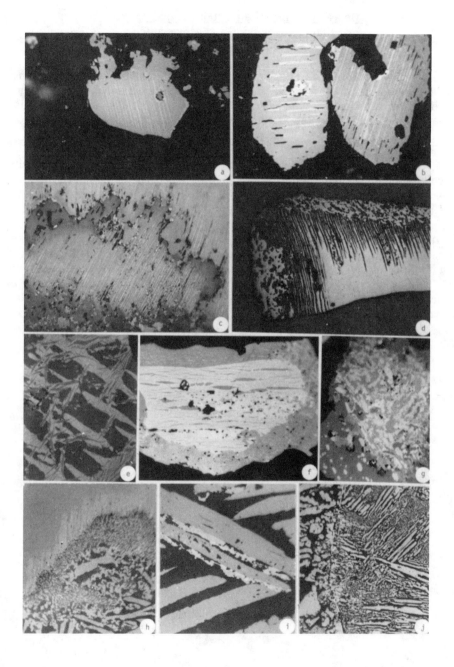

Figure 21

Metasomatic ilmenite reactions in mafic rocks: titanite, perovskite, and aenigmatite.

(a-b) Titanite ("sphene", $CaTiSiO_5$), replacement of Ilm_{ss} is illustrated in both grains; rutile is an associated phase in (a), and Hem_{ss} lamellae are exsolved from ilmenite in (b). The variation in color, which is particularly evident in (a) results from variations in crystal orientation (parallel growth and twinning) and reflection pleochroism. 0.15 mm.

(c-f) The progressive decomposition of picroilmenite illustrated in this series results initially in the marginal formation of Mg-rich Mt_{ss} + perovskite; the spinels are dark gray and perovskite ($CaTiO_3$) is white. With more intense decomposition (f) Hem_{ss} + Rut are associated phases. (c, d and f) = 0.15 mm; (e) = 750 um.

(g-h) The grains in these photomicrographs illustrate the replacement of Ilm_{ss} by aenigmatite ($Na_2Fe_5TiSi_6O_{20}$) which appears as the dark gray constituent. Grain (g) is a discrete ilmenite crystal, whereas grain (h) has {111} oriented Ilm_{ss} lamellae in Mt_{ss} and an external composite ilmenite. Note that it is only the Ilm_{ss} which is selectively replaced by aenigmatite (also known as cossyrite) and that the Mt_{ss} remains unaffected. 0.15 mm.

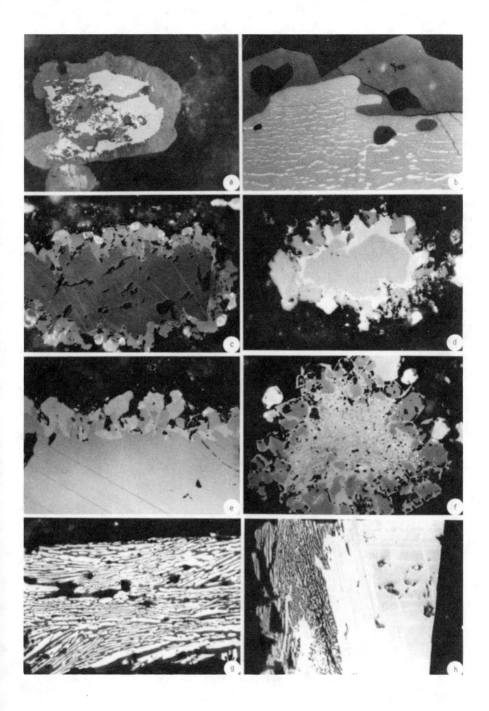

Figure 22

Oxide-silicate intergrowths: high temperature deuteric oxidation of olivine in mafic rocks

(a) A central core of partially oxidized olivine with symplectic magnetite which gradually increases in abundance towards the crystal grain boundary. The hematite diffusion rim is more diffuse than the example shown in grain (b). 0.15 mm.

(b) An oxidized olivine crystal with a well developed hematite diffusion rim and with symplectic magnetite + hematite throughout a large proportion of the crystal. 0.15 mm.

(c-e) These three crystals display varying degrees of symplectic magnetite growth. Grains (c) and (d) have marginally developed hematite whereas grain (d) is entirely magnetite with the central core of the olivine unoxidized. c and d = 0.15 mm.; e = 0.05 mm.

(f-g) These olivine crystals were oxidized in air at 800°C and 950°C respectively. Grain (f) contains hematite which has precipitated along olivine dislocations and grain (g) shows fine symplectic magnetite in subspherical units. f = 0.04 mm; g = 0.1 mm.

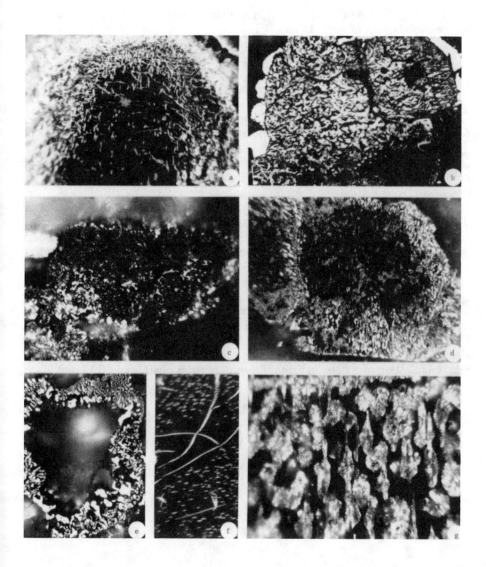

Figure 23

Oxide-silicate intergrowths in plutonic and volcanic rocks

(a) Usp-Mt$_{ss}$ "exsolved" as rods and discontinuous blebs in plagioclase. The large adjacent oxides are Ilm$_{ss}$. 0.15 mm.

(b) Symplectic arrays of Al-rich spinels in clinopyroxene. 0.15 mm.

(c) Magnetite derived from biotite and concentrated along cleavage planes. 0.15 mm.

(d) Magnetite associated with the peripheral decomposition of amphibole. 0.15 mm.

(e) Coarse symplectic magnetite in a highly oxidized crystal of olivine. Oxidation occurred at T>600 C based on the presence of associated Psb$_{ss}$. 450 um.

(f) The outer margin of this highly oxidized olivine crystal (T > 600°C) has a diffusion mantle of Hem (white) + magnesioferrite (gray). The highly reflective zone adjacent to this mantle is predominantly Hem; the core of the olivine is dark and diffuse and at high magnifications, symplectic magnetite can be resolved which is similar to that shown in grain (e). 0.15 mm.

(g-h) These olivine crystals are typical of those generally described as having undergone iddingsitization. Because iddingsite is not mono-mineralic, as is clearly apparent from the photomicrographs, the name has only descriptive connotations. The identifiable white phase is geothite and the associated darker phases (note cleavages in grain g) are smectite-related constituents. This assemblage parallels maghematization of titanomagnetite; ilmenite remains unaffected. 0.15 mm.

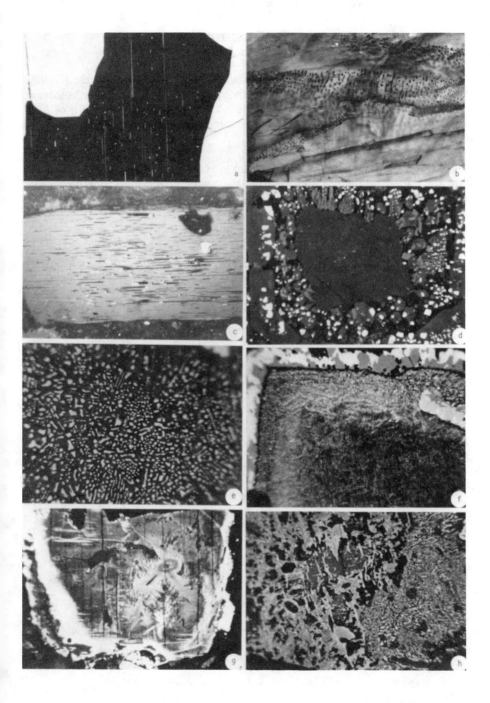

Figure 24

Oxide-silicate intergrowths in upper mantle xenoliths

(a-c) Blocky, subhedral, and euhedral Mg-Al-Fe spinel platelets along crystallographically controlled planes in orthopyroxene. These intergrowths form at the onset of mantle metasomatism and appear to derive from the instability of Al in enstatite during the introduction of metasomatic melts. 0.4 mm. Mixed transmitted and reflected light.

(d-g) Graphic intergrowths of ilmenite (black) in clinopyroxene. These intergrowths form discrete xenoliths, transported from the upper mantle by kimberlites. The origin of these xenoliths (Chapter 10) is highly controversial. 0.4 mm. Stereoscopic views of polished-thin sections.

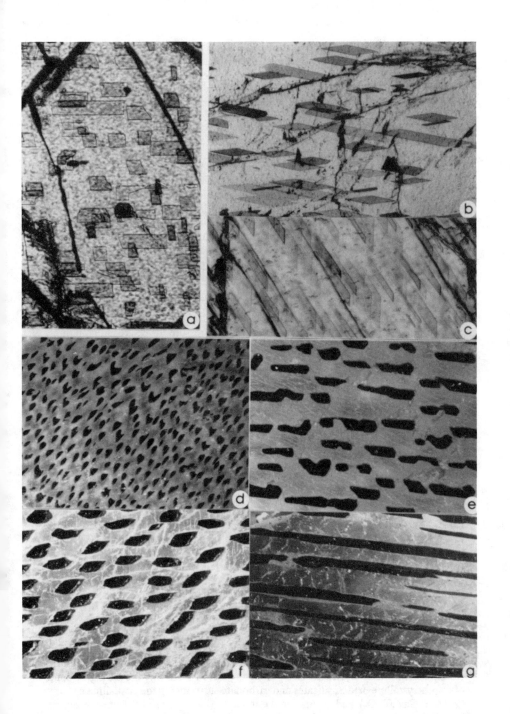

Figure 25

Oxide-silicate intergrowths in upper mantle xenoliths

(a) Nb-Cr rutile (white) - olivine (black) symplectic intergrowth of probable metasomatic origin, although other mechanisms for the assemblage have been suggested (Chapter 10). 0.4 mm.

(b) Nb-Cr rutile (white) and Mg-Cr ilmenite (gray) symplectite in olivine (black). Note the islands of rutile in ilmenite and the irregular grain boundary contacts between the two oxide minerals, suggesting possible reaction of rutile to form ilmenite. The entire assemblage is probably of metasomatic origin, as in (a); this is based primarily on the unusual composition of the rutile, and the close association of Nb-Cr rutile with metasomatic phlogopite, diopside and locally calcite. 0.4 mm.

(c) Coarse Nb-Cr rutile (white) symplectite in olivine (black). The rutile contains a fine oriented intergrowth of ilmenite, and the grain boundary contact between rutile and olivine shows a continuous gray band of armalcolite. The armalcolite is Ca-Cr and Nb-enriched and is interpreted as having formed in the presence of Ca from a metasomatic melt and by interaction with adjacent olivine. 0.4 mm.

(d) Oriented rods of rutile along apparent {111} garnet planes. This intergrowth is extremely common in some classes of eclogites (crustal and upper mantle), in amphibolite grade metamorphic rocks and in high pressure lower crustal granulites. An exsolution origin from high-Ti garnet has been proposed but the intergrowth remains enigmatic. 0.4 mm.

(e) Blebs and euhedral to subhedral crystals (white) of baddeleyite (ZrO_2) at the edge of a xenocrystic zircon. The light gray host is a complex mixture of SiO_2 (quartz or coesite), zircon, and unidentified Zr- bearing silicates. Decomposition of zircon is considered to be due to desilification of $ZrSiO_4$ on contact with highly undersaturated kimberlite. 0.4 mm.

(f) Zircon (dark gray and on the left) reacting with ilmenite (white and on the right) to form a band of diffuse baddeleyite (adjacent to the zircon), crystalline baddeleyite (white laths at the interface), and a larger irregular mass of zirconolite ($CaZrTi_2O_7$). The zirconolite is separated from ilmenite in the center of the view by dark gray calcite. Diopside is a common accessory in these reactions but is not shown; diopside and calcite and the driving force for the reaction is melt metasomatism. 0.4 mm.

(g) A higher magnification of the reaction interface between zircon and ilmenite. Cloudy microcrystalline baddeleyite is in the lower part of the photomicrograph. Baddeleyite crystals cloudy and translucent are coarser grained with distance from zircon and in association with optically and compositionally heterogeneous zirconolite (mottled gray subhedral crystal). 0.2 mm.

(h) An embayed reaction front into zircon (medium gray) with the development of a cluster of euhedral to subhedral crystals of baddeleyite and zirconolite. Note the cloudy band of microcrystalline ZrO_2 along the zircon interface, the dark gray band of cryptocrystalline oxides, silicates and carbonate surrounding the crystalline clusters (ZrO_2 + $CaZrTi_2O_7$) and separating the ilmenite (light gray right) from the zircon. The ilmenite has a very irregular replacement front and alteration to the interior of the crystal is evident along parallel, apparent (0001) rhombohedral planes. 0.4 mm.

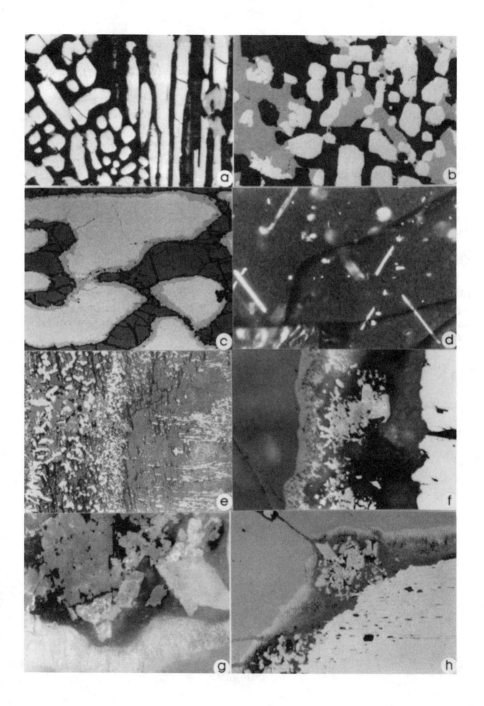

Figure 26

Spinel-silicate intergrowths in upper mantle xenoliths

(a-b) Graphic Mg-Al-Fe spinel (black) in orthopyroxene (white to gray), either approximately parallel to the pyroxene cleavage (a) or cross-cutting but still influenced by the cleavage. The clear translucent band at the spinel-pyroxene interface is pyropic garnet. Olivine and diopside are also present but are not present in these views. The xenolith is therefore, a spinel-garnet lherzolite. This is an unusual example of a possibly prograde upper mantle metamorphic reaction. 0.6 mm.

(c) Graphic and subhedral Mg-Al-Fe spinel in phlogopite after garnet (clear, translucent relic garnet persists in the lower right of the photomicrograph). This assemblage is typical of garnet harzburgites that have undergone upper mantle metasomatism. Amphibole is absent and metasomatism is inferred to have taken place at depths greater than about 120 km. 0.4 mm.

(d-f) Graphic Mg-Al-Fe spinel intergrown with enstatite and minor olivine or clinopyroxene. These fingerprint spinels are typical of harzburgites, are locally related to garnet reactions but are free of phlogopite (as in c). Spinel contacts with surrounding silicates (olivine in d) are either sharp, or irregular (diopside in e), but the parallel to subparallel finger texture into olivine, diopside, or garnet (f) is typical of these spinel intergrowths. 0.4 mm.

All these photomicrographs are in transmitted light.

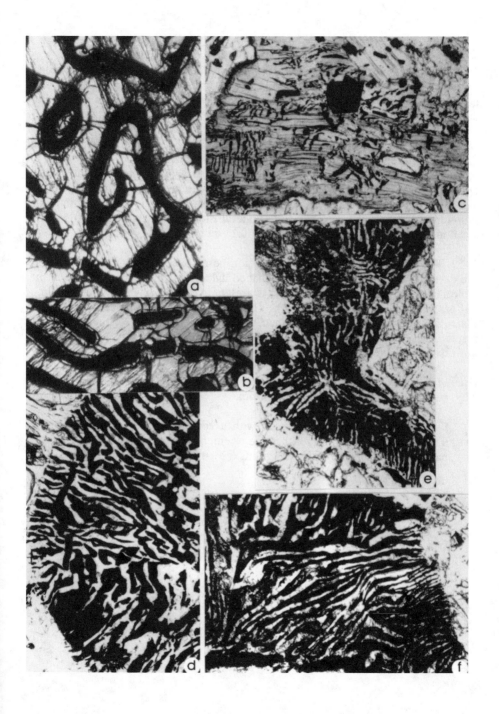

Figure 27

"Exsolution" in ilmenite$_{ss}$, hematite$_{ss}$ and rutile

(a-b) Discontinuous oriented rods of titanian chromites in kimberlitic picroilmenites. Both grains show a varied distribution of lamellae with respect to Ilm$_{ss}$ grain boundaries but the rods are uniform in size. The plane of exsolution is assumed to be parallel to (0001) ilmenite. 0.15 mm.

(c-d) Although the oriented lamellae in these kimberlitic picroilmenites have distinctly different optical contrasts, with respect to each other and with respect to their hosts, the lamellae in both grains are Mg-Al titanomagnetites; the lighter lamellae are higher in Fe$_{3+}$, and the darker lamellae are enriched in Mg and Al. The bleached zones in (c) are the result of partial oxidation; the surrounding groundmass crystals are perovskite. Peripheral rutile is present in grain (d) and a small amount is also present in grain (c). 0.15 mm.

(e-h) The host in these grains from metamorphosed beach sands is titanohematite; the darker gray lenses are Ilm$_{ss}$; the z-shaped lighter gray lamellae are rutile in the typical blitz texture. One rutile lamella is present in (e), along which abundant Ilm$_{ss}$ has exsolved; no Ilm is present in (g). The ilmenite exsolution plane is (0001) and the rutile exsolved planes are assumed to be $\{01\bar{1}1\}$ and $\{01\bar{1}2\}$; the former (e) results from exsolution *sensu stricto* whereas the latter (h) is more likely the result of an "exsolution"-like process related to oxidation. 0.08 mm.

(i) A rutile phenocryst containing sigmoidal lenses of Ilm$_{ss}$ and mantled by Ilm$_{ss}$, Mt$_{ss}$ and perovskite in kimberlite. 0.15 mm.

(j) A kimberlitic picroilmenite crystal with a core of rutile; the rutile contains irregular inclusions of Ilm$_{ss}$ rather than the oriented arrays typical of the association. The marginal and bright groundmass grains are perovskite. 0.15 mm.

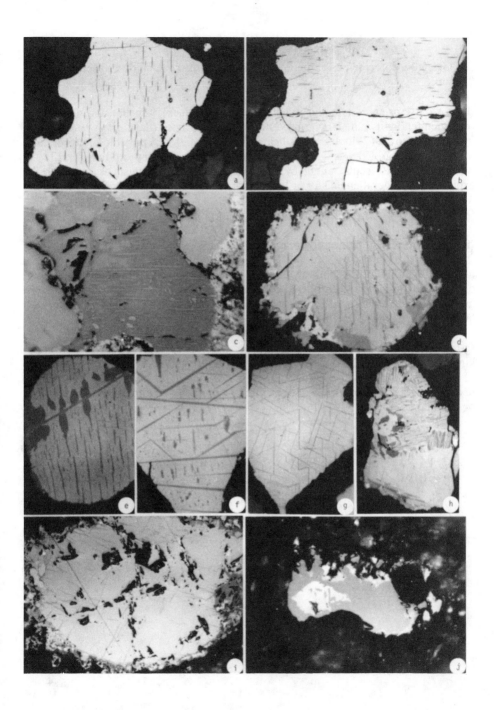

Figure 28

Upper mantle-derived metasomatic rutile

(a) Granulated, tectonized rutile and ilmenite in a discrete, silicate-absent xenolith. 0.4 mm.

(b-d) Coarse oriented lamellar ilmenite (white and black), lamellar and subgraphic (c) ilmenite (light and dark gray), and an ilmenite host (dark gray) to irregular inclusions of rutile (d) in discrete, silicate-absent xenoliths. A possible interpretation for these intergrowths is armalcolite $(Fe,Mg)Ti_2O_5$ decomposition to rutile + ilmenite, or rutile replacement by ilmenite. Neither mechanism is satisfactorily explained by mineral chemistry or low pressure experiments (Chapter 10). 0.4 mm.

(e) Chromian spinel lamellae in Nb-Cr rutile of possible exsolution origin. 0.2 mm.

(f) Transmitted light photomicrograph of coarse, fine, and ultrafine lamellar ilmenite (black) in translucent Nb-Cr rutile. This texture is uibiquitous in upper mantle rutile and implies either a high solubility of FeO (or Fe_2O_3) in rutile, and hence exsolution, or crystallographically controlled replacement of rutile by ilmenite. The latter mechanism fails to account for the multiple generations of ilmenite and the lack of high concentration of ilmenite along rutile grain boundaries. 0.2 mm.

(g-h) Deformation twinning in rutile that developed either by shearing during kimberlite eruption (g), or by hydrofracturing and shearing (h) during the injection of high-pressure metasomatic melts. The preservation of these textures and the lack of annealing, implies quenching, but an alternative interpretation is twinning induced by decompression.

Figure 29

Freudenbergite ($Na_2Fe_2Ti_6O_{16}$-$Na_2FeTi_7O_{16}$) in lower crustal granulites (a-f) and upper mantle xenoliths (g-h)

(a-b) Rutile (white) cores with successive mantles of freudenbergite (dark gray), ilmenite, perovskite, and an outermost and irregular thin margin of titanite (very dark gray). The embayments into rutile by freudenbergite suggest replacement by Na + Fe metasomatism. The ilmenite band that follows may also be due to replacement (from increasing Fe and decreasing Na). The perovskite layer marks the onset of Ca to the system, and then Si to form titanite ($CaTiSiO_5$). 0.4 mm.

(c-d) A low (0.4 mm. for c) and higher (0.12 mm. for d) magnification view of rutile, freudenbergite, ilmenite and perovskite, showing more clearly the contacts between minerals, the marked optical anisotropy (twinning) in freudenbergite (d), and the polygonal crystallization of ilmenite (lower portion of d). The apparent zonation in rutile is due to subsurface freudenbergite.

(e) A disrupted sequence in the orderly progression of low to high Ti from core to mantle but with the same minerals present as in (a) to (d). The lamellae in rutile are ilmenite. 0.4 mm.

(f) The upper portion of this crystal assemblage is rutile (core, white), freudenbergite, ilmenite, and titanite; the lower portion is rutile (core, white), ilmenite, and titanite. The bright spots around the oval silicate inclusion are sulfides. 0.4 mm.

(g-h) Rutile (white) cores replaced by freudenbergite (gray) and encased in ilmenite. These assemblages are particularly prevalent in zircon-bearing upper mantle xenoliths. Freudenbergite, along with K richterite, and jadeite are among the few source minerals for Na in the upper mantle. 0.4 mm.

The compositions of freudenbergite from the upper mantle (Chapter 10) and lower crust are relatively reduced and are dominantly $Na_2FeTi_7O_{16}$.

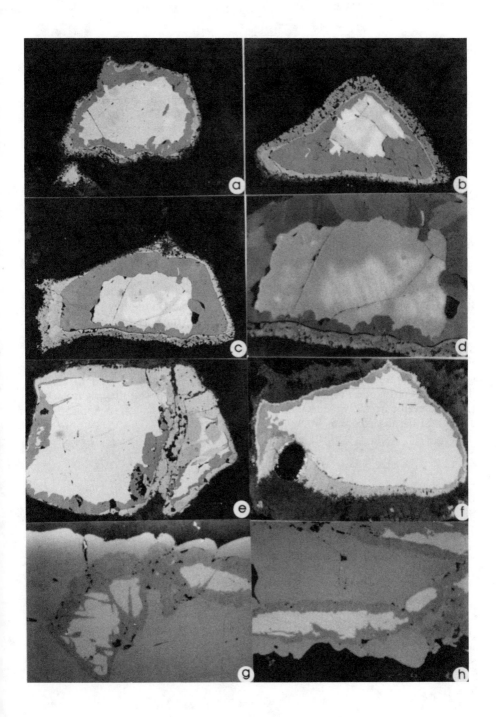

Figure 30

Armalcolite, (Fe,Mg)Ti$_2$O$_5$-pseudobrookite, Fe$_2$TiO$_5$ solid solutions in basalts

(a-b) Arm - Psb solid solution (dark gray) associated with ilmenite (light gray), ferrian rutile (white in ilmenite) and rutile (white in Arm - Psb). These textures may be interpreted either as Arm - Psb replacement of ilmenite (i.e. similar to the highly oxidized sequence illustrated in Figure 18), or of Arm - Psb reaction with liquid to form Mg-rich ilmenite (e.g. as in h). 0.4 mm.

(c) Transmitted light photomicrograph of a glass inclusion (dark gray) in basalt with pyroxene, olivine and feldspar and euhedral to subhedral laths of Arm - Psb along the walls and penetrating to the interior of the inclusion. 0.4 mm.

(d) Glass inclusions in euhedral crystals of Arm - Psb enclosed in olivine. The Arm - Psb is clearly primary and an early crystallization product of the melt. 0.4 mm.

(e-f) The case for simultaneous, or overlapping crystallization, and of Arm - Psb (gray) following ilmenite in paragenesis can each be made in these basalts. Note the inclusion of ilmenite (arrow) and the overgrowth of Arm - Psb on ilmenite (arrow) in (e). This contrasts with the euhedral crystals of Arm - Psb and ilmenite growth at the margins of Arm - Psb in (f). 0.4 mm.

(g-h) Euhedral to subhedral crystals of armalcolite (gray) adjacent to discrete ilmenite (g) or mantled by ilmenite (and accessory rutile) in (h); the bright flecks in (h) are metallic iron. These assemblages are from and are typical of Apollo 11 and 17 high-Ti basalts. 0.4 mm.

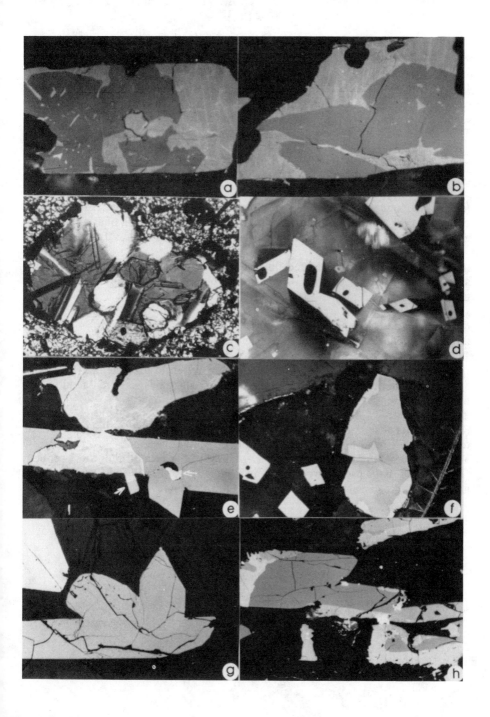

Figure 31

Armalcolite in upper mantle metasomatized xenoliths

(a-b) Armalcolite (Ca-Cr-Zb-Nb-enriched, or Cr-Zr-enriched and with minor pseudobrookite) mantles (gray) on Nb-Cr rutile (white). The lamellae in rutile in (a) are ilmenite with minor spinel; the lamellae in (b) are armalcolite and minor spinel. The enclosing silicate (black) is olivine. 0.4 mm.

(c) A blocky to subgraphic intergrowth of rutile (white) in olivine with advanced replacement of rutile by armalcolite (gray). 0.4 mm.

(d) Grain boundary, infiltration replacement of rutile by armalcolite. In some areas of this view the textural case for the primary (matrix) precipitation of armalcolite can also be made. Ilmenite is the dominant lamellar phase in rutile. 0.4 mm.

(e) Massive rutile with oriented ilmenite, and peripheral euhedral to subhedral armalcolite. Note the annulus of armalcolite on euhedral (dark gray) spinel. 0.4 mm.

(f) Irregular dark gray patches of armalcolite surrounding rutile (white) + sigmoidal ilmenite lamellae. The medium to dark gray phases are calcite + titanite. 0.4 mm.

(g) Subhedral platy laths of armalcolite (dark gray) enclosing ilmenite (light gray) and rutile (white). The lower right of the photomicrograph is massive ilmenite. 0.14 mm.

(h) A higher magnification of the view in (e) with rutile (white) and ilmenite in tones of gray (in various crystal orientations). Unlike the Ca-Cr-Nb-Zr and Cr-Zr armalcolite compositions in (a) to (d), these large stubby armalcolites are essentially $(FeMg)Ti_2O_5$, but a pseudobrookite component is also present. The former varieties are in much greater abundance than the latter. 0.15 mm.

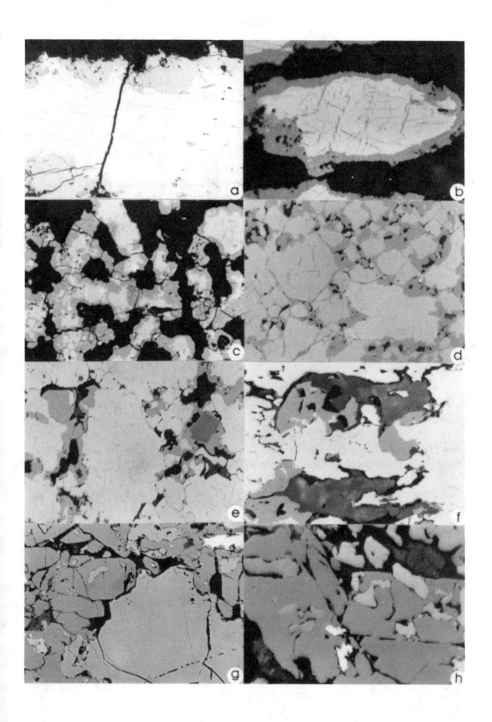

Figure 32

Lindsleyite, Ba(Ti, Cr, Fe, Zr....)21 O_38, mathiasite, K(Ti, Cr, Fe, Zr....)21 O_38, and Hawthorneite, Ba (Cr, Ti, Fe ....)12 O19 in metasomatized upper mantle xenoliths

(a) Euhedral lindsleyite enclosed in phlogopite. 0.4 mm.

(b) Lindsleyite with characteristic curvilinear fractures. 0.4 mm.

(c) Lindsleyite enclosing chromian spinel and surrounded by diopside, orthopyroxene, olivine, phlogopite and K richterite. 0.4 mm.

(d) Curvilinear cracks in lindsleyite. Mantled at one edge of the crystal is the assemblage Nb-Cr rutile (white) with sigmoidal ilmenite, and an outer rind (gray) of armalcolite. 0.4 mm.

(e) A rounded grain and a subhedral lath of mathiasite enclosed in Nb-Cr rutile. The black region is plucked mathiasite. 0.4 mm.

(f) Mathiasite (white) undergoing partial decomposition to euhedral and subhedral spinel (dark gray), ilmenite (medium gray), and rutile (white). Phlogopite, calcite, and unidentified zirconates fill the interstices (black). 0.12 mm.

(g) A complex five-phase assemblage of chromian spinel (medium gray), hawthorneite (large lighter gray areas), lindsleyite (lighter gray) ilmenite and rutile (white). 0.4 mm.

(h-i) Higher magnifications of areas from (g) showing details of the interfaces between and among spinel (Sp), hawthorneite (H), and lindsleyite (L). 0.12 mm.

The petrography and mineral chemistry of these LILE and HFSE phases are discussed in Chapter 10.

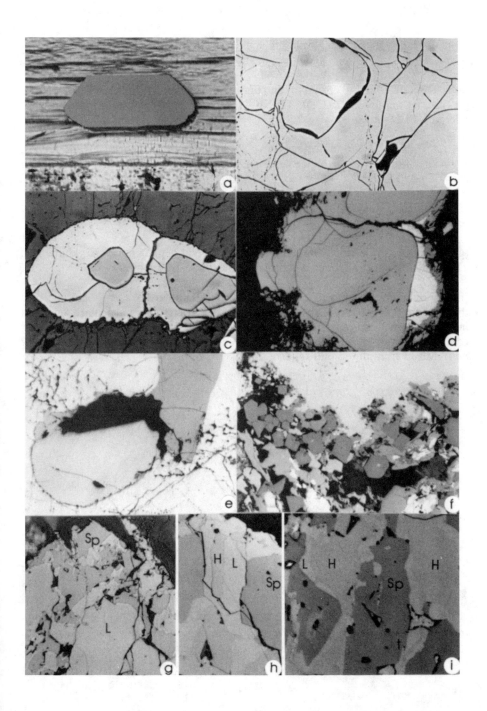

Figure 33

Exotic titanates in alkali and alkali-carbonatite associations

(a-b) Discrete euhedral crystal (a) and fluorite-type twinning (b) in strontian loparite [$SrTiO_3$, tausonite-(Na,Ca,Ce)]. $(Ti,Nb)O_3$, loparite solid solution from fenites. 90 microns. SEM images.

(c-d) Twinned (c) and intergrowth (d) strontian loparite crystals in fenites. Associated minerals are sanidine, aegirine, nepheline, lamphrophyllite, and strontian chevkinite. 0.12 mm.

(e-f) Radiating splays of kassite, $CaTi_2O_4(OH)_2$, in a carbonated peridotite xenolith. 0.12 mm.

(g) Vanadian priderite $(K,Ba)(Ti,Fe)_6O_{13}$ rosette in a phlogopite-rich kimberlite. The white dots are microbeam contamination spots. 0.12 mm.

(h) Serially zoned pyrochlore $(Na,Ca)_2Nb_2O_6(OH,F)$ from a carbonatite. The thicker and lighter band in the lower portion of the crystal is particularly enriched in Ti, as are the lighter grains throughout the crystal; the darker bands have higher concentrations of Nb, Zr, and REE and are partially to completely metamict. 0.12 mm. Mixed transmitted and reflected light.

Figure 34

Perovskite - magnetite associations
in undersaturated volcanic rocks

(a-b) Coarsely crystalline perovskite (white) in the groundmass and perovskite intergrown with Mg-Ti magnetite along the border of an ilmenite that has undergone subsolidus reduction to ilmenite + Mg magnetite lamellae. The dark rind at the edge of the ilmenite crystal is Mg-Ti magnetite. 0.12 mm.

(c) Schematic of an ilmenite and lamellar magnetite xenocryst bordered by a rim of magnetite and surrounded by alternating fingers of perovskite and magnetite. Massive, subhedral crystals of magnetite (Mt) and perovskite (Per), locally millimeters in size, form the outermost mantle. Not to scale.

(d) An intergrown network of Mg-Ti magnetite and perovskite (white, but also black plucked areas) at the margin to an ilmenite and lamellar Mg magnetite xenocryst. 0.12 mm.

(e) The core of this assemblage (upper part of the view) is a lamellar intergrowth of ilmenite, (lighter gray) and Mg magnetite (darker gray). Parallel and subparallel perovskite lamellae are present in the broad Mg titanomagnetite annulus; the outermost margin against the groundmass is an intricate intergrowth of perovskite and Mg titanomagnetite. 0.12 mm.

These ilmenite xenocrysts, and the reduction lamellae of spinel, are relatively common in melilitites and basanites; reduction is assumed to have taken place in the upper mantle prior to incorporation and subsequent reaction in the melt.

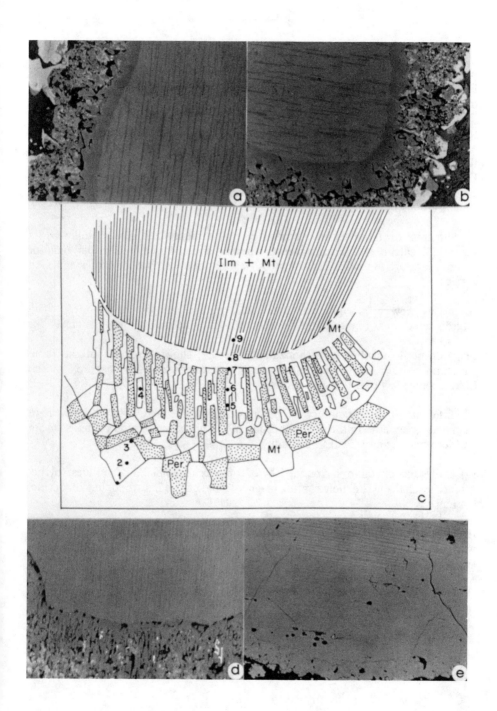

Figure 35

Magmatic carbon-oxide associations

(a) Radiating graphite (from shale incorporated into basalt) adjacent to metallic Ni-Fe (lower left) with ilmenite (small crystal, bottom central, shown at higher magnification in (b). 0.4 mm.

(b) A higher magnification of the lower central view in (a) with subhedral ilmenite enclosing metallic Ni-Fe. The gray, subspherical and radiating bleb is graphite with flecks of Ni-Fe; the bright area to the left is metallic Fe, and the mottled region is Fe + cohenite (FeC). The ilmenite has significant Ti_2O_3 in solid solution, reflecting a highly reduced state. Armalcolite, anion-deficient rutile, and non-stoichiometric spinels are also present in these basalts. 0.12 mm.

(c) Ilmenite overgrowths on a central core of serpentinized olivine, followed by perovskite, calcite, and ferromagnesian silicates, and enclosed by a second band of ilmenite against the carbonatite matrix. 2 cm. Stereoscopic view of a polished surface.

(d) This view of polygranular ilmenite is of the outmost rind in (c). The interdigitated areas (white to light gray) are calcite and perovskite. 0.4 mm.

(e) Magnetite pisolite in carbonated kimberlite, showing several of the same structural features as (c) including a highly serpentinized olivine core. 2 cm. Stereoscopic view of a polished surface.

(f) Internal structural detail of the outermost magnetite rind. The dark matrix to filamental magnetite in the central portion of the photomicrograph contains calcite and ferromagnesian silicates. 0.4 mm.

(g) Polygonal ilmenite surrounded by a complex assemblage of sulfides (pyrite, pyrrhotite, chalcopyrite, polydymite), magnetite and calcite in carbonated kimberlite. 0.4 mm.

(h) Vesicle infilling at the base of a lava flow with rims of hematite and interlayed with calcite. These infillings develop from the boiling-off of volatiles from underlying sediments. 0.4 mm.

The significance of these assemblages relates to the possible role of carbon in magmatic, immiscible oxide liquids.

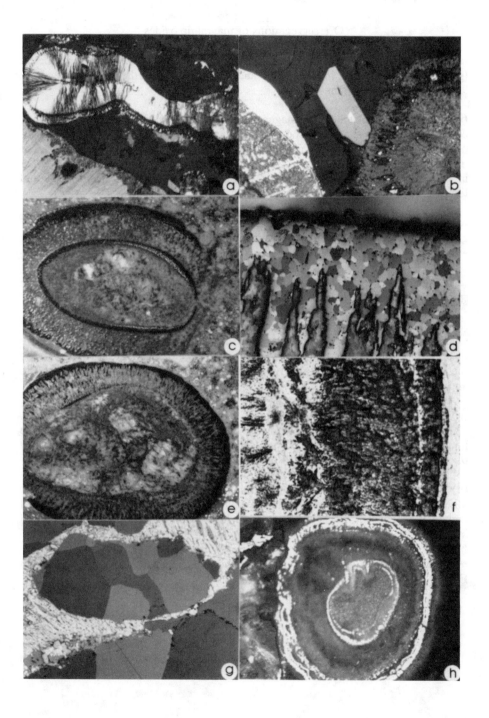

Figure 36
Ore segregations of magmatic oxides

(a) Serially zoned euhedral magnetite and the selective replacement of magnetite (gray) by hematite along {111} spinel planes in a sample from the El Laco magnetite lava flows, Chile. 0.4 mm.

(b) Selective replacement of magnetite (gray) by hematite (white). The black central region, with sharp grain boundary contacts, is an unidentified iron phosphate possibly related to vivianite, phosphoferrite, or lipscombite. Apatite is also present in this sample of the El Laco magnetite lava flow; these phosphates may have aided in depressing the magnetite liquidus to more reasonable geologically accepted temperatures (i.e. about 1200°C rather than 1600°C). 0.4 mm.

(c) This view of an El Laco sample shows abundant phosphate (black), irregular grains of magnetite (gray) and hematite (white) replacement lamellae, and a euhedral crystal of primary hematite enclosing apatite. 0.4 mm.

(d) Magnesian chromite with oriented and peripheral Mg-Cr magnetite from massive spinel-rich bands in an ultramafic intrusive. 0.4 mm.

(e-f) Complexly intergrown and exsolved chromian ulvospinel (medium gray), ulvospinel (light gray), and pleonaste-hercynite from Cr-Ti magnetite in massive oxide layers from an ultramafic complex. The large, homogeneous (gray) crystal to the left in (f) is ilmenite with a twin band of hematite (white) followed by pleonaste and magnetite. 0.4 mm.

(g) A trapped immiscible sulfide (liquid) complex in ilmenite from an anorthosite. The gray phase throughout the complex is magnetite, suggesting that the sulfide liquid had significant oxygen. Both carbon (Fig. 35) and sulfur, and possibly phosphorus (Fig. 36 b-c), play important roles in oxide liquid immiscibilities. 0.4 mm.

(h) Skeletal ilmenite (dark gray) and sulfide (pyrite, pyrrhotite) complex in a basaltic dike. These sulfide, oxide and silicate (felspar, pyroxene) assemblages are spherical and tend to concentrate in the central regions of dikes, possibly implying liquid immiscibility. 0.4 mm.

Other examples of magmatic oxide segregations in anorthosites are illustrated in Figures 5 and 16.

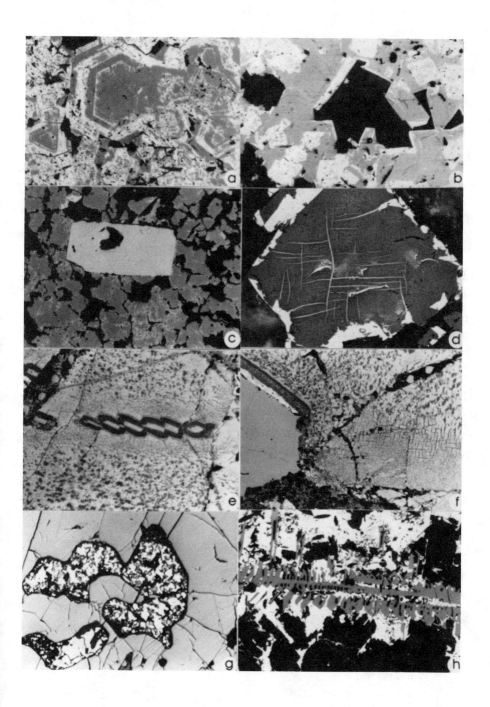

Figure 37

Metamorphic oxide ores

(a-b) Two examples of deformed garnets (light and dark gray) with dusty magnetite trails, but with coarse recrystallized magnetite (and quartz) in the surrounding matrix. Samples are typical of Archean iron formations in basement complexes. 0.4 mm.

(c) Deformed ilmenite-magnetite (in lenses and partially altered to hematite) complex from banded Proterozoic iron formations in mafic-ultramafic complexes. 0.4 mm.

(d) Euhedral gahnite (ZnAl2O4) crystal in a magnetite (white) and quartz (black) matrix from an horizon closely associated with a hydrothermal pyrite, galena and sphalerite ore body. Gahnite is a useful accessory mineral for the regional exploration of some metamorphic hydrothermal ore deposits. 0.4 mm.

(e) Highly tectonized quartzo-feldspathic vein in a massive, metamorphosed magnetite and hematite deposit of Proterozoic age. 0.4 mm.

(f) Magnetite (dark gray) with hematite (white) replacement lamellae along {111} spinel planes in association with euhedral to subhedral laths of rutile (white with cloudy internal reflections) in a chlorite schist. Magnetite + rutile is the stable oxide assemblage at low metamorphic grades. 0.4 mm.

(g) Banded and highly deformed goethite after pyrite from a metamorphosed hydrothermal sulfide deposit. 0.4 mm.

(h) Radiating fibrous goethite (dark gray) on euhedral magnetite (partially replaced by hematite-white) in an open vein in the El Laco magnetite lava flow. 0.4 mm.

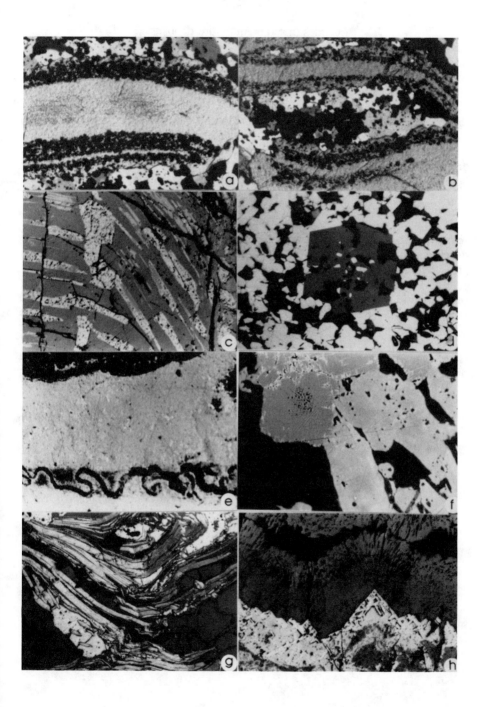

Figure 38

Oxides in sedimentary ore deposits

(a) Banded Precambrian iron formation showing both magnetite (gray) and hematite (white) in a quartz (black) matrix. Note the band of hematite normal to the horizontal magnetite fabric. 0.4 mm.

(b) Well sorted oxide (white), quartz, zircon and monazite (black and dark gray) Tertiary beach sand. The magnetite is partly to completely altered to hematite, the ilmenite has decomposed to cryptocrystalline Ti-rich phases (e.g. arizonite, proarizonite, pseudorutile) and hematite, but rutile remains optically unaffected. The up-grading of ilmenite (i.e. removal of Fe and the concentration of Ti into rutile, anatase and the cryptocrystalline Ti phases) makes these deposits economic for titanium. 2.0 mm.

(c-d) Oolitic hematite ore, partially deformed and with fibrous crystalline hematite developing in the low pressure orbicular cores. The matrix is dominantly quartz, but greenalite and siderite are also present. (c) = 0.4 mm.; (d) = 0.2 mm.

(e-f) Oolitic hematite-carbonate ores with either closely spaced oxide bands and minor carbonate (e), or with approximately equal proportions of oxide and carbonate (f). 0.4 mm.

(g-h) Oolitic magnetite and pyrite in a carbonaceous shale. Magnetite forms the outer bands interspersed with identifiable graphite and irresolvable dark matrix (g); the cores (white in g) are dominantly marcasite. (g) = 0.2 mm.; (h) = 0.4 mm.

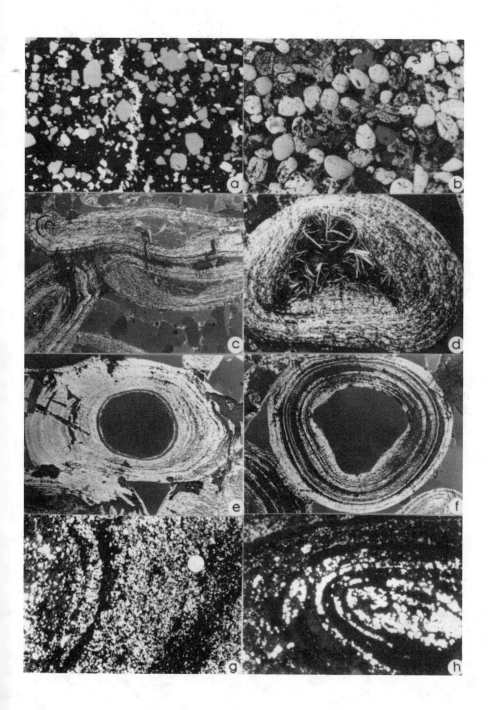

Figure 39. Oxide minerals in ocean floor rocks

(a) Cr-Al spinel xenocrysts in a fine grained basalt matrix with high concentrations of fine and extremely fine titanomagnetite in the groundmass. 0.4 mm.

(b) Cr-Al spinel core mantled by titanomagnetite. 0.15 mm.

(c) Subskeletal titanomagnetite. 0.2 mm.

(d) Subskeletal, skeletal, rod and bead titanomagnetite. 0.15 mm.

(e) High concentrations of bead titanomagnetite grading into a submicroscopic population in a plagioclase-controlled variolitic texture. 0.5 mm.

(f) Subhedral titanomagnetite adjacent to a strongly zoned feldspar phenocryst with glass inclusions and high concentrations of dusty oxides. 0.8 mm. Mixed transmitted and reflected light.

(g-h) Variolitic to subvariolitic micostructures in glass with high contents of finely dispersed oxides. 0.8 mm. Mixed transmitted and reflected light.

(i-j) Altered glass in subspherical, colloform structures that may include goethite. Titanomaghemite is visible as an oxidation product of titanomagnetite, confirming a low T origin ($<150°C$). 0.5 mm.

(k-l) Incipient to advanced replacement of titanomagnetite by titanomaghemite; (i) has previously undergone high temperature oxidation exsolution, trellis ilmenite is present along {111} spinel planes. 0.2 mm.

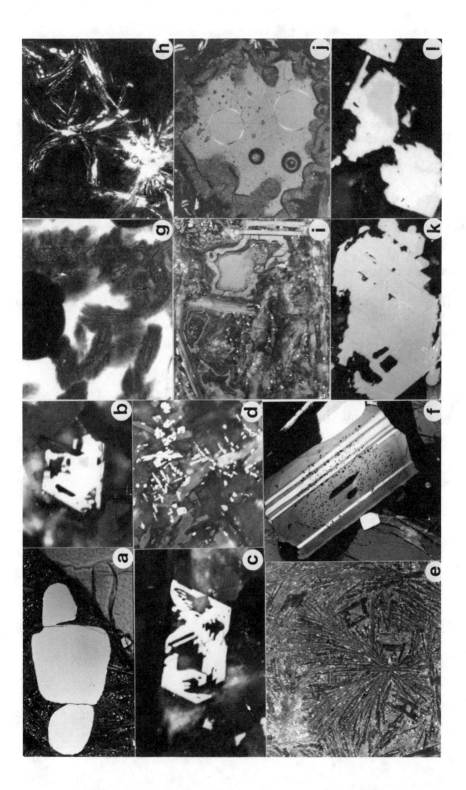

Figure 39. Oxide minerals in ocean floor rocks, continued

(m) Coarse grained equilibrated ilmenite enclosed in titanomagnetite with trellis ilmenite. 0.4 mm.

(n) Subsolidus reduced ilmenite (dark gray) with titanomagnetite (light gray) along (0001) rhombohedral planes. 0.4 mm. [Damage to photograph in upper right corner.]

(o) Titanite (gray) replacing titanomagnetite (white). 0.2 mm.

(p) Oriented rutile and titanomagnetite in Cr-Al spinel. 0.15 mm.

(q) Cr-Al spinel in serpentinite rimmed by magnetite and "ferritchromit". 0.4 mm.

(r) Highly distorted orthopyroxene containing lamellar magnetite along cleavage planes in serpentized peridotite. 0.5 mm. Mixed transmitted and reflected light.

(s-t) Network magnetite in serpentine. 0.8 mm. (s) is in reflected light and (t) is in transmitted light.

(u) Immiscible pyrite and pyrrhotite (white) blebs in association with ultrafine oxides, plagioclase and glass. 0.5 mm.

(v) Goethite (gray) replacing pyrite (white). 0.15 mm.

(w-x) Complex sulfide assemblages in cataclastic gabbro with hosts of bornite and chalcopyrite and lamellar pyrite. These assemblages are prone to magnetite replacement and goethite alteration. 0.5 mm.

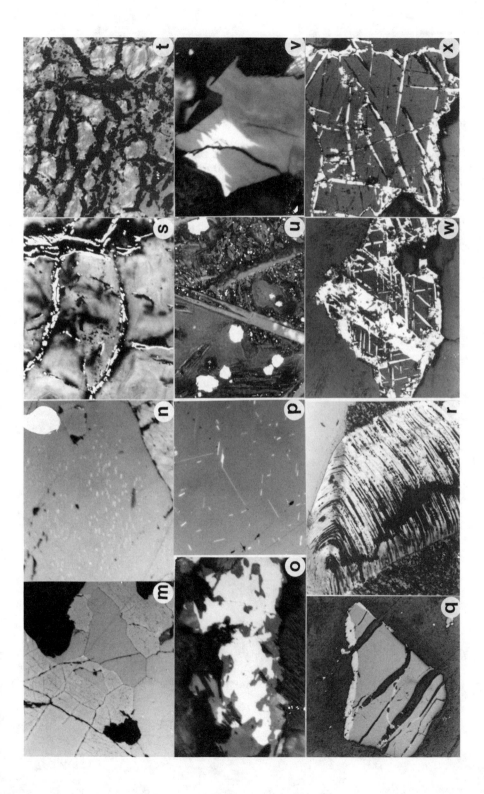

Figure 40

Biogenic magnetite

(a-f) Examples of magnetotactic magnetite morphologies. Crystals are typically 50 to 100 nm across.

(g) Magnetotactic spirillum showing a chain of magnetite crystals in the interior of the bacterium. Crystals are approximately 100 nm across and have either subcubic or pseudohexagonal cross-sections. SEM image magnified 38,500x. (Photo by John Stolz; used with permission).

(h) Magnetite framboid, possibly after colony bacterial pyrite or greigite in a deep sea sediment. Framboid is 15 microns across.

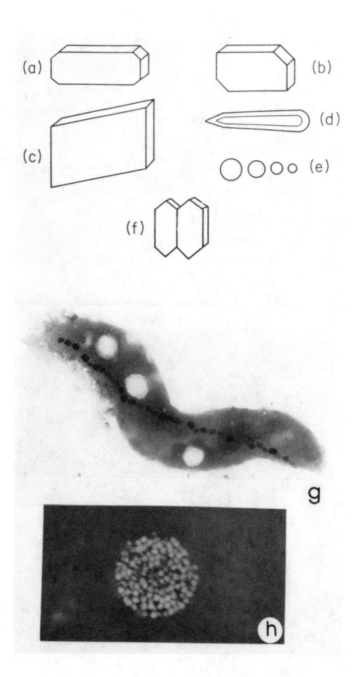

Chapter 6   M. S. Ghiorso & R. O. Sack

THERMOCHEMISTRY OF THE OXIDE MINERALS

INTRODUCTION

In this chapter we survey the thermochemical properties and energetic features of several important groups of rock-forming metal oxides. Rather than attempt an exhaustive review of the literature, we will focus on methods of construction of thermodynamic models that account for *first order* crystal chemical constraints on the energetics of these solid solutions. We discuss a few specific models for naturally occurring oxide compositions, and demonstrate that these models provide the framework for the development of an internally consistent set of end-member and mixing thermochemical properties for these mineral phases. Finally, we propose a recommended set of internally consistent values for the thermodynamic properties of end-member oxides, and indicate several directions for future research on oxide thermochemistry.

The oxides display a wide variety of crystal chemical features that make their thermodynamic description at once fascinating and treacherous. First, the crystal structures of the oxide minerals are compact. This serves to exacerbate the energetic effects of size mismatch that arise from cation substitutions. These size-mismatch contributions lead to large positive enthalpies of mixing which are manifested in extensive miscibility gaps between mineral end-members. Secondly, the oxide minerals exhibit strongly temperature- and compositionally-dependent cation ordering. Energetically, these effects are seen as "excess" entropies of mixing, and as second-order phase transitions between symmetrical high-temperature structures and their structural derivatives. The interaction of cation ordering and size mismatch often leads to *highly asymmetrical miscibility gaps* along binary joins of these solid solutions. Thirdly, the oxides display complex magnetic phase transitions which result from configurational entropy contributions arising from randomization of magnetic spins. Unlike other rock-forming minerals, oxides (in particular the Fe-Ti oxides) exhibit these features at elevated temperatures. This results in the development of "magnetic" miscibility gaps and extensive heat capacity anomalies at temperatures as high as 600°C. Last, oxides display considerable deviation from exact stoichiometry, developing defect structures, where the density of defects is a complex function of temperature, oxygen fugacity and bulk composition of the phase. These effects are the most difficult to treat thermodynamically, owing primarily to a lack of quantitative information on equilibrium defect densities for compositions of interest at geologically relevant temperatures.

CRYSTAL CHEMICAL CONSTRAINTS ON THERMODYNAMIC MODELS

Cation substitution and size mismatch

We can illustrate the dramatic energetic effects of size mismatch resulting from cation substitution by focusing on spinel solid solutions in the system $(Fe^{2+}, Mg)(Al^{3+}, Cr^{3+}, Fe^{3+})_2O_4 - (Fe^{2+}, Mg)_2TiO_4$. Experimental data on the shape and extent of miscibility gaps in geologically important spinels has been recently reviewed by Sack and Ghiorso (1991a,b). In Figure 1 we reproduce their figure which summarizes the experimental constraints via calculated miscibility gaps along constituent binaries of the supersystem. As the majority of the miscibility gaps are broadly symmetric in composition, in the context of regular solution theory their critical temperatures may be directly related to

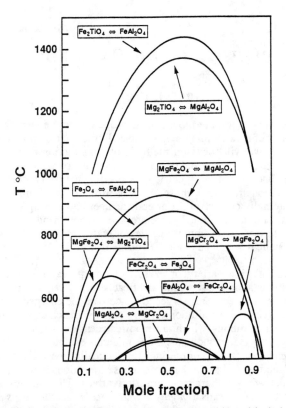

Figure 1. Comparison of calculated miscibility gaps for compositionally binary joins in the (Fe^{2+},Mg)-aluminate-chromite-titanate-ferrite spinels. Curves are labeled to indicate left and right compositions on the abscissa. After Sack and Ghiorso (1991a,b).

Table 1. Absolute differences in ionic radii for selected ions in octahedral coordination.

	r	Al^{3+}	Cr^{3+}	Fe^{2+}	Fe^{3+}	Mg^{2+}	Mn^{2+}
Al^{3+}	0.530						
Cr^{3+}	0.615	.085					
Fe^{2+}	0.780	.250	.165				
Fe^{3+}	0.645	.115	.030	.135			
Mg^{2+}	0.720	.190	.105	.060	.075		
Mn^{2+}	0.830	.300	.215	.050	.185	.110	
Ti^{4+}	0.605	.075	.010	.175	.040	.115	.225

Table 2. Absolute differences in ionic radii for selected ions in tetrahedral coordination.

	r	Al^{3+}	Fe^{2+}	Fe^{3+}
Al^{3+}	0.390			
Fe^{2+}	0.630	.240		
Fe^{3+}	0.490	.100	.140	
Mg^{2+}	0.580	.190	.050	.090

the excess enthalpy of mixing along each join ($W_H = 4 R T_c$). Along the $Fe_3O_4 - FeAl_2O_4$ join for example, the critical temperature is ~1150 K, and the excess enthalpy may be described roughly by the function: 38 $X_{Hc} X_{Mt}$ (kJ). Figure 1 should be examined in combination with Tables 1 and 2, where we have tabulated ionic radii and absolute differences in ionic radii (Shannon and Prewitt, 1969) for typical cations in octahedral and tetrahedral coordination. As an example, we read from Table 1 the difference in ionic radii of Mg^{2+} and Cr^{3+} in octahedral coordination as 0.105 Å.

To develop the discussion let us consider high-, intermediate- and low-temperature miscibility gaps along binary joins of the aluminates. We start with the high-temperature gap along the $FeAl_2O_4 - Fe_2TiO_4$ join and derive a measure of size mismatch for cation substitution in this series. Recognizing that at such high-temperatures the aluminate has essentially a random distribution of Fe^{2+} and Al^{3+} over the occupied tetrahedral and octahedral sites (Sack and Ghiorso, 1991a), we write the structural formulae[1] as $(Fe_{1/3}Al_{2/3})[Fe_{1/3}Al_{2/3}]_2O_4$ and calculate an average tetrahedral cation radius of 0.470 Å and an average octahedral cation radius of 0.613 Å. The iron titanate preserves its inverse cation distribution at elevated temperatures, giving a structural formula that may be written $(Fe)[Fe_{1/2}Ti_{1/2}]_2O_4$, with tetrahedral cation radius of 0.630 Å and average octahedral cation radius of 0.6925 Å. To define a crude measure of size mismatch we compute the quantity:

$$\Delta r = |(\Delta r)^{TET} + 2 (\Delta r)^{OCT}| \qquad (1)$$

where $(\Delta r)^{TET}$ is the difference in average radii for tetrahedral cations between the two end-members and $(\Delta r)^{OCT}$ is the corresponding quantity for octahedral cations. Our computed measure of size mismatch (Δr) for the high-temperature $FeAl_2O_4 - Fe_2TiO_4$ join is found in this manner to be 0.32 Å. Performing the same series of calculations for the other aluminate joins, gives the values reported in Table 3. We summarize the inferred relations between size mismatch, critical temperature and enthalpy of mixing (expressed as a regular solution interaction parameter) in Figure 2. It is clear that there is a good correlation between Δr and T_c, implying that the larger the size mismatch the greater the excess enthalpy of solution. In summary, we note that the energetic consequences of size mismatch accompanying cation substitution lead to positive contributions to the Gibbs energy of mixing (Lawson, 1947).

Cation-ordering phenomena

Cation ordering refers to the partitioning of two or more kinds of cations between crystallographic sites as a function of temperature, pressure and bulk composition. Two varieties are often distinguished. *Convergent* ordering refers to partitioning between symmetrically equivalent sites. *Non-convergent* ordering implies partitioning between crystallographically distinct sites. In the oxide minerals, both kinds of cation ordering occur. In solid solutions of ilmenite ($FeTiO_3$) and hematite (Fe_2O_3) for example, convergent ordering of Fe^{2+} and Ti^{4+} between otherwise crystallographically equivalent octahedral sites results in a second order phase transition between the high-temperature $R\bar{3}c$ structure and its $R\bar{3}$ subgroup. In the spinels, non-convergent ordering between tetrahedral and octahedral crystallographic sites relate *normal* $(X)[Y]_2O_4$, *inverse* $(Y)[X_{1/2}Y_{1/2}]_2O_4$ and *random* $(X_{1/3}Y_{2/3})[X_{1/3}Y_{2/3}]_2O_4$ cation distributions.

---

[1] We adopt the convention in writing the structural formulae for spinels that the cation(s) in parentheses is located on the tetrahedral site and the cation(s) in square brackets is found on the octahedral sites.

Table 3. Average tetrahedral and octahedral cation radii and size mismatch parameters for selected spinels

			FeAl$_2$O$_4$ Normal		FeAl$_2$O$_4$ Random	
			r (TET)	r (OCT)	r (TET)	r (OCT)
			0.630	0.530	0.470	0.613
Fe$_2$TiO$_4$	r (TET)	0.630	$\Delta r = 0.325$		$\Delta r = 0.319$	
	r (OCT)	0.693			$T_c \sim 1450°$ C	
Fe$_3$O$_4$	r (TET)	0.490	$\Delta r = 0.225$		$\Delta r = 0.219$	
	r (OCT)	0.713	$T_c \sim 875°$ C		$T_c \sim 875°$ C	
FeCr$_2$O$_4$	r (TET)	0.630	$\Delta r = 0.170$		$\Delta r = 0.164$	
	r (OCT)	0.615	$T_c \sim 475°$ C			

			MgAl$_2$O$_4$ Normal		MgAl$_2$O$_4$ Random	
			r (TET)	r (OCT)	r (TET)	r (OCT)
			0.580	0.530	0.453	0.593
Mg$_2$TiO$_4$	r (TET)	0.580	$\Delta r = 0.265$		$\Delta r = 0.266$	
	r (OCT)	0.663			$T_c \sim 1350°$ C	
MgFe$_2$O$_4$	r (TET)	0.535	$\Delta r = 0.223$		$\Delta r = 0.224$	
	r (OCT)	0.664	$T_c \sim 925°$ C		$T_c \sim 925°$ C	
MgCr$_2$O$_4$	r (TET)	0.580	$\Delta r = 0.170$		$\Delta r = 0.171$	
	r (OCT)	0.615	$T_c \sim 475°$			

			Fe$_3$O$_4$	
			r (TET)	r (OCT)
			0.490	0.713
Fe$_2$TiO$_4$	r (TET)	0.630	$\Delta r = 0.100$	
	r (OCT)	0.693		

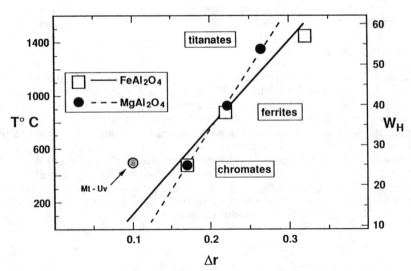

Figure 2. Correlation of size mismatch paramater (Δr, see Eqn. 1) to the critical temperature of the miscibility gaps plotted in Figure 1. The critical temperature may be interpreted as a symmetric regular solution interaction parameter via the relation: $W_H = 4 R T$. The equivalent W_H (in kJ) is indicated on the right-hand ordinate.

The degree of cation ordering may be quantified by the concept of an *ordering parameter*. The ordering parameter is usually taken to describe the difference in site mole fractions of a particular cation between two crystallographic sites. For example, ordering in $MgAl_2O_4$ spinel might be parameterized as the difference in the mole fraction of Mg on tetrahedral and octahedral sites. As the site mole fractions are uniquely determined at thermodynamic equilibrium for a given temperature, pressure and bulk composition, one sees that the ordering parameter is an implicit function of these intensive variables *in the equilibrium state*. The use of an ordering parameter allows us to distinguish between so-called "ordered" and "anti-ordered" structures. Once again, using $MgAl_2O_4$ as an example, we might consider the "ordered" structure to be the low-temperature *normal* cation distribution and the "anti-ordered" structure to be the hypothetical *inverse* cation distribution. In this case the anti-ordered distribution is never realized in nature; cation distributions in $MgAl_2O_4$ become more and more *random* at elevated temperatures (e.g., Sack and Ghiorso, 1991a). The partitioning of ordering space into physically plausible and implausible regions is a general feature of minerals that display non-convergent ordering. By contrast, the distinction between "ordered" and "anti-ordered" states is an artificial one for minerals that undergo convergent ordering, because the cation distributions in either state are equally accessible.

The thermodynamic consequences of this discussion of convergent and non-convergent ordering can be appreciated by studying Figure 3. In both panels of the figure we sketch the Gibbs free energy (G) as a function of ordering parameter (s), contoured for a range of temperatures (T). The arrows point to equilibrium values of G and s for each temperature contour. Panel (a) depicts the G-s-T relations for the case of convergent ordering, while panel (b) demonstrates the same relations for the case of non-convergent ordering. In the convergent case, the Gibbs free energy is a symmetric function of the order parameter,

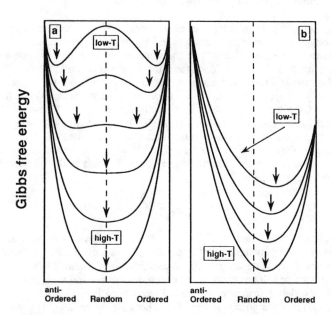

Figure 3. Sketch of the Gibbs energy of solution plotted as a function of the degree of order and temperature for the case of (a) convergent cation ordering, and (b) non-convergent cation ordering. Arrows indicate equilibrium values of the ordering state for each isotherm.

which renders the ordered and anti-ordered structures equally plausible in the low-temperature equilibrium state. At some intermediate temperature, the cation distribution eventually randomizes and the mineral undergoes a second order phase transition. At even higher temperatures there is one and only one equilibrium value of the ordering parameter (which corresponds to the structure with a random cation distribution) and the Gibbs free energy is a symmetric concave-up function about this value. In the case of non-convergent ordering, the Gibbs free energy of the anti-ordered structure is *always* more positive than that of the ordered structure, which results in an asymmetric function and the thermodynamic inaccessibility of the anti-ordered state. Equilibrium values of the ordering parameter are positive at all temperatures, and the perfectly random cation distribution is theoretically inaccessible at finite temperature.

Before we turn to the construction of thermodynamic models that account for convergent and non-convergent ordering in the oxides, the concepts of *long-range* and *short-range* order must be elucidated. Long-range order refers to a cation distribution that is periodic, characterizing the crystal as a whole. Long-range order parameters may be used to describe site occupancies in a probabilistic fashion much in the same way that mole fractions may be used to characterize the identity of cations in a particular unit cell. Short-range order refers to cation ordering that results from local interactions of nearest or next-nearest neighbor sites. The consequences of short-range ordering cannot be predicted without detailed knowledge of specific cation arrangements. Both long- and short-range ordering occur in the oxide minerals. As we will demonstrate, long-range ordering is relatively easy to incorporate into standard thermodynamic models for these phases, and fortunately is the dominant kind of ordering that occurs in the rock-forming oxides. The thermodynamic description of short-range order requires methods that rely on determination of cation occupancies through minimization of the energy of the lattice as a whole (see Burton and Kikuchi, 1984, and Chapter 8). Such methods rarely lend themselves to closed form solutions for the Gibbs energy of solution.

<u>Convergent ordering</u>. As an example for the construction of a thermodynamic model that involves long-range convergent cation ordering, we examine solid solutions along the ilmenite - hematite join . As a first step in constructing a model, we must choose an ordering parameter that describes adequately the cation distributions observed in these crystalline solutions. In Figure 4, we plot structural projections of the hematite-ilmenite lattice that identify the locations of the oxygen atoms (large gray circles) and the occupied octahedral sites (small black circles). In hematite (Fe_2O_3), Fe^{3+} atoms fill all of the octahedral positions, whereas in ilmenite ($FeTiO_3$), Fe^{2+} and Ti^{4+} are arranged in alternating layers perpendicular to the c-axis (Wechsler and Prewitt, 1984). At intermediate compositions, the distribution of Fe^{2+} and Ti^{4+} between adjacent layers is strongly temperature dependent. Labeling alternating octahedral sheets in the manner A..B..A..B, we define an ordering parameter:

$$s = X^A_{Fe^{2+}} - X^B_{Fe^{2+}} \qquad (2)$$

where $X^A_{Fe^{2+}}$ refers to mole fraction of Fe^{2+} in octahedral sites on the sheet designated A, and $X^B_{Fe^{2+}}$ refers to the equivalent quantity on the sheet designated B. For pure ilmenite, s will have the value 1 or -1, depending on our assignment of A and B labels. We adopt the convention that a value of one refers to the ordered structure and a value of negative one refers to the anti-ordered structure. We recall from our discussion of Figure 2 that the energies of the two configurations are equivalent because the ordering is convergent. With the definition of an ordering parameter, it becomes possible to write an expression for the configurational entropy of this solid solution. Let X denote the mole fraction of ilmenite in

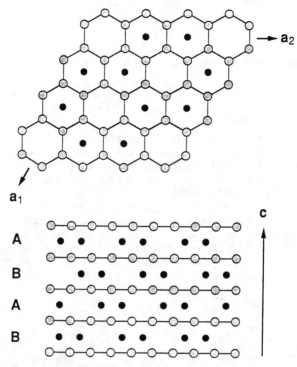

Figure 4. Structural projections of the rhombohedral oxide lattice. Gray circles are oxygen atoms, black circles denote occupied octahedral sites. The upper diagram is a basal (0001) projection, with a_1 and a_2 denoting crystallographic axes. The lower diagram is a projection parralel to the c crystallographic axis. Octahedral cation "sheets" may be viewed in this projection edge-on.

solution; it follows from Equation (2) that

$$X^A_{Fe^{3+}} = X^B_{Fe^{3+}} = 1 - X \tag{3a}$$

$$X^A_{Fe^{2+}} = X^B_{Ti^{4+}} = \frac{X + s}{2} \tag{3b}$$

$$X^B_{Fe^{2+}} = X^A_{Ti^{4+}} = \frac{X - s}{2} \tag{3c}$$

where Equation (3a) incorporates the observation (Ishikawa 1958a, b) that ferric iron exhibits minimal long-range order. The molar configurational entropy is simply

$$\overline{S}^{conf} = -R\,[\,X^A_{Fe^{3+}} \ln X^A_{Fe^{3+}} + X^B_{Fe^{3+}} \ln X^B_{Fe^{3+}} + X^A_{Fe^{2+}} \ln X^A_{Fe^{2+}}$$
$$+ X^B_{Fe^{2+}} \ln X^B_{Fe^{2+}} + X^A_{Ti^{4+}} \ln X^A_{Ti^{4+}} + X^B_{Ti^{4+}} \ln X^B_{Ti^{4+}}\,] \tag{4}$$

or alternatively,

$$\overline{S}^{conf} = -R\,[\,2(1-X)\ln(1-X) + (X+s)\ln\frac{X+s}{2} + (X-s)\ln\frac{X-s}{2}\,] \tag{5}$$

where R is the universal gas constant.

Before we proceed with the thermodynamic development, it is constructive to consider the effect of cation ordering on the functional form of the configurational entropy. To make this easier to visualize, let us define two related quantities: (1) an "ideal" entropy that arises solely from ideal mixing of cations on sites in the absence of ordering:

$$\overline{S}^{ideal} = -R\left[\,2\,(1-X)\ln(1-X) + 2X\ln\frac{X}{2}\,\right] \qquad (6)$$

and (2) an "excess" entropy arising purely from cation ordering:

$$\overline{S}^{excess} = \overline{S}^{conf} - \overline{S}^{ideal}$$

We plot $\overline{S}^{conf}$ and $\overline{S}^{excess}$ as a function of temperature and ilmenite mole fraction (X) in Figure 5. The calculations were made by computing the equilibrium state of order as a function of temperature and composition from the thermodynamic calibration of Ghiorso (1990a). Note that the high-temperature limit of the configurational entropy for pure ilmenite tends to $-2R\ln\frac{1}{2}$ (or $2R\ln 2$), and that the onset of ordering is marked by non-zero values of the "excess" entropy of mixing. There is one feature of the plot that generalizes to all crystalline solutions which exhibit cation ordering, namely, that the configurational entropy is strongly temperature dependent. Thermodynamic models that do not account for cation ordering (e.g., ones which utilize entropy formulations patterned after Eqn. 6 rather than Eqn. 5), must accommodate this "excess" entropy via temperature-dependent parameters and asymmetric higher-order Margules expansions. Because of the highly non-linear functional form of the configurational entropy, such approximations are

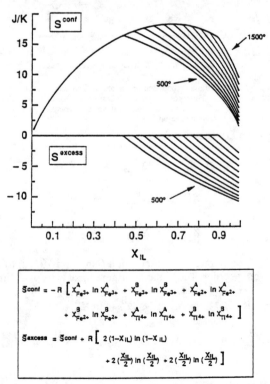

Figure 5. Molar configurational entropy ($\overline{S}^{conf}$) and molar excess entropy ($\overline{S}^{excess}$) for FeTiO$_3$ - Fe$_2$O$_3$ solutions, plotted as a function of tempetature (°C) and ilmenite content of the solid solution.

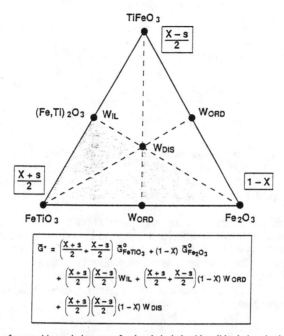

Figure 6. Vertices of composition-ordering space for rhombohedral oxide solid solutions having compositions in the binary system $FeTiO_3$-Fe_2O_3. "Mole fractions" of the species end-members are indicated by the boxed quantities associated with each vertex. These are expressed in terms of the adopted composition (X) and ordering (s) variables (see text). W_{IL}, W_{ORD}, and W_{DIS} refer to binary and ternary regular solution interaction parameters adopted to describe the enthalpy of mixing in composition-ordering space. The vibrational Gibbs energy (G^*) is the sum of the Gibbs energies of (1) the three vertices, (2) the three binary joins, and (3) the ternary interaction term. Shaded region indicates positive values of the order parameter.

rarely successful over a broad range of temperatures and compositions.

An easy way to formulate the non-configurational part of the Gibbs energy of solution is to enumerate energetic interactions between end-member species of a "composition-ordering" space (e.g., Thompson 1969, 1970). For the system under consideration the triangle displayed in Figure 6 forms a suitable representation. The vertices correspond to Fe_2O_3 and the ordered and anti-ordered end-members of ilmenite composition. The position along the ordered-anti-ordered ilmenite join denotes the extent of ordering, with the midpoint defining the hypothetical randomly ordered high-temperature structure. Coordinates within the interior of the triangle can be described in terms of "mole fractions" of the three end-member species. Formulae for computing these mole fractions are indicated on the figure as boxed quantities associated with each vertex. With reference to our composition-ordering space, the non-configurational Gibbs energy of solution (often referred to as the vibrational Gibbs energy of solution, or $\bar{G}^*$) can be modelled by adopting simple Margules parameterizations along constituent binary joins. As an example, in Figure 6 we express the excess Gibbs energy along each leg of the triangle using symmetric regular solution theory and add a "ternary" term to account for interactions between hematite and the randomly ordered ilmenite structure. A certain symmetry in interaction parameters exists because the anti-ordered and ordered structures of identical composition are energetically equivalent. The expression for $\bar{G}^*$ given in the figure is an inventory of the Gibbs energy associated with each vertex and along each join of composition-ordering space. We expand this expression and combine it with Equation (5), to generate our convergent-ordering model for the Gibbs free energy of solution along the ilmenite-hematite join:

$$\bar{G} = -T\bar{S}^{conf} + \bar{G}^*$$

$$\bar{G} = RT[2(1-X)\ln(1-X) + (X+s)\ln\frac{X+s}{2} + (X-s)\ln\frac{X-s}{2}]$$
$$+ X\bar{G}^o_{FeTiO_3} + (1-X)\bar{G}^o_{Fe_2O_3} + X(1-X)W_{ORD} + \frac{1}{4}(X^2-s^2)W_{IL}$$
$$+ (X^2-s^2)(1-X)W_{DIS} \tag{7}$$

Note that the Gibbs energy defined by Equation (7) is invariant with respect to change in sign of the order parameter. This is characteristic of all thermodynamic models meant to describe convergent ordering (e.g., Fig. 3). Equation (7) reduces to the familiar form for a two-site binary regular solution in the fully ordered case ($s = X$).

To conclude this example, let us derive an equation for the equilibrium state of order as a function of temperature and composition, examine a recent calibration of Equation (7) proposed by Ghiorso (1990a), and formulate an expression for the chemical potential of ilmenite in this solid solution series. Equilibrium values of s as a function of X and T are given by the condition of homogeneous equilibrium. This is obtained by setting the derivative, $(\partial G/\partial s)$, equal to zero, and solving the resulting implicit equation in s iteratively for specific values of X and T. From Equation (7) we obtain:

$$\left(\frac{\partial \bar{G}}{\partial s}\right) = 0 = RT\ln\frac{X+s}{X-s} - \frac{1}{2}sW_{IL} - 2s(1-X)W_{DIS} \tag{8}$$

The model parameters W_{IL} and W_{DIS} may be calibrated from experimental data on the dependence of cation ordering on composition and temperature. We illustrate Ghiorso's (1990a) calibration of Equation (7) to the data of Ishikawa (1958b) in Figure 7. The locus of T-X points that define the path of the second-order phase transition between the disordered and partially ordered structure is given by the expression:

$$\left(\frac{\partial^2 \bar{G}}{\partial s^2}\right)\bigg|_{s=0} = 0 = 2RT - \frac{1}{2}XW_{IL} - 2X(1-X)W_{DIS}$$

which demonstrates a quadratic dependence of T on X. In Figure 8 experimental data that define the T-X path of the second order phase transition are plotted (see legend for sources) and Ghiorso's (1990a) calibration of these data is indicated. Both sets of cation ordering data (e.g., Figs. 7 and 8) may be accommodated by adopting values of W_{IL} and W_{DIS} that are *constants*. Thus, there need be no excess entropy over and above that required by the configurational term (Eqn. 5). The remaining solution parameter in Equation (7) (W_{ORD}) must be calibrated from phase equilibrium data as it does not appear in the expression for homogeneous equilibrium. This brings out a general feature of all thermodynamic formulations for crystalline solutions which make provision for cation ordering. Experimental data on the state of order is not enough to completely characterize the thermodynamics of the phase. One needs additional data on (1) miscibility gaps in the system, (2) cation exchange between the phase of interest and other *well characterized* solid solutions, (3) activity-composition determinations by EMF, gas mixing or other means, or (4) calori-metric determinations of the enthalpy of mixing. If cation exchange data are utilized, then the added complication of establishing an internally consistent set of end-

231

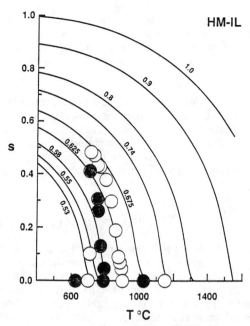

Figure 7. Ordering state of ilmenite-hematite solid solutions (s) plotted as a function of temperature (°C). The data indicated on the figure (plotted as open and solid circles) are taken from Ishikawa (1958b) and correspond to magnetic intensity measurements made on specimens with ilmenite mole fractions (X_{IL}) of 0.74, 0.675, 0.625, 0.58, 0.55 and 0.53. Calculated curves are plotted for these same mole fractions as well as X_{IL} equal to 0.80, 0.90 and 1.00. After Ghiorso (1990a).

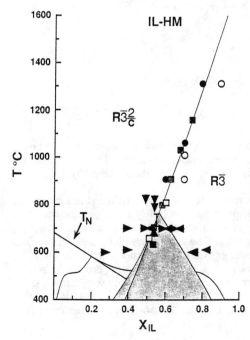

Figure 8. Phase relations plotted in terms of temperature (ordinate) and ilmenite mole fraction (abscissa) for solid solutions along the ilmenite-hematite join. Solid lines correspond to equilibrium phase boundaries between the ordered ($R\bar{3}$) and disordered ($R\bar{3}c$) structures and the region of one and two-phase stability (stipled). The darkly stipled region represents the calculated miscibility gap predicted from consideration of cation ordering alone (Ghiorso, 1990a). The extension and distortion of the miscibility gap due to magnetic ordering at low temperatures is taken from Burton and Davidson (1988) and is indicated by the lightly stipled region. The filled squares correspond to the position of the second-order phase transition as measured by Ishikawa (1958b). Circles refer to R3c - R3 reversals determined by Nord and Lawson (1989); filled circles denote the disordered structure and open circles the partially ordered structure. The open squares indicate the location of the second order transition determined by Hoffman (1975) while the "I-beam" symbol refers to a reversal bracket obtained by Burton (1982). Solid arrow-brackets denote the position of the miscibility gap as deduced by Lindsley (1973) and Burton (1984). Right-brackets represent approach to equilibrium from the right, with similar meaning attached to left, top and bottom brackets. After Ghiorso (1990a).

member, or standard state, properties (e.g., Berman 1988) must be addressed. In the system FeTiO$_3$ - Fe$_2$O$_3$ extensive information is available concerning the shape and extent of the high-temperature miscibility gap (see Fig. 8 and Chapter 3). Utilizing these data, Ghiorso (1990a) obtained a value of W$_{ORD}$ to generate the modelled gap displayed in Figure 8.

In order to derive activity-composition relations for end-member components in mineral solid solutions that exhibit cation ordering phenomena, we must utilized Darken's equation, which establishes a relationship between end-member chemical potentials and derivatives of the *molar* Gibbs energy of solution. As an example we formulate an expression for the chemical potential of ilmenite which is consistent with our model equation for the molar Gibbs energy of solution (Eqn. 7). The chemical potential of ilmenite (μ_{IL}) is given by:

$$\mu_{IL} = \bar{G} + (1 - X_{FeTiO_3})\left(\frac{\partial \bar{G}}{\partial X_{FeTiO_3}}\right)_{X_{TiFeO_3}/X_{Fe_2O_3}} \tag{9}$$

where the mole fractions refer to the end-member species of composition-ordering space as defined by the vertices of the triangle in Figure 6. The partial derivative in Equation (9) is evaluated along a pseudobinary designated by a constant mole fraction ratio of Fe$_2$O$_3$ and anti-ordered ilmenite. We could just as well have focused on the anti-ordered apex and defined the pseudobinary in terms of a constant ratio of Fe$_2$O$_3$ and ordered ilmenite. The result (μ_{IL}) would be the same because the molar Gibbs energy of the solid solution is symmetric with respect to the ordered and anti-ordered cation distributions. Evaluation of Equation (9) is complicated by the pseudobinary constraint on the partial derivative. Ghiorso (1990b) has shown that expressions of the form of Equation (9) can be reduced to equations involving simple partial derivatives with respect to composition and ordering variables. In particular, Equation (9) becomes

$$\mu_{IL} = \bar{G} + (1 - X)\left(\frac{\partial \bar{G}}{\partial X}\right)_s + (1 - s)\left(\frac{\partial \bar{G}}{\partial s}\right)_X$$

and substitution of Equation (7) results in:

$$\mu_{IL} = \bar{G}^o_{FeTiO_3} + RT \ln \frac{(X + s)^2}{4} + (1 - X)^2 W_{ORD}$$
$$+ \frac{(X - s)(2 - X - s)}{4} W_{IL} + 2(1 - X)(X - s)(1 - X - s) W_{DIS}$$

Note that the above expression is applicable to an *arbitrary* state of cation order. To obtain the chemical potential of ilmenite in its *equilibrium* ordered state, s must be determined at a particular T and X through the condition of homogeneous equilibrium (Eqn. 8).

Before leaving this example, we call attention to one feature of the solution that has general consequences. We note that the shape of the miscibility gap along the hematite-ilmenite join is strongly asymmetric (Fig. 8), yet the thermodynamic model we have formulated (e.g., Fig. 6) is completely symmetric. The asymmetry along the join develops as a consequence of projecting the symmetric gap defined in composition-ordering space (Fig. 6) onto a T-X section which specifies a state of equilibrium cation order. The general principle to be gleaned from this observation is that quite simple thermodynamic models involving temperature and pressure independent symmetric ("regular solution") formulations, can be be utilized to describe the thermodynamics of quite complex crystalline solutions, if the models are constructed utilizing end-member species which explicitly accommodate cation ordering. This generalization has been demonstrated for quite complex cases (e.g., Ghiorso 1990a, Sack and Ghiorso 1991a,b).

Non-convergent ordering. We take as an example of a thermodynamic model involving non-convergent ordering the temperature dependent distribution of Mg^{2+} and Al^{3+} over tetrahedral and octahedral sites in $MgAl_2O_4$ spinel. This example also illustrates the energetic importance of ordering phenomena to the thermodynamic properties of oxide end-members. In the spinels, cations occupy tetrahedral and octahedral sites which occur in the structure in a ratio of 1:2. At low temperatures, $MgAl_2O_4$ spinel has a *normal* cation distribution with Mg^{2+} confined largely to the tetrahedral sites and Al^{3+} occupying the octahedral sites. At elevated temperatures, the structure becomes more *random*. Several suitable choices for an ordering parameter exist. We follow Sack and Ghiorso (1991a) and define

$$s = X^{oct}_{Al^{3+}} - \frac{1}{2} X^{tet}_{Al^{3+}} \qquad (10)$$

s takes the value of unity for the *normal* cation distribution, one-third for the *random* distribution and zero for the hypothetical inverse cation distribution.

The configurational entropy of $MgAl_2O_4$ is given by:

$$\overline{S}^{conf} = -R [X^{tet}_{Mg^{2+}} \ln X^{tet}_{Mg^{2+}} + X^{tet}_{Al^{3+}} \ln X^{tet}_{Al^{3+}}$$
$$+ 2 X^{oct}_{Mg^{2+}} \ln X^{oct}_{Mg^{2+}} + 2 X^{oct}_{Al^{3+}} \ln X^{oct}_{Al^{3+}}] \qquad (11)$$

or in terms of s:

$$\overline{S}^{conf} = -R [s \ln s + (1-s) \ln (1-s) + (1-s) \ln \frac{1-s}{2} + (1+s) \ln \frac{1+s}{2}] \qquad (12)$$

The non-configurational Gibbs energy of solution can be modelled in terms of the end-member species $Mg[Al]_2O_4$ (ordered or *normal cation distribution*) and $Al[Mg_{1/2}Al_{1/2}]_2O_4$ (anti-ordered or *inverse cation distribution*), whose mole fractions may be written s and (1 − s) respectively. Analogous to our previous example we obtain

$$\overline{G}^* = s \, \overline{G}^*_{Mg[Al]_2O_4} + (1-s) \, \overline{G}^*_{Al[Mg_{1/2}Al_{1/2}]_2O_4} + s(1-s)W \qquad (13)$$

where W is a regular solution parameter describing interactions along the ordered-anti-ordered join. In Equation (12), $\overline{G}^*_{Mg[Al]_2O_4}$ is the Gibbs energy of the fully ordered end-member, and $\overline{G}^*_{Al[Mg_{1/2}Al_{1/2}]_2O_4}$ is the non-configurational part of the Gibbs energy of the hypothetical inverse structure.

Combining Equations (12) and (13) we obtain a model expression for the Gibbs energy of $MgAl_2O_4$,

$$\overline{G} = RT [s \ln s + (1-s) \ln (1-s) + (1-s) \ln \frac{1-s}{2} + (1+s) \ln \frac{1+s}{2}]$$
$$+ s \, \overline{G}^*_{Mg[Al]_2O_4} + (1-s) \, \overline{G}^*_{Al[Mg_{1/2}Al_{1/2}]_2O_4} + s(1-s)W \qquad (14)$$

The model parameters in Equation (14) may be calibrated from data on the variation of the order parameter as a function of temperature by deriving the condition of homogeneous equilibrium. Differentiating Equation (14) with respect to s yields:

$$0 = RT \ln \frac{s(1+s)}{(1-s)^2} + (\bar{G}^*_{Mg[Al]_2O_4} - \bar{G}^*_{Al[Mg_{1/2}Al_{1/2}]_2O_4}) + (1-2s)W \quad (15)$$

which is an implicit equation for s as a function of T. From our general discussion of non-convergent ordering that accompanied Figure 3, we anticipate that the difference in vibrational Gibbs energies of the ordered and anti-ordered structures (the term $\bar{G}^*_{Mg[Al]_2O_4} - \bar{G}^*_{Al[Mg_{1/2}Al_{1/2}]_2O_4}$) will be negative. Furthermore, size mismatch arguments (see above) would predict that the interaction parameter, W, would be positive. Sack and Ghiorso (1991b) obtained −18.41 kJ and 15.06 kJ respectively, in their calibration of the data displayed in Figure 9. It should be noted that at elevated temperature, the curve s(T) asymptotically approaches a constant value (1/3) that corresponds to the *random* cation distribution.

It is instructive to see how the temperature variation of the order parameter relates to macroscopic thermodynamic quantities like heat capacity, entropy and enthalpy. As an example, we will evaluate the variation in heat capacity with temperature for equilibrium ordered MgAl$_2$O$_4$ based upon our model calibration of Equation (14). The heat capacity is given by differentiating the Gibbs energy of solution,

$$\bar{C}_P = -T \left(\frac{\partial^2 \bar{G}}{\partial T^2}\right)_P = -T \left(\frac{\partial^2 \bar{G}}{\partial T^2}\right)_{P,s} - T \left(\frac{\partial^2 \bar{G}}{\partial T \partial s}\right)_P \left(\frac{\partial s}{\partial T}\right)_P$$

Substitution of Equation (14) yields

$$\bar{C}_P = \bar{C}^*_{P,Mg[Al]_2O_4} s - \bar{C}^*_{P,Al[Mg_{1/2}Al_{1/2}]_2O_4}(1-s)$$

$$-T\left[\bar{S}^*_{Al[Mg_{1/2}Al_{1/2}]_2O_4} - \bar{S}^*_{Mg[Al]_2O_4} + R \ln \frac{s(1+s)}{(1-s)^2}\right]\left(\frac{\partial s}{\partial T}\right)_P \quad (16)$$

where the derivative of the ordering parameter with respect to temperature may be evaluated by differentiating the condition of homogeneous equilibrium (Eqn. 15):

$$\left(\frac{\partial s}{\partial T}\right)_P = \frac{s(s^2-1)\left[\bar{S}^*_{Al[Mg_{1/2}Al_{1/2}]_2O_4} - \bar{S}^*_{Mg[Al]_2O_4} + R \ln \frac{s(1+s)}{(1-s)^2}\right]}{2s(s^2-1)W + RT(1+3s)} \quad (17)$$

Substitution of Equation (17) into Equation (16) permits evaluation of the heat capacity of the equilibrium ordered substance as a function of T, since s(T) is provided implicitly through Equation (15). Sack and Ghiorso (1991a,b) obtained a satisfactory calibration of Equation (14) on the assumption that the difference $\bar{C}^*_{Mg[Al]_2O_4} - \bar{C}^*_{Al[Mg_{1/2}Al_{1/2}]_2O_4}$ is constant. This assumption allows us to write the equilibrium ordered $\bar{C}_P$ as

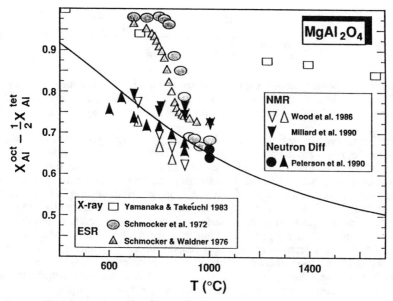

Figure 9. The ordering variable $X^{oct}_{Al^{3+}} - \frac{1}{2} X^{tet}_{Al^{3+}}$ for MgAl$_2$O$_4$, expressed as a function of temperature. Sources of data are indicated in the figure. Calibration is indicated by the solid curve. After Sack and Ghiorso (1991a).

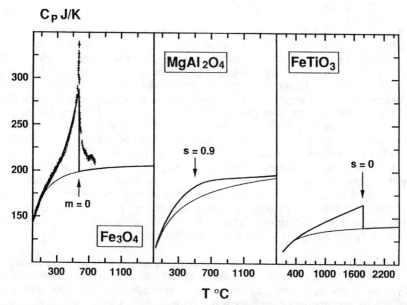

Figure 10. Comparison of heat capacity curves for three substances that show ordering phneomena. *Left panel* indicates heat capacity associated with the ferrimagnetic-paramagnetic transition in magnetite. Data are from Grønvold and Sveen (1974) and Westrum and Grønvold (1969). Heavy solid curve indicates the C$_P$ function calculated from Berman (1988). *Middle panel* indicates heat capacity associated with the non-convergent order-disorder transition in spinel. Heavy solid curve indicates the C$_P$ function calculated from Sack and Ghiorso (1991a). *Right panel* indicates heat capacity associated with the predicted convergent order-disorder transition in ilmenite. Heavy solid curve indicates the C$_P$ function calculated from Ghiorso (1990a). Thin solid curves on all three panels denote "vibrational" (non-configurational) contributions to the heat capacity (Berman, 1988).

$$\bar{C}_P^0 = \bar{C}_{P,Mg[Al]_2O_4}^* + \frac{R\,s\,(1-s^2)\left[\ln\frac{s(1+s)}{(1-s)^2}\right]^2}{2s(s^2-1)\frac{W}{RT} + (1+3s)} \quad (18)$$

where the superscript zero denotes a quantity in the standard state[2]. Equation (18) is evaluated in Figure 10. Model parameters for $MgAl_2O_4$ are from Sack and Ghiorso (1991b) and the non-configurational heat capacity is taken from Berman (1988).

The configurational contribution to the heat capacity of $MgAl_2O_4$ peaks at a value of ~20 J/K at temperatures around 500°C (Fig. 10). The configurational contribution is fairly symmetric about this maximum and does not fall to zero even at quite elevated temperatures. Most significantly, the maximum in configurational heat capacity is not associated with an extreme randomization of the cation distribution, and in fact occurs at ordering parameter values of about 0.9 (Fig. 9). As the configurational entropy and enthalpy are simply integrals of the heat capacity with respect to temperature, they too show their greatest change at low temperatures. Thus, the "thermodynamic response" to cation ordering in $MgAl_2O_4$ is premonitory to major redistribution of cations. This is a general feature of crystalline solutions that exhibit non-convergent ordering and is in direct contrast to the convergent case, where the heat capacity curve is strongly asymmetric and has a maximum at the temperature where the structure achieves a fully random cation distribution. As an example of the latter, we have plotted in Figure 10 the heat capacity of $FeTiO_3$ calculated from the calibration obtained in the previous example and the lattice C_P of Berman (1988).

<u>Magnetic contributions</u>

The oxide minerals exhibit a wide variety of magnetic phenomena that arise through the temperature dependence of the cooperative alignment of electron spins on transition metal ions in the crystal structures. We will refer collectively to the spin arrangements that give rise to these phenomena as "magnetic ordering". Although of energetic consequence in other groups of the rock-forming minerals, magnetic ordering is of particular significance in the oxides, because it persists in these phases to quite elevated temperatures. Above room temperature, four kinds of magnetic order are commonly found. Paramagnetic structures have randomly oriented electron spins and no magnetic order. Ferromagnetic structures posses an ordering derived from parallel alignment of spins between neighboring atoms, whereas antiferromagnetic and ferrimagnetic structures are ordered as a consequence of antiparallel alignment of spins. The magnetic ordering in antiferromagnets results in a zero net magnetic moment for the crystal because the spin moments of neighboring ions exactly cancel. In ferrimagnets, the spin moments do not balance exactly and the solid possess a net magnetic moment. Hematite and its solid solutions are antiferromagnets at temperatures as high as 675°C while the magnetite-rich spinels exhibit ferrimagnetic behavior at temperatures as high as 575°C.

As the ferromagnetic, antiferromagnetic or ferrimagnetic spin structure is transformed with increasing temperature to the completely random paramagnetic state, the solid gains entropy. The net entropy increase due to complete randomization of electron spins is easy to calculate, its temperature dependence is not (Gopal, 1966; Landau and Lifshitz, 1984). Let us take as an example hematite. Below 675°C, hematite is a pseudo-antiferromagnet

---

[2]In this chapter we define the standard state as unit activity of the pure substance in its equilibrium state of cation and magnetic order at any temperature and pressure.

(Dzyaloshinsky, 1958) with a spin arrangement that corresponds approximately to alternating (0001) sheets of octahedrally coordinated Fe^{3+} with spin moments oriented perpendicular to both the 2-fold and 3-fold axes, in a "positive" and "negative" direction on adjacent sheets. Given this structure, a magnetic order parameter may be defined in terms of the net magnetic moment of any individual octahedral sheet. If for example, all electron spins on all the Fe^{3+} ions in a particular sheet are oriented along the "positive" direction, we might take the magnetic order parameter to be unity. The antiparallel orientation of spins on an adjacent sheet would be described by an order parameter value of negative one, while the random orientation of spins characteristic of the paramagnetic state would correspond to a value of zero. Intermediate levels of spin alignment, corresponding to only partial attainment of complete antiferromagnetic order, are identified with absolute values of the magnetic order parameter in the inclusive range (0,1). It is clear that in the absence of an external magnetic field, the thermodynamic properties of antiferromagnetically ordered hematite must be symmetric with respect to change in sign of the magnetic order parameter. In this respect magnetic ordering resembles the convergent cation ordering discussed previously, and we should expect the Gibbs energy of a magnetic substance to follow the general form displayed in panel (a) of Figure 3. In particular, the transition from a magnetically ordered substance to a paramagnetic state should be manifested as a second-order phase transition.

Returning to hematite, let us investigate the origin of the configurational entropy associated with the disordering of a magnetic structure. In the completely ordered antiferromagnetic case there is one and only one electron ground state associated with each Fe^{3+} ion in the structure. This ground state corresponds to the spin-up or spin-down configuration with unpaired electrons (the so-called high-spin state). As there are five valence electrons in Fe^{3+} which occupy five d-orbitals, this configuration can be symbolized as ↑↑↑↑↑ or ↓↓↓↓↓, where the upward and downward pointing arrows denote the orientation of the spin moments. In the high-temperature paramagnetic structure, spin orientation is random, and for each Fe^{3+} ion the following electron spin configurations are possible:

↑↑↑↑↑   ↑↑↑↑↓   ↑↑↑↓↓   ↑↑↓↓↓   ↑↓↓↓↓   ↓↓↓↓↓

Note that spin-configurations are counted on the basis of the number of up or down moments without regard to which electron orbitals they occupy. The total magnetic configurational entropy is simply proportional to the logarithm of the ratio of configurations possible for the paramagnetic state over that possible for the magnetically ordered substance. In the case of hematite, with two Fe^{3+} ions per formulae unit, the total magnetic entropy is 2 R ln 6, or about 30 J/K-mol.

From this discussion of the origin of magnetic entropy, we can surmise that the magnetic order parameter is related to the probability of finding, on average, the electron spin moment of a Fe^{3+} atom in a random or non-random configuration. This configuration depends intimately on the nearest neighbor electron configurations of surrounding Fe^{3+} ions. This is a classic example of short-range order, where the value of the order parameter, or the probability of finding an atom in a particular state, depends on the local environment of the atom. Contrast this case to that of convergent cation ordering in hematite-ilmenite solid solutions. In the latter, the order parameter is related to the probability of finding a Fe^{2+}, Fe^{3+} or Ti^{4+} ion on a particular cation site in the structure, and is determined solely in terms of temperature, pressure and bulk composition, and not upon whether the octahedral sites surrounding the ion of interest are occupied by Fe^{2+}, Fe^{3+}, Ti^{4+}, or some combination thereof. Because of the short-range nature of magnetic ordering, it is not possible to write a closed form expression to describe the temperature dependence of magnetic entropy in a manner analogous to that of the configurational entropy of cation ordering. However, macroscopic models can be invoked which have been successful in elucidating the thermodynamics of magnetic phenomena. In particular, Dzyaloshinsky (1958, 1965) has had great success in application of Landau's theory (e.g., Landau and Lifshitz, 1980) of second order critical phenomena to magnetic phase transitions in hematite and chromite.

To complete our discussion of the thermodynamics of magnetic transitions we will develop a Landau-type model for the high-temperature antiferromagnetic-paramagnetic phase transition in hematite. Taking m to be the magnetic order parameter as defined in the previous paragraph, we write the Gibbs energy of hematite as

$$\bar{G}_{Hm} = \bar{G}_{Hm,Para} + A(T) m^2 + B m^4 \tag{19}$$

where, following Dzyaloshinsky (1958), we have truncated the Taylor expansion in m at the quartic term, and have taken the coefficient $A(T)$ to be temperature dependent. In Equation (19) B is a constant and $\bar{G}_{Hm,Para}$ is the Gibbs energy of paramagnetic Fe_2O_3 above the magnetic transition temperature, T_N. From the condition of homogeneous equilibrium,

$$\left(\frac{\partial \bar{G}_{Hm}}{\partial m}\right)_{T,P} = 0 = 2 A(T) m + 4 B m^3$$

we may readily solve for m,

$$m = \pm \sqrt{-\frac{A(T)}{2 B}} \qquad (T < T_N) \quad (20a)$$

$$m = 0 \qquad (T > T_N) \quad (20b)$$

and rewrite Equation (19) for Fe_2O_3 in its equilibrium state of magnetic order:

$$\bar{G}^o_{Hm} = \bar{G}_{Hm,Para} - \frac{[A(T)]^2}{4 B} \tag{21}$$

In most applications of Landau theory (e.g., Tolédano and Tolédano, 1987) the term $A(T)$ is modelled as a linear function of temperature, $\propto (T - T_N)$, which approximates the behavior of the more complex function in the temperature interval close to the critical point (T_N). This approximation leads to the expected quadratic temperature dependence of the magnetic contribution to the Gibbs energy near the critical point (Eqn. 21) and a square root dependence of m on T (Eqn. 20a) below T_N. Unfortunately, such a simple functional form for $A(T)$ cannot be used to reproduce measurements of magnetic heat capacity or magnetic susceptibility at temperatures even moderately removed from T_N. It is not difficult however, to choose a functional form for $A(T)$ that more closely approximates magnetic heat capacity measurements down to temperatures several hundred degrees below the magnetic phase transition. Berman and Brown (1985) have found that data on the high-temperature magnetic heat capacities of both hematite (Grønvold and Samuelsen, 1975; Grønvold and Westrum, 1959) and magnetite (Westrum and Grønvold, 1969; Grønvold and Sveen, 1974) can be modelled with a "λ"-type heat capacity function of the form:

$$C_{P,\lambda} = T(l_1 + l_2 T)^2 \tag{22}$$

where l_1 and l_2 are constants. An example of their fit to the data of Westrum and Grønvold (1969) and Grønvold and Sveen (1974) on the ferrimagnetic-paramagnetic transition in

Fe$_3$O$_4$ is shown in Figure 10. As an aside, the post-transition C$_P$ tail shown by the data plotted in Figure 10 is probably due to the preservation of magnetically ordered clusters in the paramagnetic phase. These decompose above the transition and contribute heat capacity. Berman and Brown (1985) accommodate this extra heat capacity by ascribing a "first order" enthalpy/entropy to the transition (ΔH_t and ΔS_t). This permits the integral thermodynamic properties of Fe$_3$O$_4$ (and also Fe$_2$O$_3$, which displays similar magnetic phenomena) to be accurately calculated with only minor errors in the vicinity of the transition. Adopting the formulation of Berman and Brown (1985) allows us to establish the temperature dependence of the second term in Equation (21) through the identity:

$$-\frac{[A(T)]^2}{4B} = \overline{G}_{Hm}^o - \overline{G}_{Hm,Para}$$

$$= -\int_{T_r}^{T}\int_{T_r}^{T'} \frac{C_{P,\lambda}}{T''} dT'' dT' - \left(-\int_{T_r}^{T_N}\int_{T_r}^{T'} \frac{C_{P,\lambda}}{T''} dT'' dT' + \Delta H_t - T\Delta S_t\right) \quad (23)$$

where T_r is a reference temperature (usually 298.15 K). Substitution of Equation (22) into Equation (23) yields, upon integration

$$-\frac{[A(T)]^2}{4B} = -(T_r - T)^2 [\frac{l_2^2}{2}(\frac{1}{2}T_r^2 + \frac{1}{3}TT_r + \frac{1}{6}T^2) + \frac{l_1 l_2}{3}(2T_r + T) + \frac{l_1^2}{2}]$$

$$+ (T_r - T_N)^2 [\frac{l_2^2}{2}(\frac{1}{2}T_r^2 + \frac{1}{3}T_N T_r + \frac{1}{6}T_N^2) + \frac{l_1 l_2}{3}(2T_r + T_N) + \frac{l_1^2}{2}]$$

$$- \Delta H_t + T\Delta S_t \quad (24)$$

Equation (24) permits us to stipulate the function A(T) in terms of the Landau coefficient B. The latter may be determined by raising Equation (20a) to the fourth power and substituting Equation (24):

$$m^4 = \frac{[A(T)]^2}{4B^2} = B(T_r - T)^2 [\frac{l_2^2}{2}(\frac{1}{2}T_r^2 + \frac{1}{3}TT_r + \frac{1}{6}T^2) + \frac{l_1 l_2}{3}(2T_r + T) + \frac{l_1^2}{2}]$$

$$- B(T_r - T_N)^2 [\frac{l_2^2}{2}(\frac{1}{2}T_r^2 + \frac{1}{3}T_N T_r + \frac{1}{6}T_N^2) + \frac{l_1 l_2}{3}(2T_r + T_N) + \frac{l_1^2}{2}]$$

$$+ B(\Delta H_t - T\Delta S_t) \quad (25)$$

Recognizing that at the reference temperature (298 K), hematite exhibits nearly complete magnetic order ($m \sim 1$), we obtain from Equation (25):

$$\frac{1}{B} = -(T_r - T_N)^2 \left[\frac{l_2^2}{2} \left(\frac{1}{2} T_r^2 + \frac{1}{3} T_N T_r + \frac{1}{6} T_N^2\right) + \frac{l_1 l_2}{3} (2 T_r + T_N) + \frac{l_1^2}{2} \right]$$

$$+ \Delta H_t - T_r \Delta S_t \qquad (26)$$

Equations (24) and (26) permit us to calculate the variation in equilibrium values of the magnetic order parameter with temperature. The temperature dependence of related quantities, like magnetic susceptibility may also be obtained (Tolédano and Tolédano, 1987).

Crystalline defects

The oxide minerals are known to exhibit extensive deviations from ideal stoichiometry. Certain examples come to mind immediately, like wüstite ($Fe_{1-x}O$) and its solid solutions (see Chapters 2 and 3). However, the phenomenon is of considerable importance to characterization of the thermodynamic properties of the other groups, particularly the spinels. The importance of the problem can be illuminated by focusing on a simple, well characterized example. Magnetite exhibits deviation from ideal Fe_3O_4 stoichiometry as a function of temperature and oxygen fugacity. It is an excellent example to consider because the relation between cation/oxygen ratio, T and $\log_{10} f_{O_2}$ has been carefully quantified by Dieckmann and Schmalzried (1977a,b). At lower oxygen fugacities, as along the breakdown curve of magnetite to wüstite, the phase has a slight cation excess. This may be interpreted as an excess in the ratio of ferrous/ferric iron over the nominal value. At higher oxygen fugacities, the phase becomes cation deficient. Along the hematite-magnetite buffer at temperatures in excess of 900°C, this deficiency reaches 1% (i.e., $Fe_{2.97}O_4$) and may be interpreted as an excess of ferric over ferrous iron.

Aragón and McCallister (1982) have utilized the results of Dieckmann and Schmalzried (1977a,b) to calibrate a point defect equilibrium model for the thermodynamics of magnetite in the T-$\log_{10} f_{O_2}$ range relevant to geologic processes. Their formulation is constructed on the assumptions that (1) point defects are cationic of the Frenkel type and are distributed randomly, (2) electroneutrality is preserved by changes in the Fe^{2+}/Fe^{3+} ratio, and (3) cations are statistically distributed between octahedral and tetrahedral sites. Aragón and McCallister (1982) write an equilibrium reaction between normal cations (C; Fe^{2+} or Fe^{3+} ions), cation vacancies (V) and cation interstitials (I; excess Fe) in the form of a Frenkel disorder reaction:

$$C \rightleftharpoons V + I \qquad (27)$$

which has an equilibrium constant K_1. On the assumption that defect concentrations are very small, the law of mass action for Equation (27) may be written as

$$K_1 = \left(\frac{n_V}{n_{Fe} + n_V}\right)\left(\frac{n_I}{n_{Fe} + n_V}\right) \qquad (28)$$

where n_V denotes the moles of vacancies, n_I the moles of interstitial cations, and n_{Fe} the total moles of Fe (including interstitials). An oxidation reaction, during which 3/4 of a cation site is created by addition of one oxygen atom, is described by:

$$Fe^{2+} + \frac{1}{4}O_2 = Fe^{3+} + \frac{1}{2}O^{2-} + \frac{3}{8}V \qquad (29)$$

to which Aragón and McCallister (1982) ascribe the equilibrium constant K_2, and write the law of mass action as:

$$K_2 = \frac{n_{Fe^{3+}}}{n_{Fe^{2+}}} f_{O_2}^{-1/4} \left(\frac{n_V}{n_{Fe} + n_V} \right) \qquad (30)$$

Imposing a constraint corresponding to electrical neutrality,

$$n_V - n_I = \frac{1}{8}(n_{Fe^{3+}} - 2 n_{Fe^{2+}}) , \qquad (31)$$

and defining $\quad \delta = \dfrac{n_V - n_I}{n_{Fe} + n_V} \qquad (32)$

for an arbitrary magnetite of formula unit $Fe_{3(1-\delta)}O_4$, Equations (28) and (30) may be solved simultaneously for the intrinsic f_{O_2} of magnetite of specified stoichiometry (δ):

$$f_{O_2}^{1/4} = \left[2 + \frac{24\, \delta}{1 - 8\, \delta - \dfrac{\delta}{2} - \sqrt{\dfrac{\delta^2}{4} + K_1}} \right] \frac{1}{K_2} \left[\frac{\delta}{2} + \frac{1}{2} \sqrt{\delta^2 + 4 K_1} \right]^{3/8} \qquad (33)$$

Aragón and McCallister's (1982) calibration of Equation (33) is reproduced in Figure 11 along with reference O_2 buffer curves determined by Myers and Eugster (1983). It can be seen that the intrinsic $\log_{10} f_{O_2}$ - T curve for stoichiometric Fe_3O_4 ($\delta = 0$) is steeper than that of the QFM buffer at all temperatures and crosses the QFM buffer curve at ~1000°C. Stoichiometric deviations become substantial at oxygen fugacities approaching that of the HM buffer, especially at higher temperatures.

While it might be argued that the effect of vacancies and interstitials on the activity composition relations of the stoichiometric component are of secondary consequence, it is evident that stoichiometric adjustments as large as those displayed by magnetite will have a profound effect on bulk thermodynamic properties of the end-member phases. As an

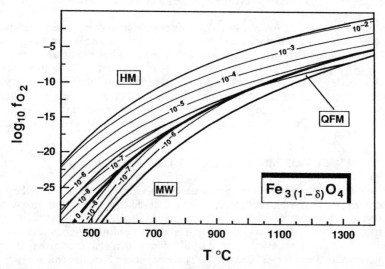

Figure 11. T- $\log_{10} f_{O_2}$ diagram indicating the stability field of magnetite with variable stoichiometry. The light curves are drawn for isopleths of δ ($Fe_{3-\delta}O_4$). The heavy curves labeled HM, QFM and MW refer to the hematite-magnetite, quartz-fayalite-magnetite and magnetite-wüstite oxygen buffers (Myers and Eugster, 1983), respectively. The T-$\log_{10} f_{O_2}$ relation for magnetite of ideal stoichiometry is indicated by the heavy curve labeled $\delta = 0$. The calculations are based upon the model of Aragón and McCallister (1982).

example, consider the common practice of utilizing measured $\log_{10} f_{O_2}$ - T relations for oxygen buffers to establish internally consistent thermodynamic data for end-member iron oxides. Let us say that we adopt thermodynamic properties ($\Delta H^°_{f,298}$, $S^°_{298}$, etc.) for fayalite and quartz and use data on the QFM buffer to constrain the enthalpy of formation of magnetite (e.g., Berman, 1988). Along QFM, magnetite is essentially stoichiometric (Fig. 11), and these phase relations can be consistently combined with other calorimetric data on the stoichiometric phase to yield internally consistent values with respect to fayalite and quartz. Now let us extend the analysis to extract a value of the enthalpy of formation of hematite from the known properties of magnetite and the T-$\log_{10} f_{O_2}$ relations of the HM buffer. On the assumption that magnetite remains stoichiometric, an internally consistent enthalpy of formation for hematite is obtained. Berman (1988) for example, retrieves a value of -825,627 kJ/mol. From our discussion of Figure 11 we see that there are at least two problems with this method of analysis: (1) the magnetite shows about a 1% deviation from ideal stoichiometry along its oxidation boundary, and (2) by analogy with magnetite, the hematite probably demonstrates a cation excess ($Fe_{2+\varepsilon}O_3$) along its reduction boundary. Thus, internally consistent thermodynamic properties for hematite obtained in this manner are suitable only for interpolating and possibly extrapolating the HM buffer. We can obtain a better estimate for the enthalpy of formation of hematite by examining Fe^{2+}-Ti partitioning data between hematite-ilmenite and magnetite-ulvöspinel solid solutions:

$$Fe_2O_3 + Fe_2TiO_4 = FeTiO_3 + Fe_3O_4 \qquad (34)$$
$$\text{rhm ox} \quad \text{spinel} \quad \text{rhm ox} \quad \text{spinel}$$

Utilizing data on the compositions of coexisting oxides, equilibrated in the range of f_{O_2}s where the spinel solutions remain essentially stoichiometric (Aragón and McCallister, 1982), Ghiorso (1990a) obtains estimates for the enthalpies of formation of both Fe_2O_3 and Fe_2TiO_4 that are internally consistent with Berman's (1988) values for Fe_3O_4 and $FeTiO_3$. His value for hematite is -822,000 kJ/mol, about 3.5 kJ/mol more positive than that derived by Berman (1988) from the HM buffer curve. This discrepancy presumably reflects the 1% stoichiometric deviation in magnetite along its oxidation breakdown curve.

In addition to the work of Aragón and McCallister (1982) on point defect equilibria in magnetite and magnetite-ulvöspinel solid solutions, Lehmann and Roux (1986) have developed thermodynamic models that include the effects of cation vacancies for spinels in the system $(Fe^{2+},Mg)(Al,Fe^{3+})_2O_4$. Much additional experimental work needs to be done before such complex treatments can be expanded to encompass the full range of spinel compositions found in nature. Of the other oxide groups, only the wüstites (in equilibrium with metallic iron) have been modelled with explicit provision for cation vacancies (Srečec et al., 1987). The whole field forms a fruitful and exciting area for future experimental and theoretical work.

SOLUTION MODELS FOR THE OXIDE MINERALS

In this section we summarize thermodynamic models for solid solutions of the cubic, rhombohedral and orthorhombic oxides. Attention is focused on models applicable to oxide compositions found in nature. Additional discussion of thermochemical models for the oxides, particularly solid solutions of the spinels along limiting binaries and activity-composition relations for magnesiowüstite in equilibrium with metallic iron, may be found in the chapter by Wood and others (Chapter 7). Thermodynamic formulations incorporating magnetic ordering are discussed by Burton in Chapter 8. Petrological applications of the models presented below are discussed by Sack and Ghiorso (Chapter 9), Frost and Lindsley (Chapter 12) and Frost (Chapter 13).

Cubic oxides: spinels

Thermodynamic models for multicomponent cubic spinels must at a minimum provide for (1) extensive miscibility gaps in many of the constituent binaries (Fig. 1), (2) the dependence of the energetics of cation ordering on composition, (3) the degree of long-range ordering of di- and trivalent cations between tetrahedral and octahedral sites, and (4) non-zero Gibbs energies of reciprocal reactions involving Fe^{2+} and Mg^{2+} end-member components. In addition, as the range of spinel compositions found in nature is extreme, a comprehensive, internally consistent formulation should encompass the system $(Fe^{2+}, Mg)(Al, Fe^{3+}, Cr)_2O_4 - (Fe^{2+}, Mg)_2TiO_4$. The construction of such a model poses a series of interesting challenges. The first attempt to devise a thermodynamic model that covers the composition space of natural spinels was the macroscopic approximation of Sack (1982). Sack (1982) utilized a two-site "Temkin" type model for the configurational entropy which he described in terms of mixing between "fictive" end-member spinel components with completely normal and inverse cation distributions. Fictive end-member components were chosen to correspond to the limiting low temperature cation distributions typically ascribed to actual unary spinels, and it was assumed that Fe^{2+} and Mg are completely disordered between tetrahedral and octahedral sites. To compensate for his neglect of the temperature and compositional dependence of cation ordering, Sack (1982) proposed an asymmetric regular solution model for the excess Gibbs energy of mixing. The resulting expression for the total Gibbs energy was calibrated based upon Fe^{2+}-Mg partitioning data and miscibility gap features. Although it reproduced most of the phase equilibrium constraints on which it was calibrated, the macroscopic approximation of Sack (1982), like all macroscopic approximations, was inadequate in several respects. Most severe was its failure to reproduce activity-composition relations for components in low concentrations. Equally troublesome were the large vibrational entropies of Fe^{2+}-Mg exchange and reciprocal reactions required to reproduce exchange data, and the need to incorporate third degree terms in the description of the vibrational Gibbs energy in order to describe the asymmetry in miscibility gaps. All of these shortcomings were manifestations of an incorrect description of the configurational entropy.

As alternatives, there have been several attempts to extend a microscopic formulation for unary spinels (Navrotsky and Kleppa, 1967) to binary and multicomponent spinels (e.g., O'Neill and Navrotsky, 1984; Lehmann and Roux, 1986; Hill and Sack, 1987; O'Neill and Wall, 1987). Notable among these has been the formulation, and its extension, of O'Neill and Navrotsky (1983, 1984). By careful analysis of cation-distribution data, they showed that the vibrational Gibbs energy of many unary spinels is at least a quadratic function of a long-range ordering variable. O'Neill and Navrotsky (1984) then attempted to extend these results to binary spinels, assuming that the vibrational Gibbs energy for such spinels is a second degree function of composition and ordering variables, but that the coefficients of cross-terms in composition-ordering variables are negligible. Because it does not deal with compositional effects on ordering, such an approach is inadequate for binary or multicomponent spinels.

We summarize here two recent formulations of activity-composition relations for natural cubic spinels: the model of Sack and Ghiorso (1991a,b) for spinels in the system $(Fe^{2+}, Mg)(Al, Cr, Fe^{3+})_2O_4 - (Fe^{2+}, Mg)_2TiO_4$ and the model of Andersen (1988; Andersen and Lindsley, 1988; Lindsley et al., 1990) for spinels in the system $(Fe^{2+}, Mg)(Fe^{3+})_2O_4 - (Fe^{2+}, Mg)_2TiO_4$. Additional models applicable to subsystems of the natural composition space are summarized in the chapter by Wood and others (see Chapter 7).

Cation ordering between normal and inverse structural states is accounted for in the Sack and Ghiorso (1991a,b) model by considering a solution of 12 end-member species (Fig. 12a), whose mole fractions can be computed in terms of four independent compositional variables and three internal ordering variables (s_1, s_2, and s_4). Modeling each join between vertices of composition-ordering space with temperature independent, symmetric regular solution parameters results in 66 coefficients. However, not all of these coefficients may be independently determined, because ternary (*sensu stricto*) interactions were ignored in Sack and Ghiorso's (1991a,b) formulation. The nature of these parameter dependencies may be illustrated (Fig. 12b) by focusing on the $(Fe^{2+},Mg)(Al,Cr)_2O_4$ quaternary system[3]. Solution parameters along the $MgAl_2O_4$-$FeAl_2O_4$ join represent Fe^{2+}-Mg interactions on tetrahedral sites. These parameters must also apply to the $MgCr_2O_4$-$FeCr_2O_4$ join. Similarly, those along the $FeCr_2O_4$-$FeAl_2O_4$ join must also apply to the $MgCr_2O_4$-$MgAl_2O_4$ join. If no strictly ternary interactions exist (top panel, Fig. 12b), thermodynamic potentials in the unshaded portion of the composition square are defined from parameters calibrated for the shaded region (this includes the pure end-member $MgCr_2O_4$!). This simplification would no longer occur if ternary interaction parameters were included in the model (lower panel): there is no particular reason why the ternary term in the $FeCr_2O_4$-$FeAl_2O_4$-$MgAl_2O_4$ system should be equivalent to that in the $FeCr_2O_4$-$MgCr_2O_4$-$MgAl_2O_4$ system ($W_{123} \neq W_{134}$). Similar arguments applied to the strictly symmetric, binary formulation for the supersystem (Sack and Ghiorso, 1991a,b), reduce the number of independent interaction parameters to 28 (8 independent species). In addition to these parameters, the thermodynamic potentials of 8 end-member species are required to calibrate the model. The model calibration of Sack and Ghiorso (1991a,b) is provided in Table 4. Expressions for calculating values of the internal ordering variables as a function of temperature and bulk composition, and equations for computing chemical potentials of end-member compositions are provided in Tables 5 and 6.

The formulation of Sack and Ghiorso (1991a,b) is consistent with (1) cation site-ordering constraints in $MgAl_2O_4$, $MgFe_2O_4$ and $FeCr_2O_4$-Fe_3O_4 spinels, (2) data defining miscibility gaps, activity-composition relations and heat of solution measurements in the supersystem, (3) distribution coefficients for the exchange of Al and Cr between cubic oxides and $(Al,Cr)_2O_3$ and $Na(Al,Cr)Si_2O_6$ solid solutions, (4) Fe^{2+}-Mg partitioning data between olivine and aluminate, ferrite, titanate, mixed aluminate-ferrite, and both synthetic and natural chromian spinels, (5) the model for rhombohedral oxides calibrated by Ghiorso (1990a), and (6) the end-member thermodynamic properties tabulated by Berman (1988). The model does not account for short-range order in the titanates or first-order phase transitions accompanying cubic-tetragonal transformations at low temperature. Nor does it correctly account for the excess entropy associated with the paramagnetic-ferrimagnetic phase transition in magnetite-rich solid solutions. The thermodynamic consequences of stoichiometric deviations due to development of high-temperature defects (e.g., Aragòn and McCallister 1982; Lehmann and Roux, 1986) have not been included in the model.

Andersen (1988) constructed two different models for the thermodynamics of spinels in the system $(Fe^{2+},Mg)(Fe^{3+})_2O_4$ - $(Fe^{2+},Mg)_2TiO_4$. The simpler of these, based upon a modified Akimoto-type formulation for the configurational entropy is recommended by Lindsley et al. (1990). Construction of the model is very straightforward. Two independent compositional variables (bulk mole fraction of Ti and Mg) are chosen and site mole fractions are defined uniquely in terms of these variables. Tetrahedral-octahedral cation ordering is not considered. The usual expression for the configurational entropy is

---

[3] Cation distributions refer to normal structures, i.e. Mg and Fe^{2+} are located on tetrahedral sites and Al and Cr are located on octahedral sites.

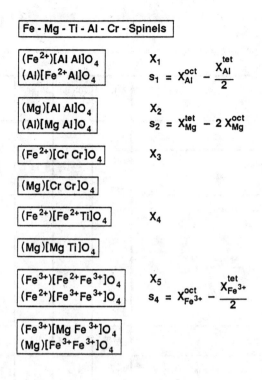

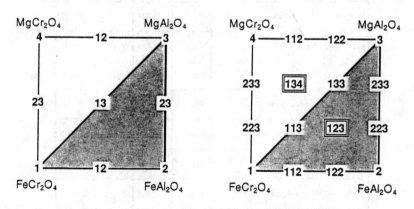

Figure 12. Details of the solution model for $(Fe^{2+},Mg)(Al,Cr,Fe^{3+})_2O_4$-$(Fe^{2+},Mg)_2TiO_4$ Fd3m spinels developed by Sack and Ghiorso (1991b). (a) Vertices of composition-ordering space. Note that the composition/ordering vertices define an 8- rather than 12-dimensional space of linearly-independent end-member species. Macroscopic variables are defined on the right. X_1 is a dependent variable. (b) Relations between the energetic parameters of dependent ($MgCr_2O_4$) and independent species in the $(Fe^{2+},Mg)(Al,Cr)_2O_4$ subsystem of $Fd3m$ spinels with normal cation ordering states. The left panel demonstrates that the thermodynamic properties of $MgCr_2O_4$ and solid solutions in the white triangle are completely defined from the end-member properties of the independent species and the regular solution interaction parameters along the 1-2, 2-3, and 1-3 joins. By contrast, the inclusion of ternary interaction parameters into the energetic model (right panel) breaks the energetic symmetry, as W_{123} need not necessarily be equal to W_{134}. After Ghiorso and Sack (1991).

Table 4. Values of model parameters for spinel solid solutions (kJ) from Sack and Ghiorso (1991a, b).

Hercynite	$\Delta \bar{H}^{*}_{1'1}$	−36.401	$\Delta \bar{S}^{*}_{1'1}$	0.000
	$W_{1'1}$	18.828		
Spinel	$W_{2'2}$	15.062		
Magnetite	$\Delta \bar{H}^{*}_{5'5}$	26.150	$\Delta \bar{S}^{*}_{5'5}$	0.000
	$W_{5'5}$	0.000		
Reciprocal Terms	$\Delta \bar{H}^{o}_{24}$	27.405	$\Delta \bar{S}^{o}_{24}$	0.000
	$\Delta \bar{H}^{o}_{25}$	33.681	$\Delta \bar{S}^{o}_{25}$	0.000
	$\Delta \bar{H}^{o}_{EX}$	−15.062	$\Delta \bar{S}^{o}_{EX}$	0.000
	$\Delta \bar{H}^{o}_{X}$	10.042	$\Delta \bar{S}^{o}_{X}$	0.000
Fe^{2+} - Mg join(s)	W^{OCT}_{FeMg}	8.368	W^{TET}_{FeMg}	8.368
Hercynite - Chromite join	$W_{13'}$	47.279	$W_{1'3'}$	24.686
Hercynite - Ulvospinel join	W_{14}	87.027	$W_{1'4}$	51.882
Hercynite - Magnetite join	W_{15}	41.840	$W_{15'}$	48.953
	$W_{1'5}$	60.250	$W_{1'5'}$	29.288
Spinel - Picrochromite join	$W_{23'}$	40.585		
Spinel - Magnesiotitanate join	W_{24}	52.718	$W_{2'\underline{4}}$	45.606
Spinel - Magnesioferrite	$W_{25'}$	64.015	$W_{2'\underline{5}}$	60.250
Chromite - Ulvospinel join	$W_{3'4}$	41.840		
Picrochromite - Magnesiotitanate join	$W_{\underline{3'4}}$	43.514		
Chromite - Magnetite join	$W_{3'5}$	41.840	$W_{3'5'}$	0.000
Ulvospinel - Magnetite join	W_{45}	25.104	$W_{45'}$	8.368
Magnesiotitanate-Magnesioferrite join	$W_{\underline{45'}}$	7.531		

Table 5. Conditions for homogeneous equilibrium in spinel solid solutions after Sack and Ghiorso (1991a, b).

$$\overline{\frac{\partial \overline{G}}{\partial s_1}} = R\,T \ln \left[\frac{(X_2 + s_1)\,(2 - X_2 - 2X_3 + s_1 - 2s_2 - 2s_4)}{(X_2 - s_1)\,(2X_3 + 2X_4 - X_2 - s_1 + 2s_2 + 2s_4)} \right] + \frac{1}{2}(\Delta \overline{G}_X^0 + \Delta \overline{G}_{EX}^0) + (\Delta \overline{G}_{24}^0 + W_{24} - W_{14} - W_{FeMg}^{OCT})$$

$$+ (\Delta \overline{G}_X^0 - W_{FeMg}^{TET} - W_{FeMg}^{OCT})\,s_1 + (W_{FeMg}^{TET} + W_{FeMg}^{OCT} + W_{1'1} - W_{22})\,s_2 + (W_{FeMg}^{TET} + W_{FeMg}^{OCT} - \Delta \overline{G}_X^0 + W_{13'} - W_{23'})\,X_3$$

$$+ W_{22} + W_{15} + W_{15'} - W_{22'5} - W_{25'})\,s_4 + (W_{FeMg}^{OCT} - W_{FeMg}^{TET})\,X_2 + \frac{1}{2}(\Delta \overline{G}_{EX}^0 - \Delta \overline{G}_X^0)\,X_4 + (W_{1'1} - W_{22} + W_{2'5} - W_{15})\,X_5$$

$$+ (W_{FeMg}^{TET} - W_{24} + W_{14} - \Delta \overline{G}_{24}^0)\,X_4 + \frac{1}{2}(\Delta \overline{G}_{EX}^0 - \Delta \overline{G}_X^0)\,X_4 + (W_{1'1} - W_{22} + W_{2'5} - W_{15})\,X_5$$

$$\overline{\frac{\partial \overline{G}}{\partial s_2}} = R\,T \ln \left[\frac{(1 - X_3 - X_4 - X_5 + s_2)\,(X_3 + X_4 - \frac{X_2}{2} - \frac{s_1}{2} + s_2 + s_4)}{(1 - X_3 - X_4 - X_5 - s_2)\,(1 - \frac{X_2}{2} - X_3 + \frac{s_1}{2} - s_2 - s_4)} \right] + (W_{1'1} + \Delta \overline{G}_{1'1}^*) + \frac{1}{2}(W_{1'1} - W_{22} + W_{FeMg}^{TET})$$

$$+ W_{FeMg}^{OCT} - \Delta \overline{G}_X^0)\,s_1 + (W_{15} + W_{15'} - W_{15} - W_{15'})\,s_4 + \frac{1}{2}(W_{FeMg}^{TET} - W_{FeMg}^{OCT} - \Delta \overline{G}_{EX}^0 + W_{1'1} - W_{22})\,X_2$$

$$- 2\,W_{1'1}\,s_2 + (\Delta \overline{G}_{24}^0 + W_{24} - W_{14})\,X_2 + (W_{13'} - W_{13'} - W_{1'1})\,X_3 + (W_{14} - W_{14} - W_{11})\,X_4$$

$$+ (W_{15} - W_{15} - W_{1'1})\,X_5$$

$$\overline{\frac{\partial \overline{G}}{\partial s_4}} = R\,T \ln \left[\frac{(X_5 + s_4)\,(2X_3 + 2X_4 - X_2 - s_1 + 2s_2 + 2s_4)}{(X_5 - s_4)\,(2 - X_2 - 2X_3 + s_1 - 2s_2 - 2s_4)} \right] + (\Delta \overline{G}_{55}^* + W_{15'} - W_{15}) + \frac{1}{2}(W_{FeMg}^{TET} + W_{FeMg}^{OCT} + W_{22})$$

$$- W_{1'1} - \Delta \overline{G}_X^0 + W_{15} + W_{15'} - W_{2'5} - W_{25'})\,s_1 + (W_{15} + W_{15'} - W_{15} - W_{15'})\,s_2 - 2\,W_{55}\,s_4$$

$$+ \frac{1}{2}(W_{FeMg}^{TET} - W_{FeMg}^{OCT} + W_{22} - W_{1'1} - \Delta \overline{G}_{EX}^0 + W_{15'} + W_{15} - W_{2'5} - W_{25'})\,X_2 + (\Delta \overline{G}_{24}^0 - \Delta \overline{G}_{25}^0 + W_{45'} - W_{45'})\,X_2$$

$$+ (W_{3'5'} - W_{3'5} + W_{15} - W_{15'})\,X_3 + (W_{15} + W_{45'} - W_{45} - W_{15'})\,X_4 + (W_{55} + W_{15} - W_{15'})\,X_5$$

Table 6. Chemical potentials of hercynite, spinel, chromite, ulvöspinel, magnetite, picrochromite, qandilite, and magnesioferrite in spinel solid solutions after Sack and Ghiorso (1991a, b).

$$\mu_{FeAl_2O_4} = \overset{o}{\mu}_{FeAl_2O_4} + RT \ln \left[\frac{(X_4 + s_2 + X_3 + s_4 - \frac{X_2 + s_1}{2})(1 - X_3 - X_4 - X_5 + s_2)^2}{s_2^0 (1 + s_2^0)^2} \right] + [\frac{1}{2}(W_{FeMg}^{TET} - W_{FeMg}^{OCT} + W_{11} - W_{22} - \Delta\bar{G}_{EX}^o) + W_{24} - W_{14} + \Delta\bar{G}_{24}^o] X_2 + (W_{13} - W_{13} - W_{11}) X_3$$

$$+ (W_{14} - W_{11}) X_4 + (W_{15} - W_{11}) X_5 + \frac{1}{2}(W_{FeMg}^{TET} + W_{FeMg}^{OCT} + W_{11} - W_{22} - \Delta\bar{G}_{EX}^o) s_1 - 2 W_{11} (s_2 - \overset{o}{s}_2) + (W_{15'} - W_{15} + W_{15'} - W_{15'}) s_4 + \frac{1}{4}(W_{FeMg}^{TET} + W_{FeMg}^{OCT} + \Delta\bar{G}_X^o) X_2^2$$

$$+ \frac{1}{2}(W_{23} - W_{13} + W_{FeMg}^{TET} - W_{FeMg}^{OCT} + \Delta\bar{G}_{EX}^o) X_2 X_3 + W_{34} - W_{24} - W_{14} - W_{FeMg}^{TET} - \Delta\bar{G}_{24}^o) X_2 X_3 + \frac{1}{2}(W_{15'} - W_{15} - W_{22}) - \Delta\bar{G}_{25}^o] X_2 X_5$$

$$+ \frac{1}{2}(W_{FeMg}^{TET} - W_{FeMg}^{OCT}) X_2 s_1 + \frac{1}{2}(W_{24} - W_{14} - W_{FeMg}^{TET} - \Delta\bar{G}_{24}^o) X_2 s_2 + \frac{1}{2}(W_{45'} - W_{45} + W_{25} + W_{11} - W_{22} - W_{15} + W_{FeMg}^{OCT} - W_{FeMg}^{TET} + \Delta\bar{G}_{EX}^o) X_2 s_4$$

$$+ \Delta\bar{G}_{25}^o - \Delta\bar{G}_{24}^o] X_2 s_4 + W_{13} X_3^2 + (W_{14} - W_{34} + W_{13'}) X_3 X_4 + (W_{15} - W_{35} + W_{13'}) X_3 X_5 + \frac{1}{2}(W_{23'} - W_{13} - W_{FeMg}^{TET} - W_{FeMg}^{OCT}) X_3 s_1 + (W_{13'} - W_{13} + W_{11}) X_3 s_2$$

$$+ (W_{15} - W_{15} + W_{35} - W_{35'}) X_3 s_4 + W_{14} X_4^2 + (W_{14} + W_{15} - W_{45}) X_4 X_5 + \frac{1}{2}(-W_{14} + W_{24} - W_{FeMg}^{TET} + \Delta\bar{G}_{24}^o - \frac{1}{4}(\Delta\bar{G}_X^o - \Delta\bar{G}_{EX}^o)) X_4 s_1 + (W_{14} - W_{14} + W_{11}) X_4 s_2 + (W_{15} - W_{15} + W_{45} - W_{45'}) X_4 s_4$$

$$+ W_{15} X_5^2 + \frac{1}{2}(W_{15'} - W_{15} - W_{11}) X_5 s_2 + (W_{15'} - W_{15} + W_{22} - W_{55'}) X_5 s_4 + (W_{15'} - W_{15} - \Delta\bar{G}_X^o) s_1^2 + \frac{1}{2}(W_{22} - W_{11} - W_{FeMg}^{TET} - W_{FeMg}^{OCT} - \Delta\bar{G}_X^o) s_1 s_2$$

$$+ \frac{1}{2}(W_{22} + W_{22} - W_{15} - W_{15} + W_{11} - W_{22} - W_{FeMg}^{TET} - W_{FeMg}^{OCT} + \Delta\bar{G}_{EX}^o) s_1 s_4 + W_{11} [s_2^2 - (\overset{o}{s}_2)^2] + (W_{15} - W_{15'} + W_{15} - W_{15}) s_2 s_4 + W_{55} s_4^2$$

Note below $\overset{o}{s}_1 = 2\overset{o}{s}_2 - 1$

$$\mu_{MgAl_2O_4} = \overset{o}{\mu}_{MgAl_2O_4} + RT \ln \left[\frac{(\frac{X_2 + s_1}{2})(1 - X_3 - X_4 - X_5 + s_2)^2}{s_2^0 (1 + s_2^0)^2} \right] - (s_2)^2 W_{22} + 2 s_2^0 W_{22} + [W_{14'} - W_{24} - \Delta\bar{G}_{24}^o + \frac{1}{2}(\Delta\bar{G}_X^o + \Delta\bar{G}_{EX}^o)] (1 - X_2 - X_3 - X_4 - X_5) + \frac{1}{2}(-W_{FeMg}^{TET} - W_{FeMg}^{OCT} + W_{14} + W_{11} - W_{22}) X_2$$

$$+ (W_{13'} + W_{24} - W_{34} - W_{23} + W_{FeMg}^{TET} + W_{14} - W_{24} - W_{23'} + W_{FeMg}^{OCT} + W_{11} - W_{22}) s_1 + [(W_{14'} - W_{24} - W_{11} - \Delta\bar{G}_{24}^o) + \frac{1}{2}(\Delta\bar{G}_X^o + \Delta\bar{G}_{EX}^o)] X_4 + [(W_{14} - W_{24} - W_{15} + W_{25} - W_{22} - \Delta\bar{G}_{25}^o - \Delta\bar{G}_{24}^o) + \frac{1}{2}(\Delta\bar{G}_X^o + \Delta\bar{G}_{EX}^o)] X_5$$

$$+ \frac{1}{2}(\Delta\bar{G}_X^o + \Delta\bar{G}_{EX}^o) s_4 + \frac{1}{4}(W_{FeMg}^{TET} + W_{FeMg}^{OCT} + \Delta\bar{G}_{EX}^o) X_2^2 + \frac{1}{2}(W_{23'} - W_{13'} + W_{FeMg}^{TET} - W_{FeMg}^{OCT} + \Delta\bar{G}_{EX}^o) X_2 X_3 + \frac{1}{2}(W_{24} - W_{14} - W_{FeMg}^{TET} - \Delta\bar{G}_{24}^o) X_2 X_4 - \frac{1}{2}(W_{45'} - W_{45} - \Delta\bar{G}_{24}^o) X_2 X_4$$

$$- \frac{1}{2}(\Delta\bar{G}_X^o + \Delta\bar{G}_{EX}^o) X_2 X_5 + \frac{1}{2}(W_{FeMg}^{TET} - W_{FeMg}^{OCT}) X_2 s_1 + \frac{1}{2}(W_{FeMg}^{TET} - W_{FeMg}^{OCT} - \Delta\bar{G}_{25}^o) X_2 X_5 + [(W_{FeMg}^{OCT} - W_{FeMg}^{TET} + W_{22} - W_{11} + \Delta\bar{G}_{EX}^o) - W_{24} + W_{14} - W_{FeMg}^{TET} - \Delta\bar{G}_{24}^o) + \frac{1}{4}(\Delta\bar{G}_{EX}^o - \Delta\bar{G}_X^o)] X_2 X_4$$

$$+ \frac{1}{2}(W_{15'} - W_{22} + W_{22} - W_{11}) - \Delta\bar{G}_{25}^o] X_2 X_5 + \frac{1}{2}(W_{FeMg}^{OCT} - W_{FeMg}^{TET}) X_2 s_1 + \frac{1}{2}(W_{FeMg}^{TET} - W_{FeMg}^{OCT} + W_{22} - W_{11} + \Delta\bar{G}_{EX}^o) X_2 s_2 + (W_{14} - W_{22} - W_{11} + \Delta\bar{G}_{EX}^o) X_2 s_4 + [W_{45} - W_{45'} + \Delta\bar{G}_{25}^o - \Delta\bar{G}_{24}^o$$

$$+ \tfrac{1}{2}(W_{\text{FeMg}}^{\text{OCT}} - W_{\text{FeMg}}^{\text{TET}} - W_{15} + W_{15'} + W_{22} + W_{25} + W_{11} - W_{22} + \Delta\bar{G}_{\text{EX}}^{0})]X_{2}4 + W_{13}^{2}X_{3}^{2} + (W_{14} - W_{34} + W_{13})X_{3}X_{4} + (W_{15} - W_{35} + W_{13})X_{3}X_{5} + \tfrac{1}{2}(W_{23} - W_{13} - W_{\text{FeMg}}^{\text{TET}} - W_{\text{FeMg}}^{\text{OCT}} + \Delta\bar{G}_{X}^{0})X_{3}s_{1}$$

$$+ (W_{13'} - W_{13} + W_{11})X_{3}s_{2} + (W_{15'} - W_{15} + W_{35'} - W_{35})X_{3}s_{4} + W_{14}^{2}X_{4}^{2} + \tfrac{1}{2}[-W_{14} + W_{24} - W_{\text{FeMg}}^{\text{TET}} + \Delta\bar{G}_{24}^{0}]X_{4}s_{1} + (W_{15} + W_{14} - W_{45})X_{4}X_{5} + \tfrac{1}{2}(\Delta\bar{G}_{X}^{0} - \Delta\bar{G}_{\text{EX}}^{0})X_{4}s_{2}$$

$$+ (W_{15} + W_{45} - W_{45'})X_{4}s_{4} + W_{15}^{2}X_{5}^{2} + \tfrac{1}{2}(W_{15} - W_{22} + W_{11})X_{5}s_{1} + (W_{15} - W_{15'} + W_{\text{FeMg}}^{\text{TET}} - W_{\text{FeMg}}^{\text{OCT}} - \Delta\bar{G}_{X}^{0})^{2}s_{1}^{2}$$

$$+ \tfrac{1}{2}(W_{22} - W_{11} - W_{\text{FeMg}}^{\text{OCT}} + \Delta\bar{G}_{X}^{0})s_{1}s_{2} + \tfrac{1}{2}(W_{25} + W_{22} - W_{15'} - W_{\text{FeMg}}^{\text{OCT}} + W_{11} + W_{11} - W_{22} + \Delta\bar{G}_{\text{EX}}^{0})s_{1}s_{4} + W_{11}^{2}s_{2}^{2} + (W_{15} + W_{15'} - W_{15})s_{2}s_{4} + W_{55}^{2}s_{4}^{2}$$

$$\mu_{\text{FeCr}_2\text{O}_4} = \mu_{\text{FeCr}_2\text{O}_4}^{0} + RT\ln\left[(X_{4} + s_{2} + X_{3} + s_{4} - \tfrac{X_{2} + s_{1}}{2})(X_{3})^{2}\right] + W_{13}(1 - X_{2} - X_{3} - X_{4} - X_{5}) + [W_{13'} + W_{24} - W_{34} + \Delta\bar{G}_{24}^{0} + \tfrac{1}{2}(W_{13} - W_{23} + W_{\text{FeMg}}^{\text{TET}} - W_{\text{FeMg}}^{\text{OCT}} - \Delta\bar{G}_{\text{EX}}^{0})]X_{2} - W_{13}X_{3} + (W_{34} - W_{14})X_{4}$$

$$+ (W_{35} - W_{15})X_{5} + \tfrac{1}{2}(W_{\text{FeMg}}^{\text{TET}} + W_{\text{FeMg}}^{\text{OCT}} + W_{13'} - W_{23} - \Delta\bar{G}_{X}^{0})s_{1} + (W_{13'} - W_{13} - W_{11})s_{2} + (W_{35'} - W_{35} + W_{15} - W_{15})s_{4} + \tfrac{1}{4}(W_{\text{FeMg}}^{\text{TET}} - W_{\text{FeMg}}^{\text{OCT}})X_{2}^{2} + [W_{34} - W_{34} - \Delta\bar{G}_{24}^{0}$$

$$+ \tfrac{1}{2}(W_{23} - W_{13} + W_{\text{FeMg}}^{\text{TET}} - W_{\text{FeMg}}^{\text{OCT}} + \Delta\bar{G}_{\text{EX}}^{0})]X_{2}X_{3} + [\tfrac{1}{2}(W_{24} - W_{14} - W_{\text{FeMg}}^{\text{TET}} + \Delta\bar{G}_{24}^{0})]X_{2}X_{4} + \tfrac{1}{2}(W_{15} - W_{22} + W_{22} - W_{11}) - \Delta\bar{G}_{25}^{0}]X_{2}X_{5} + \tfrac{1}{2}(W_{\text{FeMg}}^{\text{TET}} - W_{\text{FeMg}}^{\text{OCT}})X_{2}s_{1}$$

$$+ [W_{14} - W_{24} - \Delta\bar{G}_{24}^{0} + \tfrac{1}{2}(W_{22} - W_{11} - W_{\text{FeMg}}^{\text{TET}} + \Delta\bar{G}_{\text{EX}}^{0})]X_{2}s_{2} + [W_{45'} - W_{45} + \Delta\bar{G}_{25}^{0} - W_{\text{FeMg}}^{\text{TET}} - W_{\text{FeMg}}^{\text{OCT}} + \Delta\bar{G}_{\text{EX}}^{0})]X_{2}s_{4} + W_{14}X_{4}^{2}$$

$$+ (W_{34} + W_{13})X_{3}X_{4} + (W_{15} + W_{13} - W_{35})X_{3}X_{5} + \tfrac{1}{2}(W_{23} - W_{13} - W_{\text{FeMg}}^{\text{TET}} - W_{\text{FeMg}}^{\text{OCT}} + \Delta\bar{G}_{X}^{0})X_{3}s_{1} + (W_{13'} - W_{13} + W_{11})X_{3}s_{2} + (W_{15'} - W_{15} + W_{35'} - W_{35})X_{3}s_{4} + W_{14}X_{4}^{2}$$

$$+ (W_{14} + W_{15} - W_{45})X_{4}X_{5} + [\tfrac{1}{2}(W_{24} - W_{14} - W_{\text{FeMg}}^{\text{TET}} - W_{\text{FeMg}}^{\text{OCT}} - \Delta\bar{G}_{\text{EX}}^{0})]X_{4}s_{1} + (W_{14} - W_{14} + W_{11})X_{4}s_{2} + (W_{15'} - W_{15} + W_{45} - W_{45})X_{4}s_{4} + W_{15}X_{5}^{2} + \tfrac{1}{2}(W_{15} - W_{25} + W_{22} - W_{11})X_{5}s_{1}$$

$$+ (W_{15} - W_{15} + W_{11})X_{5}s_{2} + (W_{15'} - W_{15} - W_{22} - W_{\text{FeMg}}^{\text{TET}} - W_{\text{FeMg}}^{\text{OCT}} + \Delta\bar{G}_{X}^{0})s_{1}s_{2} + \tfrac{1}{2}(W_{15'} + W_{15} - W_{15})s_{1}s_{4} + W_{11}^{2}s_{2}^{2}$$

$$+ \tfrac{1}{2}(W_{25} + W_{22} - W_{15'} - W_{\text{FeMg}}^{\text{OCT}})s_{2}s_{4} + W_{55}^{2}s_{4}^{2}$$

$$\mu_{\text{Fe}_2\text{TiO}_4} = \mu_{\text{Fe}_2\text{TiO}_4}^{0} + RT\ln\left[(X_{4} + s_{2} + X_{3} + s_{4} - \tfrac{X_{2} + s_{1}}{2})(1 - s_{2} - X_{3} - s_{4} - \tfrac{X_{2} - s_{1}}{2})X_{4}\right] + W_{14}(1 - X_{2} - X_{3} - X_{4} - X_{5}) + (W_{14} + \tfrac{1}{2}[W_{14} - W_{24} + W_{\text{FeMg}}^{\text{TET}} + \Delta\bar{G}_{24}^{0} + \tfrac{1}{2}(\Delta\bar{G}_{X} - \Delta\bar{G}_{\text{EX}}^{0})])X_{2} + (W_{34} - W_{34} - W_{13})X_{3}$$

$$- W_{14}X_{4} + (W_{45} - W_{15})X_{5} + \tfrac{1}{2}[W_{14} - W_{24} + W_{\text{FeMg}}^{\text{TET}} - \Delta\bar{G}_{24}^{0} + \tfrac{1}{2}(\Delta\bar{G}_{\text{EX}} - \Delta\bar{G}_{X}^{0})]s_{1} + (W_{14} - W_{14} - W_{11})s_{2} + (W_{15'} - W_{15} + W_{45'} - W_{45})s_{4} + \tfrac{1}{4}(W_{\text{FeMg}}^{\text{TET}} + W_{\text{FeMg}}^{\text{OCT}} + \Delta\bar{G}_{X}^{0})X_{2}^{2} + [W_{34} - W_{14} - \Delta\bar{G}_{24}^{0}]$$

$$+ \tfrac{1}{2}(W_{23} - W_{13'} + W_{\text{FeMg}}^{\text{TET}} - W_{\text{FeMg}}^{\text{OCT}} - \Delta\bar{G}_{\text{EX}}^{0})]X_{2}X_{3} + \tfrac{1}{2}[W_{24} - W_{14} - W_{\text{FeMg}}^{\text{TET}} - \Delta\bar{G}_{24}^{0} - \tfrac{1}{2}(\Delta\bar{G}_{\text{EX}} - \Delta\bar{G}_{X}^{0})]X_{2}X_{4} + [\tfrac{1}{2}(W_{22} - W_{11} + W_{15} - W_{25}) - \Delta\bar{G}_{25}^{0}]X_{2}X_{5} + \tfrac{1}{2}(W_{\text{FeMg}}^{\text{TET}} - W_{\text{FeMg}}^{\text{OCT}})X_{2}s_{1}$$

$$+ [W_{14} - W_{24} + W_{\text{FeMg}}^{\text{TET}} + \Delta\bar{G}_{\text{EX}}^{0}]X_{2}s_{2} + [W_{45} - W_{45} + \Delta\bar{G}_{25}^{0} - \Delta\bar{G}_{24}^{0} + \tfrac{1}{2}(W_{11} - W_{22} + W_{25} + W_{22} - W_{15} - W_{15} + W_{\text{FeMg}}^{\text{TET}} + \Delta\bar{G}_{\text{EX}}^{0})]X_{2}s_{4}$$

$$+ W_{13} X_3^2 + (W_{13} + W_{14} - W_{34}) X_3 X_4 + (W_{13} + W_{15} - W_{35}) X_3 X_5 + \frac{1}{2}(W_{23} - W_{13} - W_{FeMg}^{TET} - W_{FeMg}^{OCT} + \Delta \bar{G}_X^0) X_3 s_1 + (W_{13'} + W_{11} - W_{13}) X_3 s_2 + (W_{15} - W_{15} + W_{35} - W_{35'}) X_3 s_4 + W_{14} X_4^2$$

$$+ (W_{14} + W_{15} - W_{45}) X_4 X_5 + \frac{1}{2}[W_{24} - W_{14} - W_{FeMg}^{TET} - \Delta \bar{G}_{24}^0 + \frac{1}{2}(\Delta \bar{G}_X^0 - \Delta \bar{G}_{EX}^0)] X_4 s_1 + (W_{11} + W_{14} - W_{14'}) X_4 s_2 + (W_{15} - W_{15} + W_{45} - W_{45'}) X_4 s_4 + W_{15} X_5^2 + \frac{1}{2}(W_{22} - W_{11} + W_{15} - W_{25}) X_5 s_1$$

$$+ (W_{11} + W_{15} - W_{15'}) X_5 s_2 + (W_{15} - W_{15} - W_{55}) X_5 s_4 + \frac{1}{4}(W_{FeMg}^{TET} + W_{FeMg}^{OCT} - \Delta \bar{G}_X^0)^2 s_1^2 + \frac{1}{2}(W_{22} - W_{11} - W_{22} + W_{25} + W_{25} - W_{15} - W_{FeMg}^{TET} - W_{FeMg}^{OCT} + \Delta \bar{G}_X^0) s_1 s_2$$

$$+ \Delta \bar{G}_X^0) s_1 s_4 + W_{11} s_2^2 + (W_{15'} - W_{15} + W_{15} - W_{15'}) s_2 s_4 + W_{55} s_4^2$$

$$\mu_{Fe_3O_4} = \mu_{Fe_3O_4}^o + RT \ln \left[\frac{(X_5 - s_4)(1 - s_2 - X_3 - s_4 - \frac{X_2 - s_1}{2})(X_5 + s_4)}{(1 - s_4)_4^o{}^2 (1 + s_4)_4^o} \right] + W_{15}(1 - X_2 - X_3 - X_4 - X_5)$$

$$+ \frac{1}{2}(W_{25} + W_{11} - W_{22} - W_{15}) s_1 + (W_{15} - W_{15} - W_{11}) s_2 + (W_{55} - W_{15} + W_{15}) s_4 + \frac{1}{4}(W_{FeMg}^{TET} + W_{FeMg}^{OCT} + \Delta \bar{G}_{EX}^0) X_2^2 + [\frac{1}{2}(W_{23'} - W_{13'} - W_{FeMg}^{TET} - W_{FeMg}^{OCT} + \Delta \bar{G}_{EX}^0) X_2 s_1$$

$$+ \frac{1}{2}[W_{24} - W_{14} - W_{FeMg}^{TET} - \Delta \bar{G}_{24}^0 + \frac{1}{2}(\Delta \bar{G}_{EX}^0 - \Delta \bar{G}_X^0)] X_2 X_4 + [\frac{1}{2}(W_{15} + W_{22} - W_{11} - W_{25}) - \Delta \bar{G}_{25}^0 + \frac{1}{2}(W_{FeMg}^{TET} - W_{FeMg}^{OCT} + \Delta \bar{G}_{EX}^0)] X_2 X_5 + [\frac{1}{2}(W_{FeMg}^{TET} + W_{FeMg}^{OCT} + \Delta \bar{G}_{EX}^0)] X_2 X_5$$

$$- \Delta \bar{G}_{24}^0] X_2 s_2 + \frac{1}{2}[W_{25} + W_{22} - W_{15} - W_{15} + W_{FeMg}^{TET} - W_{FeMg}^{OCT} + W_{11} - W_{22} + \Delta \bar{G}_{EX}^0) + W_{45} - W_{45'} + \Delta \bar{G}_{23}^0 - \Delta \bar{G}_{24}^0] X_2 s_4 + W_{13} X_3^2 + (W_{13} + W_{14} - W_{34}) X_3 X_4 + (W_{13} + W_{15} - W_{35}) X_3 X_5$$

$$+ \frac{1}{2}(W_{23'} - W_{13'} - W_{FeMg}^{TET} - W_{FeMg}^{OCT} + \Delta \bar{G}_X^0) X_3 s_1 + (W_{11} - W_{13} + W_{13}) X_3 s_2 + (W_{15} - W_{15} + W_{35} - W_{35'}) X_3 s_4 + W_{14} X_4^2 + (W_{14} + W_{15} - W_{45}) X_4 X_5 + \frac{1}{2}[W_{24} - W_{14} - W_{FeMg}^{TET} + \Delta \bar{G}_{24}^0$$

$$+ \frac{1}{2}(\Delta \bar{G}_X^0 - \Delta \bar{G}_{EX}^0)] X_4 s_1 + (W_{11} + W_{14} - W_{14}) X_4 s_2 + (W_{15} - W_{15} + W_{45} - W_{45'}) X_4 s_4 + W_{15} X_5^2 + \frac{1}{2}(W_{15} + W_{22} - W_{11} - W_{25}) X_5 s_1 + (W_{11} + W_{15} - W_{15'}) X_5 s_2 + (W_{15'} - W_{15} - W_{55}) X_5 s_4$$

$$+ \frac{1}{4}(W_{FeMg}^{TET} + W_{FeMg}^{OCT} - \Delta \bar{G}_X^0)^2 s_1^2 + \frac{1}{2}(W_{22} - W_{11} - W_{22'} + W_{25} + W_{25} - W_{15} - W_{FeMg}^{TET} - W_{FeMg}^{OCT} + \Delta \bar{G}_X^0) s_1 s_2 + W_{11} s_2^2 + (W_{15'} - W_{15} + W_{15} - W_{15'}) s_2 s_4$$

$$+ W_{55}[s_4^2 - (s_4^o)^2]$$

$$\mu_{MgCr_2O_4} = \mu_{FeCr_2O_4} + \mu_{MgAl_2O_4} - \mu_{FeAl_2O_4}$$

$$\mu_{Mg_2TiO_4} = \mu_{Fe_2TiO_4} + 2\mu_{MgAl_2O_4} - 2\mu_{FeAl_2O_4}$$

$$\mu_{MgFe_2O_4} = \mu_{Fe_3O_4} + \mu_{MgAl_2O_4} - \mu_{FeAl_2O_4}$$

combined with an asymmetric regular solution model for the excess Gibbs energy of solution. The regular solution parameters are allowed to be temperature dependent. The model is calibrated from (1) exchange data between spinels and rhombohedral oxides and spinels and (Fe^{2+},Mg)-olivines, and (2) miscibility gap features along the Fe$_3$O$_4$-Fe$_2$TiO$_4$ join. Like the model of Sack and Ghiorso (1991a,b) it does not account for magnetic phase transitions or defect equilibria.

In Figure 13 we plot activity-composition relations for the system Fe$_2$TiO$_4$ - Fe$_3$O$_4$ and

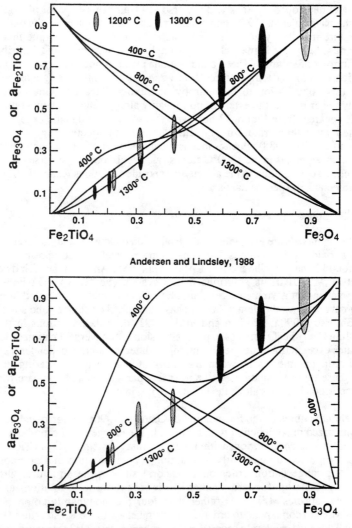

Figure 13. Calculated activity-composition relations for Fe$_2$TiO$_4$-Fe$_3$O$_4$ spinels at 400°, 800° and 1300°C compared with experimental determinations of Fe$_3$O$_4$ activity. *Ellipses* indicate the ranges of Fe$_3$O$_4$ activity consistent with the interpretation by Katsura et al. (1975) of the phase equilibrium data of Webster and Bright (1961) (*shaded*, 1200°C) and of Taylor (1964) (*solid*, 1300°C). In the upper panel, curves are calculated from the model of Sack and Ghiorso (1991a,b). In the lower panel, curves are calculated from the model of Andersen and Lindsley (1988).

compare the formulation of Sack and Ghiorso (1991a,b) to that of Andersen and Lindsley (1988). Results indicate general agreement between the two models at 800°C and noticeable disagreement at 400° and 1300°C. The Andersen and Lindsley (1988) formulation includes an excess entropy of solution that is responsible for the pronounced temperature dependence of their activity-composition relations. Adoption of this excess entropy is motivated by the need to account simultaneously for two-oxide Fe-Ti exchange experiments over the temperature range 600°-1000°C and the inferred extent of the low temperature miscibility gap along the binary (e.g., Price, 1981; Lindsley, 1981). The effect of this excess entropy renders their formulation inconsistent with the analysis of activity-composition relations at 1200° and 1300°C reported by Katsura et al. (1975). The thermodynamic model developed by Sack and Ghiorso (1991a,b) (upper panel of Fig. 13) does not violate the Katsura et al. (1975) constraints but also does not support the existence of a low-temperature miscibility gap (with $T_C > 450°C$). These authors argue that the low-temperature miscibility gap in magnetite-ulvöspinel arises from magnetic contributions to the excess entropy of solution. These contributions result in non-zero excess entropies of mixing which develop at temperatures *below the Curie point of magnetite* (~575°C). Neglecting this contribution to the entropy of mixing, Sack and Ghiorso (1991a,b) consequently under-predict the extent of non-ideality along the join at temperatures below 600°C. It is apparent from Figure 13 however, that when the magnetic excess entropy is approximated by a temperature-dependent Margules parameter (e.g., the model of Andersen and Lindsley, 1988) which is then applied at temperatures substantially above the paramagnetic transition (as in the 1300°C curves calculated from the Andersen and Lindsley (1988) model), its contribution to the activity-composition relations results in disagreement with high-temperature experimental data.

Rhombohedral oxides

Two solution models are available for the evaluation of activity-composition relations in rhombohedral oxide solid solutions. Both span approximately the compositional range of the oxides found in nature. The model of Andersen (1988; Andersen and Lindsley, 1988; Lindsley et al., 1990) treats solutions in the system $(Fe^{2+},Mg)TiO_3$-Fe_2O_3 using temperature-dependent asymmetric regular solution theory and is calibrated largely from partitioning data between the rhombohedral phase and spinel or olivine solid solutions. The model of Ghiorso (1990a; Ghiorso and Sack, 1991) describes solutions in the system $(Fe^{2+},Mg,Mn^{2+})TiO_3$-$Fe_2O_3$, makes explicit provision for convergent cation ordering, and accommodates excess Gibbs energies of mixing with temperature independent, symmetric regular solution parameters between the joins of composition-ordering space. We summarize these two formulations in the next few paragraphs and compare the derived activity-composition relations along the $FeTiO_3$-Fe_2O_3 join.

The essential features of the formulation of Ghiorso (1990a; Ghiorso and Sack, 1991) may be summarized in conjunction with Figure 14. The energetics of rhombohedral oxide solid solutions having compositions within the quaternary system Fe_2O_3-$FeTiO_3$-$MgTiO_3$-$MnTiO_3$, are modeled by explicitly accounting for temperature and compositionally dependent long-range cation order on octahedral sites in the structure. Ordering is accounted for in a straightforward manner by viewing these solid solutions as mixtures between seven "species" (Fig. 14) rather than four thermodynamic components. The species are chosen to correspond to ordered end-members with the $R\bar{3}$ structure ($Fe^{2+}TiO_3$, $MgTiO_3$, $MnTiO_3$) and randomly ordered end-members with the $R\bar{3}c$ structure. Mole fractions of each species are then defined in terms of three independent variables which denote the bulk mole fractions of ilmenite (X_{IL}), geikielite (X_{GK}) and pyrophanite (X_{PY}) and three independent internal variables (s,t, and u) which parameterize the ordering state at a

253

Table 7. Values of model parameters for rhombohedral oxide solid solutions (kJ) from Ghiorso (1990a).

Ilmenite	ΔH_{DIS}^{IL*}	17	ΔS_{DIS}^{IL*}	0
Geikielite	ΔH_{DIS}^{GK*}	17	ΔS_{DIS}^{GK*}	0
Pyrophanite	ΔH_{DIS}^{PY*}	17	ΔS_{DIS}^{PY*}	0
Ilmenite - Geikielite join	$W_{IL,GK}^{dis}$	25	$W_{IL,GK}^{ord}$	5
	$\Delta W_{IL,GK}$	12	$\Delta W_{GK,IL}$	8
Ilmenite - Pyrophanite join	$W_{IL,PY}^{dis}$	2.2	$W_{IL,PY}^{ord}$	2.2
	$\Delta W_{IL,PY}$	17	$\Delta W_{PY,IL}$	17
Ilmenite - Hematite join	$W_{IL,HM}^{dis}$	5	$\Delta W_{IL,HM}$	−19
Geikielite - Hematite join	$W_{GK,HM}^{dis}$	5	$\Delta W_{GK,HM}$	−19
Geikielite - Pyrophanite join	$W_{GK,PY}^{dis}$	25	$W_{GK,PY}^{ord}$	5
	$\Delta W_{GK,PY}$	8	$\Delta W_{PY,GK}$	12
Pyrophanite - Hematite join	$W_{PY,HM}^{dis}$	5	$\Delta W_{PY,HM}$	−19

Fe - Mg - Mn - Ti Rhombohedral Oxides

$(Fe^{3+})(Fe^{3+})O_3$ X_{HM}

$(Fe^{2+})(Ti)O_3$
$(Fe_{1/2}^{2+}Ti_{1/2})(Fe_{1/2}^{2+}Ti_{1/2})O_3$ X_{IL}
$s = X_{Fe^{2+}}^A - X_{Fe^{2+}}^B$

$(Mg)(Ti)O_3$
$(Mg_{1/2}Ti_{1/2})(Mg_{1/2}Ti_{1/2})O_3$ X_{GK}
$t = X_{Mg}^A - X_{Mg}^B$

$(Mn)(Ti)O_3$
$(Mn_{1/2}Ti_{1/2})(Mn_{1/2}Ti_{1/2})O_3$ X_{PY}
$u = X_{Mn}^A - X_{Mn}^B$

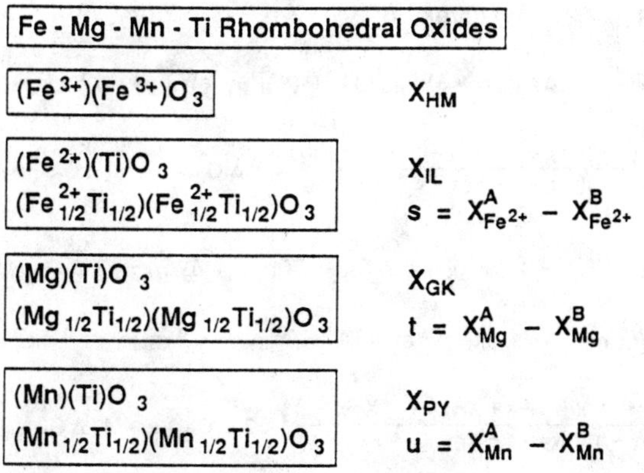

Figure 14. Vertices of composition-ordering space for rhombohedral oxide solid solutions having compositions in the quaternary system FeTiO₃-MgTiO₃-MnTiO₃-Fe₂O₃. Macroscopic variables are defined on the right. X_{HM} is a dependent variable. After Ghiorso and Sack (1991).

particular temperature and bulk composition. The excess Gibbs energy is constructed by treating mixing along each binary "species" join as a symmetric regular solution with temperature-independent coefficients. The 21 binaries in this system (i.e., 7 × 6/2) imply 21 regular solution parameters. Three of these, which correspond to interactions between ordered and disordered species of the same bulk composition, are set to zero. This leaves 18 solution parameters that must be calibrated in addition to the thermodynamic properties of the 7 end-member species. The results of Ghiorso's (1990; Ghiorso and Sack, 1991) calibration are reported in Table 7. Expressions for calculating equilibrium values of s, t and u as function of T and composition and for computing activity-composition relations for end-member compositions are provided in Tables 8 and 9.

Ghiorso's (1990; Ghiorso and Sack, 1991) solution theory for the rhombohedral oxides reproduces data on (1) the temperature dependence of ordering and the path of the second order $R\bar{3}$-$R\bar{3}c$ phase transition along the ilmenite-hematite join (e.g., Figs. 7 and 8), (2) the shape of the miscibility gap along the ilmenite-hematite join, (3) exchange equilibria between olivine and oxide solid solutions along the ilmenite-geikielite join, (4) experimental data on Fe^{2+}-Ti exchange between rhombohedral and cubic oxides, and (5) direct measurements of solution properties along the pyrophanite-ilmenite join. The model does not include the effects of the second-order paramagnetic-antiferromagnetic phase transition and is consequently inapplicable below ~550°C.

Andersen's (1988; Andersen and Lindsley, 1988; Lindsley et al. 1990) solution model for $FeTiO_3$-$MgTiO_3$-Fe_2O_3 oxides is constructed on the assumption that the phase remains fully ordered over the temperature-composition range of interest, and that the excess Gibbs energy of mixing may be expressed in terms of temperature-dependent asymmetric regular

Table 8. Conditions of homogeneous equilibrium for rhombohedral oxide solid solutions: $(Fe^{2+}, Fe^{3+}, Mg, Mn)_2O_3$ from Ghiorso (1990a).

$$\left(\frac{\partial \bar{G}}{\partial s}\right) = RT\frac{1}{2}\ln\frac{(X_{II}+s)(X_{II}+s+X_{Gk}+t+X_{PY}+u)}{(X_{IL}-s)(X_{II}-s+X_{Gk}-t+X_{PY}-u)} - 2\Delta G_{DIS}^{IL\,*} s - 4(\Delta G_{DIS}^{IL\,*}+\Delta W_{IL,HM})s$$

$$(1-X_{IL}-X_{GK}-X_{PY}) + 4(\Delta W_{IL,GK}-\Delta G_{DIS}^{IL\,*})sX_{GK} + 4(\Delta W_{IL,PY}-\Delta G_{DIS}^{IL\,*})sX_{PY}$$

$$+ (W_{IL,GK}^{ord}-W_{IL,GK}^{dis}-\Delta W_{IL,GK}-\Delta W_{GK,IL})t + (W_{IL,PY}^{ord}-W_{IL,PY}^{dis}-\Delta W_{IL,PY}-\Delta W_{PY,IL})u$$

$$\left(\frac{\partial \bar{G}}{\partial t}\right) = RT\frac{1}{2}\ln\frac{(X_{Gk}+t)(X_{II}+s+X_{Gk}+t+X_{PY}+u)}{(X_{Gk}-t)(X_{II}-s+X_{Gk}-t+X_{PY}-u)} - 2\Delta G_{DIS}^{GK\,*} t - 4(\Delta G_{DIS}^{GK\,*}+\Delta W_{GK,HM})t$$

$$(1-X_{IL}-X_{GK}-X_{PY}) + 4(\Delta W_{GK,IL}-\Delta G_{DIS}^{GK\,*})tX_{IL} + 4(\Delta W_{GK,PY}-\Delta G_{DIS}^{GK\,*})tX_{PY}$$

$$+ (W_{IL,GK}^{ord}-W_{IL,GK}^{dis}-\Delta W_{IL,GK}-\Delta W_{GK,IL})s + (W_{GK,PY}^{ord}-W_{GK,PY}^{dis}-\Delta W_{GK,PY}-\Delta W_{PY,GK})u$$

$$\left(\frac{\partial \bar{G}}{\partial u}\right) = RT\frac{1}{2}\ln\frac{(X_{PY}+u)(X_{II}+s+X_{Gk}+t+X_{PY}+u)}{(X_{PY}-u)(X_{II}-s+X_{Gk}-t+X_{PY}-u)} - 2\Delta G_{DIS}^{PY\,*} u - 4(\Delta G_{DIS}^{PY\,*}+\Delta W_{PY,HM})u$$

$$(1-X_{IL}-X_{GK}-X_{PY}) + 4(\Delta W_{PY,IL}-\Delta G_{DIS}^{PY\,*})uX_{IL} + 4(\Delta W_{PY,GK}-\Delta G_{DIS}^{PY\,*})uX_{GK}$$

$$+ (W_{IL,PY}^{ord}-W_{IL,PY}^{dis}-\Delta W_{IL,PY}-\Delta W_{PY,IL})s + (W_{GK,PY}^{ord}-W_{GK,PY}^{dis}-\Delta W_{GK,PY}-\Delta W_{PY,GK})t$$

Table 9. Chemical potentials of ilmenite, hematite, geikielite and pyrophanite in rhombohedral oxide solid solutions. From Ghiorso (1990a) and Ghiorso and Sack (1991).

$$\mu_{FeTiO_3} = \mu^\circ_{FeTiO_3} + RT\left[\ln(X_{IL} + s + t) + \ln(X_{IL} + s + t + X_{GK} + t + X_{PY} + u) - 2RT\ln(1 + s^\circ)\right] + \Delta G^{IL^*}_{DIS}(1 - s^2) - (1 - s^{\circ 2}) + \Delta G^{GK^*}_{DIS} t^2 + \Delta G^{PY^*}_{DIS} u^2 + W^{dia}_{IL,GK} X_{GK}(1 - X_{IL}) + W^{dia}_{IL,PY} X_{PY}(1 - X_{IL})$$
$$+ W^{dis}_{IL,HM}\left[((1 - X_{IL} - X_{GK} - X_{PY})(1 - X_{IL})\right] - W^{dis}_{GK,PY} X_{GK} X_{PY} - W^{dis}_{PY,HM} X_{PY}(1 - X_{IL} - X_{GK} - X_{PY}) - (\Delta W_{IL,HM} + \Delta G^{IL^*}_{DIS}) 4 s (1 - s) (1 - X_{IL} - X_{GK} - X_{PY})$$
$$+ (\Delta W_{GK,HM} + \Delta G^{GK^*}_{DIS}) 4 t^2 (1 - X_{IL} - X_{GK} - X_{PY}) + (\Delta W_{PY,HM} + \Delta G^{PY^*}_{DIS}) 4 u^2 (1 - X_{IL} - X_{GK} - X_{PY}) + (\Delta W_{IL,GK} - \Delta G^{IL^*}_{DIS}) X_{GK} + (\Delta W_{IL,PY} - \Delta G^{IL^*}_{DIS}) 4 s (1 - s) X_{PY}$$
$$+ (\Delta W_{GK,IL} - \Delta G^{GK^*}_{DIS}) 2 t^2 ((1 - X_{IL}) - X_{IL}] - \Delta W_{GK,HM} \Delta W_{PY,GK} + (W^{dia}_{IL,PY} - W^{dia}_{IL,GK} - \Delta W_{IL,GK} - \Delta W_{GK,IL})(1 - s) s$$
$$+ (W^{ord}_{IL,PY} - W^{ord}_{IL,PY} - \Delta W_{IL,PY} - \Delta W_{PY,IL})(1 - s) u - (W^{ord}_{GK,PY} - W^{dia}_{GK,PY} - \Delta W_{GK,PY} - \Delta W_{PY,GK}) t u$$

$$\mu_{Fe_2O_3} = \mu^\circ_{Fe_2O_3} + 2RT\ln(1 - X_{IL} - X_{GK} - X_{PY}) + \Delta G^{IL^*}_{DIS} s^2 + \Delta G^{GK^*}_{DIS} t^2 + \Delta G^{PY^*}_{DIS} u^2 - W^{dia}_{IL,GK} X_{IL} X_{GK} - W^{dia}_{IL,PY} X_{IL} X_{PY} + W^{dis}_{IL,HM} X_{IL}(X_{IL} + X_{GK} + X_{PY}) - W^{dis}_{GK,PY} X_{GK} X_{PY}$$
$$+ W^{dis}_{GK,HM} X_{GK}(X_{IL} + X_{GK} + X_{PY}) + W^{dis}_{PY,HM} X_{PY}(X_{IL} + X_{GK} + X_{PY}) + (\Delta W_{IL,HM} + \Delta G^{IL^*}_{DIS}) 2 s^2 ((1 - X_{IL} - X_{GK} - X_{PY})) + (\Delta W_{GK,HM} + \Delta G^{GK^*}_{DIS}) 4 t^2 X_{IL}$$
$$- (X_{IL} + X_{GK} + X_{PY})] + (\Delta W_{PY,HM} + \Delta G^{PY^*}_{DIS}) 2 u^2 (1 - X_{IL} - X_{GK} - X_{PY}) - (\Delta W_{IL,GK} - \Delta G^{IL^*}_{DIS}) 4 s^2 X_{GK} - (\Delta W_{GK,IL} - \Delta G^{GK^*}_{DIS}) 4 t^2 X_{IL}$$
$$- (\Delta W_{GK,PY} - \Delta G^{PY^*}_{DIS}) 4 t^2 X_{PY} - (\Delta W_{PY,IL} - \Delta G^{PY^*}_{DIS}) 4 u^2 X_{IL} - (\Delta W_{PY,GK} - \Delta G^{PY^*}_{DIS}) 4 u^2 X_{GK} - (W^{ord}_{IL,PY} - W^{dia}_{IL,PY} - \Delta W_{IL,PY} - \Delta W_{PY,IL}) s u$$
$$- (W^{ord}_{GK,PY} - W^{dia}_{GK,PY} - \Delta W_{GK,PY} - \Delta W_{PY,GK}) t u$$

$$\mu_{MgTiO_3} = \mu^\circ_{MgTiO_3} + RT\left[\ln(X_{GK} + t) + \ln(X_{IL} + s + X_{GK} + t + X_{PY} + u) - 2RT\ln(1 + t^\circ)\right] + \Delta G^{IL^*}_{DIS} s^2 + \Delta G^{GK^*}_{DIS}[(1 - t)^2 - (1 - t^{\circ})^2] - W^{dia}_{IL,GK} X_{GK}(1 - X_{GK}) X_{IL} + W^{dia}_{IL,PY} X_{IL} (1 - X_{PY})$$
$$- W^{dis}_{IL,HM} (1 - X_{IL} - X_{GK} - X_{PY}) X_{IL} + W^{dis}_{GK,PY} X_{GK} (1 - X_{PY}) + W^{dis}_{GK,HM} X_{GK}(1 - X_{GK})(1 - X_{IL} - X_{GK} - X_{PY}) + W^{dis}_{PY,HM} X_{PY}(1 - X_{IL} - X_{GK} - X_{PY}) + (\Delta W_{IL,HM} + \Delta G^{IL^*}_{DIS}) 4 s^2 (1 - X_{IL} - X_{GK} - X_{PY})$$
$$- (\Delta W_{GK,HM} + \Delta G^{GK^*}_{DIS}) 4 t (1 - t) (1 - X_{IL} - X_{GK} - X_{PY}) + (\Delta W_{PY,HM} + \Delta G^{PY^*}_{DIS}) 4 u^2 (1 - X_{IL} - X_{GK} - X_{PY}) + (\Delta W_{IL,GK} - \Delta G^{IL^*}_{DIS}) 2 s^2 ((1 - X_{GK}) - X_{GK}] - \Delta W_{IL,PY} - \Delta W_{PY,GK})$$
$$+ (\Delta W_{GK,IL} - \Delta G^{GK^*}_{DIS}) 4 t^2 X_{PY} + (\Delta W_{GK,PY} - \Delta G^{PY^*}_{DIS}) 4 t (1 - t) X_{PY} + (\Delta W_{PY,IL} - \Delta G^{PY^*}_{DIS}) 4 u^2 X_{IL} + (\Delta W_{PY,GK} - \Delta G^{PY^*}_{DIS}) 4 u (1 - u) X_{GK} - (W^{dia}_{IL,PY} - W^{dia}_{IL,PY} - \Delta W_{IL,GK} - \Delta W_{GK,IL}) s (1 - t)$$
$$+ (W^{ord}_{IL,PY} - W^{dia}_{IL,PY} - \Delta W_{IL,PY} - \Delta W_{PY,IL}) s u + (W^{ord}_{GK,PY} - W^{dia}_{GK,PY} - \Delta W_{GK,PY} - \Delta W_{PY,GK})(1 - t) u$$

$$\mu_{MnTiO_3} = \mu^\circ_{MnTiO_3} + RT\left[\ln(X_{PY} + u) + \ln(X_{IL} + s + X_{GK} + t + X_{PY} + u) - 2RT\ln(1 + u^\circ)\right] + \Delta G^{IL^*}_{DIS} s^2 + \Delta G^{GK^*}_{DIS} t^2 + \Delta G^{PY^*}_{DIS}[(1 - u)^2 - (1 - u^\circ)^2] - W^{dia}_{IL,GK} X_{GK} X_{IL} + W^{dia}_{IL,PY} X_{IL} (1 - X_{PY})$$
$$- W^{dis}_{IL,HM} (1 - X_{IL} - X_{GK} - X_{PY}) X_{IL} + W^{dis}_{GK,PY} X_{GK} (1 - X_{PY}) + W^{dis}_{GK,HM} X_{GK}(1 - X_{IL} - X_{PY}) - (\Delta W_{PY,HM} + \Delta G^{PY^*}_{DIS}) 4 u(1 - u)(1 - X_{IL} - X_{GK} - X_{PY}) + W^{dis}_{PY,HM}(1 - X_{PY})(1 - X_{IL} - X_{GK} - X_{PY}) + (\Delta W_{IL,HM} + \Delta G^{IL^*}_{DIS}) 4 s^2 (1 - X_{IL} - X_{GK} - X_{PY})$$
$$- (\Delta W_{GK,HM} + \Delta G^{GK^*}_{DIS}) 4 t^2 (1 - X_{IL} - X_{GK} - X_{PY}) - \Delta W_{IL,GK} 4 s^2 X_{GK} + (\Delta W_{IL,PY} - \Delta G^{PY^*}_{DIS}) 2 s^2 [(1 - X_{PY}) - X_{PY}]$$
$$+ (\Delta W_{GK,IL} - \Delta G^{GK^*}_{DIS}) 4 t^2 X_{IL} + (\Delta W_{GK,PY} - \Delta G^{PY^*}_{DIS}) 2 t^2 [(1 - X_{PY}) - X_{PY}] + (\Delta W_{PY,IL} - \Delta G^{PY^*}_{DIS}) 4 u (1 - u) X_{IL} + (\Delta W_{PY,GK} - \Delta G^{PY^*}_{DIS}) 4 u (1 - u) X_{GK} - (W^{dia}_{IL,GK} - W^{dia}_{IL,GK} - \Delta W_{IL,GK} - \Delta W_{GK,IL}) s t$$
$$+ (W^{ord}_{IL,PY} - W^{dia}_{IL,PY} - \Delta W_{IL,PY} - \Delta W_{PY,IL}) s (1 - u) + (W^{ord}_{GK,PY} - W^{dia}_{GK,PY} - \Delta W_{GK,PY} - \Delta W_{PY,GK}) t (1 - u)$$

solution parameters. Their models are consistent with (1) the shape of the miscibility gap along the ilmenite-hematite join, (2) exchange equilibria between olivine and oxide solid solutions along the ilmenite-geikielite join, and (3) experimental data on Fe^{2+}-Ti and Fe^{2+}-Mg exchange between rhombohedral and cubic oxides. Like the Ghiorso (1990a) model the formulation of Andersen and Lindsley (1988) suffers by not accounting for magnetic ordering at temperatures below ~550°C.

A comparison of calculated activity-composition relations for $FeTiO_3$-Fe_2O_3 solid solutions is shown in Figure 15. The left-most panels correspond to the calibration of Ghiorso (1990a). Calculated activity-composition relations from Andersen and Lindsley (1988) are shown on the right panels. The two formulations agree quite nicely at 800°C, but differ markedly at both lower and higher temperatures. For compositions of practical interest ($X_{IL} > 0.9$), the calculated activity of ilmenite differs insignificantly between the two formulations. The real discrepancy arises in the temperature dependence of the activity of Fe_2O_3 in ilmenite-rich solid solutions. The extreme variation in temperature displayed by the activity-composition relations of Andersen and Lindsley (1988) is due to a large excess entropy Margules parameter which presumably functions to proxy for ordering state variations across the join.

<u>Orthorhombic oxides</u>

Brown and Navrotsky (1989) have developed a thermodynamic model for non-convergent cation ordering in the pseudobrookite-type compound karroite ($MgTi_2O_5$). They demonstrate that in the low temperature ordered structure, Mg^{2+} occupies the M1 octahedral sites and Ti^{4+} the M2 sites (there are twice as many M2 and M1 sites in the structure, see Chapter 2). At elevated temperatures, the distribution becomes more random. Solid solutions between karroite and ferro-pseudobrookite ($FeTi_2O_5$) behave similarly, with Fe^{2+} and Mg^{2+} occupying M1 sites at low temperatures tending towards a random distribution over M1 and M2 at elevated T (Wechsler, 1977). Natural examples of these solid solutions are found on the moon (armalcolite). Under extremely oxidizing conditions in terrestrial igneous and metamorphic rocks, ternary solutions of pseudobrookite (Fe_2TiO_5) and the Fe^{2+} and Mg^{2+} end-members may be found. The thermodynamics of cation ordering are somewhat complicated for these compounds in that the low-temperature cation distribution of Fe_2TiO_5 most likely has Fe^{3+} on the M1 and both Fe^{3+} and Ti^{4+} on the M2 sites (Chapter 2).

Although thermodynamic models for the Fe_2TiO_5-$(Fe^{2+},Mg)_2TiO_5$ orthorhombic oxides have not been calibrated, in principle their formulation follows directly from the concepts and examples proposed above for the cubic spinels and rhombohedral phases. We propose the following formulation. In the system Fe_2TiO_5-$(Fe^{2+},Mg)_2TiO_5$ there are six end-members of composition ordering space:

$(Fe^{3+})^{M1}(Fe^{3+},Ti)_2^{M2}O_5$ $(Ti)^{M1}(Fe^{2+},Ti)_2^{M2}O_5$ $(Ti)^{M1}(Mg,Ti)_2^{M2}O_5$

$(Ti)^{M1}(Fe^{3+})_2^{M2}O_5$ $(Fe^{2+})^{M1}(Ti)_2^{M2}O_5$ $(Mg)^{M1}(Ti)_2^{M2}O_5$

which may be described in terms of two independent compositional variables:

$$x = X_{Fe^{2+},total} \text{ and } y = X_{Mg,total}$$

and three independent ordering variables:

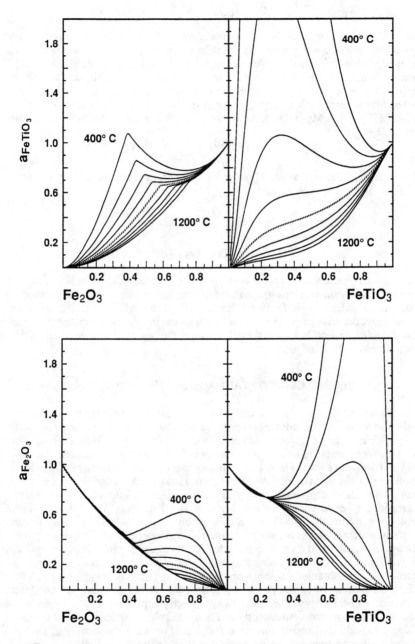

Figure 15. Calculated activity-composition relations for $FeTiO_3$-Fe_2O_3 rhombohedral oxides at 400° to 1200°C in 100° increments. The 800°C isotherm is indicated by the dashed curves. The upper and lower left panels are from Figure 12 of Ghiorso (1990a), while the right panels are calculated from the model of Andersen and Lindsley (1988).

$$s = \frac{3}{2}X^{M1}_{Fe^{3+}} - X^{M2}_{Fe^{3+}}, \quad t = X^{M1}_{Fe^{2+}} - X^{M2}_{Fe^{2+}} \quad \text{and} \quad u = X^{M1}_{Mg} - X^{M2}_{Mg}$$

Site mole fractions may be derived from these five definitions, and the configurational entropy of solution is consequently expressed in terms of x, y, s, t, and u. The vibrational Gibbs energy of solution will be formed as a 2nd-degree Taylor expansion in all five independent variables, which results in 21 terms. As usual the Taylor coefficients are of four types. Three of the 21 parameters are standard state contributions from the fully ordered end-member of each ternary apex (i.e., the low-temperature cation ordering state of Fe_2TiO_5, $FeTi_2O_5$ and $MgTi_2O_5$). Another three are exchange reactions between ordered and anti-ordered pairs:

$$(Fe^{3+})^{M1}(Fe^{3+},Ti)^{M2}_2O_5 = (Ti)^{M1}(Fe^{3+})^{M2}_2O_5$$

one for each end-member apex. Three more arise from reciprocal ordering reactions between two of the three apices, i.e.,

$$(Ti)^{M1}(Fe^{2+},Ti)^{M2}_2O_5 + (Mg)^{M1}(Ti)^{M2}_2O_5 =$$

$$(Ti)^{M1}(Mg,Ti)^{M2}_2O_5 + (Fe^{2+})^{M1}(Ti)^{M2}_2O_5$$

The remaining 12 parameters would correspond to regular solution-type interaction terms (Ws) accounting for interactions of the four different cations on both the M1 and M2 sites (4 × 3/2, taken twice). It is likely that in the calibration of this model, a number of equivalences would arise between the intra-site Ws reducing the total number of parameters to a manageable number. We leave the calibration to the interested reader armed with more data than we have at our disposal.

RECOMMENDED STANDARD STATE PROPERTIES

We recommend the use of an internally consistent set of thermochemical properties for the end-member oxides in order that equilibria may be evaluated between them and the widest possible range of silicate phases. For these purposes the data base of Berman (1988) is most applicable in that his formulation incorporates the heat capacity functions of Berman and Brown (1985). These heat capacity expressions are constructed to facilitate extrapolation of the derived constants to igneous and mantle conditions (i.e., high T). Berman (1988) also incorporates magnetic phase transitions into his expressions for the high-temperature heat capacities of both hematite and the ferrite spinels. For convenience we reproduce Berman's (1988) oxide end-member properties in Tables 10 and 11. Berman's algorithms for computing apparent Gibbs energies of formation[4] from the elements at elevated T and P are also provided. To the derived values reported by Berman (1988) we have added internally consistent values for a number of cubic and rhombohedral oxides. These are obtained in the work of Ghiorso (1990a) and Sack and Ghiorso (1991a,b) via analysis of exchange equilibria between "Berman calibrated" and "Berman uncalibrated" end-members. These values are directly dependent on the solution theory adopted by the authors and consequently on the myriad phase equilibrium, cation ordering, miscibility gap and calorimetric constraints considered by them in constructing the models. These end-member properties should only be used in conjunction with their formulations for activity composition relations of cubic spinel and rhombohedral oxide solid solutions.

---

[4]An apparent Gibbs energy of formation is the difference of the Gibbs energy of the mineral *evaluated at the temperature and pressure of interest* and the sum of the Gibbs energies of the constituent elements evaluated at 298.15 K and 1 bar.

Table 10. Standard state thermodynamic properties of selected oxide minerals:
Enthalpies of formation, entropies of formation and lattice and magnetic heat capacity functions ($C_p^\bullet$) for temperatures greater than 298 K and pressures of 10^5 Pascals. From Berman (1988), Ghiorso (1990a) and Sack and Ghiorso (1991a, b).

	ΔH_f^o (kJ)	S^o (J/K)	k_0	k_1 ($\times 10^{-2}$)	k_2 ($\times 10^{-5}$)	k_3 ($\times 10^{-7}$)	T_λ (K)	l_1 ($\times 10^2$)	l_2 ($\times 10^5$)	$\Delta_t H$ (J)
ANATASE TiO_2			78.10	0.0	-31.251	32.912				
CHROMITE $FeCr_2O_4$	-1445.490	142.676	236.874	-16.796	0.000	-16.765				
CORUNDUM AL_2O_3	-1675.700	50.82	155.02	-8.284	-38.614	40.908				
GEIKIELITE $MgTiO_3$	-1572.56	74.56	146.20	-4.160	-39.998	40.233				
HEMATITE FE_2O_3	-822.	87.40	146.86	0.0	-55.768	52.563	955	-7.403	27.921	1287
HERCYNITE $FeAl_2O_4$	-1947.681	115.362	235.190	-14.370	-46.913	64.564				
ILMENITE $FeTiO_3$	-1231.947	108.628	150.00	-4.416	-33.237	34.815				
magnesioferrite $MgFe_2O_4$	-1406.465	122.765	196.66	0.0	-74.922	81.007	665	15.236	-53.571	931
magnesiotitanate Mg_2TiO_4	-2161.998	120.170	226.11	-13.801	-17.011	4.128				
$MgTi_2O_5$			232.58	-7.555	-56.608	58.214				
MAGNETITE Fe_3O_4	-1117.403	146.114	207.93	0.0	-72.433	66.436	848	-19.502	61.037	1565
PERICLASE MgO	-601.500	26.951	61.11	-2.962	-6.212	0.584				
PEROVSKITE $CaTiO_3$			150.49	-6.213	0.0	-43.010				
PICROCHROMITE $MgCr_2O_4$	-1783.640	106.020	201.981	-5.519	-57.844	57.729				
PseudoBROOKITE Fe_2TiO_4			261.35	-15.307	0.0	-23.466				
RUTILE TiO_2	-944.750	50.460	77.84	0.0	-33.678	40.294				
SPINEL $MgAl_2O_4$	-2300.313	84.535	235.90	-17.666	-17.104	4.062				
ULVÖSPINEL Fe_2TiO_4	-1488.500	185.447	249.63	-18.174	0.000	-5.453				
WUSTITE $Fe_{0.947}O$			65.18	-2.803	0.0	-1.737				

$$298\,K < T < T_\lambda \quad C_P = k_0 + k_1 T^{-0.5} + k_2 T^{-2} + k_3 T^{-3} + T(l_1 + l_2 T)^2$$

$$T_\lambda < T \quad C_P = k_0 + k_1 T^{-0.5} + k_2 T^{-2} + k_3 T^{-3}$$

$$\Delta G_{f,\,app}^o = \Delta H_f^o + \int_{T_r}^{T} C_P(T,P_r)\,dT - T\left[S^o + \int_{T_r}^{T} \frac{C_P(T,P_r)}{T}\,dT\right] + \int_{P_r}^{P} V^o(T,P)\,dP$$

Table 11. Standard state thermodynamic properties of selected oxide minerals: Volumes for temperatures greater than 298 K and pressures of 10^5 Pascals. From Berman (1988).

	V_{T_r,P_r}^o (J/K)	v_1 ($\times 10^{-6}$)	v_2 ($\times 10^{-12}$)	v_3 ($\times 10^{-6}$)	v_4 ($\times 10^{-10}$)
CORUNDUM AL_2O_3	2.558	-0.385	0.375	21.342	47.180
HEMATITE FE_2O_3	3.027	-0.479	0.304	38.310	1.650
ILMENITE $FeTiO_3$	3.170	-0.584	1.230	27.248	29.968
MAGNETITE Fe_3O_4	4.452	-0.582	1.751	30.291	138.470
PERICLASE MgO	1.125	-0.622	1.511	37.477	3.556
RUTILE TiO_2	1.882	-0.454	0.584	25.716	15.409
SPINEL $MgAl_2O_4$	3.977	-0.489	0.0	21.691	50.528

$$V_{T,P}^o = V_{T_r,P_r}^o [1 + v_1(P - P_r) + v_2(P - P_r)^2 + v_3(T - T_r) + v_4(T - T_r)^2]$$

$T_r = 298.15\,K,\,P_r = 1\,bar$

OXIDE GEOTHERMOMETERS

Thermochemical models for oxide solid solutions may be combined with internally consistent end-member standard state properties to construct mineral geothermometers, geobarometers and oxygen fugacity sensors. These formulations form the core of applications of oxide thermochemistry which are discussed in the remainder of this volume. Although it is beyond the scope of the present chapter to summarize the experimental and field constraints that bear on the formulation and application of silicate-oxide geothermometry/geobarometry to natural systems (cf. Chapters 7 and 9), it is appropriate to summarize here the construction of oxide-oxide geothermometers and oxygen barometers.

Perhaps the best known of all mineral geothermometers is the Fe-Ti oxide geothermometer/oxygen barometer of Buddington and Lindsley (1964). The theoretical foundations of this method are predicated on the establishment of Fe^{2+}-Ti exchange equilibrium between spinel and coexisting rhombohedral oxide. Thermodynamic formulations of the Fe-Ti oxide geothermometer/oxygen barometer have been proposed by Lindsley and coworkers (Andersen and Lindsley, 1988; Lindsley et al., 1990) and by Ghiorso and Sack (1991). From the compositions of coexisting oxides, and an assumption regarding the pressure at which the two-oxides last equilibrated, temperature is calculated by zeroing the Gibbs energy change of the Fe^{2+}-Ti exchange reaction given by Equation (34). Once a temperature is obtained, oxygen fugacity is estimated from the relation

$$\log_{10} f_{O_2} = -\frac{\Delta G^o_{HM}}{2.303\ R\ T} - 4 \log_{10} a^{rhm\ ox}_{Fe_2O_3} + 6 \log_{10} a^{spinel}_{Fe_3O_4}$$

where ΔG^o_{HM} is the standard state Gibbs energy change of the reaction

$$6\ Fe_2O_3 = 4\ Fe_3O_4 + O_2 \qquad (35)$$
$$\text{rhm ox} \qquad \text{spinel} \quad \text{gas}$$

and a^i_j refers to the activity of the jth end-member in the ith phase. Andersen and Lindsley (1988) and Lindsley et al. (1990) utilize activity composition relations and standard state properties for the exchange reaction (Eqn. 34) derived by Andersen (1988) to formulate their geothermometer. Ghiorso and Sack (1991) utilize the standard state properties provided in Tables 10 and 11 and the solution theory developed by Ghiorso (1990a) and Sack and Ghiorso (1991a,b) to construct their version of the geothermometer. The two approaches are compared in Figure 16, where T-$\log_{10} f_{O_2}$ relations are provided for coexisting binary $FeTiO_3$-Fe_2O_3 and Fe_2TiO_4-Fe_3O_4 oxides. Using a diverse suite of volcanic rocks, Ghiorso and Sack (1991) provide a detailed comparison of the two versions of the Fe-Ti oxide geothermometer. They show that the two geothermometers yield fairly comparable results in the temperature range 750°-850°C and at oxygen fugacities within one $\log_{10}$ unit of the quartz-fayalite-magnetite (QFM) oxygen buffer.

DIRECTIONS FOR FUTURE RESEARCH

The two most serious shortcomings of the thermodynamic models summarized in this chapter are the assumptions that the majority of oxide minerals approximate ideal stoichiometry and that long-range cation order dominates the configurational entropy under geologic temperatures and pressures. The former assumption can be demonstrated to be seriously flawed for very magnesian magnesioferrite-magnetite solid solutions and for magnetite-rich spinels at high oxygen fugacities (> NNO) and temperatures. Neglect of the

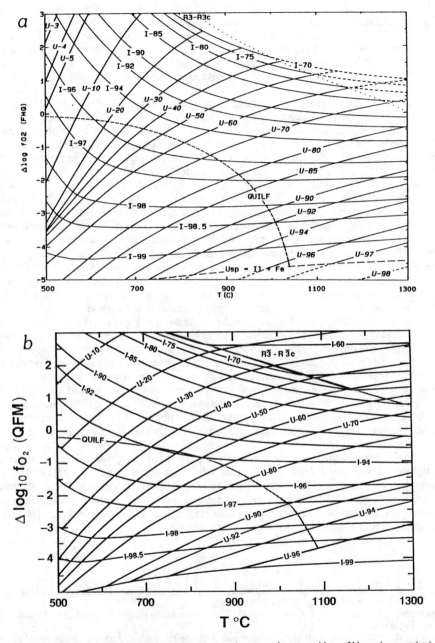

Figure 16. Graphical Fe-Ti oxide geothermometer/oxygen barometer for compositions of binary (*sensu stricto*) coexisting rhombohedral and cubic oxides. Relative oxygen fugacity, defined as the calculated $\log_{10} f_{O_2}$ minus that of the quartz-fayalite-magnetite buffer, is plotted on the ordinate. Composition of the spinel is indicated in mol % cent ulvöspinel (U) while that of the rhombohedral phase is given in mol % ilmenite (I). The path of the second order phase transition in the rhombohedral oxides is denoted by the curve labeled R3-R3c. The curve labeled QUILF refers to the $\log_{10} f_{O_2}$-T path defined by equilibrium between quartz, fayalite and both oxides. (a) From Andersen and Lindsley (1988, Fig. 2). The stability of the assemblage ulvöspinel + ilmenite + iron (Usp = Il + Fe) is also shown. (b) After Ghiorso and Sack (1991).

defect equilibria in constructing solution models for the oxides is equivalent to defining a fictive standard state that cannot be simply related to calorimetric constraints (Sack and Ghiorso, 1991a). Extension of the models to make provision for deviations from ideal stoichiometry would require much additional data on the level and nature of point defects in the oxides as a function of temperature, pressure and composition. Such data are necessary for further advance and will require an extensive experimental effort.

The assumption that long-range cation order dominates the configurational entropy is implicit in all of the "closed-form" models discussed above. More sophisticated thermodynamic treatments that involve lattice energy minimization techniques (Burton, 1984, 1985; Chapter 8) reveal that the effects of short-range cation order are important for a detailed understanding of phase relations at low to moderate geologic temperatures. Short-range cation order is a high-temperature manifestation of symmetry reducing phase transitions in the titanate spinels (Sack and Ghiorso, 1991a), and is likely the cause of the development of ferri-ferrous clusters that have a profound influence on the magnetic properties of intermediate hematite-ilmenite and ulvöspinel-magnetite solid solutions (Ishikawa and Akiomoto, 1957; Banerjee et al., 1967). Unfortunately, it is unclear how the effects of short-range ordering might be incorporated into a closed form (analytical) description of the configurational entropy. Including ternary or higher order Margules parameters in the excess Gibbs energy as a proxy for short-range order contributions would cure the symptoms but not the disease. To calibrate higher order terms would require more extensive and more highly precise data than are currently available. Additionally, the low-order polynomial form of the excess Gibbs energy is unlikely to provide an adequate approximation for the necessarily complex functional form of the short-range order entropy contribution. This is an exciting area for future research which will involve not only experimental documentation of short-range ordering phenomena, but also the need to develop mathematical formalisms to describe short-range ordering in these complex solid solutions. As descriptions of short-range cation ordering are akin to problems involving the development of magnetic order, we should anticipate the development of thermodynamic models that couple the latter with both long-range and short-range cation ordering. Such developments must precede the complete interpretation of oxide phase equilibria under metamorphic conditions (Ghiorso and Sack, 1991).

Further improvements in thermodynamic models for the oxides depends on the acquisition of additional data on (1) cation site ordering in spinels, rhombohedral and orthorhombic oxides and (2) low-temperature (<50 K) heat capacities and magnetic phase transitions, for both end-member compositions and *their solid solutions*. Though they may seem mundane and old fashioned, these data are fundamental and prerequisite for further progress on the thermochemistry of the oxide minerals.

REFERENCES

Andersen, D.J. (1988) Internally consistent solution models for Fe-Mg-Mn-Ti oxides. Ph.D. dissertation, State University of New York (Stony Brook), 202 p.
Andersen, D.J., Lindsley, D.H. (1988) Internally consistent solution models for Fe-Mg-Mn-Ti oxides: Fe-Ti oxides. Am. Mineral. 73, 714-726.
Aragón, R., McCallister, R.H. (1982) Phase and point defect equilibria in the titanomagnetite solid solution. Phys. Chem. Minerals 8, 112-120.
Banerjee, S.K., O'Reilly, W., Gibb, T.C., Greenwood, N.N. (1967) The behavior of ferrous ions in iron-titanium spinels. J. Phys. Chem. Solids 28, 1323-1335.
Berman, R.G. (1988) Internally-consistent thermodynamic data for minerals in the system Na_2O-K_2O-CaO-MgO-FeO-Fe_2O_3-Al_2O_3-SiO_2-TiO_2-H_2O-CO_2. J. Petrol. 29, 445-522.

Berman, R.G., Brown, T.H. (1985) Heat capacities of minerals in the system $Na_2O-K_2O-CaO-MgO-FeO-Fe_2O_3-Al_2O_3-SiO_2-TiO_2-H_2O-CO_2$: representation, estimation, and high temperature extrapolation. Contrib. Mineral. Petrol. 89, 168-183.

Brown, N.E., Navrotsky, A. (1989) Structural, thermodynamic, and kinetic aspects of disordering in the pseudobrookite-type compound karrooite, $MgTi_2O_5$. Am. Mineral. 74, 902-912.

Buddington, A.F., Lindsley, D.H. (1964) Iron-titanium oxide minerals and synthetic equivalents. J. Petrol. 5, 310-357.

Burton, B. (1982) Thermodynamic analysis of the systems $CaCO_3-MgCO_3$, α-Fe_2O_3, and Fe_2O_3-$FeTiO_3$. Ph.D. dissertation, State University of New York (Stony Brook).

Burton, B. (1984) Thermodynamic analysis of the system Fe_2O_3-$FeTiO_3$. Phys. Chem. Minerals 11, 132-139.

Burton, B.P. (1985) Theoretical analysis of chemical and magnetic ordering in the system Fe_2O_3-$FeTiO_3$. Am. Mineral. 70, 1027-1035.

Burton, B.P., Davidson, P.M. (1988) Multicritical phase relations in minerals. In: Ghose, S., Coey, J.M.D., Salje, E. (eds) Structural and Magnetic Phase Transitions in Minerals. Springer-Verlag, New York, Advances in Phys. Geochem. 7, 60-90.

Burton, B.P., Kikuchi, R. (1984) The antiferromagnetic-paramagnetic transition in α-Fe_2O_3 in the single prism approximation of the cluster variation method. Phys. Chem. Minerals 11, 125-131.

Dieckmann, R., Schmalzried, J. (1977a) Defects and cation diffusion in magnetite (I). Berichte der Bunsengesellschaft für Physikalische Chemie 81, 344-347.

Dieckmann, R., Schmalzried, J. (1977b) Defects and cation diffusion in magnetite (II). Berichte der Bunsengesellschaft für Physikalische Chemie 81, 414-419.

Dzyaloshinsky, I. (1958) A thermodynamic theory of "weak" ferromagnetism of antiferromagnetics. J. Phys. Chem. Solids 4, 241-255.

Dzyaloshinsky, I. (1965) Theory of helicoidal structures in antiferromagnets. III. Soviet Physics JETP 20, 665-671.

Ghiorso, M.S. (1990a) Thermodynamic properties of hematite-ilmenite-geikielite solid solutions. Contrib. Mineral. Petrol. 104, 645-667.

Ghiorso, M.S. (1990b) Application of the Darken equation to mineral solid solutions with variable degrees of order-disorder. Am. Mineral. 75, 539-543.

Ghiorso, M.S., Sack, R.O. (1991) Fe-Ti oxide geothermometry: Thermodynamic formulation and the estimation of intensive variables in silicic magmas. Contrib. Mineral. Petrol. (in press)

Gopal, E.S.R. (1966) Specific Heats at Low Temperatures. Plenum Press, New York, 240 p.

Grønvold, F., Samuelsen, E.J. (1975) Heat capacity and thermodynamic properties of α-Fe_2O_3 in the region 300-1050 K. Antiferromagnetic transition. J. Phys. Chem. Solids 36, 249-256.

Grønvold, F., Sveen, A. (1974) Heat capacity and thermodynamic properties of synthetic magnetite (Fe_3O_4) from 300 to 1050 K. Ferrimagnetic transition and zero point entropy. J. Chem. Thermodynamics 6, 859-872.

Grønvold, F., Westrum, E.F., Jr. (1959) α-Ferric oxide: Low temperature heat capacity and thermodynamic functions. J. Am. Chem. Soc. 81, 1780-1783.

Hill, R.L., Sack, R.O. (1987) Thermodynamic properties of Fe-Mg titaniferous magnetite spinels. Canadian Mineral. 25, 443-464.

Hoffman, K.A. (1975) Cation diffusion processes and self-reversal of thermoremanent magnetization in the ilmenite-hematite solid solution series. Geophys. J. Royal Astronomical Soc. 41, 65-80.

Ishikawa, Y. (1958a) Electrical properties of $FeTiO_3$-Fe_2O_3 solid solution series. J. Phys. Soc. Japan 13, 37-42.

Ishikawa, Y. (1958b) An order-disorder transformation phenomena in the $FeTiO_3$-Fe_2O_3 solid solution series. J. Phys. Soc. Japan 13, 828-837.

Ishikawa, Y., Akimoto, S. (1957) Magnetic properties of the $FeTiO_3$-Fe_2O_3 solid solution series. J. Phys. Soc. Japan 12, 1083-1098.

Katsura, T., Wakihara, M., Shin-Ichi, H., Sugihara, T. (1975) Some thermodynamic properties in spinel solid solutions with the Fe_3O_4 component. J. Solid State Chem. 13, 107-113.

Landau, L.D., Lifshitz, E.M. (1980) Statistical Physics. Part I (v. 5 of Course of Theoretical Physics). Pergamon Press, New York, 3rd ed., revised by E.M. Lifshitz and L.P. Pitaevskii, 460 p.

Landau, L.D., Lifshitz, E.M. (1984) Electrodynamics of Continuous Media (v. 8 of Course of Theoretical Physics). Pergamon Press, New York, 2nd ed., revised by E.M. Lifshitz and L.P. Pitaevskii, 544 p.

Lawson, A.W. (1947) On simple binary solid solutions. J. Chemical Physics 15, 831-842.

Lehmann, J., and Roux, J. (1986) Experimental and theoretical study of $(Fe^{2+},Mg)(Al,Fe^{3+})_2O_4$ spinels: Activity-composition relationships, miscibility gaps, vacancy contents. Geochim. Cosmochim. Acta 50, 1765-1783.

Lindsley, D.H. (1973) Delimitation of the hematite-ilmenite miscibility gap. Geol. Soc. Am. Bull. 84, 657-662.

Lindsley, D.H. (1981) Some experiments pertianing to the magnetite-ulvöspinel miscibility gap. Am. Mineral. 66, 759-762.

Lindsley, D.H., Frost, B.R., Andersen, D.J., Davidson, P.M. (1990) Fe-Ti oxide-silicate equilibria: Assemblages with orthopyroxene. In: Spencer, R.J., Chou, I.-M. (eds) Fluid-Mineral Interactions: A Tribute to H. P. Eugster. Geochemical Soc., Special Publications 2, 103-119.

Millard, R.L., Peterson, R.C., Hunter, B.K. (1991) Temperature dependence of cation disorder in $MgAl_2O_4$ spinel using aluminum-27 and oxygen-17 magic-angle spinning nuclear magnetic resonance spectroscopy. Am. Mineral. (in press).

Myers, J., Eugster, H.P. (1983) The system Fe-Si-O: oxygen buffer calibrations to 1,500 K. Contrib. Mineral. Petrol. 82, 75-90.

Navrotsky, A., Kleppa, O.J. (1967) The thermodynamics of cation distributions in simple spinels. J. Inorganic Nuclear Chem. 30, 479-498.

Nord, G.L. Jr., Lawson, C.A. (1989) Order-disorder transition-induced twin domains and magnetic properties in ilmenite-hematite. Am. Mineral. 74, 160-176.

O'Neill, H.St.C., Navrotsky, A. (1983) Simple spinels: crystallographic parameters, cation radii, lattice energies, and cation distributions. Am. Mineral. 68, 181-194.

O'Neill, H.St.C., Navrotsky, A. (1984) Cation distributions and thermodynamic properties of binary spinel solid solutions. Am. Mineral. 69, 733-753.

O'Neill, H.St.C., Wall, V.J. (1987) The olivine-spinel oxygen geobarometer, the nickel precipitation curve and the oxygen fugacity of the upper mantle. J. Petrol. 28, 1169-1192.

Peterson, R.C., Lager, G.A., Hitterman, I.P.N.S. (1991) A Time-Of-Flight neutron diffraction study of $MgAl_2O_4$ at temperatures up to 1273 K. Am. Mineral., (in press).

Price, G.D. (1981) Subsolidus phase relations in the titanomagnetite solid solution series. Am. Mineral. 66, 751-758.

Sack, R.O. (1982) Spinels as petrogenetic indicators: Activity-composition relations at low pressures. Contrib. Mineral. Petrol. 79, 169-182.

Sack, R.O., Ghiorso, M.S. (1991a) An internally consistent model for the thermodynamic properties of Fe-Mg- titanomagnetite - aluminate spinels. Contrib. Mineral. Petrol. 106, 474-505.

Sack, R.O., Ghiorso, M.S. (1991b) Chromian spinels as petrogenetic indicators: thermodynamics and petrological applications. Am. Mineral. (in press).

Schmocker, U., Boesch, Waldner, F. (1972) A direct determination of cation disorder in $MgAl_2O_4$ spinel by ESR. Physical Letters 40A, 237-238.

Schmocker, U., Waldner, F. (1976) The inversion parameter with respect to the space group of $MgAl_2O_4$ spinels. J. Physics C: Solid State Physics 9, L235-237.

Shannon, R.D., Prewitt, C.T. (1969) Effective ionic radii in oxides and fluorides. Acta Cryst. 25B, 928-9.

Srečec, I., Ender, A., Woermann, E., Gans, W., Jacobsson, E., Eriksson, G., Rosén, E. (1987) Activity-composition relations of the magnesiowüstite solid solution series in equilibrium with metallic iron in the temperature range 1050-1400 K. Phys. Chem. Minerals 14, 492-498.

Taylor, R.W. (1964) Phase equilibrium in the system $FeO-Fe_2O_3-TiO_2$ at 1300°C. Am. Mineral. 49, 1016-1030.

Thompson, J.B., Jr. (1969) Chemical reactions in crystals. Am. Mineral. 54, 341-375.

Thompson, J.B., Jr. (1970) Chemical reactions in crystals: Corrections and clarification. Am. Mineral. 55, 528-532.

Tolédano, J.-C., Tolédano, P. (1987) The Landau theory of phase transitions (Lecture Notes in Physics, Vol. 3), World Scientific Publishing Co., Teaneck, NJ., 451 p.

Webster, A.H., Bright, N.F.H. (1961) The system iron-titanium-oxygen at 1200°C and oxygen partial pressures between 1 atm and $2X10^{-14}$ atm. J. Am. Ceramic Soc. 44, 110-116.

Wechsler, B.A. (1977) Cation distribution and high-temperature crystal chemistry of armalcolite. Am. Mineral. 62, 913-920.

Wechsler, B.A., Prewitt, C.T. (1984) Crystal structure of ilmenite ($FeTiO_3$) at high temperature and at high pressure. Am. Mineral. 69, 176-185.

Westrum, E.F., Jr., Grønvold, F. (1969) Magnetite (Fe_3O_4) heat capacity and thermodynamic properties from 5 to 350 K, low temperature transition. J. Chem. Thermodynamics 1, 543-557.

Wood, B.J., Kirkpatrick, R.J., Montez, B. (1986) Order-disorder phenomena in $MgAl_2O_4$ spinel. Am. Mineral. 71, 999-1006.

Yamanaka, T., Takéuchi, Y. (1983) Order-disorder transition in $MgAl_2O_4$ spinel at temperatures up to 1700°C. Zeits. Kristallogr. 165, 65-78.

Chapter 7　　　　　　　　　　　B. J. Wood, J. Nell & A. B. Woodland

MACROSCOPIC AND MICROSCOPIC THERMODYNAMIC PROPERTIES OF OXIDES

INTRODUCTION

The purpose of this chapter is to elucidate the links between phase equilibria in oxide systems (Chapter 3) and the use of oxide compositions to extract petrogenetic information (P, T, fO_2, etc.) for rocks (Chapters 9, 10, 11 and 12). In order to make this progression it is generally necessary to have an understanding of the mixing properties of complex minerals and our purpose here is to review how these are obtained.

Petrogenetic indicators (geothermometers/barometers, etc.) rely on the equilibrium thermodynamic properties of phases which commonly coexist in rocks. These properties may be loosely divided into those characteristic of pure end-member phases (Fe_3O_4, $FeTiO_3$, MgO, etc.) and those due to solid solution in, for example, Fe_3O_4 - Fe_2TiO_4 spinel, "FeO"-MgO oxide, and so on. End-member properties are derived from a combination of calorimetric and phase equilibrium data on the pure phases. These may be illustrated with respect to the system Fe-O, for which the free energies of formation of wüstite, magnetite and hematite from metallic iron and oxygen gas are derived from phase equilibria (log fO_2-temperature) measurements of the equilibria (Chapter 3):

$$\text{Fe} + \frac{y}{2} O_2 = \text{FeO}_y \qquad (\text{IW, } y>1) \qquad (1)$$
$$\text{metal} \qquad \text{wüstite}$$

$$\frac{6}{4-3x} \text{FeO}_x + O_2 = \frac{2}{4-3x} \text{Fe}_3O_4 \qquad (\text{WM, } x>y) \qquad (2)$$
$$\text{wüstite} \qquad \text{magnetite}$$

$$4 \text{Fe}_3O_4 + O_2 = 6 \text{Fe}_2O_3 \qquad (\text{MH}) \qquad (3)$$

Note that wüstite has been formally given the stoichiometry FeO_y even though it is strictly cation-deficient and that a formula such as $Fe_{1-x}O$ would be crystallochemically more correct (Chapters 2 and 3). The formalism adopted here is for convenience in what follows and is thermodynamically acceptable even if unusual. Knowing the positions of curves (1), (2), (3) in log fO_2-T space gives the free energies of formation of all three oxides in the experimental temperature range, provided the stoichiometry of wüstite is known. An important additional constraint is provided by the convergence of reactions (1) and (2) at about 830 K (O'Neill, 1988). The resultant set of free energies may then be extrapolated in temperature and pressure using calorimetrically measured heat capacities and volumes derived from X-ray diffraction (e.g., Robie et al., 1978). The construction of large internally-consistent thermodynamic data bases for end-member oxides and silicates using such data has been the subject of several recent reviews (e.g., Berman, 1988; Holland and Powell, 1990). We will not, therefore, treat the methodology in more detail, turning instead to the problems of multicomponent phases.

Consider a phase j consisting of i components each of which has a pure end-member molar free energy (G_i^o) and which mix to yield a phase with molar free energy G_j:

$$G_j = \sum_i X_i G_i^o + G_{mix} \qquad (4)$$

In Equation (4) X_i is the mole fraction of component i and G_j is made up of a combination of the properties of the pure end-members and a mixing term G_{mix}. Formally, G_{mix} can be broken down into enthalpy and entropy terms:

$$G_{mix} = H_{mix} - T S_{mix} \qquad (5)$$

or it may be regarded as a sum of the activities of the constituent components:

$$G_{mix} = \sum_i X_i \, RT \, ln \, a_j^i \qquad (6)$$

Equations (5) and (6) embody the experimental approaches that may be taken towards understanding the mixing properties of phases. Direct determination of free energy involves measuring the activities of one or more components in the mixed phase. Indirect approaches involve measurement of enthalpy or entropy of mixing in order to characterise the properties of the phase. We continue with review of the experimental techniques in general use and a description of their application to a number of oxide systems.

ACTIVITY MEASUREMENTS

Equilibration with metal

To illustrate the principle of activity measurement, let us consider Equilibrium (1), involving iron metal coexisting with pure wüstite. Along the IW boundary pure metal coexists with pure wüstite and by definition, using the standard states of pure wüstite and pure metal at the P and T of interest, we have:

$$G_{wu} = G^o_{FeO_y} \; ; \; RT \, ln \, a^{Wu}_{FeO_y} = 0 \; ; \; a^{Wu}_{FeO_y} = 1 \; .$$

If we add magnesium to the system Fe-O then the wüstite becomes an MgO-FeO solid solution while the metal remains essentially pure Fe. Hence the activity of FeO_y in wüstite is no longer 1.0 while the activity of Fe in the metal is still 1.0. The result is that the fO_2 of equilibrium is shifted to values below the IW curve.

At IW we have:

$$\frac{y}{2} RT \, ln \, fO_2^o = -\Delta G^o_{IW}$$

while at some fO_2 below IW where the oxide phase is no longer pure FeO_y we have:

$$\frac{y}{2} RT \, ln \, fO_2' - \frac{y}{2} RT \, ln \, a^{Wu}_{FeO_y} = -\Delta G^o_{IW}$$

which yields:

$$ln \, a^{Wu}_{FeO_y} = \frac{y}{2} (ln \, fO_2' - ln \, fO_2^o) \qquad (7)$$

Therefore, by measuring the shift in equilibrium fO_2 as a function of magnesium content we obtain the relationship between composition and a_{FeOy}.

In practice, oxygen fugacity shifts of this type are measured in two ways, either by a conventional phase equilibrium technique or by emf measurements. The phase equilibrium method would involve placing samples of Fe metal and magnesiowüstite solid solution in a furnace and equilibrating them with gas mixtures (CO-CO_2 or H_2-CO_2) of known composition and fO_2. After quenching, the compositions of the magnesiowüstites would be measured and the equilibrium compositions at fixed fO_2 reversed by approach from high and low mole fractions of FeO_y.

An emf cell, illustrated in Figure 1, involves placing the magnesiowüstite-iron mixture in an inert atmosphere on one side of an oxygen-ion electrolyte. A reference assemblage (in this case generally iron-wüstite) is placed on the other side, also in an inert atmosphere isolated from the inside assemblage. Oxygen ion electrolytes are generally made of yttria-doped ZrO_2 or calcia-doped ZrO_2, although many other possibilities exist (see, for example, Subbarao, 1980). The difference in μO_2 between the inside (unknown) and outside (reference) assemblages is given by the emf difference across the leads positioned as shown in Figure 1 and yields, for oxygen fugacity difference:

$$\text{emf} = \frac{RT}{4F} \ln \frac{fO_2^{\text{inside}}}{fO_2^{\text{ref}}} \qquad (8)$$

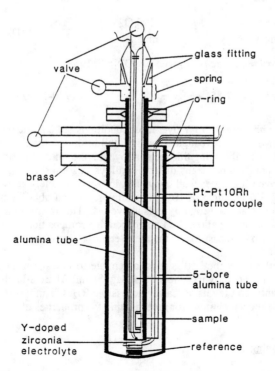

Figure 1. Experimental arrangement for measuring the fO_2 of an unknown sample relative to a reference buffer using an oxygen-ion electrolyte.

where R is the gas constant and F is Faraday's constant. The emf method is extremely precise (better than 0.01 log fO_2 units) and works extremely well, particularly for measuring oxygen buffers (O'Neill, 1988, for example).

A potential disadvantage is that electronic conductivity becomes significant at high temperatures and at low fO_2 values (generally well below IW) so that the fO_2-T range of applicability may be limited. Also, when dealing with solid solutions dilute in Fe-endmembers, measured emf values often drift with time (Srecec et al., 1987; O'Neill et al., 1989) because of changing composition of the solid solution. The latter effect arises from slight oxygen leakage across the cell which oxidises metallic Fe and increases the iron content of the coexisting solid solution.

Equilibration with noble metals

Equilibration of oxide solid solutions with Fe (or Ni or Co) metal is an extremely powerful method of activity determination. It is limited, however, to oxygen fugacities below the Metal-Oxide buffer and cannot, therefore, be used extensively to investigate solid solutions stable at higher fO_2's such as those containing Fe_3O_4 or Fe_2O_3.

The metal-equilibration technique may, however, be extended to more oxidised solutions by making use of the very low activity coefficients (about 10^{-3}; Hultgren et al., 1973) of Fe and other transition elements in dilute solution in platinum or palladium. Consider, for example, the equilibrium (metastable above 830 K) formed by the intersection of (1) and (2):

$$3 \text{ Fe} + 2 \text{ O}_2 = \text{Fe}_3\text{O}_4 \quad \text{(IM)} \tag{9}$$
$$\text{metal} \qquad\qquad \text{magnetite}$$

The equilibrium oxygen fugacity is given by (O'Neill, 1988):

$$\log fO_2 = 7.85 - \frac{28450}{T} \qquad (900\text{-}1600 \text{ K}) \tag{10}$$

but is metastable with respect to IW over most conditions of geologic interest. Fe_3O_4 solid solutions may, however, be investigated using Equilibrium (9) by dissolving a few percent Fe into platinum and equilibrating this metal alloy with the solid solution at known fO_2. The very low activity coefficient means that these alloys in solid solution mixtures can be in equilibrium at oxygen fugacities well *above* IW, in fact completely above the stability field of wüstite. Using 5% Fe and γ of 10^{-3} as examples, we can obtain equilibrium between Fe_3O_4 and the alloy at about 7 log fO_2 units above IM. This means that, provided activity-composition relations for the alloys are known, activity-composition relations in M_3O_4- and M_2O_3-type oxides can be measured by the metal-equilibration technique.

Although the alloy method is most amenable to investigating transition element oxide activities, a recent study has shown that Mg, Si and Al are all soluble in Pd metal to measurable concentration levels at readily accessible fO_2 (Chamberlin et al., 1990). This provides a potential new technique for investigating the properties of nontransition-element oxides.

Interphase partitioning

A third widely-used activity measuring device is that of determining the equilibrium partitioning of elements between two phases. If the mixing properties of one are known, those of the second may be unequivocally derived. This may be illustrated with respect to the data shown in Figure 2a for Al-Cr partitioning between $MgCr_2O_4$-$MgAl_2O_4$ spinels and Cr_2O_3-Al_2O_3 oxides at 1323 K (Oka et al., 1984).

The partitioning of Cr and Al between spinel and sesquioxide phases may be represented by the equilibrium:

$$Al_2O_3 + MgCr_2O_4 = MgAl_2O_4 + Cr_2O_3 \quad (11)$$
$$\text{oxide} \quad \text{spinel} \quad \text{spinel} \quad \text{oxide}$$

for which we have, at any given pressure and temperature:

$$-\Delta G^o_{P,T} = RT \ln\left(\frac{a^{Sp}_{MgAl_2O_4} \cdot a^{oxide}_{Cr_2O_3}}{a^{Sp}_{MgCr_2O_4} \cdot a^{oxide}_{Al_2O_3}}\right) \quad (12)$$

Assuming that $MgCr_2O_4$-$MgAl_2O_4$ spinels can be treated as two-site Cr-Al mixtures, the relationship between activity and composition is expressed by (e.g., Wood and Fraser, 1976, p. 102; Thompson, 1967):

$$a^{Sp}_{MgAl_2O_4} = X^2_{MgAl_2O_4} \cdot \gamma_{MgAl_2O_4}$$
$$a^{Sp}_{MgCr_2O_4} = X^2_{MgCr_2O_4} \cdot \gamma_{MgCr_2O_4} \quad (13)$$

(Note that it does not strictly matter whether the assumption of two-site mixing is correct or not, but that the closer it is to correct, the better-behaved will be the compositional dependence of γ. Also note that γ is often raised to the same power as X, but that this is a matter of convenience only and we have opted for the form of (13) in order to make the succeeding expressions a little simpler). Substituting (13) into (12):

$$\frac{\Delta G^o}{2.303 RT} = -\log\left[\frac{a^{Ox}_{Cr_2O_3} \cdot X^{Sp2}_{MgAl_2O_4}}{a^{Ox}_{Al_2O_3} \cdot X^{Sp2}_{MgCr_2O_4}}\right] - \log \gamma_{MgAl_2O_4}$$

$$+ \log \gamma_{MgCr_2O_4} ;$$

$$= -\log Q - \log \gamma_{MgAl_2O_4} + \log \gamma_{MgCr_2O_4} \quad (14)$$

Differentiating Equation (14) at constant P and T and applying the Gibbs-Duhem equation yields:

$$d \log \gamma_{MgCr_2O_4} = (1 - X_{MgCr_2O_4}) \, d \log Q$$

Integration by parts leads to:

$$\log \gamma_{MgCr_2O_4} = \log Q \, (1 - X_{MgCr_2O_4}) + \int_1^{X_{MgCr_2O_4}} \log Q \, dX_{MgCr_2O_4}$$

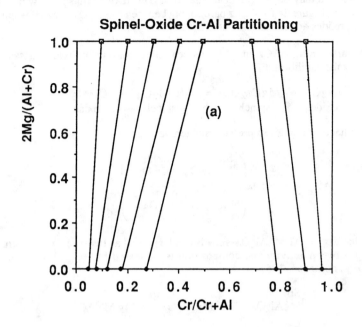

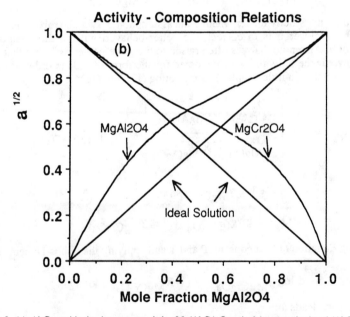

Figure 2. (a) Al-Cr partitioning between coexisting Mg(Al,Cr)$_2$O$_4$ spinel (upper points) and (Al,Cr)$_2$O$_3$ oxide (lower points) at 25 kbar and 1050°C (Oka et al., 1984). (b) Activities of MgCr$_2$O$_4$ and MgAl$_2$O$_4$ derived from the partitioning data. Note that √activity is used for the two-site solution for ease of comparison with ideality.

$$\log \gamma_{MgAl_2O_4} = -X_{MgCr_2O_4} \log Q + \int_0^{X_{Mg_2CrO_4}} \log Q \, dX_{MgCr_2O_4} \quad (15)$$

Given activity-composition relations for the sesquioxide (Jacob, 1978; Chatterjee et al., 1982), the measurement of equilibrium values of $X^{Sp}_{MgCr_2O_4}$, $X^{Ox}_{Cr_2O_3}$, etc. over the whole join enables Q to be calculated and the integrals in Equation (15) to be evaluated graphically. This leads to the activity-composition relations shown in Figure 2b.

A variation of the interphase partitioning approach involves making use of the immiscibility that is common in oxide and silicate systems. In, for example, $FeAl_2O_4$ - Fe_3O_4 there is a large, nearly symmetric, solvus whose crest is at 860°C (Turnock and Eugster, 1962) (Fig. 3a). The activity of Fe_3O_4 is equal on each side of the solvus as is $a_{FeAl_2O_4}$ in each phase:

$$RT \ln a^a_{Fe_3O_4} = RT \ln a^b_{Fe_3O_4}$$

$$RT \ln a^a_{FeAl_2O_4} = RT \ln a^b_{FeAl_2O_4} \quad (16)$$

Given the compositions of the two phases a reasonable estimate of the activity coefficients may be obtained by assuming that the solid solution obeys an appropriate nonideal model such as that of symmetric solution (Thompson, 1967). The result, however, depends on order-disorder relations, which, in the case of spinels, are complex. For example, one could treat $Fe^{2+}(Fe^{3+})_2O_4$ - $Fe^{2+}(Al^{3+})_2O_4$ as a normal solid solution with (Al-Fe^{3+}) mixing on the two octahedral sites only.

This would yield:
$$a_{Fe_3O_4} = X^2_{Fe_3O_4} \cdot \gamma_{Fe_3O_4} \quad (17)$$

and, for symmetric solution:
$$RT \ln \gamma_{Fe_3O_4} = W(X_{FeAl_2O_4})^2$$

where W is the nonideal interaction parameter between Fe_3O_4 and $FeAl_2O_4$. Magnetite has a dominantly inverse configuration, however, so that Fe^{2+} is actually octahedral and Fe^{3+} both tetrahedral and octahedral, i.e., $Fe^{3+}(Fe^{2+}Fe^{3+})_2O_4$. So there should actually be three site mixing between (essentially) inverse Fe_3O_4 and normal $FeAl_2O_4$, yielding:

$$a_{Fe_3O_4} = X^3_{Fe_3O_4} \cdot \gamma_{Fe_3O_4} \quad (18)$$

In the first case, the condition of equilibrium on the solvus becomes (with symmetric solution):

$$RT \ln \left(\frac{X^b_{Fe_3O_4}}{X^a_{Fe_3O_4}} \right)^2 = W \left[X^{a2}_{FeAl_2O_4} - X^{b2}_{FeAl_2O_4} \right] \quad (19)$$

while in the second case we obtain:

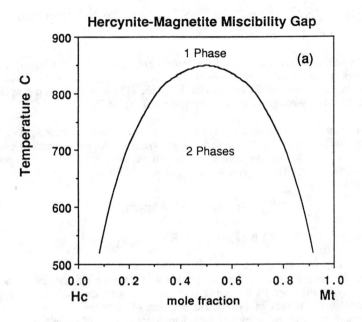

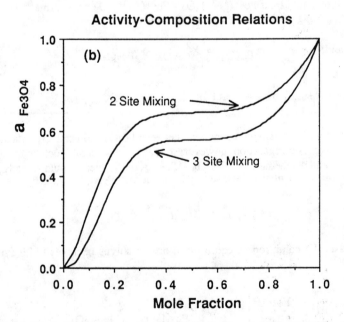

Figure 3. (a) Sketch of the solvus on the join $FeAl_2O_4$-Fe_3O_4 (Turnock and Eugster, 1962). (b) Activity-composition relations for Fe_3O_4 component calculated from the solvus assuming either two-site or three-site mixing (see text).

$$RT \, ln \left(\frac{X^b_{Fe_3O_4}}{X^a_{Fe_3O_4}} \right)^3 = W \left[X^{a2}_{FeAl_2O_4} - X^{b2}_{FeAl_2O_4} \right] \qquad (20)$$

The two assumptions yield quite different activity-composition relations as shown at the temperature of the crest of the solvus in Figure 3b. In practise the extent of disorder is somewhere between the two end-member assumptions (Mason and Bowen, 1981), demonstrating the importance of knowing order-disorder relations in these phases in order to derive correct activity-composition relations.

ENTHALPY MEASUREMENTS

Enthalpy measurements of oxides and silicates are generally made by measuring the heat evolved or absorbed during dissolution in acid at low temperatures or in borate melts at high temperatures (950-1200 K approx.). The difference between the weighted sum of end-member enthalpies of solution and that of the solid solution gives H_{mix}. Relatively few measurements have been made on oxide solid solutions of geologic interest, however, principally because the latter generally contain Fe. It is extremely difficult to prevent some oxidation of such samples during thermal equilibration in the calorimeter or, more important, during dissolution in the solvent. The oxidation of Fe(II) to Fe(III) is generally exothermic and during dissolution causes spurious and scattered results.

Where available, however, enthalpy of solution measurements provide an important complement to activity measurements in that they aid in determination of the temperature dependence of the latter. From Equations (5) and (6) we have, for a binary solid solution:

$$X_1 \, RT \, ln \, a_1 + X_2 \, RT \, ln \, a_2 = H_{mix} - T \, S_{mix} \qquad (21)$$

If atoms (1) and (2) are completely randomly mixed on one site per formula unit (a = Xγ), then the major contribution to the entropy of mixing is configurational:

$$S_{config} = -\sum_i X_i \, R \, ln \, X_i$$

$$S_{mix} = S_{config} + S^{xs}$$

which yields:

$$RT \, X_1 \, ln \, \gamma_1 + RT \, X_2 \, ln \, \gamma_2 = H_{mix} - TS^{xs} \qquad (22)$$

From (22) we see that if the weighted sum of activity coefficients is equal to H_{mix} then there is no excess entropy of vibrational or order-disorder origin. Davies and Navrotsky (1981) used this approach to compare enthalpies of mixing on the NiO-MgO join with measured values of activity coefficient. Their results indicated a substantial negative excess entropy of mixing which they ascribed to short range ordering of Mg and Ni atoms, i.e., despite the simple structure, Ni and Mg atoms are probably not randomly mixed on the octahedral sublattice.

SPECIFIC SYSTEMS

FeO - MgO

Although magnesiowüstites are not volumetrically significant minerals in near-surface rocks, they are probably very important in the lower mantle, being produced by the breakdown of the spinel form of $(Mg,Fe)_2SiO_4$ to oxide plus $(Mg,Fe)SiO_3$ perovskite at pressures greater than 230 kbar (Ito and Takahashi, 1989).

The solid solutions are not binary, since magnesiowüstites always contain significant Fe^{3+} and may in fact be richer in Fe^{3+} than Fe^{2+} (Valet et al., 1975). Their activity-composition relations have been determined several times by the phase equilibrium method of equilibrating with metallic Fe (Schmahl et al., 1960; Hahn and Muan, 1962; Berthet and Perrot, 1970). A recent comprehensive study by Srecec et al. (1987) used combined emf and phase equilibrium methods. These authors measured activities in the temperature range 1050-1400 K and obtained, for ternary MgO-FeO-$Fe_{2/3}^{3+}O$, the following activity coefficients at Fe-saturation (T in K):

$$\ln \gamma_{FeO} = \left(\frac{521}{T} + 5.062\right)X^2_{Fe_{2/3}O} + \left(\frac{-719}{T} + 3.041\right)X_{Fe_{2/3}O}X_{MgO}$$

$$+ \left(\frac{525}{T} + 0.48\right)X^2_{MgO} \qquad (23)$$

$$\ln \gamma_{MgO} = \left(\frac{386}{T} + 0.593\right)X^2_{FeO} + \left(\frac{6211}{T} - 3.587\right)X_{FeO}X_{Fe_{2/3}O}$$

$$+ \left(\frac{-33630}{T} + 36.21\right)X^2_{Fe_{2/3}O} \qquad (24)$$

$$\ln \gamma_{Fe_{2/3}O} = \left(\frac{781}{T} - 0.48\right)X^2_{FeO} + \left(\frac{698}{T} + 2.582\right)X_{FeO}X_{MgO}$$

$$+ \left(\frac{1209}{T} + 2.262\right)X^2_{MgO} \qquad (25)$$

The content of $Fe_{2/3}O$ at Fe saturation is a strong function of $Fe/(Fe+Mg) = X_{Fe}$ and is represented by (Srecec et al., 1987):

$$X_{Fe_{2/3}O} = 0.0375\,X_{Fe} - 0.1008\,X^2_{Fe} + 0.2155\,X^3_{Fe} \qquad (26)$$

Other rocksalt structure oxides

A large number of activity and phase equilibrium data are available for other binary oxides with the rocksalt structure. These were summarized by Davis and Navrotsky (1983), who fitted them – as far as possible – to a one-site symmetric solution model:

$$a_{MO} = X_{MO}\gamma_{MO}\ ; \qquad RT\ln \gamma_{MO} = W(1-X_{MO})^2 .$$

Values of W were found to correlate very well with the volume mismatch between the two end-members (Table 1) defined as follows:

$$\Delta V = \frac{V_2 - V_1}{V_{\overline{12}}} \qquad (27)$$

where V_2 and V_1 are the molar volumes of larger and smaller components respectively and $V_{\overline{12}}$ is the mean of the two. Size mismatch is generally the most important contributing factor to nonideality because of the strain energy stored in the lattice when large ions are substituted onto small lattice positions or vice versa. One of the principal exceptions to the correlation is the join MgO-NiO which exhibits negative excess enthalpies and entropies of mixing due to Ni-Mg ordering (Davies and Navrotsky, 1981). Davies and Navrotsky argue that clustering of Mg next-nearest neighbors and Ni may have a tendency to stabilise the open-shell d^8 configuration of Ni^{2+}.

Spinel structure oxides

Most geologically important spinels are of the 2-3 $(A^{2+})[B_2^{3+}]O_4$ or 4-2 $(A^{4+})[B_2^{2+}]O_4$ type with cations occupying one tetrahedral, denoted ( ), and two octahedral, denoted [ ], sites per formula unit. If A ions occupy the tetrahedral sites then the configuration is normal, while if one of the B cations is tetrahedral, $(B^{3+})[A^{2+}B^{3+}]O_4$ for example, then the configuration is called inverse. The normal-inverse relationships of a number of important spinels are given in Table 2. In practise, most end-member spinels do not adhere absolutely to the idealised configurations and there is octahedral-tetrahedral disorder of A and B cations. This impacts activity-composition relations through configurational entropy of mixing, which depends on the extent of disorder. We will continue by addressing activity-composition relations for specific systems, deferring a detailed discussion of order-disorder relations until later.

Table 1. Nonidealities of rocksalt structure oxides
(Davies and Navrotsky, 1981)

System	W (kJ)	ΔV	T (K)
CoO-FeO	3.214	0.0317	1473
CoO-MgO	4.799	0.0412	1100-1300
CoO-NiO	4.934	0.0577	1000-1300
CaO-CdO	0	0.0661	1273
FeO-MnO	5.243	0.0911	1423
CoO-MnO	5.464	0.128	1273
MgO-MnO	18.277	0.169	1473
MnO-NiO	12.328	0.1854	1200
CaO-SrO	23.762	0.2167	1073-1473
CaO-MnO	13.734	0.2295	1375
CaO-FeO	33.231	0.319	1123-1318
CdO-NiO	40.673	0.3468	1323
CaO-CoO	30.899	0.355	1373-1573
CaO-MgO	60.611	0.395	2288-2703
CaO-NiO	46.091	0.4106	1373-1968

Table 2
Cation configuration in spinels

Formula	Cation distribution
$MgAl_2O_4$	Normal
$MgCr_2O_4$	Normal
$FeAl_2O_4$	Normal
$FeCr_2O_4$	Normal
Mg_2SiO_4	Normal
Fe_3O_4	Inverse
Fe_2TiO_4	Inverse

Fe_3O_4 - Fe_2TiO_4

This join is one of the most important mineralogically because of its importance as a petrogenetic indicator (Chapter 12) and because of the permanent magnetization carried by Fe_3O_4-rich spinels in many rocks (Chapter 4). Direct activity measurements have been carried out by Katsura et al. (1975) and by Woodland (1988) using phase equilibria and emf measurements respectively. Several other published phase equilibria studies may also be used to constrain activities as may the presence of a solvus at temperatures below about 550°C (Vincent et al., 1957; Lindsley, 1981).

Fe_3O_4 - Fe_2TiO_4 spinels coexist, at low fO_2 with metallic Fe and at higher fO_2 with wüstite FeO_y. Thus, activities are measured by the fO_2 displacement of Equilibrium (2) (WM) and metastable Equilibrium (9) (IM) as the Fe_3O_4 component is diluted by Fe_2TiO_4. The activity of Fe_2TiO_4 may then be obtained from the Gibbs-Duhem equation in the normal manner (e.g., Darken and Gurry, 1953). The method is complicated slightly by the fact that the stoichiometry of wüstite and hence of reaction (2) is fO_2-dependent, becoming poorer in O as fO_2 decreases. This is circumvented by treating wüstite as if it were a solid solution of Fe and O. Since the activity of Fe, relative to metal is 1.0 at the IW boundary (fO_2^o), the activity of Fe in FeO_y at any other fO_2 may be obtained from the Gibbs-Duhem equation as follows:

$$\log a_{Fe}^{fO_2} = -\frac{1}{2} \int_{fO_2^o}^{fO_2} y \, d \log fO_2$$

Thus, we actually use metastable Equilibrium (9):

$$\underset{\text{metal}}{3 \text{ Fe}} + 2 O_2 = \underset{\text{spinel}}{Fe_3O_4}$$

under conditions where wüstite is stable, by determining the activity of Fe as a function of fO_2 and stoichiometry y (from Vallet and Raccah, 1965).

Figure 4 shows emf activity measurements of Woodland (1988) at 800° and 900°C and phase equilibria data at 1000°-1300°C (Schmahl et al., 1960; Webster and Bright, 1961; Taylor, 1964; Katsura et al., 1976) from which we have derived values of $a_{Fe_3O_4}$ by the above method. As seen in Figure 4 and noted by Katsura et al. (1975), the high temperature data approximate $a_{Fe_3O_4}$ equal to $X_{Fe_3O_4}$. This should not be taken as near-ideal behavior, however, because the mixing of $Fe^{3+}(Fe^{2+}Fe^{3+})O_4$ and $Fe^{2+}(Fe^{2+}Ti^{4+})O_4$, both inverse spinels, would yield an ideal $a_{Fe_3O_4}$ of $X_{Fe_3O_4}^2$. Figure 4 shows that large excess free energies are needed to explain the activity data. The latter are best fit by taking explicit account of order-disorder and activity-composition relations in the manner discussed later. We find, however, that in this temperature range the macroscopic activity data can be reasonably well fit with a subregular model as follows:

$$a_{Fe_3O_4} = X_{Fe_3O_4}^2 \gamma_{Fe_3O_4} \tag{28}$$

$$RT \ln \gamma_{Fe_3O_4} = [10580 + 63060 \, X_{Fe_3O_4}] \, (1-X_{Fe_3O_4})^2 \text{ Joules} \tag{29}$$

$$RT \ln \gamma_{Fe_2TiO_4} = [42110 + 63060 \, X_{Fe_2TiO_4}] \, (1-X_{Fe_2TiO_4})^2 \text{ Joules}$$

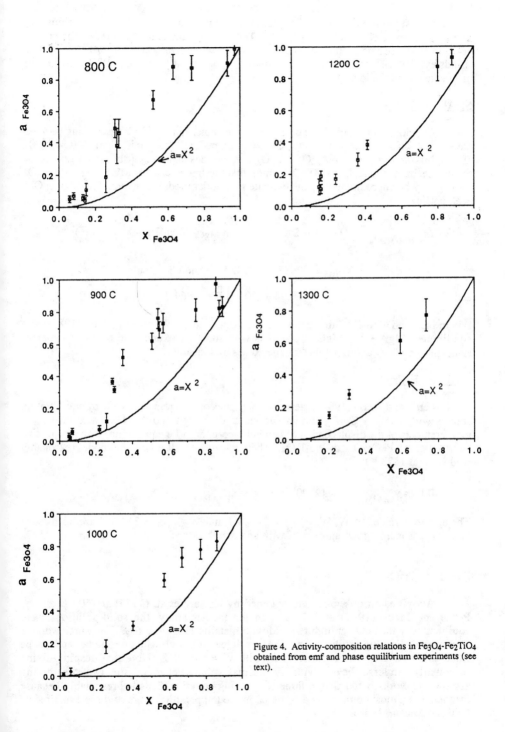

Figure 4. Activity-composition relations in Fe_3O_4-Fe_2TiO_4 obtained from emf and phase equilibrium experiments (see text).

It must be emphasized that because of their empirical nature these relationships may only be used between 800° and 1300°C. They do not, for example, predict the correct position of the solvus on the Fe_2TiO_4-Fe_3O_4 join which has a crest at, or below, 550°C (Lindsley, 1981), the crest being calculated to be slightly above 700°C towards the magnetite end of the join.

$MgAl_2O_4$ - Cr_2O_4

Activities on this join have been obtained by determining Al-Cr partitioning between the spinel solid solutions and Al_2O_3 - Cr_2O_3 mixtures as illustrated in Figure 2 (Oka et al., 1984). Mixing properties of Al_2O_3 - Cr_2O_3 sesquioxides had previously been measured by the emf technique (Jacob, 1978). Assuming that the free energy of $MgAl_2O_4$ - $MgCr_2O_4$ mixing may be approximated by the two-site subregular model, as in eqn. (28) above, Oka et al. obtain:

$$RT \ln \gamma_{MgAl_2O_4} = [W_{Sp} + 2X_{MgAl_2O_4}(W_{MgCr} - W_{Sp})](1 - X_{MgAl_2O_4})^2$$

where

$$W_{Sp} = 19686 + 0.0182\,P + 0.463\,T \quad \text{Joules}$$

and

$$W_{MgCr} = 23894 + 0.0504\,P + 1.964\,T \quad \text{Joules}$$

P is in bars and T in K. Similar results were obtained by Webb and Wood (1986) who equilibrated $MgAl_2O_4$ - $MgCr_2O_4$ spinels with sodic pyroxenes and obtained apparent spinel properties by assuming that the pyroxenes mix ideally.

$FeAl_2O_4$ - $FeCr_2O_4$

Petric and Jacob (1982) determined activities on this join at 1373 K by equilibrating spinels with Al_2O_3 - Cr_2O_3 solid solutions as discussed above. They also made emf measurements of the displacement of hercynite-corundum-Fe equilibrium with progressive substitution of $FeCr_2O_4$ for $FeAl_2O_4$. Their results also fit a two site model reasonably well with (at 1373 K):

$$RT \ln \gamma_{FeAl_2O_4} = [22900 + 17608\, X_{FeAl_2O_3}](1 - X_{FeAl_2O_4})^2$$

The apparent values of W, $W_{FeAl_2O_4}$ = 22900 J and $W_{FeCr_2O_4}$ = 31700 J are, as may be seen, in reasonably good agreement with analogous values for the $MgCr_2O_4$ - $MgAl_2O_4$ join.

Fe_3O_4 - $FeCr_2O_4$

Activities on this join were obtained by Katsura et al. (1975) at 1500 K and by Petric and Jacob (1982) at 1673 K. In the former study, the solid solutions were equilibrated with FeO_y at high fO_2 and with metallic Fe at low fO_2. The latter workers employed Pt-Fe alloy equilibration. The data agree quite well with one another and can be fitted to a two-site symmetric solution with W = 11000 J. Note that despite mixing (nominally) inverse Fe_3O_4 with the normal spinel $FeCr_2O_4$, the two-site mixing assumption works better than a three-site mixing equation. This is presumably because magnetite is almost completely disordered in the temperature range of these experiments (Wu and Mason, 1981).

Fe_3O_4 - $FeAl_2O_4$

There is a solvus on this join (Turnock and Eugster, 1962) which may be used to constrain activity-composition relations to some extent (Fig. 3). Petric et al. (1981) have also made direct activity measurements using the Pt-Fe alloy technique at 1573 K. The solvus is essentially symmetric and in the temperature range 873-1133 K fits the two-site mixing model (Eqn. 17) with a W value of 37000 J/mol. This model also fits, within experimental uncertainty, the activity data at 1573 K, although a better fit is obtained if explicit provision is made for the temperature dependence of octahedral-tetrahedral disorder (Nell and Wood, 1989).

$FeAl_2O_4$ - Fe_2TiO_4; $MgAl_2O_4$ - Mg_2TiO_4

Both these systems have well developed, slightly asymmetric miscibility gaps which may be used to estimate the extent of nonideality. If for simplicity we assume mixing of normal $FeAl_2O_4$ with inverse Fe_2TiO_4 (Table 2) then activity-composition relations obey Equation (26). The asymmetric solvi constrain excess properties also to be asymmetric, which we obtain by fitting the solvus using the method of Thompson and Waldbaum (1969) and representing activity by equations of the form of (28) and (29):

$MgAl_2O_4$ - Mg_2TiO_4

$$RT \ln \gamma_{MgAl_2O_4} = (43000 + 34400 \, X_{MgAl_2O_4})(1-X_{MgAl_2O_4})^2 \text{ J}$$

$$RT \ln \gamma_{Mg_2TiO_4} = (60200 - 34400 \, X_{Mg_2TiO_4})(1-X_{Mg_2TiO_4})^2 \text{ J}$$

$FeAl_2O_4$ - Fe_2TiO_4

$$RT \ln \gamma_{FeAl_2O_4} = (48500 + 27600 \, X_{FeAl_2O_4})(1-X_{FeAl_2O_4})^2 \text{ J}$$

$$RT \ln \gamma_{Fe_2TiO_4} = (62300 - 27600 \, X_{Fe_2TiO_4})(1-X_{Fe_2TiO_4})^2 \text{ J}$$

The solvi and apparent activity-composition relations are very similar on the two joins, but, as demonstrated earlier (Fig. 3), these are model-dependent and cannot be well constrained without a microscopic activity model.

RHOMBOHEDRAL OXIDES

Rhombohedral oxides contain cations in octahedral coordination and exhibit either long range order of cations, space group $R\bar{3}$ ($FeTiO_3$, $MgTiO_3$) or disorder (α-Fe_2O_3, α-Al_2O_3, Cr_2O_3) with space group $R\bar{3}c$.

Al_2O_3 - Cr_2O_3

Mixing properties on this join have been determined by Jacob (1978) using the emf technique to determine the conditions of equilibrium with Cr metal. Jacob's data were amplified by Chatterjee et al. (1982) who showed that the solid solution could be treated as subregular in the range 1073-1600 K:

$$a_{Cr_2O_3} = X^2_{Cr_2O_3} \gamma_{Cr_2O_3}$$

$$RT \ln \gamma_{Cr_2O_3} = [W_{Cr_2O_3} + 2X_{Cr_2O_3}(W_{Al_2O_3} - W_{Cr_2O_3})](1-X_{Cr_2O_3})^2$$

and

$$W_{Cr_2O_3} = 37484 + 0.0386\,P + 4.334\,T \quad \text{Joules}$$

$$W_{Al_2O_3} = 31729 + 0.0006\,P + 4.719\,T \quad \text{Joules}$$

with P in bars and T in K.

Al_2O_3 - Fe_2O_3

This system has been studied at atmospheric pressure by Muan and Gee (1956), Muan (1958) and Atlas and Sumida (1958). There is a large, slightly asymmetric miscibility gap which can be fitted by assuming a two site subregular model ($a = X^2 \gamma$) with the following approximate parameters:

$$RT \ln \gamma_{Fe_2O_3} = (68000 - 29200\,X_{Fe_2O_3})(1-X_{Fe_2O_3})^2 \quad J$$

$$RT \ln \gamma_{Al_2O_3} = (53400 + 29200\,X_{Al_2O_3})(1-X_{Al_2O_3})^2 \quad J$$

At high temperature (>1591 K) and high fO_2 (>0.03 bar), there is, additionally, a stability field of the 1:1 ordered phase $FeAlO_3$ (Muan, 1958).

Fe_2O_3 - Mn_2O_3

Immiscibility on this join (Muan and Somiya, 1962) yields a strongly asymmetric solvus which, if treated as a two-site asymmetric solution, yields approximately:

$$RT \ln \gamma_{Fe_2O_3} = (-9470 + 110630\,X_{Fe_2O_3}](1-X_{Fe_2O_3})^2 \quad J$$

$$RT \ln \gamma_{Mn_2O_3} = (45845 - 110630\,X_{Mn_2O_3}](1-X_{Mn_2O_3})^2 \quad J$$

$FeTiO_3$ - $MnTiO_3$

O'Neill et al. (1989) have determined the properties of Fe-Mn ilmenites in the temperature range 1050-1300 K by emf measurements on the assemblage iron-rutile-ilmenites$_{ss}$. They found that Fe-Mn mixing is attended by small positive nonidealities which may be represented by a symmetrical (one-site) model with:

$$W = 2200 \pm 300 \quad J/mol$$

$FeTiO_3$ - $MgTiO_3$

The properties of Fe-Mg ilmenites may be estimated from the measured partitioning of iron and magnesium between coexisting ilmenite and olivine (Andersen and Lindsley,

1979, 1981; Bishop, 1976). At high temperatures ($FeSi_{0.5}O_2$ - $MgSi_{0.5}O_2$) olivines exhibit small, positive, symmetric deviations from ideality (Wiser and Wood, 1991) with, on a one-site basis:

$$ln\, \gamma_{MgSi_{0.5}O_2} = \frac{445}{T}(1 - X_{MgSi_{0.5}O_2})^2$$

The partitioning data (Fig. 5) may be treated in the same way as described earlier to obtain Al-Cr mixing in spinels. Alternatively, a plot of $RT\, ln\, Q$

$$\left[Q = \frac{a^{ol}_{MgSi_{0.5}O_2} \cdot X^{ilm}_{FeTiO_3}}{a^{ol}_{FeSi_{0.5}O_2} \cdot X^{ilm}_{MgTiO_3}}\right] \quad \text{versus} \quad X^{ilm}_{MgTiO_3} \quad \text{(Fig. 5)}$$

shows the extent of nonideality. If $RT\, ln\, Q$ were a linear function of X_{MgTiO_3} then, from the relationship between Q and the equilibrium constant, it is easily found that the solution would be symmetric with

$$W = 1/2 \frac{d(RT\, ln\, Q)}{dX_{MgTiO_3}}$$

The data at 800° and 900°C (Fig. 5) yield $W \approx 4300$ J/mole on the symmetric model. It might be argued from Figure 5 that $RT\, ln\, Q$ is a nonlinear function of X_{MgTiO_3}, in which case the solution is asymmetric or more complex. In view of the relatively large errors in Q at very low and very high values of X_{MgTiO_3}, however, it is difficult to justify a model more complex than the symmetric one.

Fe_2O_3 - $FeTiO_3$

The hematite-ilmenite join is petrologically the most important of the rhombohedral oxides. It exhibits several complexities, the most important of these being the transformation from octahedrally disordered R3c structure at pure Fe_2O_3 to ordered $R\bar{3}$ at the $FeTiO_3$ end. As may be seen from Figure 6 (Chapters 6 and 8) the second order transformation is both composition- and temperature-dependent and leads to a miscibility gap at temperatures below 700°C.

Activity-composition estimates may be made in several different ways. We commenced by using the large number of available data on the compositions of coexisting Fe-Ti spinel and rhombohedral oxide solid solutions (Webster and Bright, 1961; Buddington and Lindsley, 1964; Katsura et al. 1976; Spencer and Lindsley, 1981; Hammond et al., 1982; Andersen and Lindsley, 1988). These yielded information on the equilibria:

$$\begin{array}{cccc} FeTiO_3 + Fe_3O_4 & = & Fe_2O_3 + Fe_2TiO_4 \\ \text{rhomb} \quad \text{sp} & & \text{rhomb} \quad \text{sp} \end{array} \quad (30)$$

and

$$\begin{array}{cc} 4\, Fe_3O_4 + O_2 = & 6\, Fe_2O_3 \\ \text{sp} & \text{rhomb} \end{array} \quad (31)$$

All of the experimental results refer to ilmenite-rich ($R\bar{3}$) solid solutions, so we are able, from Equation (30), to use our measurements of Fe_3O_4-Fe_2TiO_4 properties (Fig. 4) to obtain activities of Fe_2O_3-$FeTiO_3$ solutions. Given reasonable estimates of uncertainties in the data, which are up to ±2 mol % in composition and ±0.2 log units in fO_2 for the high

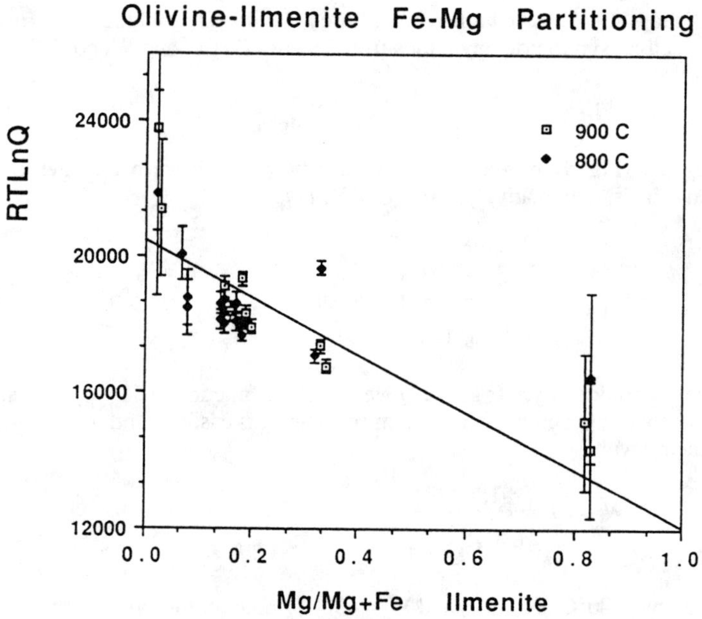

Olivine-Ilmenite Fe-Mg Partitioning

Figure 5. Plot of $RT \ln Q \left[Q = \dfrac{a^{ol}_{MgSi_{0.5}O_2} \cdot X^{ilm}_{FeTiO_3}}{a^{ol}_{FeSi_{0.5}O_2} \cdot X^{ilm}_{MgTiO_3}} \right]$ as a function of $MgTiO_3$ content of ilmenite. Negative slope indicates ilmenite is nonideal (see text).

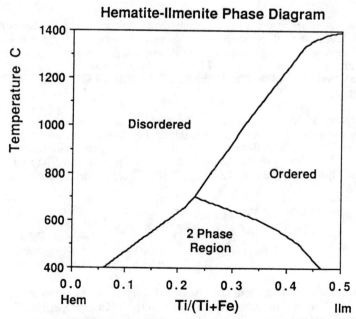

Figure 6. Order-disorder and miscibility relations in $FeTiO_3$ - Fe_2O_3 (after Burton, 1984).

temperature results of Webster and Bright (1961), we find that $a_{Fe_2O_3}$ can be described by a two-site symmetric model:

$$RT \ln a_{Fe_2O_3} = RT \ln X^2_{Fe_2O_3} + W_{hem\text{-}ilm}(1-X_{Fe_2O_3})^2$$

with $\quad W_{hem\text{-}ilm} = 30.0 \pm 8.0$ kJ/mol

This simple model, taken in conjunction with the Fe_3O_4-Fe_2TiO_4 results, is adequate to describe all of the results on the exchange Equilibrium (30) and the redox Equilibrium (31). We must emphasize again, however, that it refers only to ordered ilmenite-rich solid solutions.

An additional estimate of nonideality in $FeTiO_3$-rich solutions may be obtained from the form of the miscibility gap in Figure 6. Although this is not a normal solvus because of its intersection with the second-order transformation, we can use the right-hand limb to make an estimate of $W_{hem\text{-}il}$ using Equation (19) and assuming a metastable mirror-image left-hand limb. The result is about 32 kJ/mol. This must be an overestimate because the true solvus in (R$\bar{3}$) structure should lie metastably within the observed miscibility gap.

Finally, Ghiorso (1990) (Chapter 6) has developed a much more detailed model of $FeTiO_3$-Fe_2O_3-$MgTiO_3$ solid solutions with provision for intermediate ordering states. Although he didn't have our spinel activity data he obtained a result in very good agreement with ours: $W_{hem\text{-}il}$ of 24 kJ/mol in the ordered R$\bar{3}$ structure.

MICROSCOPIC PROPERTIES

As has been emphasized several times in this discussion, the entropy of mixing, S_{mix}, provides the temperature dependence of activity in a solid solution. The dominant contribution is configurational and it is necessary to know the partitioning of atoms between any nonequivalent sites in the structure in order to estimate S_{conf} with confidence. In simple rocksalt structure oxides it is generally assumed that the cations mix randomly on one site per formula unit. That this is not always true is evident from (Mg,Ni)O solid solutions which exhibit large negative deviations from ideal S_{mix} probably due to Mg-Ni clustering (Davies and Navrotsky, 1981). Deviations from ideal S_{mix} are also apparent in magnesiowüstites (Mg,Fe)O$_y$ in which significant proportions of Fe^{3+} occur. Dissolution of Fe^{3+} in the structure occurs with incorporation of a nominal negatively charged vacancy as follows:

$$3\,Fe^{2+} + 1/2\,O_2 = 2\,Fe^{3+} + V''_{Fe} + FeO$$

where V''_{Fe} denotes the vacancy.

Several experimental techniques (neutron diffraction, Mössbauer spectroscopy, IR spectroscopy) have shown that the vacancies are not randomly distributed in the structure, but tend to associate with Fe^{3+} as short-range ordered clusters. Furthermore, some of the Fe^{3+} enters tetrahedral interstices rather than the "normal" octahedral positions. Thus, the configurational entropy of Fe^{2+}-Fe^{3+}-V''_{Fe} mixing is far from the ideal value and activity-composition relations are correspondingly more complex. Although these are yet not completely described, Brynestad and Flood (1958) suggested that dissolution of Fe^{3+} in magnesiowustite could be explained by complete Fe^{3+}-V''_{Fe}-Fe^{3+} association and more recent data (Valet et al., 1975) are in reasonable agreement with such a model.

Given that microscopic order-disorder relations may be important even in the "simplest" compounds, it is clear that these problems are magnified in spinels and rhombohedral oxides where long-range ordering is present. In our earlier discussion we assumed ideal or near-ideal configurational entropies for these phases and used simple models (symmetric or asymmetric) to describe nonideal behaviour. Now we wish to consider some examples of how configurational entropies may actually be determined.

Most of the techniques available for intraphase partitioning measurements have been described extensively elsewhere. *Reviews in Mineralogy* Volume 18 (1988) contains, for example, discussions of vibrational, Mössbauer, optical, NMR, EXAFS and EPR techniques, all of which provide information on order-disorder relations. Although most of these methods may be applied at high temperature, they are much more readily and accurately applicable at room temperature and below on quenched samples. The same applies to X-ray, neutron and electron diffraction methods. Since, however, it is difficult to quench-in intraphase partitioning in oxides (e.g., Wood et al., 1986), an in-situ, high temperature technique is really required. In some cases this may be provided by measurement of thermopower and electrical conductivity, a simple in-situ method that can be used at high temperatures under controlled pO_2. It is applicable to Fe-bearing samples provided the electrical conduction occurs by electron-hopping. Since this is the technique with which we have been most closely associated, we will continue by describing its application to magnetite and to more complex Fe-spinels.

Electrical conduction in Fe_3O_4

At temperatures above about 300 K, magnetite conducts by the small polaron mechanism (Wu and Mason, 1981), characterised by a small activation energy (0.12 eV) and low carrier mobilities of about 0.1 cm^2/V·s (Dieckmann et al., 1983; Austin and Mott, 1969). In this mechanism, the carrier (either an electron, n-type, or hole, p-type) is trapped or localised by a lattice distortion. Conduction occurs by thermally activated hopping which in Fe_3O_4 involves an electron hopping from Fe^{2+} to its nearest neighbor Fe^{3+}. The electrical properties thermoelectric power (Q) and conductivity (σ) depend on the concentration of donors (Fe^{2+}) and acceptors (Fe^{3+}) and on how these are distributed between octahedral and tetrahedral sites in the lattice.

In a small polaron conductor the thermoelectric power is given by (e.g., Chaiken and Beni, 1976):

$$Q = \frac{(\mu/e - S)}{T}$$

where e is the electric charge, T the temperature, μ the chemical potential and S the vibrational entropy associated with electron transport. From the definition of chemical potential we can introduce the normal extensive variables as follows:

$$\frac{\mu}{T} = -\left(\frac{\partial S}{\partial N}\right)_{E,V}$$

where N is the number of particles and S, E, V have their usual meanings. The entropy of the conducting system may be obtained from the Boltzmann relationship ($S = k \ln W$), which yields:

$$Q = -\left(\frac{S}{T} + \frac{k}{e}\left(\frac{\partial \ln W}{\partial N}\right)\right)$$

The permutability (W) is calculated for a system with a fixed carrier concentration, i.e., N_A sites ($Fe^{2+} + Fe^{3+}$) and N randomly distributed electrons (Fe^{2+}). In systems where electron-electron repulsion is sufficiently large to prevent double occupancy of a jump site (RT << coulomb on-site interaction), the N electrons distributed over N_A sites may be either spin-up or spin-down and W is given by:

$$W = \frac{N_A! \, 2^N}{N! \, (N_A - N)!}$$

which leads to the following expression for Q:

$$Q = -\frac{k}{e} ln\left[\frac{2(1-c)}{c}\right] - \frac{S}{T} \tag{32}$$

in Equation (32) c is the ratio of electrons to total sites (N/N_A) which is equivalent to [Fe^{2+}/($Fe^{2+} + Fe^{3+}$)]. The entropy of transport term is predicted to be very small (~10 μV/K) if conduction takes place between sites of similar energy (Austin and Mott, 1969; Emin, 1975) so that for most practical purposes it may be neglected, yielding:

$$Q = -\frac{k}{e} ln\left[\frac{2\,Fe^{3+}}{Fe^{2+}}\right] \tag{33}$$

From Equation (33) we obtain a predicted thermopower of -119 μV/K for pure Fe_3O_4 if all possible hopping mechanisms (i.e., tetrahedral-tetrahedral, octahedral-tetrahedral and tetrahedral-octahedral) are equally probable. In fact, the observed thermopower at 600°C (Fig. 7) is about -85 μV/K (Wu and Mason, 1981). This implies that Fe^{3+}/Fe^{2+} on the conducting sites is lower than the stoichiometric value and, since Fe_3O_4 is dominantly inverse at low temperatures (e.g., O'Neill and Navrotsky, 1983), it implies that conductivity is restricted to octahedral-octahedral hopping.

Several other lines of evidence support the inference that hopping is confined to octahedral sites. The Mossbauer spectrum of Fe_3O_4 (Kündig and Hargrove, 1969) indicates electron delocalization on the octahedral sites with an effective charge of $Fe^{2.5+}$. Thermally activated hopping on octahedral sites was also found in the Mossbauer spectra of Zn-Ti ferrites by Lotgering and Van Diepen (1977). Amthauer and Rossman (1984) pointed out that electron delocalization as observed using the Mossbauer effect is only found in minerals in which Fe^{2+} and Fe^{3+} sites form infinite chains (e.g., ilvaite) sheets (cronstedite) or three-dimensional networks (as in the octahedral sublattice of magnetite). Molecular orbital calculations (Sherman, 1987) also show that Fe^{2+}-Fe^{3+} charge transfer is energetically favoured where the two sites involved share a common face or edge. This occurs with neighboring octahedra in spinels but neighboring tetrahedra are isolated from one another and octahedra share only corners with tetrahedra.

Thus, experimental and theoretical evidence indicate that electrical conduction in magnetite takes place by octahedral electron hopping. This model was used by Wu and Mason to calculate Fe^{2+} - Fe^{3+} distribution in magnetite (Fig. 7b) with the anticipated result of increasing disorder (lower Q) with increasing temperature.

Measurement technique

Thermopower measurements are preferably made on single crystal samples, but in the case of solid solutions large homogeneous single crystals are difficult to obtain. High density sintered polycrystalline slabs are therefore generally used. The sample pellet is cut

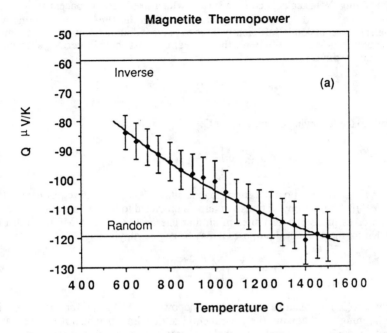

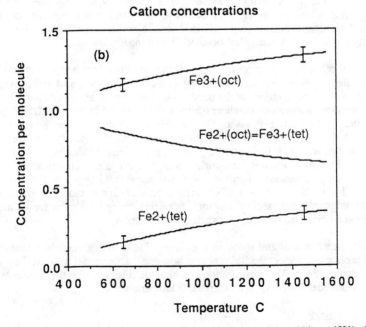

Figure 7. (a) Thermopower (Q) in Fe_3O_4 as a function of temperature (Wu and Mason, 1981). (b) Cation distributions calculated from Q assuming octahedral electron hopping (see text).

to a bar, four holes drilled in it, and then mounted on four thermocouples. The thermocouples (typically $Pt_{94}Rh_6$-$Pt_{70}Rh_{30}$) are threaded through a 13-hole ceramic and the sample pulled up tight to ensure good electrical contact. (A small amount of Pt paint around the thermocouple bead may improve contact). The sample assembly is then lowered into the furnace and closed in a 15°-20°C temperature gradient. After annealing at high temperature and controlled fO_2, selected to maintain Fe^{2+}-Fe^{3+} stoichiometry, measurements may begin.

The thermocouples provide measurements of temperature and their common $Pt_{94}Rh_6$ leads are used to measure the thermoelectric voltage ΔV of the sample. The four thermocouples provide six ΔT and corresponding ΔV pairs which, when corrected for the small thermoelectric voltage of the leads provides thermopower as follows :

$$Q = \lim_{\Delta T \to 0} \frac{DV}{DT} \qquad \mu V/K$$

Electrical conductivity (d.c.) is measured by passing a small (10^{-4} A) current between the outer thermocouple leads and measuring the voltage drop between the inner leads. The advantage of a four-point technique is that it circumvents possible non-ohmic electrode effects that may exist at the contact between the current-carrying leads and the much less conducting sample material. We then have:

$$\sigma = \frac{i \cdot l}{A \cdot V} \qquad (\Omega\ cm)^{-1}$$

where i is the current, l is the distance between the two inner electrodes, A is the cross-sectional area of the sample and V is the measured voltage drop between the inner electrodes.

<u>Fe_3O_4 - $MgFe_2O_4$</u>

The measurements on Fe_3O_4 can obviously be extended to any spinel in which the conduction mechanism is the same. Figure 8a shows thermopower in different Fe_3O_4-$MgFe_2O_4$ solid solutions measured in-situ at 600°-1400°C. As can be seen, there is a monotonic decrease in Q with increasing $MgFe_2O_4$ content, an observation that can be related to increasing Fe^{3+}/Fe^{2+} ratio. Since, however, there are two independent cation exchange equilibria:

$$\underset{oct}{Fe^{3+}} + \underset{tet}{Fe^{2+}} = \underset{oct}{Fe^{2+}} + \underset{tet}{Fe^{3+}}$$

$$\underset{oct}{Fe^{3+}} + \underset{tet}{Mg^{2+}} = \underset{oct}{Mg^{2+}} + \underset{tet}{Fe^{3+}}$$

thermopower alone is not adequate to characterise cation partitioning completely. Additional information may be obtained from electrical conductivity. The electrical conductivity of a small polaron conductor may be related as follows to the expression for diffusion (D) in a three-dimensional lattice (e.g., Dieckmann and Schmalzreid, 1977):

$$D = \frac{a^2 \Gamma}{6} \tag{34}$$

where Γ is the jump rate and a the jump distance. At high temperatures (T>Debye temperature) the jump rate follows an Arrhenius relationship (Emin, 1982; Dieckmann et al., 1983):

$$\Gamma = [\nu \cdot Z \cdot \beta \exp(-E_H/kT)] \cdot P \tag{35}$$

where ν is the lattice vibration frequency, z the coordination number, ß is the fraction of available jump sites (i.e., 1–c), E_H is the activation energy and P is the probability that a jump will be successfully executed when a jump opportunity occurs. In the adiabatic limit of small polaron transport, P is essentially 1.0 (Emin, 1982; Tuller and Nowick, 1977). Equations (34) and (35) may now be combined with the Nernst-Einstein equation for mobility (m) of a charge carrier (m = De/kT) to give:

$$m = \frac{(1-c) \cdot \nu \cdot g \cdot e \cdot a^2}{kT} \exp\left(\frac{-E_H}{kT}\right)$$

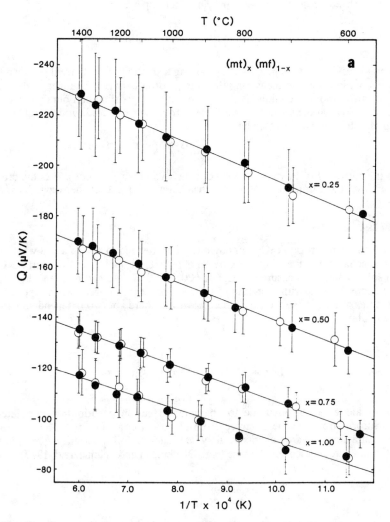

Figure 8. (a) Thermopower (Q) in Fe$_3$O$_4$ (mt) - MgFe$_2$O$_4$ (mf) solid solution. Solid and open symbols refer to measurements taken during "down" and "up" temperature cycles, respectively. [See nexy page.] (b) Conductivity ($\Omega \cdot$cm)$^{-1}$ for the same join plotted as $ln(\sigma T)$ vs. 1/T. (c) Calculated cation distributions at 1000°C assuming octahedral electron hopping. (Solid lines are best-fit O'Neill-Navrotsky model; see text).

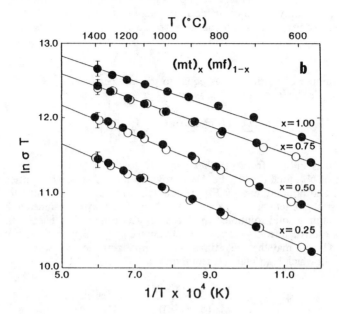

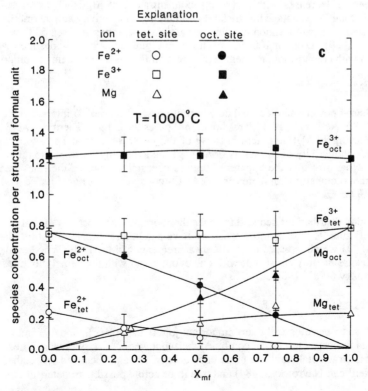

Figure 8, continued. See previous page for legend.

where g is z/6, i.e., unity for an octahedral hopping mechanism. The electrical conductivity (σ) is obtained from the mobility through the relation:

$$\sigma = m \cdot e \cdot N$$

Because the number of particles N is equal to the number of ($Fe^{2+} + Fe^{3+}$) positions multiplied by c [$Fe^{2+}/(Fe^{2+} + Fe^{3+})$], σ is given by:

$$\sigma = \frac{N_A \cdot c(1-c) \cdot g \cdot a^2 \cdot e^2 \nu}{kT} \exp\left(\frac{E_H}{kT}\right) \tag{36}$$

The quantity c is known from the thermopower so that measurement of σ enables determination of N_A, total ($Fe^{2+} + Fe^{3+})_{oct}$ provided jump distance (a), lattice vibration frequency (ν) and hopping energy E_H are known. Mason (1987) argued that a and ν should not be strong functions of composition and that they may be assumed to be constant for Fe_3O_4-containing solid solutions which are not very dilute. In this case, the hopping energy may be estimated from the apparent activation energy (E_a) obtained from a plot of ln σT versus 1/T. N_A may then be estimated by normalizing the solid solution data to that for pure Fe_3O_4 in which N_A (total octahedral iron) is 2.0:

$$N_A(ss) = \frac{2 \sigma_{ss}(c \cdot (1-c))_{mt}}{\sigma_{mt}(c \cdot (1-c))_{ss}} \exp\left(\frac{E_a(ss) - E_a(mt)}{kT}\right) \tag{37}$$

Figure 8b shows the compositional dependence of σ for the Fe_3O_4 - $MgFe_2O_4$ join. As can be seen, σ decreases with increasing concentration of $MgFe_2O_4$ as a result of the decrease in total Fe on octahedral sites. By combining conductivity and thermopower data we can calculate apparent cation distributions at all temperatures between 600° and 1400°C (Nell et al., 1989). Data for 1000°C are shown in Figure 8c. Note that despite large shifts in Q and σ with composition, the degree of inversion Fe^{3+}_{tet} is independent of composition.

Fe_3O_4 - $FeAl_2O_4$

Thermopower and electrical conductivity data, at 25 mol % intervals (Mason and Bowen, 1981; Nell et al., 1989), are shown in Figures 9a and 9b. Thermopower increases with increasing $FeAl_2O_4$ content because of the decrease in total Fe^{3+} concentration. Similarly, conductivity decreases as $FeAl_2O_4$ is added owing (Eqn. 36) to the decline in total Fe. Note also that the apparent hopping energies (slopes of the lines in Fig. 9b) increase monotonically with increasing $FeAl_2O_4$ content from 0.12 eV (at Fe_3O_4) to 0.40 eV (at 25% Fe_3O_4).

Cation distributions deduced from the thermopower /conductivity data at 1000°C are shown in Figure 9c. For completeness we have added an estimate of cation disorder in pure $FeAl_2O_4$ which is based on NMR measurements of $MgAl_2O_4$ (Wood et al., 1986). The results are broadly consistent with one another and with the spinel model of O'Neill and Navrotsky (1983, 1984).

Fe_3O_4 - $FeCr_2O_4$

This is a petrologically very important join (Chapters 6, 9, and 10) and one which provides an important test of the combined thermopower/conductivity technique. Cr^{3+}, by virtue of its large octahedral site preference energy, remains localized on octahedral sites (e.g., O'Neill and Navrotsky, 1984) and it is in principle possible to determine Fe^{2+}-Fe^{3+}

distribution by *either* measuring Q (Eqn. 33) or σ (Eqn. 36). Data for Q and cation distributions derived from them at 1400°C are shown in Figures 10a and 10b, respectively. We tested the correspondence between Q and σ by combining them to calculate the total ($Fe^{2+}+Fe^{3+}$) concentration on octahedral sites from Equation (37). This result may then be compared with that anticipated from the assumption that all Cr is on octahedral sites,

i.e., $$(Fe^{2+} + Fe^{3+})^{oct}_{theor} = (2 - 2\, X_{FeCr_2O_4})$$

We obtain excellent agreement at 75% Fe_3O_4 and 50% Fe_3O_4, but physically impossible results at 25% Fe_3O_4. Inspection of the composition dependence of activation energy (Fig. 10c) reveals a rapid increase with $X_{FeCr_2O_4}$ at about 60% $FeCr_2O_4$, similar to that found in Fe_3O_4 - $MgCr_2O_4$ by Verwey et al. (1947). This is almost certainly due to a change of conduction mechanism at high Fe_3O_4 dilution.

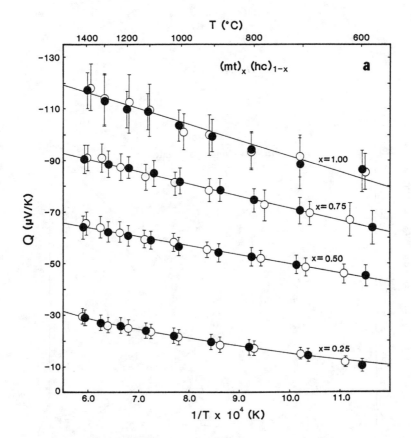

Figure 9. (a) Q in Fe_3O_4 (mt) - $FeAl_2O_4$ (hc) solid solutions (symbols as in Fig. 8). (b) Conductivity as a function of composition and temperature for Fe_3O_4-$FeAl_2O_4$. (c) Calculated cation distributions at 1000°C. Solid lines are best-fit O'Neill-Navrotsky values (see text).

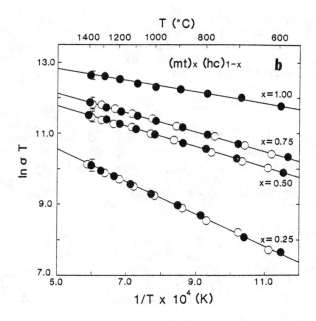
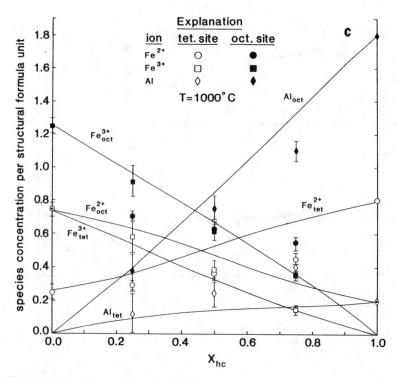

Figure 9, continued. See previous page for legend.

Thus, we find the thermopower/conductivity results to be consistent with the octahedral Fe^{2+}-Fe^{3+} hopping mechanism over a large part of the Fe_3O_4 - $FeCr_2O_4$ join, with a change in conduction mechanism (possibly involving Cr^{3+}) near the $FeCr_2O_4$ end. Cation distributions based on the results which are consistent with octahedral hopping are shown in Figure 10b. As might be expected, cation distributions become more normal (increasing Fe^{2+}_{tet}) with increasing $FeCr_2O_4$ content.

Fe_3O_4 - Fe_2TiO_4

The thermopower technique was used by Trestman-Matts et al. (1983) to measure apparent octahedral-tetrahedral partitioning of Fe^{2+} and Fe^{3+} in titanomagnetites. We present their data in Figure 11 as a plot of $RT \ln K_{CD}$ versus the concentration of Fe^{3+} in tetrahedral sites, where

$$K_{CD} = \frac{(Fe^{2+})_{oct} \cdot (Fe^{3+})_{tet}}{(Fe^{3+})_{oct} \cdot (Fe^{2+})_{tet}}$$

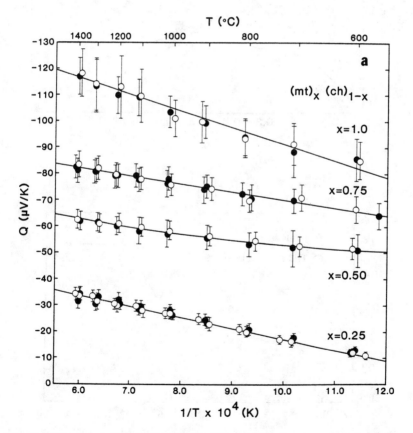

Figure 10. (a) Thermopower for Fe_3O_4 (mt) - $FeCr_2O_4$ (ch) solid solutions (symbols as in Figure 8). (b) Cation distributions at 1400°C obtained from Q and σ data at $X_{mt} > 0.4$ (see text). (c) Compositional dependence of activation energy on the join Fe_3O_4-$FeCr_2O_4$. Note apparent rapid increase close to $FeCr_2O_4$.

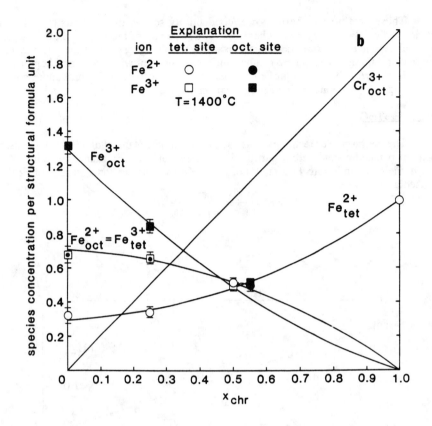

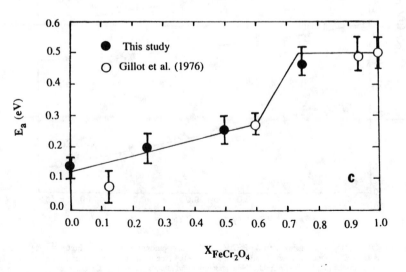

Figure 10, continued. See previous page for legend.

(Note that the sample described as containing 58 mol % Fe_2TiO_4 by Trestman-Matts et al. actually contains 49% Fe_2TiO_4.) As can be seen from Figure 11, Fe^{2+}-Fe^{3+} partitioning is essentially a linear function of $X^{tet}_{Fe^{3+}}$ over a very wide range of composition and temperature. This is a result consistent with the O'Neill-Navrotsky (1983, 1984) spinel model.

More complex spinels

Each of the binaries discussed so far has a maximum of three cations distributed over octahedral and tetrahedral sites. With two independent partitioning equilibria, measurements of thermopower and conductivity are sufficient to characterise distributions, provided the octahedral hopping mechanism applies. The technique may be extended to spinels in which four cations disorder (Fe_3O_4 - $MgAl_2O_4$ for example) by using the Fe^{2+}-Fe^{3+} distribution to measure the partial molar entropy of Fe_3O_4 and applying the Gibbs-Duhem equation (e.g., Darken, 1950) to obtain the partial molar entropy of $MgAl_2O_4$. $S_{MgAl_2O_4}$, combined with Fe^{2+}-Fe^{3+} distribution, gives Mg-Al distribution between octahedral and tetrahedral sites.

Integrating order-disorder and activity relations

O'Neill and Navrotsky (1983, 1984) have shown from lattice energy arguments that the enthalpy of disordering of a spinel $A^{2+}B_2^{3+}O_4$ or $A^{4+}B_2^{2+}O_4$ should be a quadratic function of the degree of inversion X:

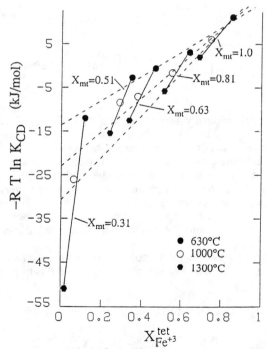

Figure 11. Plot of $-RT \ln K_{CD}$ $\left(K_{CD} = \frac{(Fe^{2+})_{oct} \cdot (Fe^{3+})_{tet}}{(Fe^{3+})_{oct} \cdot (Fe^{2+})_{tet}} \right)$ versus tetrahedral Fe^{3+} content for Fe_3O_4 (mt) - Fe_2TiO_4 spinels from Trestman-Matts et al. (1983) at 630°, 1000° and 1300°C. The O'Neill-Navrotsky model requires $RT \ln K_{CD}$ to be a linear function of $X^{tet}_{Fe^{3+}}$. The data agree well with the model, except at very low values of $X^{tet}_{Fe^{3+}}$.

$$\Delta H_E = \alpha_E X + \beta_E X^2 \tag{38}$$

In Equation (38) X is the number of atoms of B on A sites or of A atoms on B sites per formula unit. The partitioning of, for example, Mg^{2+} and Al^{3+} between octahedral and tetrahedral sites may be considered in terms of the exchange equilibrium:

$$Mg^{2+}_{tet} + Al^{3+}_{oct} = Mg^{2+}_{oct} + Al^{3+}_{tet} \tag{39}$$

which, in the O'Neill-Navrotsky model has a partition coefficient given by:

$$-RT \ln \left(\frac{X^{tet}_{Al^{3+}} \cdot X^{oct}_{Mg^{2+}}}{X^{tet}_{Mg^{2+}} \cdot X^{oct}_{Al^{3+}}} \right) = \alpha_{Mg-Al} + 2\beta \, n^{oct}_{Mg}$$

where n^{oct}_{Mg} is the degree of inversion. O'Neill and Navrotsky (1983, 1984) showed that this type of equation fitted the available data on partitioning in a number of simple spinels quite well. They also suggested that values of β of -20 kJ/mol and -60 kJ/mol were appropriate for 2-3 and 4-2 spinels, respectively.

In the case of Fe^{2+}-Fe^{3+} disorder in Fe_3O_4 (Wu and Mason, 1981) a temperature-dependence of the exchange free energy is needed to explain the approach to complete disorder at high temperatures. This yields:

$$-RT \ln \left(\frac{X^{tet}_{Fe^{3+}} \cdot X^{oct}_{Fe^{2+}}}{X^{tet}_{Fe^{2+}} \cdot X^{oct}_{Fe^{3+}}} \right) = \alpha_{Fe^{2+}-Fe^{3+}} + 2\beta \, n^{oct}_{Fe^{2+}} - T\sigma_{Fe^{2+}-Fe^{3+}} \tag{40}$$

In Figure 12 we show our measured values of Fe^{2+}-Fe^{3+} partition coefficient for several binary 2-3 spinels (Nell et al., 1989). As can be seen, the data indicate a good correlation between total degree of inversion (total 3+ ions on tetrahedral sites) and the left hand side of Equation (40). This indicates close adherence to the O'Neill-Navrotsky model. In Table 3 we compare values of α and β obtained from our studies with those presented by O'Neill and Navrotsky (1984). As can be seen, agreement is, if not perfect, very good indeed. We therefore have a good basis on which to calculate cation distributions in complex spinels. In such cases, there are several exchange equilibria of the form of (39) and the degree of inversion X is the total number of trivalent ions on tetrahedral sites.

Integration of microscopic with macroscopic properties is most readily achieved by separating the free energy of the solid solution into configurational entropy and vibrational free energy terms:

$$G = G^* - TS_{conf} \tag{41}$$

The configurational entropy is then obtained from cation partitioning using:

$$S_{conf} = -R \sum_i X^{tet}_i \ln X^{tet}_i - 2R \sum_i X^{oct}_i \ln X^{oct}_i$$

while the vibrational free energy is expanded as a Taylor series in order and compositional parameters as discussed by Sack (1982), Hill and Sack (1987) and Nell and Wood (1989).

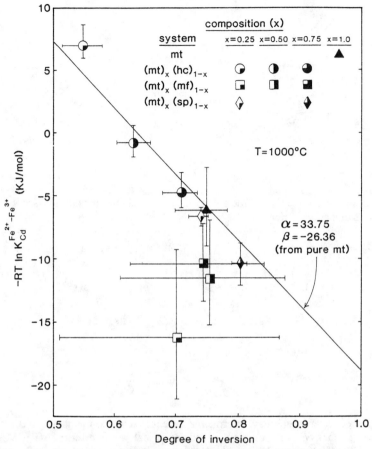

Figure 12. Similar plot to Figure 11 for solid solutions of Fe$_3$O$_4$ (mt), FeAl$_2$O$_4$ (hc), MgFe$_2$O$_4$ (mf) and MgAl$_2$O$_4$ (sp). In this case $-RT \ln K_{CD}$ is plotted against total 3 + (Fe^{3+} + Al^{3+}) on tetrahedral sites. Note that most solid solutions agree reasonably well with the O'Neill-Navrotsky model.

Table 3. Free energy parameters for spinel disorder (kJ)

Cation pair	O'Neill and Navrotsky (1984)		Nell and Wood (1989)	
	α	β	α	β
Fe^{2+}-Fe^{3+}	16.0 +0.0058T	-20.0	33.75	-26.36
Fe^{2+}-Al	52.0 +0.0058T	-20.0	50.98	-32.85
Mg-Fe^{3+}	20.0	-20.0	20.96	-19.13
MgAl	56.0	-20.0	38.19	-25.62

If, for example, we treat the solid solution Fe$_3$O$_4$ - MgAl$_2$O$_4$ - MgFe$_2$O$_4$ - FeAl$_2$O$_4$, then there are two compositional (r_1, r_2) and three order parameters (s_1, s_2, s_3) required to fully characterise it (Nell and Wood, 1989). One unique set would be:

$$r_1 = 1 - \left(X_{Mg}^{tet} + 2X_{Mg}^{oct} \right)$$

$$r_2 = \frac{1}{2}\left(X_{Al}^{tet} + 2X_{Al}^{oct}\right)$$

$$s_1 = X_{Al}^{oct} - \frac{1}{2}X_{Al}^{tet}$$

$$s_2 = X_{Fe^{3+}}^{oct} - \frac{1}{2}X_{Fe^{3+}}^{tet}$$

$$s_3 = 2X_{Mg}^{oct}$$

The vibrational part of the Gibbs free energy G^* may then be expanded as a second degree Taylor series in the five parameters:

$$\begin{aligned}G^* =\ & g_0 + g_{r_1}r_1 + g_{r_2}r_2 + g_{s_1}s_1 + g_{s_2}s_2 + g_{s_3}s_3 + \\ & g_{r_1r_1}r_1^2 + g_{r_2r_2}r_2^2 + g_{s_1s_1}s_1^2 + g_{s_2s_2}s_2^2 + g_{s_3s_3}s_3^2 + \\ & g_{r_1r_2}r_1r_2 + g_{s_1s_2}s_1s_2 + g_{s_1s_3}s_1s_3 + g_{s_2s_3}s_2s_3 + \\ & g_{r_1s_1}r_1s_1 + g_{r_1s_2}r_1s_2 + g_{r_1s_3}r_1s_3 + g_{r_2s_1}r_2s_1 + \\ & g_{r_2s_2}r_2s_2 + g_{r_2s_3}r_2s_3\end{aligned}$$

Each of the coefficients (g_{ij}) may, in principle, be equated to end-member free energies μ_i^o, symmetric interaction parameters W_{ab} and reciprocal solution terms ΔG_r^o (Wood and Nicholls, 1978). This is done by initially deriving numerical values of the coefficients from the macroscopic and microscopic properties of the end-members and binaries. The latter may also, in general, be fairly easily described in terms of the μ_i^o, W_{ab} and ΔG_r^o, so that it is possible to work out the relationships between these more familiar parameters and the g_{ij} coefficients.

For example, with the order and compositional parameters used above (Nell and Wood, 1989), the condition of internal equilibrium is:

$$\left(\frac{\partial G}{\partial s_1}\right)_{P,T,r_1,r_2,s_2,s_3} = \left(\frac{\partial G}{\partial s_2}\right)_{P,T,r_1,r_2,s_1,s_3} = \left(\frac{\partial G}{\partial s_3}\right)_{P,T,r_1,r_2,s_1,s_2} = 0$$

The second of these relates to internal Fe^{2+}-Fe^{3+} partitioning and yields:

$$RT\ ln\frac{(Fe_{tet}^{2+})\cdot(Fe_{oct}^{3+})}{(Fe_{oct}^{2+})\cdot(Fe_{tet}^{3+})} = g_{s_2} + 2g_{s_2s_2}s_2 + g_{r_1s_2}r_1 + g_{r_2s_2}r_2 \\ + g_{s_1s_2}s_1 + g_{s_2s_3}s_3 \quad (42)$$

In pure magnetite this reduces further to an equation of the same form as the O'Neill-Navrotsky model:

$$RT \ln \frac{(Fe^{2+}_{tet}) \cdot (Fe^{3+}_{oct})}{(Fe^{2+}_{oct}) \cdot (Fe^{3+}_{tet})} = (g_{s_2} + g_{r_1 s_2}) + 2g_{s_2 s_2} s_2 \qquad (43)$$

The parameters in (43) can be obtained by fitting the cation distributions in pure magnetite. They are then related to the familiar free energy and solution parameters by treating Fe_3O_4 as a mixture of completely normal $Fe^{2+}(Fe^{3+})_2O_4$ (n) and completely inverse $Fe^{3+}(Fe^{2+}Fe^{3+})_2O_4$ (i) (Nell and Wood, 1989). In the case of pure partially disordered magnetite we would have:

$$RT \ln \frac{(Fe^{2+}_{tet}) \cdot (Fe^{3+}_{oct})}{(Fe^{2+}_{oct}) \cdot (Fe^{3+}_{tet})} = (\mu^o_n - \mu^o_i + W_{ni} + \Delta G^o_{ni}) - 2s_2(W_{ni} + \Delta G^o_{ni})$$

where W_{ni} refers to the interaction parameter between normal (n) and inverse (i) Fe_3O_4 and ΔG^o_{ni} to the reciprocal reaction between these two components. The relationship between recognizable solution properties and the g_{ij} coefficients is clear. By considering each of the end-members and binaries in turn and adding-in macroscopic properties a complete solution model is built up (Hill and Sack, 1987; Nell and Wood, 1989).

SUMMARY

We have described the methods generally used to measure the macroscopic properties of oxides and have given equations that describe the activity-composition relations in most of the oxide solid solutions of mineralogic interest. We have also shown how these are related to microscopic properties and discussed some methods of determining the latter in spinel solid solutions. Finally, we have introduced one way of combining macroscopic and microscopic measurements into a self-consistent model for a complex solid solution with cation disorder. Specific descriptions of the calibration of these models for oxides may be found in Hill and Sack (1987), Nell and Wood (1989) and Ghiorso (1990).

REFERENCES

Amthauer, G. and G.R. Rossman (1984) Mixed valence of iron in minerals with cation clusters. Phys. Chem. Minerals 11, 37-51.

Andersen, D.J. and D.H. Lindsley (1979) The olivine-ilmenite thermometer. Proc. 10th Lunar Planet. Sci. Conf., 493-507.

Andersen, D.J. and D.H. Lindsley (1981) A valid Margules formulation for an asymmetric ternary solution: Revision of the olivine-ilmenite thermometer, with applications. Geochim. Cosmochim. Acta 45, 847-853

Andersen, D.J. and D.H. Lindsley (1988) Internally consistent solution models for Fe-Mg-Mn-Ti oxides: Fe-Ti oxides. Am. Mineral. 73, 714-726.

Atlas, L.M. and W.K. Sumida (1958) Solidus, subsolidus and subdissociation phase equilibria in the system Fe-Al-O. J. Am. Ceram. Soc. 41, 150-156.

Austin, I.G. and N.F. Mott (1969) Polarons in crystalline and non-crystalline materials. Advances in Physics 18, 41-108.

Berman, R.G. (1988) Internally-consistent thermodynamic data for minerals in the system Na_2O-K_2O-CaO-MgO-FeO-Fe_2O_3 - Al_2O_3 -SiO_2 -TiO_2 -H_2O-CO_2. J. Petrol. 29, 445-522.

Berthet, A. and P. Perrot (1970) Equilibres dans le système Fe-Mg-O à 850°C. Mém. Sci. Rev. Mét. 67, 747-753.

Bishop, F.C. (1976) Partitioning of Fe^{2+} and Mg between ilmenite and some ferromagnesian silicates. Ph.D. dissertation, Univ. Chicago, Chicago, IL, 137 pp.

Brynestad, J. and H. Flood (1958) The redox equilibrium in wüstite and solid solution of wüstite and magnesium oxide. Zeits. Elektrochem. Ber. Bunsenges. Phys. Chem. 62, 953-958.

Buddington, A.F. and D.H. Lindsley (1964) Iron-titanium oxide minerals and synthetic equivalents. J. Petrol. 5, 310-357.

Burton, B. (1984) Thermodynamic analysis of the system Fe_2O_3-$FeTiO_3$. Phys. Chem. Mineral. 11, 132-139.

Chaikin, P.M. and G. Beni (1976) Thermopower in the correlated hopping regime. Phys. Rev. B13, 647-651.

Chamberlin, L., J. Beckett and E. Stolper (1990) A new experimental technique for the direct determination of oxide activities in mineral solid solutions. Geol. Soc. Am. Abstr. Prog. A70.

Chatterjee, N.D., H. Leistner, L. Terhart, K. Abraham, and R. Klaska (1982) Thermodynamic mixing properties of corundum-eskolaite, α-$(Al,Cr^{3+})_2O_3$, crystalline solutions at high temperatures and pressures. Am. Mineral. 67, 725-735.

Darken, L.S. (1950) Application of the Gibbs-Duhem equation to ternary and multicomponent systems. J. Am. Chem. Soc. 72, 2909-2914.

Darken, L.S. and R.W. Gurry (1953) Physical Chemistry of Metals, McGraw-Hill, New York.

Davies, P.K. and A. Navrotsky (1981) Thermodynamics of solid solution formation in NiO-MgO and NiO-ZnO. J. Solid State Chem. 38, 264-276.

Davies, P.K. and A. Navrotsky (1983) Quantitative correlations of deviations from ideality in binary and pseudobinary solid solutions. J. Solid State Chem. 46, 1-22.

Dieckmann, R. and H. Schmalzried (1977) Defects and cation diffusion in magnetite II. Ber. Bunsenges. Phys. Chem. 81, 414-419.

Dieckmann, R., C.A. Witt, and T.O. Mason (1983) Defects and cation diffusion in magnetite (V): Electrical conduction, cation distribution and point defects in Fe_3O_4. Ber. Bunsenges Phys. Chem. 87, 495-503.

Emin, D. (1975) Thermoelectric power due to electronic hopping motion. Phys. Rev. Lett. 35, 882-885.

Ghiorso, M.S. (1990) Thermodynamic properties of hematite-ilmenite-geikielite solid solutions. Contrib. Mineral. Petrol. 104, 645-667.

Gillot, B., J.F. Ferriot, and A. Rousset (1976) Electrical conductivity of magnetites substituted for aluminium and chromium. J. Phys. Chem. Solids, 37, 857-862.

Hahn, W.C. and A. Muan (1962) Activity measurements in oxide solid solutions: The system "FeO"-MgO in the temperature interval 1100° to 1300°C. Trans. AIME 224, 416-420.

Hammond, P.L., L.A. Tompkins, S.E. Haggerty, K.J. Spencer, and D.H. Lindsley (1982) Revised data for co-existing magnetite and ilmenite near 1000°C, NNo and FMQ buffers. Geol. Soc. Am. Abstr. Prog. 14, 506.

Hill, R.L. and R.O. Sack (1987) Thermodynamic properties of Fe-Mg titaniferous magnetite spinels. Can. Mineral. 25, 443-464.

Holland, T.J.B. and R. Powell (1990) An enlarged and updated internally consistent thermodynamic data set with uncertainties and correlations in the system K_2O-Na_2O-CaO-MgO-MnO-FeO-Fe_2O_3-Al_2O_3-TiO_2-SiO_2-C-H_2-O_2. J. Met. Geol. 8, 89-124.

Hultgren, R.P., D. Desai, D.T. Hawkins, M. Gleiser and K.K. Kelley (1973) Selected values of the thermodynamic properties of binary alloys. American Society for Metals, Metals Park, Ohio.

Ito, E. and E. Takahashi (1989) Postspinel transformations in the system Mg_2SiO_4-Fe_2SiO_4 and some geophysical implications. J. Geophys. Res. 94, 10637-10646.

Jacob, K.T. (1978) Electrochemical determination of activities in Cr_2O_3-Al_2O_3 solid solutions. J. Electrochem. Soc. 125, 175-179.

Katsura, T., K. Kitayama, R. Aoyagi, and S. Sasajima (1976) High temperature experiments related to Fe-Ti oxide minerals in volcanic rocks. Kazan (Volcanoes) 1, 31-56.

Katsura, T., M. Wakihara, S.I. Hara, and T. Sugihara (1975) Some thermodynamic properties of spinel solid solutions with the Fe_3O_4 component. J. Solid State Chem. 14, 107-113.

Kündig, W. and R.S. Hargrove (1969) Electron hopping in magnetite. Solid State Commun. 7, 223-227.

Lindsley, D.H. (1981) Some experiments pertaining to the magnetite-ulvospinel miscibility gap. Am. Mineral. 66, 759-762.

Lotgering, F.K. and A.M. Van Diepen (1977) Electron exchange between Fe^{2+} and Fe^{3+} ions on octahedral sites in spinels studied by means of paramagnetic Mossbauer spectra and susceptibility measurements. J. Phys. Chem. Solids 38, 565-572.

Mason, T.O. (1987) Cation intersite distributions in iron-bearing minerals via electrical conductivity/Seebeck effect. Phys. Chem. Minerals 14, 156.

Mason, T.O. and H.K. Bowen (1981) Electronic conduction and thermopower of magnetite and iron-aluminate spinels. J. Am. Ceram. Soc. 64, 237-242

Muan, A. (1958) On the stability of the phase Fe_2O_3 - Al_2O_3. Am. J. Sci. 256, 413-422.

Muan, A. and C.L. Gee (1956) Phase equilibrium studies in the system iron oxide-Al_2O_3 in air and at 1 atm O_2 pressure. J. Am. Ceram. Soc. 39, 207-214.
Muan, A. and S. Somiya (1962) The system iron oxide-manganese oxide in air. Am. J. Sci. 260, 230-240.
Nell, J. and B.J. Wood (1989) Thermodynamic properties in a multicomponent solid solution involving cation disorder: Fe_3O_4-$MgFe_2O_4$-$FeAl_2O_4$-$MgAl_2O_4$ spinels. Am. Mineral. 74, 1000-1015.
Nell, J., B.J. Wood, and T.O. Mason (1989) High temperature cation distributions in Fe_3O_4-$MgAl_2O_4$-$MgFe_2O_4$-$FeAl_2O_4$ spinels from thermopower and conductivity measurements,. Am. Mineral. 74, 339-351.
Oka, Y., P. Steinke, and N.D. Chatterjee (1984) Thermodynamic mixing properties of $Mg(Al,Cr)_2O_4$ spinel crystalline solution at high temperatures and pressures. Contrib. Mineral. Petrol. 87, 196-204.
O'Neill, H.St.C. (1988) Systems Fe-O and Cu-O: Thermodynamic data for the equilibria Fe-"FeO", Fe-Fe_3O_4, "FeO"-Fe_3O_4, Fe_3O_4-Fe_2O_3, Cu-Cu_2O, and Cu_2O-CuO from emf measurements. Am. Mineral. 73, 470-486.
O'Neill, H.St.C. and A. Navrotsky (1983) Simple spinels: Crystallographic parameters, cation radii, lattice energies and cation distributions. Am. Mineral. 68, 181-194.
O'Neill, H.St.C. and A. Navrotsky (1984) Cation distributions and thermodynamic properties of binary spinel solid solutions. Am. Mineral. 69, 733-753.
O'Neill, H.St.C., M.I. Pownceby, and V.J. Wall (1989) Activity-composition relations in $FeTiO_3$-$MnTiO_3$ ilmenite solid solutions from EMF measurements at 1050-1300K. Contrib. Mineral. Petrol. 103, 216-222.
Petric, A. and K.T. Jacob (1982) Thermodynamic properties of Fe_3O_4-FeV_2O_4 and Fe_3O_4-$FeCr_2O_4$ spinel solid solutions. J. Am. Ceram. Soc. 65, 117-123.
Petric, A., K.T. Jacob, and C.B. Alcock (1981) Thermodynamic properties of Fe_3O_4-$FeAl_2O_4$ spinel solid solutions. J. Am. Ceram. Soc. 64, 632-639.
Robie, R.A., B.S. Hemingway, and J.R. Fisher (1978) Thermodynamic properties of minerals and related substances at 298.15K and 1 bar (10^5 Pa) pressure and at higher temperature. U.S. Geol. Surv. Bull. 1452, 456 pp.
Sack, R.O. (1982) Spinels as petrogenetic indicators: Activity-composition relations at low pressure. Contrib. Mineral. Petrol. 71, 169-186.
Schmahl, V.N.G., B. Frisch, and E. Hargarter (1960) Zur Kenntnis der Phasenverhaltnisse im System Fe-Ti-O bei 1000°C. Zeits. Anorganische Allgemeine Chemie 305, 40-54.
Sherman, D.M. (1987) Molecular orbital (SCF-XαSW) theory of metal-metal charge transfer processes in minerals. II Applications to $Fe^{2+} \rightarrow Ti^{4+}$ charge transfer transitions in oxides and silicates. Phys. Chem. Mihneral. 14, 364-367.
Spencer, K.J. and D.H. Lindsley (1981) A solution model for co-existing iron-titanium oxides. Am. Mineral. 66, 1189-1201.
Srecec, I., A. Ender, E. Woermann, W. Gans, E. Jacobsson, G. Eriksson, and E. Rosen (1987) Activity-composition relations of the magnesiowustite solid solution series in equilibrium with metallic iron in the temperature range 1050-1400°C. Phys. Chem. Minerals 14, 492-498.
Subbarao, E.C. (ed.) (1980) Solid Electrolytes and Their Applications. Plenum Press, New York.
Taylor, R.W. (1964) Phase equilibria in the system FeO-Fe_2O_3-TiO_2 at 1300°C. Am. Mineral. 49, 1016-1030.
Thompson, J.B., Jr. (1967) Thermodynamic properties of simple solutions. In: Researches in Geochemistry, Vol. II, P. H. Abelson (ed), John Wiley and Sons, New York, 340-361.
Thompson, J.B., Jr. and D.R. Waldbaum (1969) Mixing properties of sanidine crystalline solutions III. calculations based on two-phase data. Am. Mineral. 54, 811-838.
Trestman-Matts, A., S.E. Dorris, S. Kumarakrishnam, and T.O. Mason (1983) Thermoelectric determination of cation distributions in Fe_3O_4-Fe_2TiO_4. J. Am. Ceram. Soc. 66, 829-834.
Tuller, H.L.A and A.S. Nowick (1977) Small polaron electron transport in reduced CeO_2 single crystals. J. Phys. Chem. Solids 38, 859-867.
Turnock, A.C. and H.P. Eugster (1962) Fe-Al oxides: Phase relations below 1000°C. J. Petrol. 3, 553-565.
Valet, P-M., W. Pluschkell and H-J. Engell (1975) Gleichgewichte von MgO-FeO-Fe_2O_3 Muschkristallen mit Sauerstoff. Arch. Eisenhütterwes. 46, 383-388.
Vallet, P. and P. Raccah (1965) Contribution a l'étude des propriétés thermodynamiques du protoxyde de fer dolide. Mem. Scientifiques et Revue de Metallurgie 62, 1-29.
Verwey, E.J., P.W. Haayman, and F.C. Romeijn (1947) Physical properties and cation arrangement of oxides with spinel structures II. Electronic conductivity. J. Chem. Phys. 15, 181-187.
Vincent, E.A., J.B. Wright, R. Chevallier, and S. Mathieu (1957) Heating experiments on some natural titaniferous magnetites. Mineral. Mag. 31, 624-655.
Webb, S.A.C. and B.J. Wood (1986) Spinel-pyroxene-garnet relationships and their dependence on Cr/Al ratio. Contrib. Mineral. Petrol. 92, 471-480.

Webster, A.H. and N.F.H. Bright (1961) The system iron-titanium-oxygen at 1200°C and oxygen partial pressures between 1 atmosphere and 2×10^{-14} atmospheres. J. Am. Ceram. Soc. 44, 110-116.

Wiser, N. and B.J. Wood (1991) Experimental determination of activities in Fe-Mg olivines at 1400 K. Contrib. Mineral. Petrol., in press.

Wood, B.J. and D.G. Fraser (1976) Elementary Thermodynamics for Geologists. Oxford Univ. Press, Oxford, 303 pp.

Wood, B.J., R.J. Kirkpatrick, and B. Montez (1986) Order-disorder phenomena in $MgAl_2O_4$ spinel. Am. Mineral. 71, 999-1006.

Wood, B.J. and J. Nicholls (1978) The thermodynamic properties of reciprocal solid solutions. Contrib. Mineral. Petrol. 66, 389-400.

Woodland, A.B. (1988) Fe-Ti and Fe-Al oxides as indicators of oxygen fugacity in rocks. Ph.D. dissertation, Northwestern Univ., Evanston, IL.

Wu, C.C. and T.O. Mason (1981) Thermopower measurements of cation distribution in magnetite. J. Am. Ceram. Soc. 64, 520-522.

Chapter 8
Benjamin P. Burton
THE INTERPLAY OF CHEMICAL AND MAGNETIC ORDERING

TERMINOLOGY

The term "ordering" will be used for any deviation from randomness that requires an order parameter to describe it. "Chemical ordering" includes such phenomena as atomic clustering, exsolution, or order-disorder phenomena. "Magnetic ordering" includes transitions between paramagnetic, ferromagnetic, antiferromagnetic or ferrimagnetic states (Banerjee, Chapter 3). Coupling[1] between two or more order parameters that are associated with critical points leads to what are called *multicritical phenomena* (Blume et al., 1971; Allen and Cahn, 1982; references in Pynn and Skjeltorp, 1983) so there is a large overlap between this paper and Burton and Davidson (1988) who considered some aspects of multicritical phase relations in minerals. The essential difference between this paper and the previous one is the specific focus on magnetic ordering as opposed to ordering in general.

Ordering is generally associated with a critical temperature (T_c), and the amount of energy absorbed or released in an ordering transition is approximately kT_c, $\Delta H_{ORD} \sim kT_c$, where k is Boltzmann's constant and ΔH_{ORD} is the difference in enthalpy between the ordered and disordered states. Thus, if a system exhibits both chemical and magnetic ordering transitions at comparable T_c's then at least two order parameters will be required for a satisfactory description. In such cases energetic coupling between order parameters is always present, and there may be symmetrical coupling as well (Salje, 1985 and Salje et al., 1985; Davidson and Burton, 1987).

Some systems such as spinel exhibit what Thompson (1968) called "nonconvergent" ordering which is not associated with a critical temperature because ordering does not lead to a change in symmetry. The characteristic temperature for order-disorder phenomena of this type has been called a "crossover" temperature (Salje, 1985; Salje et al., 1985). The case of "nonconvergent" ordering highlights an interesting difference between magnetic and atomic order-disorder phenomena: there is a critical temperature and a change in symmetry associated with antiferromagnetic ordering in $MgFe_2O_4$ ($Fe[Mg,Fe]_2O_4$; where [ ] symbolizes the octahedral sites). There is no phase transition or change in symmetry, however, that is associated with Mg^{2+}-Fe^{3+} or Mg^{2+}-Fe^{2+} order-disorder between tetrahedral and octahedral sites. Superficially, this appears to contradict the analogy between an atomic order-disorder transition and an antiferromagnetic transition. What it really indicates, however, is the additional possibility for symmetry breaking in the magnetic system that arises because of the indistinguishability of electrons; i.e., the configurations ↑[↓] and ↓[↑] (↑ = "spin-up" and ↓ = "spin-down" respectively) are indistinguishable and therefore identical in energy. The analogous atomic configurations Fe[Mg] and Mg[Fe] are clearly different in energy so there is no symmetry to break.

---

[1]In this context "coupling" means that changes in one order parameter cause changes in the other.

For magnetic transitions the term "Curie temperature" is typically used for paramagnetic-ferromagnetic or paramagnetic-ferrimagnetic transitions, and "Néel temperature" is used for paramagnetic-antiferromagnetic transitions, but sometimes the latter is also applied to paramagnetic-ferrimagnetic transitions (Goodenough, 1963). Fortunately, these distinctions are not important for the discussion presented below, so the term "critical temperature" will be used for all chemical or magnetic order-disorder transitions.

INTRODUCTION

Magnetic ordering makes significant contributions to the thermodynamics of hematite (Fe_2O_3), magnetite (Fe_3O_4) and solid solutions based on them; particularly hematite-ilmenite (Fe_2O_3-$FeTiO_3$) and magnetite-ulvöspinel (Fe_3O_4-Fe_2TiO_4). It is generally appreciated that cation ordering in these and other ionic solutions can cause dramatic changes in magnetic properties. For example, T_c for $MgFe_2O_4$ is 679K when it is slowly cooled, but 612K when it is quenched from 1200-1400° C (Swatsky et al., 1969), and this variation is due to changes in the chemical order parameter for degree of inversion (concentration of Mg^{2+} on tetrahedral sites). It is less widely appreciated, however, that magnetic ordering can perturb the solution properties. Specific phase diagram features that may arise as a result of magnetic ordering are: (1) kinks in transus lines, i.e. phase boundaries of the type A/(A+B) where A and B are structurally unrelated phases; (2) kinks in miscibility gaps where magnetic ordering induces a critical end point or tricritical point (Allen and Cahn, 1982; Burton and Davidson, 1988); (3) "cones" on miscibility gaps where the critical line for magnetic ordering enters the cone at a tricritical point.

All three effects listed above cause changes in the solubilities of solute atoms/ions in magnetically ordered solid solutions. The sign of the solubility change will depend upon the slope of the line of critical temperatures (λ-line) for magnetic ordering: if the slope of $T_c(X)$ is negative then the solubility of B in A-rich solid solution will decrease (Fig. 1a); if the slope is positive then it will increase (Fig. 1b) relative to what it would be without magnetic ordering (dotted lines of metastable A_p+B extensions in Fig. 1). Another effect is the perturbation of order parameters that are associated with chemical ordering such as the degree of "inversion" in a spinel. This may be an important effect in the system Fe_3O_4-$MgFe_2O_4$ because both end members have relatively high temperature magnetic transitions (Table 1).

MAGNETIC HEAT CAPACITIES

Before discussing the effects of magnetic ordering on phase relations in hematite- or magnetite- based solutions it will be helpful to review some of the experimental data on the magnetic transitions in Fe_2O_3 and Fe_3O_4. It is useful to consider the specific heat data (Grønvold and Samuelsen, 1975; Grønvold and Sveen, 1974; Hemingway, 1990) and particularly $C_m(T)$ the magnetic contribution to the specific heat. From $C_m(T)$ one can estimate magnetic contributions to the free energy, both above and below T_c. Figures 2 are plots of $C_m(T)$ for hematite and magnetite respectively. In hematite, the transition is between a paramagnetic high temperature state and an essentially antiferromagnetic low temperature state that

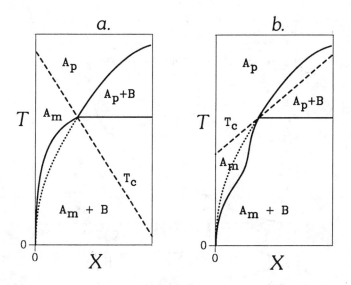

Figure 1. Schematic drawings of the intersection between a magnetic transition and a transus line (a phase boundary of the form A/A+B where A and B need not be structurally related): A_p = paramagnetic A; A_m = magnetically ordered A; (a) $dT_c/dX < 0$, magnetic ordering reduces the solubility of B in A_m; (b) $dT_c/dX > 0$, magnetic ordering enhances the solubility of B in A_m.

Table 1: Magnetic transition temperatures

System	Type of Transition	T_c Kelvins	References
Fe_2O_3	P - ~A A - A'	955 ~259	Grønvold and Samuelsen (1975) References in Lindsley (1976)
$FeTiO_3$	P - A	56±2	References in Lindsley (1976)
Fe_3O_4	P - f	849.5 843-854	Grønvold and Sveen (1974) References in Lindsley (1976)
Fe_2TiO_4	P - A A - A'	99.1 56.0	Todd and King (1953)
$FeCr_2O_4$	P - A	88	Connolly and Copenhaber (1972)
$MgFe_2O_4$	P - A	679* 612**	Swatsky et al. (1969)
$FeAl_2O_4$	P - A	8	Connolly and Copenhaber (1972)
Fe_2O_3-$FeTiO_3$	P - A	~ 950 - 895X ~ 895 - 622X	Ishikawa and Akimoto (1957) Westcott-Lewis and Parry (1971)
Fe_3O_4-Fe_2TiO_4	P - f	~ 853 - 725X	Nagata (1962)

P = paramagnetic; A = antiferromagnetic; ~A = antiferromagnetic with "weak" parasitic ferromagnetism; f = ferrimagnetic; * cooled from 1250° C at 4° C/hr; ** Quenched from between 1250-1400° C into H_2O.

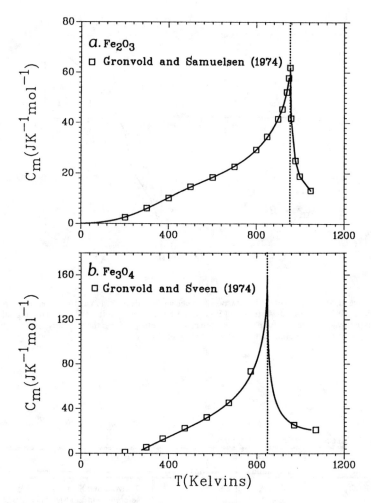

Figure 2. Magnetic heat capacity curves, $C_m(T)$ for: (a) hematite (Fe_2O_3) as estimated by Grønvold and Samuelsen (1975); (b) magnetite (Fe_3O_4) Grønvold and Sveen (1974). The Fe_3O_4 data markers are from the partial data set in Table 3 of Grønvold and Sveen (1974); the curve was digitized from their Figure 2 which is evidently based on their complete data set.

exhibits a weak parasitic magnetic moment (Dzyaloshinsky, 1957 and 1958). In magnetite the transition is between a paramagnetic high-temperature state and a ferrimagnetic state. Both curves exhibit the characteristic λ-shape of a continuous second-order, transition[2] (or a very weakly first-order transition). The presence of

---

[2] Here, the term "second-order transition" is used for a transition that would be described as "λ-transformation with no first order break" in the classification scheme given by Carpenter (1985; column 4 of his Fig. 2) which is a modified version of that in Thompson and Perkins (1981). Because the difference between what

significant high-temperature "tails" on the $C_m(T)$ curves indicates that short-range order (SRO) above T_c makes an important contribution to the thermodynamics.

Separation of C_m from other contributions to the specific heat is a very difficult problem (cf. Grønvold and Samuelsen, 1975; Grønvold and Sveen, 1974; Haas, 1988). For example the calculation by Burton and Kikuchi (1984) suggests that the Grønvold and Samuelsen (1975) values for $C_m(T)$ are too large by a factor of about two, and the thermodynamic analysis by Haas (1988) leads to a similar conclusion.

The magnetic contributions to the enthalpy and entropy are:

$$\Delta H_m(T) = \int_0^T C_p^{mag} dT$$

$$\Delta S_m(T) = \int_0^T \frac{C_p^{mag}}{T} dT$$

which can be conveniently broken up into low- and high-temperature parts that correspond to long-range-order (LRO) and short-range-order (SRO) contributions, respectively.

For $T < T_c$:

$$\Delta H^{LRO} = \int_{T=0}^{T=T_c} C_m^{LRO} dT$$

and

$$\Delta S^{LRO} = \int_{T=0}^{T=T_c} \frac{C_m^{LRO}}{T} dT$$

For $T > T_c$:

$$\Delta H^{SRO} = \int_{T=T_c}^{T=\infty} C_m^{SRO} dT$$

and

$$\Delta S^{SRO} = \int_{T=T_c}^{T=\infty} \frac{C_m^{SRO}}{T} dT$$

In Fe_2O_3 the total magnetic contribution to the enthalpy is about 9.1 kJ/mol of which $\Delta H_{SRO} \sim 1.3$ kJ/mol, and in Fe_3O_4 the corresponding numbers are $\Delta H_m \sim 19.2$ kJ/mol, and $\Delta H_{SRO} \sim 2.5$ kJ/mol (based on the analysis of Haas, 1988). The fraction

---

they refer to as "second-order" and what they call "λ-transformation with no first order break" is just the presence of short-range order above T_c, the distinction seems unnecessary.

$\Delta H_{SRO}/\Delta H_m$ is an important parameter for modeling the magnetic contribution to the free energy (see below).

THEORETICAL METHODS

Two distinct theoretical approaches have been widely used to add the magnetic contribution to a free energy model:

(1) The magnetic degrees of freedom are included by *increasing the number of components*. For example, the chemically binary system A-B with A magnetic and B nonmagnetic can be modeled as a quasibinary (ternary) system with components A↑, A↓ and B (where A↑ is read A spin-up; e.g. Meijering, 1963; Burton, 1985; Burton and Davidson, 1988). Similarly, if both A and B are magnetic the quasibinary system is quaternary with components A↑, A↓, B↑, and B↓. Note that this approach requires energy parameters (J's) for various magnetic interactions (e.g., between nearest neighbors).

(2) The magnetic contribution can be treated phenomenologically as an excess contribution to the heat capacity. The procedure is to fit a purely empirical function to experimental magnetic specific heat (C_m) data. Relevant thermodynamic functions such as the magnetic contributions to enthalpy and entropy (ΔH_m, ΔS_m) are obtained by integration of the empirical function. This approach was pioneered by Inden (1975, 1976a,b, 1981 and 1982), Hillert et al. (1967), and Hillert and Jarl (1978).

Method (1) is most appropriate when the objective is to construct a physically rigorous model that yields fundamental understanding of the phase transition behavior. Method (2) is preferable when the objective is to fit experimental data precisely.

Method (1): Meijering's "regular pseudo-ternary model"

This is the simplest model of type (1) above, and it has been reviewed in Burton and Davidson (1988). The present discussion will therefore be restricted to the two cases that are most likely to appear in oxide mineral systems (Fig. 3). The components are A↑, A↓ and B (A spin-up, A-spin down and B, respectively). The chemical interactions between A and B are those of a symmetric regular solution, and magnetic interactions between A↑ and A↓ are assumed to be ferromagnetic (assuming antiferromagnetic coupling yields identical results at zero applied magnetic field). The free energy function for this model can be written as:

$$G = W_J(X_{A\uparrow}X_{A\downarrow}) + W(X_{A\uparrow}+X_{A\downarrow})X_B + RT(X_{A\uparrow}\ln X_{A\uparrow} + X_{A\downarrow}\ln X_{A\downarrow} + X_B\ln X_B)$$

Where $W_J > 0$ is the magnetic interaction parameter[3], W is the regular solution parameter, and X_i is the mol fraction of component i. Note that this is equivalent to a

---

[3]Usually, magnetic interaction parameters are symbolized as $J_{i,j}(r)$ for the magnetic interaction between r'th nearest neighbor sites i and j, and $J_{\uparrow\uparrow}(r) = J_{\downarrow\downarrow}(r) = -J_{\uparrow\downarrow}(r) = J_{\downarrow\uparrow}(r)$. However, Meijering (1963) used a parameter equivalent to W_J above.

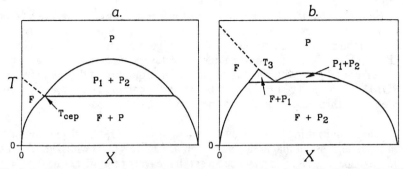

Figure 3. Two of the four phase diagram topologies that can be obtained in the Meijering (1963) "pseudoternary model:" P = paramagnetic; F = ferromagnetic; T_{cep} = a critical end point; T_3 = a tricritical point. In (a) the λ-line of magnetic transitions intersects the miscibility gap ar a critical end point, and in (b) the it intersects a small cone shaped two-phase field at a tricritical point.

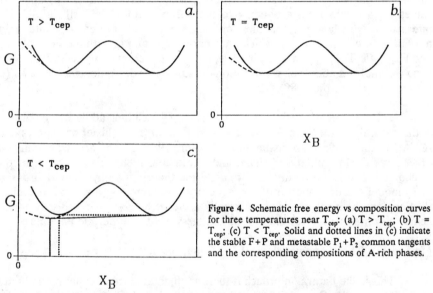

Figure 4. Schematic free energy vs composition curves for three temperatures near T_{cep}: (a) $T > T_{cep}$; (b) $T = T_{cep}$; (c) $T < T_{cep}$. Solid and dotted lines in (c) indicate the stable F+P and metastable P_1+P_2 common tangents and the corresponding compositions of A-rich phases.

ternary system that contains three symmetric binary regular solutions: one between A↑ and B, that is identical to that between A↓ and B, and a third between A↑ and A↓. Thus, ferromagnetic ordering is formally analogous to phase separation in this Ising type formulation. Similarly, antiferromagnetism is formally analogous to an order-disorder transition. The λ-line of second-order magnetic transitions is just the line of consolute temperatures for the A↑-A↓ "miscibility gap" at constant X_B.

Meijering (1963) enumerated four different phase diagram topologies that can arise for different ratios of W_J/W, but the two shown in Figure 3 (for which $0 \le W_J/W \le 1.06$) are the only ones that seem likely to be realized in rock-forming systems. The two cases that arise for larger values of W_J/W would imply miscibility gaps at temperatures that are too low for chemical equilibration to be achieved. In Figure 3a ($W_J/W = 1$), the λ-line intersects the miscibility gap at a *critical end point* (T_{cep}) and in Figure 3b ($W_J/W = 0.75$) it intersects it at a *tricritical point* (T_3; Allen and Cahn, 1982; Burton and Davidson, 1988 and references therein). Note that both figures exhibit the characteristic reduction in solubility of the nonmagnetic component below T_c. Figures 4 are schematic free energy vs. composition curves that correspond to the phase relations represented in Figure 3a at: (a) $T_c(A) > T > T_{cep}$; (b) $T = T_{cep}$; (c) $T < T_{cep}$. The dashed line indicates the free energy for a magnetically ordered crystal (A_F) with a stable or metastable equilibrium degree of magnetic order ($\partial G/\partial X_A = \partial G/\partial X_A = \partial G/\partial X_B = 0$). From the common tangent construction, it is evident why this leads to a kink in the A-rich phase boundary at T_{cep}.

An obstacle to using this approach on Fe_2O_3- or Fe_3O_4-based solutions is that one requires a large set of magnetic interaction parameters (J_r's) which generally can not be looked up. That is, one may need a set that covers r significant near-neighbor (nn) superexchange interactions ($J_1, J_2,...J_r$; Goodenough, 1963) and it may be necessary to consider both Fe^{3+}-O-Fe^{3+} and Fe^{3+}-O-Fe^{2+} type interactions. For example, Samuelsen and Shirane (1970) report significant values for J_1-J_4 in Fe_2O_3, but similar studies of magnetite (Glasser and Milford, 1963; Alperin et al., 1966) only report values for a single J corresponding to the tetrahedral-[octahedral] Fe^{3+}-O-$[Fe^{3+}]$ interaction. Note however, that these only account for Fe^{3+}-O-Fe^{3+} type interactions, not Fe^{3+}-O-Fe^{2+} interactions which are clearly important in both the Fe_2O_3-$FeTiO3$ and Fe_3O_4-Fe_2TiO_4 systems. Ultimately, it becomes necessary to make some simplifying assumptions about the composition or many-body dependencies of the J's (e.g., Burton and Davidson, 1988).

Treating the magnetic degree of freedom by including additional components is the natural approach to follow in a model based on a cluster variation method (CVM; Kikuchi, 1951) approximation (Burton, 1985; Burton and Davidson, 1988) or a Monte Carlo simulation (Binder and Heermann, 1988). These are the methods of choice if one is attempting to formulate the theory in a physically rigorous way, or to model the dependence of magnetic order parameters (e.g. sublattice magnetization) on chemical ordering. Typically however, they have the disadvantage of achieving physical rigor at the cost of goodness of fit to experimental data.

Method (2): The phenomenological $C_m(T)$ approach

This is the natural approach if there is an abundance of experimental data to fit, and if a precise fit to the data is the prime objective. Typically, one combines a regular-type solution model for chemical interactions with magnetic interaction parameters that are obtained by fitting magnetic specific heat data. Separate functions are used for the high- and low-temperature parts of the specific heat, and two specific sets of functions have been widely used. The first was proposed by Inden (1975, 1976a,b):

$$C_m^{LRO} = K^{LRO} R \ln \frac{(1+\tau^3)}{(1-\tau^3)} \quad for \quad \tau = \frac{T}{T_c} < 1$$

$$C_m^{SRO} = K^{SRO} R \ln \frac{(1+\tau^{-5})}{(1-\tau^{-5})} \quad for \quad \tau > 1$$

where K^{LRO} and K^{SRO} are phenomenological constants. Integration of these equations leads to

For $\tau < 1$:

$$\Delta H^{LRO}(T) = K^{LRO} R T_c \{ (1-\tau)\ln(1-\tau)$$

$$+ \tau\ln(\frac{1+\tau^3}{1+\tau+\tau^2}) + \ln(\frac{1+\tau}{\sqrt{1+\tau^2+\tau^4}})$$

$$+ \sqrt{3}\cdot\arctan(\frac{2\tau-1}{\sqrt{3}}) - \sqrt{3}\cdot\arctan(\frac{2\tau+1}{\sqrt{3}}) + 2\sqrt{3}\cdot\arctan(\frac{1}{\sqrt{3}}) \}$$

and

$$\Delta S^{LRO} = \frac{K^{LRO}\pi^2 R}{12}$$

and

For $\tau > 1$:

$$\Delta H^{SRO}(T) = K^{SRO} R T_c \{ (\tau+1)\ln(\tau+1) - (\tau-1)\ln(\tau-1)$$

$$+ \tau\ln(\frac{1-\tau+\tau^2-\tau^3+\tau^4}{1+\tau+\tau^2+\tau^3+\tau^4})$$

$$- \cos(\frac{\pi}{5})\ln[\tau^4 - 2\tau^2\cos(\frac{2\pi}{5}) + 1]$$

$$- \cos(\frac{3\pi}{5})\ln[\tau^4 + 2\tau^2\cos(\frac{\pi}{5}) + 1]$$

$$+ 2\sin(\frac{\pi}{5})[\arctan(\frac{\tau-\cos(\frac{\pi}{5})}{\sin(\frac{\pi}{5})}) - \arctan(\frac{\tau+\cos(\frac{\pi}{5})}{\sin(\frac{\pi}{5})})]$$

$$+ 2\sin(\frac{3\pi}{5})[\arctan(\frac{\tau-\cos(\frac{3\pi}{5})}{\sin(\frac{3\pi}{5})}) - \arctan(\frac{\tau+\cos(\frac{3\pi}{5})}{\sin(\frac{3\pi}{5})})]$$

$$+ \ln(\frac{5}{4}) + \cos(\frac{\pi}{5})\ln[4\sin^2(\frac{\pi}{5})] + \frac{3\pi}{5}\sin(\frac{\pi}{5})$$

$$+ \cos(\frac{3\pi}{5})\ln[4\sin^2(\frac{3\pi}{5})] - \frac{\pi}{5}\sin(\frac{3\pi}{5}) \}$$

and

$$\Delta S^{LRO}(\infty) = \frac{K^{SRO}\pi^2 R}{20}$$

Inden (1976a) defined a quantity (f) equal to the fraction of magnetic specific heat that is due to SRO:

$$f = \Delta H_{SRO}/\Delta H_m$$

which is evidently structure-specific: $f = 0.4$ for fcc metals and $f = 0.28$ for bcc metals. Considering the discussion of magnetic specific heats above, $f \sim 0.14$ for hematite, and $f \sim 0.13$ for magnetite.

Because the expressions listed above are rather complicated, another formulation (Hillert and Jarl, 1978) has become somewhat more popular. It is incorporated in the "THERMOCALC" databank/modeling code which is generally available

(Sundman et al., 1985). In this formulation the magnetic contribution to Gibbs energy is approximated as:

$$G_m^{Ord} = \phi(\tau) RT \ln(\beta+1) \qquad (1)$$

where $\tau = T/T_c$, β is the average magnetic moment per atom in Bohr magnetons, and the reference state is the random ($T \to \infty$) paramagnetic state.

$$\phi(\tau \le 1) = 1 - [\frac{79\tau^{-1}}{140f} + \frac{474}{497}(\frac{1}{f}-1)(\frac{\tau^3}{6}+\frac{\tau^9}{135}+\frac{\tau^{15}}{600})]/A \qquad (2)$$

$$\phi(\tau > 1) = -(\frac{\tau^{-5}}{10}+\frac{\tau^{-15}}{315}+\frac{\tau^{-25}}{1500})/A \qquad (3)$$

where

$$A = \frac{518}{1125} + \frac{11692}{15975}(\frac{1}{f}-1) \qquad (4)$$

and f above is equal to "f" in Inden's formulation. Note that both τ and β above are in general functions of bulk composition and other order parameters (e.g., site occupancies). This approach was used by Bergman and Ågren (1985; 1986a,b) to model chemical and magnetic ordering in the systems MnO-NiO, MnO-CoO and NiO-CoO (see below).   Added in proof: See also Sundman, B. (1991) An assessment of the FeO system. J. Phase Equil. 12, 127-140.

EFFECTS ON PHASE DIAGRAM TOPOLOGIES

Metallurgical systems

The interplay of chemical and magnetic ordering in metallurgical systems has been extensively studied by both theoretical and experimental techniques (Zener, 1955; Meijering, 1963; Inden, 1975, 1981, 1982; Hillert and Jarl, 1978; Miodownik, 1982 and references therein; Sanchez and Lin, 1984). Historically, theoretical predictions (Zener, 1955, and Meijering, 1963) preceded experimental verification (see the reviews by Inden, 1982, Miodownik, 1982, and Nishizawa et al., 1987, and references therein), but verification has been comprehensive. Table 1 in Miodownik (1982) lists over 20 binary and ternary Fe-, Co-, or Ni-based systems in which magnetic ordering causes experimentally measurable changes in phase boundaries. For example the solubility of Mo in face centered cubic Co (Fig. 5) is significantly perturbed by ferromagnetic ordering. Experiments reveal a kink where the λ-line intersects the phase boundary. This example represents an experimental proof of the major point made in the introduction: in solutions between magnetic and nonmagnetic components, the solubility of the nonmagnetic component will be reduced by magnetic ordering.

Ceramic and mineral systems

There appear to be only four theoretical studies of oxide systems that include magnetic as well as chemical interactions. (NiO-MnO, CoO-NiO and CoO-MnO, Bergman and Ågren, 1985, 1986a and 1986b, respectively; Fe_2O_3-$FeTiO_3$, Burton,

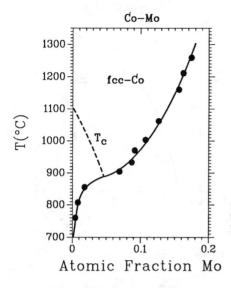

Figure 5. Displacement of the Co/(Co+Co$_3$Mo) phase boundary as a result of magnetic ordering in fcc Co. Data points are from Ko and Nishizawa (1979).

1985; Burton and Davidson, 1988). Of these, only hematite-ilmenite is an important rock forming system, and there do not appear to be any experimental studies that verify or contradict the theoretical results. In the case of magnetite-ulvöspinel there have been several theoretical studies (Price, 1981; Trestman-Matts et al., 1983; O'Neill and Navrotsky, 1984) but none of them included the magnetic contribution to the free energy.

Rock salt structure systems. Bergman and Ågren (1985, 1986a,b; Figs. 6) modeled the systems MnO-NiO, MnO-CoO and NiO-CoO with a combination of regular, or subregular, models for chemical ordering plus "Method (2)" above for the magnetic ordering. The predicted miscibility gaps in these systems are at sufficiently low temperatures that they will probably not be experimentally realized; so these examples are included more for completeness than for mineralogical significance. Dotted lines in Figures 6b and 6c indicate phase boundaries that were added by this author to ensure compliance with the phase rule.

Hematite-ilmenite. This is the only important rock forming system for which there is a model that includes both chemical and magnetic contributions the free energy. The model was originally presented in Burton (1985) and updated in Burton and Davidson (1988). The "single prism approximation" (SPA) of the CVM was used because it covers the first four nearest neighbor (nn) cation-cation pairs, and can therefore include both chemical and magnetic interactions for up to fourth nn's. The "single prism" cluster of four cation sites that is used in the SPA is depicted in Figure 7c. Model components are Fe↑, Fe↓ and Ti^{4+} ions without making any distinction between Fe^{3+} and Fe^{2+}.

Chemical ordering in Fe$_2$O$_3$-FeTiO$_3$ gives rise to both an order-disorder transition in ilmenite-rich solutions (Ishikawa, 1958; references in Burton and Davidson, 1988; Ghiorso, 1990; and Chapter 6), and to exsolution at low temperatures (e.g., Rumble, 1976). The order-disorder transition is between a less ordered hematite-rich

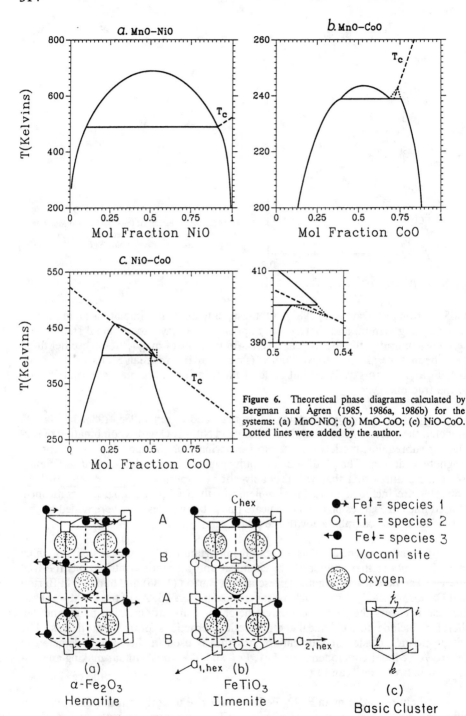

Figure 6. Theoretical phase diagrams calculated by Bergman and Ågren (1985, 1986a, 1986b) for the systems: (a) MnO-NiO; (b) MnO-CoO; (c) NiO-CoO. Dotted lines were added by the author.

Figure 7. Schematic representations of: (a) the crystal structure of antiferromagnetic Fe_2O_3; (b) paramagnetic $FeTiO_3$; (c) the largest cluster used in the Burton (1985) and Burton and Davidson (1988) SPA calculations of the hematite ilmenite phase diagram.

corundum structure phase (space group $R\bar{3}c$; Fig. 7a) and a low temperature, ilmenite-rich ilmenite structure phase (space group $R\bar{3}$; Fig. 7b), and it is apparently second-order in character (Ishikawa, 1958).

To obtain a qualitatively correct fit to the observed chemical ordering phenomena it was assumed that:

(1) First-nn interactions (between basal layers; indices j-k in Fig. 7c) favor Fe-Ti ordering;
(2) Second-nn interactions (within basal layers; indices i-j, k-l in Fig. 7c) favor Fe-Fe and Ti-Ti nn's.

Assumption (1) above guarantees that the ilmenite structure phase will be stable at low temperatures, and (2) guarantees that there will be a miscibility gap between hematite- and ilmenite-rich solutions. There are parameterizations other than (2) above that guarantee a miscibility gap, but they either lead to topologically equivalent phase diagrams, or to diagrams that contradict experiment. Specifically, if one combines a "long-range" regular solution term with attractive 1'st and 2'nd nn Fe-Ti interactions then the $R\bar{3}c$-$R\bar{3}$ transition is predicted to be first-order, and ilmenite is predicted to have the lithium niobate structure, both of which are contrary to experiment (Ishikawa, 1958).

As noted above, there is an essentially paramagnetic-antiferromagnetic transition in the hematite-rich region that follows an approximately linear trend: $T_c(X) \sim 950 - 1790X$ (Ishikawa and Akimoto, 1958), where X is mol fraction Ti_2O_3. To approximate this result, scaled[4] values for pure Fe_2O_3 (Samuelsen and Shirane, 1970) were used, and it was assumed that the magnetic contribution to the total energy is zero in any four-site cluster that contains one or more Ti^{4+} ion. This assumption leads to a $T_c(X)$ trend that is somewhat less negative in slope than the experimental one (Ishikawa and Akimoto, 1957; chain-dotted line Fig. 8b).

The predicted effect of magnetic ordering on the solubility of ilmenite in hematite solution is enhanced relative to what it would be if there were no $R\bar{3}c$-$R\bar{3}$ transition (e.g., if there were a symmetrical miscibility gap between hematite and ilmenite). The $R\bar{3}c$-$R\bar{3}$ order-disorder transition causes a distortion of the $R\bar{3}c$ + $R\bar{3}$ field such that more ilmenite is soluble in hematite solution than vice versa, but below the three phase isothermal line this asymmetry is removed. At the three phase line the difference in composition between the predicted equilibrium from chemical interactions alone and that with chemical plus magnetic interactions is $\Delta X = 0.075 = 15$ mol% $FeTiO_3$.

Magnetite-Ulvöspinel. As noted above there have been several studies of this system (Vincent et al., 1957; Price, 1981; Lindsley, 1981; Trestman-Matts et al., 1983; O'Neill and Navrotsky, 1984; Chapters 6 and 7). All these studies suggest the presence of a miscibility gap at low temperatures (Fig. 9), but none of the calculations included the magnetic contribution to the free energy. The strongest constraint on

---

[4]The Samuelsen and Shirane (1970) values were divided by a constant factor so that the calculated value for $T_c(Fe_2O_3)$ would equal the experimental one.

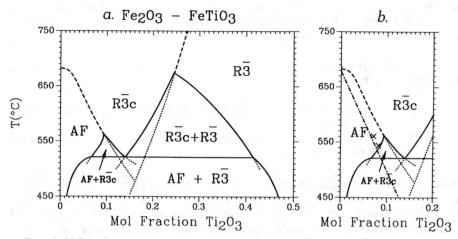

Figure 8. (a) Part of a calculated phase diagram for the system hematite-ilmenite, Fe_2O_3-$FeTiO_3$. $R\bar{3}c$ = disordered hematite-ilmenite solid solution; $R\bar{3}$ = ordered ilmenite structure solution; AF = antiferromagnetically ordered hematite-rich solid solution. Note how the interplay of antiferromagnetic ordering with the miscibility gap is predicted to cause a small AF + $R\bar{3}$ two phase field. (b) The hematite-rich region of (a) with the experimental antiferromagnetic $T_c(X)$ curve indicated by data points and a dash-dotted line.

the maximum allowed value for the consolute temperature (T_{cons}) is provided by the experiments of Lindsley (1981) which indicate 833K < T_{cons} < 853K = $T_c(Fe_3O_4)$. Thus the change in energy that is associated with chemical ordering ($\Delta H_{chem} \sim kT_{cons}$) is approximately equal to the change in energy that is associated with magnetic ordering ($\Delta H_{mag} \sim kT_c$). Clearly, the magnetic contribution to the free energy of mixing is about as large as the chemical one that causes low temperature immiscibility.

Figure 10 illustrates how including the magnetic term changes the free energy of mixing. The magnetic term is of the Hillert and Jarl (1978) type (Equations 1-4) with β = 5.0, f = 0.13 and T_c = 853 - 725X_{Usp}. The chemical interaction parameters are from model (2) of O'Neill and Navrotsky (1984): α = 16 + 5.76T/1000 (kJ/mol)[5]; β = -20 kJ/mol; W = 9.2 kJ/mol. Molar site occupancies in Fe_3O_4-Fe_2TiO_4 are approximately:

Species	Tet	Oct	Sum
Fe^{2+}	1-b	X+b	1+X
Fe^{3+}	b	2-2X-b	2-2X
Ti^{4+}	0	X	X
Sum:	1	2	3

---

[5] The caption to Figure 1 in O'Neill and Navrotsky (1984) gives α (Fe^{2+}-Fe^{3+}) = 16 + 0.0057T (kJ/mol) as above, but their Table A1 gives α (Fe^{2+}-Fe^{3+}) = 16 - 0.00576T (kJ/mol). Neither value seems to give tielines that are consistent with the ON2 solvus (Fig. 9).

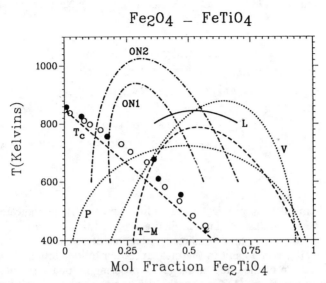

Figure 9. Several proposed miscibility gaps for the system magnetite-ulvöspinel (Fe_3O_4-Fe_2TiO_4): V = Vincent et al. (1957); P = Price (1981); L = Lindsley (1981); T-M = Trestman-Matts et al. (1983); ON1, ON2 = curves (1) and (2) in Figure 13 of O'Neill and Navrotsky (1984). The straight dashed line connects $T_c(Fe_3O_4)$ to $T_c(Fe_2TiO_4)$. Solid circles are $T_c(X)$ data from Akimoto and Katsura (1959), and open circles are data from Yama-ai et al. (1961; in Nagata, 1962). Of these, only V and L are based on phase equilibria experiments. V used natural samples and L used synthetics.

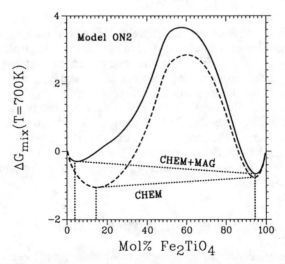

Figure 10. Free energy of mixing curves for the system Fe_3O_4-Fe_2TiO_4 calculated with the chemical interaction parameters of model ON2 (Fig. 9) and Hillert and Jarl (1978) magnetic parameters: $T_c(K) = 853-725X$; $\beta = 5.0$; $f = 0.13$. The dashed curve was calculated with chemical interactions only. The solid curve was calculated with chemical plus magnetic interactions.

[O'Neill and Navrotsky (1984); Trestman-Matts et al. (1983); Navrotsky (1987)], and therefore, the free energy of mixing is:

$$\Delta G_{mix} = \Delta G_{chem}(b, X, T) + \Delta G_{mag}(X, T)$$
$$- (1-X)\Delta G_{chem}(b, X=0, T) - X\Delta G_{chem}(b, X=1, T)$$
$$- (1-X)\Delta G_{mag}(X=0, T) - X\Delta G_{mag}(X=1, T)$$

where

$$\Delta G_{chem} = b(\alpha + \beta b) + WX(1-X)$$
$$+ RT\{b\ln b + (1-b)\ln(1-b) + (X+b)\ln(X+b) + (2-2X-b)\ln(2-2X-b) + X\ln X\}$$

and

ΔG_{mag} is given by Equations 1-4.

By itself, the chemical term predicts a miscibility gap at 700K as indicated by the double-well shape of the dashed curve in Figure 10. Addition of the magnetic term (solid curve Fig. 10), leads to a predicted reduction of about 10 mol % in the solubility of Fe_2TiO_4 component in the magnetite-rich solid solution. This is only a crude approximation of the effect of magnetic ordering on phase relations in this system. As noted above, the ΔG_{mag} depends upon site occupancies (b) as well as composition. Also, the site preference energies (α's and β's) in the O'Neill and Navrotsky (1984) formulation implicitly include a magnetic contribution. That is, Fe^{3+}-O-$[Fe^{3+}]$ configurations are lower in energy than Fe^{2+}-O-$[Fe^{3+}]$ or Fe^{3+}-O-$[Fe^{2+}]$ configurations, and therefore magnetic interactions add to a tetrahedral site preference for Fe^{3+} relative to Fe^{2+}.

SUMMARY

The interplay of chemical and magnetic ordering is a well established phenomenon that has been experimentally verified in several metallurgical systems (e.g. Miodownik, 1982; Inden, 1982). The essential requirement for significant coupling between magnetic and chemical ordering phenomena is that ΔH_{mag} be large enough, or equivalently, that T_c for the magnetic transition be high enough (e.g., $T_c \sim T_{cons}$ in a system with a miscibility gap). Clearly, this requirement is satisfied in the hematite-ilmenite and magnetite-ulvöspinel systems; the key question is: will kinetics prevent full equilibration at the temperatures of interest? These are significant effects but they will not be observed by accident. For experimental work the temperatures may be too low to achieve equilibrium, but in regionally metamorphosed rocks they most probably are not.

OPPORTUNITIES FOR FUTURE RESEARCH

A list of geologically important systems in which magnetic ordering may be significant is compiled in Table 2. Most of these systems have miscibility gaps between magnetite and a component that has no magnetic transition (e.g., $MgAl_2O_4$) or a magnetic transition at a very low temperature (e.g., Fe_2TiO_4, $T_c \sim 99K$). In the cases of hematite-ilmenite and magnetite-ulvöspinel there are data on the variations

Table 2: Systems in which magnetic ordering is expected to significantly affect the phase relations.

System	Expected Effect	References and (data type)
hematite - ilmenite Fe_2O_3 - $FeTiO_3$	T_λ + "cone" on the hematite-rich side of the Hem-Ilm misc. gap	References in Lindsley (1976) Burton (1985) Burton and Davidson (1988)
magnetite - ulvöspinel Fe_3O_4 - Fe_2TiO_4	T_{cep} or T_λ + "cone" on the magnetite-rich side of the Mt-Usp misc. gap	Lindsley (1976) Lindsley (1981) Price (1981) (misc. gap)
magnetite - chromite Fe_3O_4 - $FeCr_2O_4$	T_{cep} on the Mt-rich side of the Mt-Chro. misc. gap	Petric et al. (1981) Petric and Jacob (1982) O'Neill and Navrotsky (1984) Ghiorso and Sack Chapter 6 (misc. gap)
magnetite - spinel Fe_3O_4 - $MgAl_2O_4$	T_{cep} on the Mt-rich side of the Mt-Spinel misc. gap	O'Neill and Navrotsky (1984) (misc. gap)
magnetite - hercynite Fe_3O_4 - $FeAl_2O_4$	T_{cep} on the Mt-rich side of the Mt-Hc misc. gap	Turnock and Eugster, 1962 Nell et al. (1989) Ghiorso and Sack Chapter 6 Wood and Nell Chapter 7 (misc. gap)
magnetite - magnesioferrite Fe_3O_4 - $MgFe_2O4$	coupling between the degree of inversion and magnetic ordering	Swatsky et al. (1969) $T_c(MgFe_2O_4)$ is a strong function of the degree of inversion

of magnetic critical temperatures as functions of composition, and these trends roughly follow Vegard's law (approximately straight lines between the two T_c's). Thus, it is not hard to estimate where the λ-lines will intersect the miscibility gaps, and therefore, where kinks in the solubility curves are to be expected. Another interesting problem is the effect of magnetic ordering on degree of inversion which may be important in the system Fe_3O_4-$MgFe_2O_4$.

REFERENCES

Akimoto, A and Katsura, T. (1959) Magneto-chemical study of the generalized titanomagnetite in volcanic rocks. J. Geomag. Geoelectricity 3, 69-90.
Allen, S.M. and Cahn, J.W. (1982) Phase diagram features associated with multicritical points in alloy systems. Bulletin of Alloy Phase Diagrams V3, #3, 287-295.
Alperin, H.A., Steinsvoli, O., Nathans, R. and Shirane, G. (1967), Magnon scattering of polarized neutrons by the diffraction method: measurements on magnetite, Phys. Rev. 154, 508-154.
Bergman, B. and Ågren, J. (1985) thermodynamic assessment of the system MnO-NiO, J. Am. Ceram. Soc. 68 [8] 444-450.
---- and ---- (1986a) Immiscibility in the CoO-NiO system, J. Am. Ceram. Soc., 69 [10] C-248 - C-250.
---- and ---- (1986b) thermodynamic assessment of the system CoO-MnO, J. Am. Ceram. Soc., 69 [12] 877-881.
Binder, K. and Heermann, D.W. (1988) "Monte Carlo Simulation in Statistical Physics, An Introduction." Springer Series in Solid-State Sciences 80, Springer-Verlag, New York.
Blume, M., Emery, V.J. and Griffiths, R.B. (1971) Ising model for the λ-transition and phase separation in 3He-4He mixtures. Phys. Rev. A4, #3, 1071-1077.

Burton, B. P. and Davidson, P. M. (1988) Multicritical phase relations in minerals, In "Structural and Magnetic Phase Transitions in Minerals" S. Ghose, J.M.D. Coey and E. Salje, Editors, Advances in Physical Geochemistry V7, 60-90. Springer-Verlag.

----- (1985) Theoretical analysis of chemical and magnetic ordering in the system Fe_2O_3-$FeTiO_3$. Am. Mineral. 72, 329-336.

----- and R. Kikuchi (1984) The antiferromagnetic-paramagnetic transition in αFe_2O_3 in the single prism approximation of the cluster variation method. Phys. Chem. Minerals 11, 125-131.

Carpenter, M.A. (1985) Order-disorder transformations in mineral solid solutions, In S.W. Kieffer and A. Navrotsky, Editors, Rev. Mineral. 14, 187-223.

Connolly, T.F. and Copenhaber, E.J. (1972) "Solid State Physics Literature Guide", Plenum Press.

Davidson, P.M. and Burton, B.P. (1987) Order-disorder in omphacitic pyroxenes: a model for coupled substitution in the point approximation. Am. Mineral., 72, 337-344.

Dzyaloshinskii, I.E. (1957) A thermodynamic theory of "weak" ferromagnetism of antiferromagnetic substances, Soviet Physics JETP 5, 1259-1272 [translated from Zhu. Eksper. Theor. Fiziki 32, 1547-1562.

Dzyaloshinskii, I.E. (1958) A thermodynamic theory of "weak" ferromagnetism of antiferromagnetics, J. Physics Chemistry Solids, 4, 241-255.

Ghiorso, M.S. (1990) Thermodynamic properties of hematite-ilmenite-geikielite solid solutions. Contrib. Mineral. Petrol., 104, 645-667.

Goodenough, J.B. (1963) "Magnetism and the Chemical Bond." Wiley-Interscience, New York.

Glasser, M.L. and Milford, F.J. (1963) Spin wave spectra of magnetite, Phys. Rev., 130, 1783-1789.

Grønvold, F. and Samuelsen, E.J. (1975) Heat capacity and thermodynamic properties of α-Fe_2O_3 in the region 300 to 1050 K Antiferromagnetic transition, J. Phys. Chem. Solids, 36, 249-256.

Grønvold, F. and Sveen, A. (1974) Heat capacity and thermodynamic properties of synthetic magnetite (Fe_3O_4) from 300 to 1050 K. Ferrimagnetic transition and zero point entropy. J. Chem. Thermo., 6, 859-872.

Haas, J.L., Jr. (1988) Recommended standard electrochemical potentials and fugacities of oxygen for the solid buffers and thermodynamic data in the systems iron-silicon-oxygen, nickel-oxygen, and copper-oxygen. Preliminary report to CODATA Task Group on Chemical Thermodynamic Tables, U.S. Geol. Survey, Reston, VA.

Hemingway, B.S. (1990) Thermodynamic properties of bunsenite, NiO, magnetite, Fe_3O_4, and hematite, Fe_2O_3, with comments on selected oxygen buffer reactions. Am. Mineral., 75, 781-790.

Hillert, M., Wada, T. and Wada, H. (1967) The α-γ equilibrium in Fe-Mn, Fe-Mo, Fe-Ni, Fe-Sb, Fe-Sn and Fe-W systems, J. of the Iron and Steel Institute, 205, 539-546.

Hillert, M. and Jarl, M. (1978) A model for alloying effects in ferromagnetic metals, CALPHAD 2, #3, 227-238.

Inden, G. (1975) Determination of chemical and magnetic interchange energies in bcc alloys, Zeit. Metallkunde, B66, H.11, 648-653.

---- (1976a) Approximate description of the configurational specific heat during a magnetic order-disorder transformation, Project Meeting, CALPHAD V, June 21-25, III.4-1 - III.4-12, Max-Planck Institute fur Eisenforschung G.m.b.H., Duesseldorf, F.R.G.

---- (1976b) Computer calculation of the free energy contributions due to chemical and/or magnetic ordering, Project Meeting, CALPHAD V, June 21-25, IV.1-1 - IV.1-35, Max-Planck Institute fur Eisenforschung G.m.b.H., Duesseldorf, F.R.G.

----- (1981) The role of magnetism in the calculation of phase diagrams, Physica, 103B, 82-100.

----- (1982) The effect of continuous transformations on phase diagrams. Bull. Alloy Phase Diagrams 2, 412-422.

Ishikawa, Y. and Akimoto, S. (1957) Magnetic properties of the $FeTiO_3$-Fe_2O_3 solid solution series. J. Phys. Soc. Japan, 12, 1083-1098.

----- (1958) An order-disorder transformation phenomenon in the $FeTiO_3$-Fe_2O_3 solid solution series. J. Phys. Soc. Japan, 13, 828-837.

Kikuchi, R. (1951) A theory of cooperative phenomena. Phys. Rev., 81, 988-1003.

Ko, M. and Nishizawa (1979) J. Japan Inst. Met. 43(2), 126-. Cited in Miodownik (1982).

Lindsley, D.H. (1976) Experimental studies of oxide minerals. In D. Rumble, editor, Rev. Mineral. 3, L-61-L-84.

----- (1981) Some experiments pertaining to the magnetite-ulvöspinel miscibility gap. Am. Mineral. 66, 759-762.

Meijering, J. L. (1963) Miscibility gaps in ferromagnetic alloy systems. Phillips Res. Rep. 13, 318-330.
Miodownik, A. P. (1982) The effect of magnetic transformations on phase diagrams. Bull. Alloy Phase Diagrams, 2, 406-412.
Nagata, T. (1962) Magnetic properties of ferrimagnetic minerals of Fe-Ti-O system. Proc. Benedum Earth Magnetism Symp. 69-86.
Navrotsky, A. (1987) Models of crystalline solutions. In Carmichael, I.S.E. and Eugster, H.P., editors, Rev. Mineral., 17, 35-69.
Nell, J., Wood, B.J., and Mason, T.O. (1989) High-temperature cation distributions in Fe_3O_4-$MgAl_2O_4$-$MgFe_2O_4$-$FeAl_2O_4$ spinels from thermopower and conductivity measurements. Am. Mineral., 74, 339-351.
Nishizawa, T., Hasebe, M. and Ko, M. (1987) Thermodynamic analysis of solubility and miscibility gap in ferromagnetic alpha iron alloys, Acta Metallurgica, 27, 817-828.
O'Neill, H. St. C. and Navrotsky, A. (1984) Cation distributions and thermodynamic properties of binary spinel solid solutions, Am. Mineral., 69, 733-753.
Petric, A. and Jacob, K.T. (1981) Thermodynamic properties of Fe_3O_4-$FeAl_2O_4$ spinel solid solutions, J. Am. Ceram. Soc., 64, 632-639.
---- and ----, (1982) Thermodynamic properties of. Fe_3O_4-FeV_2O_4 and Fe_3O_4-$FeAl_2O_4$ spinel solid solutions, J. Am. Ceram. Soc., 65, 117-123.
Price, G.D. (1981) Subsolidus phase relations in the titanomagnetite solid solution series. Am. Mineral., 66, 751-758.
Pynn, R. and Skjeltorp, A., Editors (1983) "Multicritical Phenomena" NATO ASI Series, V106, Plenum.
Rumble, D (1976) Oxide minerals in metamorphic rocks. In D. Rumble, editor, Rev. Mineral. 3, R-1-R-24.
Salje, E. (1985) Thermodynamics of sodium feldspar I, order parameter treatment and strain induced coupling effects. Phys. Chem. Minerals, 12, 93-98.
Salje, E., Kuscholke, B., Wruck, B. and Kroll, H. (1985) Thermodynamics of sodium feldspar II, experimental results and numerical calculations. Phys. Chem. Minerals 12, 99-107.
Samuelsen, E.J. and Shirane, G. (1970) Spin waves in antiferromagnets with corundum structure, Physica (Utrect), 43, 353-374.
Sanchez, J. M. and Lin, C.H. (1984) Modeling magnetic and chemical ordering in binary alloys. Phys. Rev., B30(3), 1448-1453.
Sundman, B., Jansson, B. and Andersson, J.O. (1985) The THERMO-CALC databank system, CALPHAD 9, #2, 153-190.
Swatsky, G.A., Van Der Woude, F., and Morrish, A.H. (1969) Mössbauer study of several ferrimagnetic spinels. Phys. Rev. 187, 747-757.
Thompson, A.B. and Perkins, E.H. (1981) Lamda transitions in minerals, in "Thermodynamics of Minerals and Melts," A. Navrotsky, R.C. Newton and B.J. Wood, editors. Advances in Physical Geochemistry, 1, 35-62, Springer-Verlag, New York.
Thompson, J.B. (1968) Chemical reactions in crystals. Am. Mineral., 54, 341-374.
Todd, S.S. and King, E.G. (1953) Heat capacities at low temperatures and entropies at 298.16°K. of titanomagnetite and ferric titanate. J. Am. Chem. Soc., 75, 4547-4549.
Trestman-Matts, A., Dorris, S.E., Kumarakrishnan, S., and Mason, T.O. (1983) Thermoelectric determination of cation distributions in Fe_3O_4-Fe_2TiO_4. J. Am. Ceram. Soc., 66, 829-834.
Turnock, A.C. and Eugster, H.P. (1962) Fe-Al oxides: phase relationships below 1000°C, J. Petrology 3, 533-565.
Vincent, E.A., Wright, J.B., Chevallier, R., and Mathieu, S. (1957) Heating experiments on some natural titaniferous magnetite. Mineral. Mag. 31, 634-655.
Westcott-Lewis, M.F. and Parry, L.G. (1971) Magnetism in rhombohedral iron-titanium oxides. Aust. J. Phys. 24, 719-734.
Yama-ai, M., Nagata, T. and Akimoto, S. (1962): these data points were digitized from a graph appearing in Nagata (1962).
Zener, C. (1955) Impact of magnetism on metallurgy, J. Met., Trans. AIME, 203, 619-630.

Chapter 9   R. O. Sack & M. S. Ghiorso

CHROMITE AS A PETROGENETIC INDICATOR

INTRODUCTION

Spinels in which chromium is a non-negligible constituent, are nearly ubiquitous, accessory minerals in low-pressure, basic and ultramafic igneous and metamorphic rocks of the terrestrial planets, and their associated satellites and debris (e.g., meteorites and asteroids). The notion that chromian spinels are potentially important petrogenetic indicators has been spawned, at least in part, in recognition that systematic relationships exist between spinel chemistry and chemical zoning, bulk-rock composition or mineral assemblage, and geological environment and process (e.g., Irvine, 1965, 1967; Jackson, 1969; Hill and Roeder, 1974; El Goresy, 1976; Sack, 1982; Allan et al., 1988). Although the focus of most petrological studies has been to enumerate general relationships between spinel chemistry, rock type, and process, recent studies have placed increasing emphasis on deriving quantitative estimates of intensive variables associated with igneous/metamorphic events from analysis of spinel-bearing assemblages. Such applications include Fe-Mg exchange geothermometry, geobarometry of upper mantle/lower crustal phase transitions, and the use of composition zoning as a sliding-scale indicator of magmatic processes or reaction progress and fluid evolution in metamorphic systems. The success of such efforts depends critically on our having adequate knowledge of relationships between structure, composition, and stability for spinels in a compositional subspace with the minimum complexity needed to describe most natural varieties, $(Fe,Mg)(Al,Cr,Fe^{3+})_2O_4 - (Fe,Mg)_2TiO_4$.

In this chapter some of the composition data from nature will be used to illustrate how chromian spinels may be used as petrogenetic indicators. We will focus on the fundamental field data that provide critical tests of calibrations for the various exchange reactions that govern its composition. First we will summarize data pertaining to the Fe-Mg exchange reactions involving chromian spinels and other common, ferromagnesian crystalline solutions (e.g., olivine and orthopyroxene). These are the best constrained and most fundamental of the relevant exchange reactions. In focusing on such data it is our intention to illustrate both the magnitude of the nonidealities associated with Fe-Mg substitutions and the dependencies of Fe-Mg partitioning relations on cation ordering and the other spinel exchange substitutions (e.g., Cr-Al, Cr-Fe). Applications to petrogenesis are illustrated by comparing calibrations of the olivine-spinel Fe-Mg geothermometer to data from specific suites of meteorites and terrestrial rocks. Composition data for chromian spinel+glass assemblages in MORB-type basalts are examined to illustrate the composition dependence of the Cr-Al exchange reaction between these phases and the use of zoning in chromian spinels as an indicator of magmatic processes. We then examine the use of chromian spinels as indicators of processes in the parent body for meteorites of the howardite-eucrite-diogenite association. Finally, calculated phase relations (Sack and Ghiorso, 1991b) are compared with natural constraints on miscibility gap features to clarify phase relations in the 'spinel prism' (e.g., Irvine, 1965) under conditions appropriate to the greenschist-amphibolite facies of metamorphism.

Nomenclature and chemical species

Although there are numerous descriptors that are reserved to designate specific spinel end-member species and solid solutions (e.g., picrochromite, qandilite, picotite, etc.), such vestiges of the spinel *appellative arcanum* will not be used here. Because it is

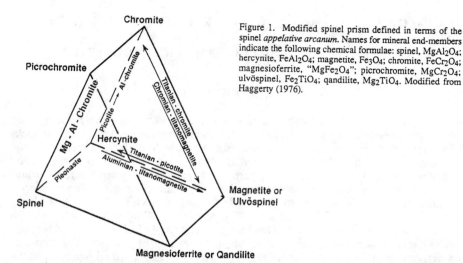

Figure 1. Modified spinel prism defined in terms of the spinel *appelative arcanum*. Names for mineral end-members indicate the following chemical formulae: spinel, $MgAl_2O_4$; hercynite, $FeAl_2O_4$; magnetite, Fe_3O_4; chromite, $FeCr_2O_4$; magnesioferrite, "$MgFe_2O_4$"; picrochromite, $MgCr_2O_4$; ulvöspinel, Fe_2TiO_4; qandilite, Mg_2TiO_4. Modified from Haggerty (1976).

not our purpose to unduly focus attention on restricted parts of spinel composition space, we will use the family name spinel preceeded by appropriate chemical descriptors, only where such modifiers do not detract from the meaning.

IRON-MAGNESIUM EXCHANGE REACTIONS

Fe-Mg partitioning relations among the common ferromagnesian silicates and oxides are among the most fundamental of petrogenetic indicators, and they must be fully characterized as a prerequisite to quantitative modelling of multicomponent systems. Following Ramberg and DeVore's (1951) pioneering study of Fe-Mg partitioning relations between olivine and orthopyroxene, calibrations of the Fe-Mg exchange reactions governing such relations have been widely used to estimate 'equilibration' temperatures. The petrogenetic significance of Fe-Mg partitioning between chromian spinel and Fe-Mg olivine was recognized early on. Irvine (1965, 1967) and others noted a pronounced positive correlation between Fe/Mg and Cr/Al ratios in chromian spinels in alpine peridotites and mafic layered intrusions. It was apparent that such correlations were crystallochemically based, because such rocks are characterised by restricted ranges of Fe/Mg ratios of olivines (Fo_{85-92}) and inferred temperatures of formation. In such rocks spinels typically have compositions close to the Al-Cr face of the so-called "spinel prism" defined by the vertices of the $(Fe^{2+},Mg)(Al,Cr,Fe^{3+})_2O_4$ composition space (Fig. 1). Irvine (1965) made the simplest possible provision for such compositional dependencies in his formulation for this Fe-Mg exchange reaction; he assumed that these spinels are ideal reciprocal solutions. In this formulation the logarithm of the distribution coefficient for the Fe-Mg exchange reaction between Fe-Mg olivine and a spinel

$$(\ln K_D^{OL-SP}; K_D^{OL-SP} \equiv [(n_{Fe^{2+}}^{SP})(n_{Mg^{2+}}^{OL})]/[(n_{Mg^{2+}}^{SP})(n_{Fe^{2+}}^{OL})])$$

is given by the sum of the logarithms of the distribution coefficients for the constituent reactions

$$\tfrac{1}{2} Mg_2SiO_4 + FeAl_2O_4 = \tfrac{1}{2} Fe_2SiO_4 + MgAl_2O_4 , \qquad (1)$$

$$\tfrac{1}{2} Mg_2SiO_4 + FeCr_2O_4 = \tfrac{1}{2} Fe_2SiO_4 + MgCr_2O_4 , \qquad (2)$$

and
$$\tfrac{1}{2} Mg_2SiO_4 + Fe_3O_4 = \tfrac{1}{2} Fe_2SiO_4 + MgFe_2O_4 , \qquad (3)$$

weighted according to the relative proportions of Al, Cr, and Fe^{3+} in the spinel. In the familar expression for the Fe-Mg exchange reaction consistent with this assumption,

$$\ln K_D^{OL-SP} = Y_{Al^{3+}} \left\{ \frac{\Delta \bar{G}^o_{(1)}}{RT} \right\} + Y_{Cr^{3+}} \left\{ \frac{\Delta \bar{G}^o_{(2)}}{RT} \right\} + Y_{Fe^{3+}} \left\{ \frac{\Delta \bar{G}^o_{(3)}}{RT} \right\}, \qquad (4)$$

the quantities $Y_{Al^{3+}}$, $Y_{Cr^{3+}}$, $Y_{Fe^{3+}}$ indicate the respective molar ratios of trivalent cations $Al^{3+}/(Al^{3+} + Cr^{3+} + Fe^{3+})$, $Cr^{3+}/(Al^{3+} + Cr^{3+} + Fe^{3+})$, and $Fe^{3+}/(Al^{3+} + Cr^{3+} + Fe^{3+})$, and the $\Delta \bar{G}^o_{(i)}$'s are the standard state Gibbs energies of Reactions (1) - (3).

Since its initial statement there have been many attempts to calibrate the formulation of the olivine-spinel Fe-Mg exchange reaction given by Equation (4) (e.g., Jackson, 1969; Evans and Frost, 1975; Fabriès, 1979; Roeder et al., 1979; Mukherjee and Viswanath, 1987). Such calibrations have tended to be based either on thermochemical data (e.g., Jackson, 1969; Roeder et al., 1979) or partitioning data derived from petrological studies combined with temperature estimates based on geological intuition (e.g., Evans and Frost, 1975; Fabriès, 1979). More recent calibrations have tended to incorporate both types of constraints as well as partitioning data obtained from experimental studies (e.g., Fujii, 1977; Mukherjee and Viswanath, 1987; Mukherjee et al., 1989). The near solidus equilibration temperatures obtained for many olivine-chromian spinel pairs from the calibration of Jackson (1969) were taken at face value in many early studies of alpine peridotites, layered intrusions, and iron meteorites (e.g., Bunch et al., 1970; Bunch and Keil, 1971; Dick, 1977), possibly because such estimates accorded with geological expectation. However, it became increasing evident that this calibration gave unrealistically high temperatures for chromian spinel-olivine assemblages in metamorphosed serpentinites (Evans and Frost, 1975) or basalts (e.g., Evans and Wright, 1972; Sigurdsson and Schilling, 1976). Subsequent attempts to resurrect this formulation of the olivine-spinel geothermometer have also been unsuccessful in describing both the temperature and composition dependencies of Fe-Mg partitioning, as documented by both experimental and petrological studies (e.g., Sack, 1982; Engi, 1983; Jamieson and Roeder, 1984; Lehmann and Roux; 1986), because this formulation fails to make explicit provision for nonidealities in the Fe-Mg substitutions in olivine and spinel or long-range cation ordering in spinel. Minor modifications and extensions of this simple formulation to include titanium substitution in spinel or provisions for nonideality in Fe-Mg substitutions (e.g., Sack, 1982; Engi, 1983) have also proved unsatisfactory in this regard. Symptomatic of the inadequacies of such modifications or extensions is the unphysical nature of the calibrations (strongly temperature-dependent parameters or asymmetric representations of the nonconfigurational, 'vibrational', Gibbs energy). These deficiencies arise from failure to account for the extent of Fe-Mg nonideality in both olivine and spinel and, in part, from lack of provision for order-disorder relations in spinels.

Fe-Mg nonideality

The central issue underlying the development of Fe-Mg exchange geothermometers is the magnitude of the nonideality associated with the Fe-Mg substitution in the common ferromagnesian oxides and silicates. The simplest case that may be considered is a hypothetical one in which Fe-Mg exchange occurs between two binary Fe-Mg minerals which each have only one type of crystallographically distinct site on which this substitution operates. Even in this case, the nonideality associated with the Fe-Mg substitutions in the constituent phases cannot be established simply from constraints on Fe-Mg partitioning betweeen them. Only the relative nonidealities may be determined from such data, unless the absolute nonideality associated with one of the phases is known or there are independent, internally consistent constraints on each of the end-member components. The analysis is further complicated for common ferromagnesian minerals by Fe-Mg substitution in crystallographically distinct sites (e.g., spinels, olivines, pyroxenes, amphiboles), ordering of Fe and Mg between such sites (e.g., spinels, pyroxenes, amphiboles), and pronounced dependencies of Fe-Mg exchange energies on the degree of operation of other exchange substitutions (e.g., spinels, Eqn. 4). Where Fe-Mg ordering is a complicating factor nonidealities associated with the Fe-Mg substitution are temperature dependent and the macroscopic deviations from ideality are reduced relative to those due to mixing on individual sites. In spinels, strongly bulk composition and temperature dependent cation ordering and pronounced differences between the energies of Reaction (1) and Reactions (2), (3), and the analogous Fe-Mg exchange reaction between olivines and titanate spinels,

$$\frac{1}{2}Mg_2SiO_4 + \frac{1}{2}Fe_2TiO_4 = \frac{1}{2}Fe_2SiO_4 + \frac{1}{2}Mg_2TiO_4 \;, \tag{5}$$

make it difficult to unambiguously determine *a priori* the nonidealities associated with the Fe-Mg substitutions on individual octahedral and tetrahedral sites. Accordingly, it is instructive to determine the magnitudes of nonidealities associated with the Fe-Mg substitutions in more nearly binary ferromagnesian minerals (olivines, orthopyroxenes, cummingtonite-grunerites) as a prelude to an inquiry into the site nonidealities associated with Fe-Mg substitutions in spinels.

It may be readily demonstrated that critical experimental and field data require positive deviations from ideal mixing of at least ~2 kJ/gfw for Fe-Mg substitutions on *individual* octahedral sites in olivines, orthopyroxenes, and amphiboles (e.g., Saxena and Ghose, 1971; Chatillon-Colinet et al., 1983, Anovitz et al., 1988; Sack and Ghiorso, 1989, 1991a, 1991b; Evans and Ghiorso, in prep.). Among these minerals, Fe-Mg olivine appears to display the greatest deviation from ideal mixing of Fe^{2+} and Mg^{2+}, largely because it exhibits less ordering of these cations between different crystallographic sites than do orthopyroxene or cummingtonite-grunerite solid solutions (e.g., Hafner and Ghose, 1971; Saxena and Ghose, 1971; Seifert, 1978; Aikawa et al., 1985; Anovitz et al., 1988; Ottonello et al., 1990). The more positive deviation from ideality displayed by olivine is manifested, for example, in the large compositional dependence to the distribution coefficient for the Fe-Mg exchange reaction between olivine and orthopyroxene (e.g., Fig. 1, Sack and Ghiorso, 1989), first documented by Ramberg and DeVore (1951). It is also manifested in a similarly large compositional dependence to the distribution coefficient for the Fe-Mg exchange reaction between olivine and cummingtonite-grunerite solid solutions (e.g., Fig. 2; Evans and Ghiorso, in prep.). Given equilibration temperatures, the distribution data may be used to establish plausible relationships between the macroscopic nonidealities associated with the minerals involved in these exchange reactions (e.g., Fig. 3; Fig. 2, Sack and Ghiorso, 1989). These relationsphips may be further refined by imposing the conditions that the calibration for the Fe-Mg exchange reaction also satisfy both (1) independent standard state constraints on the energies of the exchange reactions (e.g., Berman, 1988; Evans and Ghiorso, in prep.) and (2) standard state constraints on Fe end-member reactions involving quartz (Bohlen et al., 1980; Lattard and Evans, 1991) and the P-T-X characteristics of quartz-bearing assemblages in magnesian systems (e.g., Fig. 3; Bohlen and Boettcher, 1981; Davidson and Lindsley, 1989; Fonarev et al., 1979). Based on such an analysis for olivine + orthopyroxene-bearing assemblages Sack and Ghiorso (1989) have demonstrated that values of a macroscopic regular-solution-type

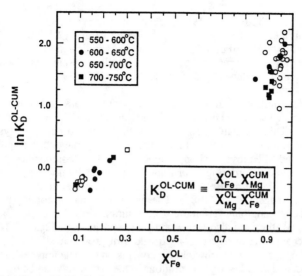

Figure 2. Logarithms of Fe-Mg distribution coefficients for natural olivine-cummingtonite pairs compared with molar Fe/(Fe + Mg) ratios of olivine. Composition data and temperature estimates are from the compilation of Evans and Ghiorso (in prep.).

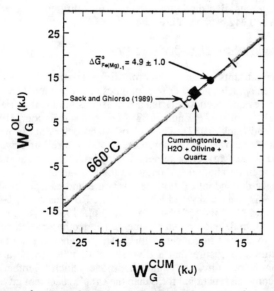

Figure 3. Ranges of macroscopic regular solution-type parameters for $(Fe,Mg)Si_{1/2}O_2$ olivine and for $(Fe,Mg)Si_{8/7}O_{22/7}(OH)_{2/7}$ cummingtonite-grunerite solid solutions (W_G^{OL} and W_G^{CUM}) that satisfy Fe-Mg distribution data (Fig. 2) and standard state constraints compared with an independent estimate for W_G^{OL}. *Shaded band* with temperature label corresponding to the average temperature estimated for the olivine and cummingtonite-bearing rocks (660°C) indicates ranges of these parameters consistent with the compositional dependence of the Fe-Mg distribution data. *Solid circle* and associated *error bar* superimposed on this band represent the mean and range of W_G^{OL} and W_G^{CUM} consistent with the independent estimate of Evans and Ghiorso (in prep.) for the standard state Gibbs energy of the Fe-Mg exchange reaction, $\Delta \bar{G}^{\circ}_{Fe(Mg)_{-1}}$. *Solid square* indicates corresponding range of these parameters obtained from employing the constraints on the standard state Gibbs energy of the grunerite Fe-endmember breakdown reaction (Lattard and Evans, 1991) and on the P-T-X characteristics of the indicated four-phase assemblage in magnesian systems (Fonarev et al., 1979). *Open circle* represents the value of W_G^{OL} obtained by Sack and Ghiorso (1989).

parameter W_G^{OL} (e.g., Thompson, 1967) describing deviations from ideality in $(Mg,Fe)Si_{1/2}O_2$ olivines must be in the range 8.4-10.5 kJ/gfw (2.1-2.6 kJ/gfw positive deviation from ideality in $Mg_{1/2}Fe_{1/2}Si_{1/2}O_2$ olivine). As indicated in Figure 3, ranges of values of W_G^{OL} derived from a corresponding analysis for olivine + cummingtonite-grunerite assemblages overlap those deduced by Sack and Ghiorso (1989).

In the subsequent discussion we adopt the estimate of Sack and Ghiorso (1989) for the nonideality associated with the Fe-Mg substitution in olivine (W_G^{OL} = 10.2 ± 0.3 kJ/gfw in $(Mg,Fe)Si_{1/2}O_2$). This estimate is consistent with the macroscopic analysis outlined above and a host of other constraints including those on long-range ordering of Fe^{2+} and Mg^{2+} between M1 and M2 sites in orthopyroxene, and activity-composition relations and enthalpies of mixing and disordering in orthopyroxene and olivine (e.g., Berman, 1988; Saxena and Ghose, 1971; Anovitz et al., 1988; Chatillon-Colinet et al., 1983; Kitayama and Katsura, 1968; Sharma et al., 1987). The analogous macroscopic parameters for orthopyroxene and cummingtonite-grunerite solid solutions (W_G^{OPX} and W_G^{CUM}) are smaller than W_G^{OL} (e.g., Fig. 3); they are also temperature dependent (e.g., Figs. 13-14, Sack and Ghiorso, 1989), because the energetic consequences of pronounced cation ordering are subsumed within them. It is noteworthy that the conclusion that site nonidealities of at least ~ 2kJ/gfw are associated with Fe-Mg substitutions in these minerals can be inferred simply from the Fe-Mg ordering and enthalpy of mixing data for orthopyroxene. To accomodate the fact that orthopyroxenes display positive enthalpies of mixing the microscopic parameters characterizing nonidealities due to the Fe-Mg substitutions on individual M1 and M2 octahedral sites (W_{M1}^{OPX} and W_{M2}^{OPX}) must each be greater than about 8.5 KJ/site to compensate for the large negative deviations from nonideality contributed by the pronounced ordering of Fe and Mg between these sites.

Cation ordering and the olivine-spinel geothermometer

Experimental constraints on activity-composition relations and Fe^{2+}-Mg^{2+} exchange reactions between spinels and other ferromagnesian phases indicate that spinels in constituent Fe-Mg binaries and pseudobinaries are nearly as nonideal as Fe-Mg olivines (e.g., Trinel-Dufour and Perrot, 1977; Engi, 1983; Jamieson and Roeder, 1984; Lehmann and Roux, 1986; Hill and Sack, 1987). The use of such constraints in calibrating the olivine-spinel Fe-Mg exchange geothermometer is not straightforward, because $(Fe,Mg)(Al,Cr,Fe^{3+})_2O_4$ - $(Fe,Mg)_2TiO_4$ spinels display both long-range and short-range ordering of cations between tetrahedral and octahedral sites, and the nature and degree of such ordering is strongly dependent on both temperature and composition (e.g., Roth, 1964; Banerjee et al., 1967; Navrotsky and Kleppa, 1967; Robbins et al., 1971; DeGrave et al., 1975; Wechsler and Navrotsky, 1984; Millard et al., 1991; Peterson et al., 1991; Sack and Ghiorso, 1991a). In developing an olivine-spinel geothermometer constraints on such ordering must be evaluated in concert with those on standard state properties and phase equilibrium features to ensure that a diverse set of phenomena are adequately accounted for in the context of a physically plausible model. Such a geothermometer will necessarily be of more complex functional form than the simple statement given in Equation (4), making explicit provision for dependencies of the Fe-Mg distribution coefficient on both composition and ordering variables.

In this section we compare the results of a recent formulation and calibration of the olivine-spinel Fe-Mg exchange geothermometer (Sack and Ghiorso, 1991b) with some of the critical field constraints. This comparison illustrates the use of Fe-Mg exchange geothermometers as petrogenetic indicators and some of the petrological manifestations of both cation ordering and Fe-Mg nonideality. The olivine-spinel Fe-Mg exchange geothermometer of Sack and Ghiorso (1991b) is derived from a thermodynamic model for

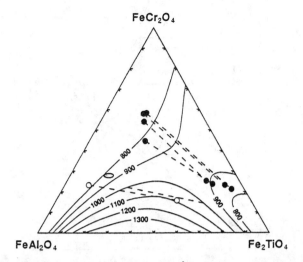

Figure 4. Calculated 800°, 900°, 1000°, 1100°, 1200°, and 1300°C miscibility gaps (*solid curves*) for FeAl$_2$O$_4$ - FeCr$_2$O$_4$ - Fe$_2$TiO$_4$ spinels together with experimental constraints and composition data for coexisting Cr- and Ti-rich spinels in lunar basalts (*solid circles* connected by *dashed lines*). *Shaded ellipse* indicates an estimate for the extent of ternary miscibility at 1000°C derived from X-ray determinative curves and synthesis experiments by Muan et al. (1972). *Open circles* connected by a *dashed line*, indicate a similarly derived estimate for the sense of a tieline for coexisting ternary spinels at 1000°C (Muan et al., 1972). A comparable calculated tieline is indicated by the remaining *dashed line*. Lunar spinel data from Agrell et al. (1970), Champness et al. (1971), Haggerty and Meyer (1970), and Taylor et al. (1971).

chromian spinels that is formulated for the simplifying assumptions (1) that cations exhibit only long-range non-convergent ordering between tetrahedral and octahedral sites, (2) that the *vibrational* Gibbs energy may be described by a Taylor expansion of only second degree in composition and ordering variables, and (3) that excess *vibrational* entropies of ordering and mixing are negligible (i.e., all coefficients of second degree in the Taylor expansion are constants independent of temperature) (Chapter 6). Experimental determinations of Fe-Mg partitioning relations between spinels and olivines (e.g., Sack, 1982; Engi, 1983; Jamieson and Roeder, 1984; Murck and Campbell, 1986; Hill and Sack, 1987) constitute only one type of constraint utilized in calibrating this model. Other active constraints on the calibration of this model include those on (1) distribution coefficients for the exchange of Al and Cr between spinels and (Al,Cr)$_2$O$_3$ solid solutions (e.g., Petric and Jacob, 1982a; Oka et al., 1984), (2) miscibility gap features in Cr-Ti-Al spinels (e.g., Fig. 4; Fig. 1, Chapter 6; Muan et al., 1972), (3) activity-composition data in the chromite-magnetite subsystem and for (Al,Cr)$_2$O$_3$ solid solutions (e.g., Katsura et al., 1975; Petric and Jacob, 1982b, Chatterjee et al., 1982), and (4) cation site-ordering constraints (e.g., Kriessman and Harrison, 1956; Epstein and Frackiewicz, 1958; Mozzi and Paladino, 1963; Roth, 1964; DeGrave et al., 1975; Millard et al., 1991; Peterson et al., 1991). The calibration is internally consistent with the thermodynamic database of Berman (1988) and the analyses of the solution properties of olivines, orthopyroxenes, and rhombohedral oxides of Sack and Ghiorso (1989) and Ghiorso (1990). Fundamental field constraints such as those summarized in Figures 5-8 were not considered in developing the calibration.

The most straightforward field test of the adequacy of olivine-spinel Fe-Mg exchange geothermometers and of its internal consistency with respect to the olivine-orthopyroxene exchange geothermometer is provided by the composition data for these minerals in ordinary H, L, and LL chondrites (e.g., Keil and Fredrickson, 1964; Bunch et al., 1967). Assuming the blocking temperatures of these two geothermometers are similar, results of a comparison between them should have a fairly unambiguous interpretation. Fe^{3+}/Fe^{2+} ratios in the ferromagnesian phases in these metal-saturated meteorites are

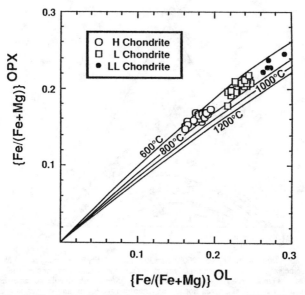

Figure 5. Molar Fe/(Fe + Mg) ratios of olivines (OL) and orthopyroxenes (OPX) of H, L, and LL chondrites (Keil and Fredrickson, 1964) plotted with 600, 800, 1000, and 1200°C isotherms for the olivine-orthopyroxene Fe-Mg exchange reaction calculated from the calibration of Sack and Ghiorso (1989).

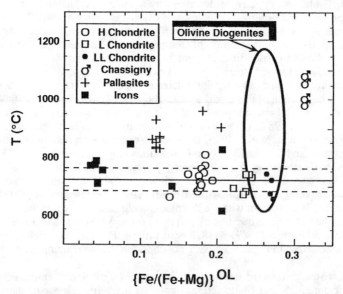

Figure 6. Temperatures calculated for various meteorites from the olivine-spinel Fe-Mg exchange geothermometer of Sack and Ghiorso (1991b) expressed as a function of molar Fe/(Fe + Mg) ratios of olivines (OL). Ellipsoid indicates range of these quantities obtained for olivine diogenites (Figs. 14-15; Sack et al., 1991). Remaining data from: Keil and Fredrickson (1964); Bunch et al. (1967, 1970); Bunch and Keil (1971); Floran et al. (1978). *Horizontal continuous* and *dashed lines* indicate the respective mean and 1σ range of olivine-spinel Fe-Mg exchange temperatures calculated for ordinary chondrites from chromite and olivine data reported by Bunch et al. (1967, p. 1574-1576, Tables 2-4) and Keil and Fredricksson (1964, p. 4394-4395, Table 3).

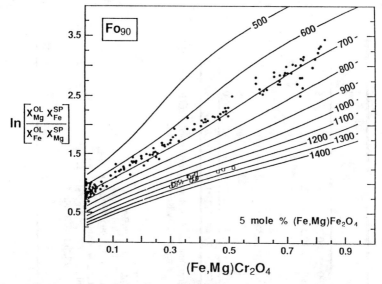

Figure 7. Fe-Mg partitioning data for olivine-chromian spinel pairs from metamorphosed green spinel-chlorite-enstatite-olivine rocks (*solid circles*; Evans and Frost, 1975; their Fig. 6) and calculated temperatures and spinel chromium contents of olivine-spinel pairs from Juan de Fuca Ridge axial valley lavas (*open squares*; J.F. Allan, pers. comm.) compared with the calibration of the olivine-spinel Fe-Mg exchange reaction of Sack and Ghiorso (1991b) for $(Fe,Mg)([Al,Cr]_{0.95}[Fe^{3+}]_{0.05})_2O_4$ spinels and Fe-Mg exchange potentials corresponding to those of $(Mg_{0.9},Fe_{0.1})_2SiO_4$ olivines. X_{Mg}^{OL}, X_{Mg}^{SP}, X_{Fe}^{OL}, and X_{Fe}^{SP} indicate the respective molar Mg/(Mg + Fe^{2+}) and Fe^{2+}/(Mg + Fe^{2+}) ratios in olivine (OL) and spinel (SP).

negligible. Orthopyroxenes have negligible Al- and Cr-contents and their Ca-contents are low (X_{Ca}^{M2} < 0.03). Finally, spinels in these meteorites are nearly end-member chromite-pichrochromite solid solutions (e.g., Table 1). Uncertainties associated with the ordering states of spinels are not an issue, because end-member chromite and picrochromite spinels have fully ordered, normal cation distributions (e.g., Chapter 6; Dunitz and Orgel, 1957). In Figures 5 and 6 we plot the relevant mineral composition data from ordinary H, L, and LL chondrites on the calibrations of the olivine-orthopyroxene and olivine-spinel Fe-Mg exchange geothermometers of Sack and Ghiorso (1989, 1991b).

From the comparisons illustrated in Figures 5 and 6 it is evident that the two geothermometers give virtually identical equilibration temperatures for the three different classes of chondrites. The average temperature obtained from the olivine-spinel geothermometer for the chondrites, 720°±40°C, is slightly lower than that deduced for ordinary H chondrites by Olsen and Bunch (1984) using the two pyroxene Ca-Mg exchange geothermometers of both Kretz (1982) and Lindsley (1983), 820°-830°C. The latter observation is consistent with the fact that, during cooling, Ca-Mg and Ca-Fe exchange reactions between coexisting pyroxenes typically freeze at higher temperatures than Fe-Mg exchange reactions, particularly those between olivine and spinel. Excellent correspondence between equilibration temperatures calculated from these thermometers has also been noted for the iron-rich olivine and orthopyroxene-bearing assemblages found in rhyolites, where such temperatures also accord with those calculated from coexisting spinel and rhombobohedral oxides (Ghiorso and Sack, 1991). Finally, the calibration of the olivine-orthopyroxene Fe-Mg exchange geothermometer displayed in Figure 5 is a reminder that the relative partitioning of cations between coexisting phases does not necessarily decrease with increasing temperature where cation ordering and site nonideality associated with the substitution of the exchanging elements are complicating factors.

Table 1. Chemical analyses chromian spinels in meteorites (wt %).

	1	2	3	4	5	6	7	8	9	10	11	12
TiO_2	2.20	2.63	2.92	0.17	1.29	6.25	0.69	0.90	0.95	0.83	2.28	4.1
Al_2O_3	5.92	5.58	5.39	5.09	3.82	5.28	21.21	9.79	10.54	7.97	7.14	9.3
Cr_2O_3	57.02	56.17	55.23	64.31	64.70	44.68	42.22	56.57	52.73	60.09	53.9	48.9
V_2O_3	0.69	0.73	0.75	0.56	0.39	0.39	n.d.	n.d.	n.d.	------	0.56	<0.02
FeO	31.04	33.05	34.05	23.39	20.31	38.12	27.67	24.62	29.58	25.47	31.9	35.7
MgO	2.60	1.97	1.73	5.70	7.00	3.78	5.54	6.11	2.87	5.83	1.80	0.59
MnO	0.96	0.69	0.66	0.63	1.95	0.61	0.49	0.58	0.73	0.49	0.62	0.59
Total	100.43	100.82	100.73	99.85	99.46	99.30	97.82	98.57	97.40	100.68	98.20	99.93

1-3. Average compositions of chromites in ordinary H (#1, 12 analyses), L (#2, 6 analyses), and LL (#3, 4 analyses) chondrites (Bunch et al. (1967, p. 1574-1576, Tables 2-4) for which analyses of olivines and orthopyroxenes are reported by Keil and Fredricksson (1964, p. 4394-4395, Table 3).
4. Average chromite composition in ten pallasites (Bunch and Keil, 1971, p. 148, Tables 1-2).
5. Average chromite composition in nine iron meteorites (Bunch et al., 1970, p.152, Table 7).
6. Average chromite composition in Chassigny (Floran et al., 1978, p. 1215, Table 2, ZnO = 0.19 wt%).
7. 'Equilibrated' magnesian, chromian spinel in olivine diogenite ALH A77256 (Sack et al., 1991, Table 1).
8. 'Unre-equilibrated', magnesian, chromian spinel in olivine diogenite ALH 84001 (Sack et al., 1991, Table 1).
9. 'Equilibrated', iron-rich, chromian spinel in olivine diogenite ALH 84001 (Sack et al., 1991, Table 1).
10. 'Unre-equilibrated' chromian spinel from ordinary diogenite Yamoto 6902 (Takeda et al., 1978, p.176, Table 4).
11. 'Equilibrated' chromian spinel from ordinary diogenite Yamato 75032 (Takeda et al., 1978, p.176, Table 4).
12. Average chromite composition in eucrites (Bunch and Keil, 1971, p. 152, Table 7).

Composition data for coexisting olivines and chromian spinels in prograde metamorphic serpentinites (Evans and Frost, 1975) also provide critical field tests of calibrations of the olivine-spinel Fe-Mg exchange geothermometer and inferences regarding the site distributions of cations in $(Fe,Mg)(Al,Cr)_2O_4$ and $(Fe,Mg)(Cr,Fe^{3+})_2O_4$ spinels. The serpentinites considered by Evans and Frost (1975) have olivines with Mg/(Mg+Fe) ~0.90 (Fo_{90}) and spinels with compositions near either the $(Fe,Mg)(Al,Cr)_2O_4$ or the $(Fe,Mg)(Cr,Fe^{3+})_2O_4$ face of the spinel prism. There is a progressive increase in the range of proportions of trivalent cations in these spinels with increasing metamorphic grade. Spinels coexisting with olivine are essentially ferrites at the lowest grades of metamorphism characterized by brucite + diopside + antigorite (or chrysotile/lizardite)-bearing assemblages. With increasing metamorphic grade they may incorporate progressively more chromium, such that by middle amphibolite facies (talc + olivine-bearing rocks) spinels may exhibit a wide range of Cr/Fe^{3+} ratios (e.g., Fig. 8). Fe^{3+}-rich spinels are Al-poor, but the most chromian spinels may incorporate up to about 25% of aluminate components. At the highest grades of metamorphism (upper amphibolite-granulite facies) $(Fe,Mg)(Al,Cr)_2O_4$ spinels with about 5 mole % of $(Fe,Mg)Fe_2O_4$ are common in enstatite + chlorite-bearing assemblages, and the aluminum content of the spinels is controlled by the thermal decomposition of chlorite. Evans and Frost (1975) have used mineral composition data from these rocks to derive a tentative 700°C Fe-Mg exchange isotherm for Fo_{90} olivines coexisting with Cr-Al spinels.

In Figure 7 we compare the calibration of the olivine-spinel Fe-Mg exchange reaction of Sack and Ghiorso (1991b) for $(Fe,Mg)([Al,Cr]_{0.95}[Fe^{3+}]_{0.05})_2O_4$ spinels and Fo_{90} olivines with the Fe-Mg partitioning data of Evans and Frost (1975) for olivine-spinel pairs in chlorite + enstatite-bearing serpentinites. Also plotted on this Figure are spinel chromium contents and calculated temperatures for olivine-spinel pairs from a suite of MORB-type lavas in small cones on the axial valley of the Juan de Fuca Ridge. These lavas have olivines with $Fo_{88.4-82.1}$ and Ti-poor spinels with between 5.6 and 13.9 mole % of ferrite components. Eruption temperatures of 1170° to 1220°C have been estimated for them based on phase equilibrium constraints (J.F. Allan, pers. comm.). Of note are the excellent correspondences between calculated temperatures and those inferred for both suites of rocks. This agreement may be interpreted as providing indirect confirmation of two inferences regarding cation distributions in aluminate spinels: (1) $FeAl_2O_4$ is significantly more normal than $MgAl_2O_4$ and (2) the thermal dependence of the ordering state of $MgAl_2O_4$ is virtually identical to that inferred from the tight reversal brackets on degrees of ordering obtained from the spectroscopic studies of Millard et al. (1991) and Peterson et al. (1991) (cf., Fig. 9, Chapter 6). The former inference is a consequence of accomodating a diverse set of phase equilibrium constraints in the context of a physically plausible model of simple formulation. Because calculated isotherms are very sensitive to the ordering state of $MgAl_2O_4$, it is not possible to simultaneously accomodate both these phase equilibrium constraints and critical field data such as those summarized in Figure 6, unless the latter inference is substantially correct. Larger thermal dependencies of the degrees of ordering of $MgAl_2O_4$ would be problematic, because they lower both the aluminate and chromate intercepts of the 700°C isotherm on Figure 6, and would thus yield temperatures too low to be consistent with the thermal decomposition of chlorites in the serpentinites examined by Evans and Frost (1975).

The calibration of the olivine-spinel geothermometer of Sack and Ghiorso (1991b) is also consistent with the composition data of Evans and Frost (1975) for olivine - $(Fe,Mg)(Cr,Fe^{3+})_2O_4$ spinels, given the temperatures generally inferred to be associated with the relevant prograde metamorphic assemblages (Fig. 8). As in the case discussed above, the calculated isotherms are very sensitive to assumptions regarding the thermal dependence of equilibrium ordering states of end-member components displaying variable degrees of cation ordering, in this case Fe$_3$O$_4$ and "MgFe$_2$O$_4$". Fe$_3$O$_4$ must have cation distributions that are substantially inverse at all temperatures to satisfy both these critical field data and the active phase equilibrium and cation ordering constraints (cf. Sack and Ghiorso, 1991a,b). The distinctly sigmoidal shape of the calculated isotherms appearing in Figure 8 reflects the change in site preference that Fe^{3+} undergoes with increasing Cr/(Cr + Fe^{3+}) (e.g., Fig. 9; Robbins et al., 1971), parallels a similar phenomenon exhibited by the molar volumes of Fe$_3$O$_4$ - FeCr$_2$O$_4$ solid solutions (e.g., Francombe, 1957), and accords with the observation of Evans and Frost (1975) that Fe-Mg exchange isotherms at intermediate Cr/(Cr + Fe^{3+}) ratios display a distinct curvature that becomes less pronounced with increasing temperature. It is most likely that the change in site preference exhibited by Fe^{3+} leads to the development of miscibility gaps in FeCr$_2$O$_4$ - Fe$_3$O$_4$ spinels at temperatures substantially above those that would be predicted from considerations based on size mismatch between end-member "species", inverse Fe$_3$O$_4$ and normal FeCr$_2$O$_4$ (cf., Fig. 2, Chapter 6). The calculated miscibility gaps are fairly symmetric, reflecting the nearly symmetric nature of the related ordering phenomenon (e.g., Fig. 9). It is noteworthy that argentian tetrahedrite fahlores,

$$\sim(Cu,Ag)^{III}_6([Cu,Ag]_{2/3}[Fe,Zn]_{1/3})^{IV}_6 Sb_4S_{13} ,$$

(e.g., Spiridonov, 1984; Ebel and Sack, 1991) displays parallel systematics with respect to distribution coefficients for the Fe-Zn exchange reaction with sphalerites, volume-composition systematics, miscibility gap features, and order-disorder phenomena (e.g., Sack et al., 1987a; Sack, 1991). Although the exact mechanisms responsible for the parallels between the physical and thermochemical properties of ferrichromite spinels and argentian fahlores are not known in detail, we suspect that such phenomena are fairly common features in ordered, multisite reciprocal mineral solutions. At the very least, it

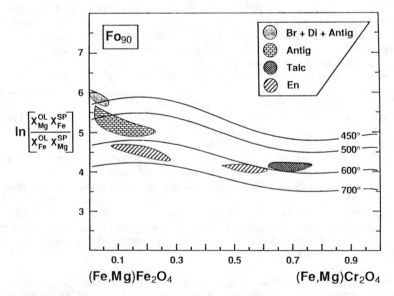

Figure 8. Fe-Mg partitioning data for olivine-ferrichromite spinel pairs from prograde metamorphic serpentinites (Evans and Frost, 1975; their Fig. 7) compared with the calibration of the olivine-spinel Fe-Mg exchange reaction of Sack and Ghiorso (1991b) for ferrite-chromite spinels and Fe-Mg exchange potentials corresponding to those of (Mg$_{0.9}$,Fe$_{0.1}$)$_2$SiO$_4$ olivines. X^{OL}_{Mg}, X^{SP}_{Mg}, X^{OL}_{Fe}, and X^{SP}_{Fe} indicate the respective molar Mg/(Mg + Fe^{2+}) and Fe^{2+}/(Mg + Fe^{2+}) ratios in olivine (OL) and spinel (SP). *Shaded areas* enclose the most densely populated regions of the graph of Evans and Frost (1975) for indicated prograde metamorphic assemblages. Abbreviations are: Br, brucite; Di, diopside; Antig, antigorite; En, orthopyroxene.

should be evident that such features are inconsistent with formulations of exchange geothermometers that do not make exhibit provision for order-disorder phenomena, nonidealities associated with substitution of the exchanging elements, and their interactions.

PETROLOGICAL APPLICATIONS

There are at least three ways in which composition data from chromian spinels may be used to obtain information relevant to the analysis of petrogenesis. First, composition data may be compared with calibrations of exchange reactions between spinels and coexisting phases to estimate equilibrium temperatures. In practice, reliable estimates of equilibration temperatures may be obtained only from the Fe-Mg exchange reactions between chromian spinel and coexisting ferromagnesian phases. Developing calibrations for exchange reactions based on the chromian spinel substitutions Cr - Al, Cr - Fe^{3+}, and Cr_2 - FeTi is problematic at present owing to the absence of constraints on standard state

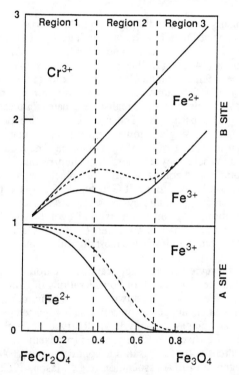

Figure 9. Calculated 25°C site distributions for chromite-magnetite solid solutions compared with approximate boundaries between regions of differing cation ordering schemes inferred from magnetic, Mössbauer, and volume-composition data (Robbins et al., 1971). Regions 1 and 3 correspond to the composition ranges for which $RTln\left[\dfrac{X^{tet}_{Fe^{3+}} X^{oct}_{Fe^{2+}}}{X^{tet}_{Fe^{2+}} X^{oct}_{Fe^{3+}}}\right]$ is either strongly negative or positive (i.e., regions in which magnetite behaves primarily as a normal and inverse component, respectively). *Solid* and *dashed curves* indicate site distributions calculated from the model of Sack and Ghiorso (1991b) and by utilizing for a value for the enthalpy of the Fe^{2+}-Fe^{3+} ordering reaction in Fe_3O_4 ~20% smaller than their estimate, respectively (cf. Sack and Ghiorso, 1991b).

properties and activity-composition relations for chromian components in most of the relevant coexisting phases (e.g., silicate melt, pyroxene), and the paucity of experimental and natural constraints with which to calibrate such exchange reactions. Second, the conditions of equilibrium corresponding to net transport reactions may be evaluated to estimate other intensive parameters (e.g., f_{O_2}, P) that pertain to chromian spinel-bearing multiphase assemblages. Chromium is a non-negligible constituent in spinels in many assemblages for which the following geologically important net transport reactions may be written:

$$O_2 + 6\ Fe_2SiO_4 = 3\ Fe_2Si_2O_6 + 2\ Fe_3O_4, \qquad (6)$$
$$\text{Olivine} \qquad \text{Orthopyroxene} \quad \text{Spinel}$$

$$CaAl_2Si_2O_8 + 2\ Mg_2SiO_4 = MgAl_2O_4 + Mg_2Si_2O_6 + CaMgSi_2O_6, \qquad (7)$$
$$\text{Plagioclase} \qquad \text{Olivine} \qquad \text{Spinel} \quad \text{Orthopyroxene} \quad \text{Clinopyroxene}$$

and
$$MgAl_2O_4 + 2\ Mg_2Si_2O_6 = Mg_3Al_2Si_3O_{12} + Mg_2SiO_4. \qquad (8)$$
$$\text{Spinel} \qquad \text{Orthopyroxene} \qquad \text{Garnet} \qquad \text{Olivine}$$

Calibrations of statements of the conditions of equilibrium corresponding to the first of these reactions have been used to define the oxygen fugacity of the mantle (e.g., Mo et al., 1982; O'Neill and Wall, 1987; Mattioli and Wood, 1988; Chapter 11) while those associated with Reactions (14) and (15) have been used to estimate the depths associated with the boundaries between the plagioclase, spinel, and garnet peridotite facies (e.g., Frost, 1976; O'Neill, 1981). Of course activity-composition relations must be known for each of the constituent minerals to produce realistic estimates for these parameters and their displacements due to Cr-substitution. The phases must also preserve 'equilibrium' compositions appropriate to a single, known temperature, or such compositions must be reconstructed from considerations of exchange equilibria. The requisite consistency tests are seldom performed. Where they are not, calculated results should be viewed with skepticism. Finally, composition zoning of chromian spinels may be used as a sliding-scale indicator of crystallization history (e.g., El Goresy, 1976), crystal fractionation, magma mixing, and assimilation in magmatic systems (e.g., Allan et al., 1988, 1989), or of subsolidus re-equilibration, reaction progress, and evolution of fluid composition in metamorphic systems (e.g., Hoffman and Walker, 1978). This last activity is arguably the most exciting of the three, but it is clearly the most difficult to quantify.

In this section we review several applications of chromian spinels as petrological indicators that illustrate the importance of considerations of exchange equilibria in the analysis of petrogenesis. We draw our first two examples from rock suites in which at least some of the chromian spinels appear to preserve their high-temperature compositional signatures: (1) olivine- and plagioclase-phyric MORB-type basalt glasses from seamounts adjacent to the East Pacific Rise; and (2) diogenites (pyroxenites) from the lower crust - upper mantle of the parent body for basaltic achondrite group (BAG) meteorites of the howardite-eucrite-diogenite (HED) association (e.g., Mason, 1962; BVSP, 1981). The inference that spinels from quenched basalts preserve their original Fe-Mg ratios is certainly not surprising. Such an inference is so for plutonic rocks in which retrograde Fe-Mg exchange reactions between olivine and spinel typically continue to temperatures below 900°-1000°C (e.g., Fig. 10) unless olivine and spinel are separated from diffusional contact as, for example, in certain monomineralic layers in the Stillwater intrusion (e.g., Irving, 1967). The example drawn from MORB-type basalts is used to demonstrate that the Cr-content of spinels crystallized from basaltic liquids is strongly dependent on the degree of fractionation of the host liquid; it is intended as a reminder that petrogenetic inferences derived from spinel chemistry and chemical zoning in lavas and hypabyssal rocks may be premature until a valid description for the thermodynamic properties of chromium-bearing

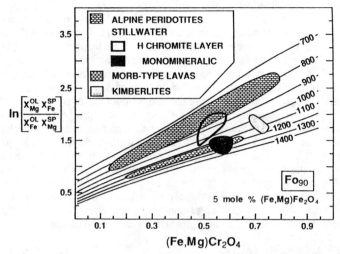

Figure 10. Representative ranges of calculated olivine-spinel temperatures and spinel Cr-contents for alpine peridotites, ocean basalt lavas, the Stillwater H Chromitite Zone, and kimberlite nodules plotted on the calibration of the olivine-spinel Fe-Mg exchange reaction of Sack and Ghiorso (1991b) for $(Fe,Mg)([Al,Cr]_{0.95} \cdot [Fe^{3+}]_{0.05})_2O_4$ spinels and Fe-Mg exchange potentials corresponding to those of $(Mg_{0.9}Fe_{0.1})_2SiO_4$ olivines. *Areas* for alpine peridotites, kimberlites, and the Stillwater H Chromitite Zone either enclose points or correspond directly to areas given in Figures 6 and 8 of Evans and Frost (1975). Area for MORB-type basalt lavas is based on analyses from J.F. Allan (per. comm.). X^{OL}_{Mg}, X^{SP}_{Mg}, X^{OL}_{Fe}, and X^{SP}_{Fe} indicate the respective molar $Mg/(Mg + Fe^{2+})$ and $Fe^{2+}/(Mg + Fe^{2+})$ ratios in olivine (OL) and spinel (SP).

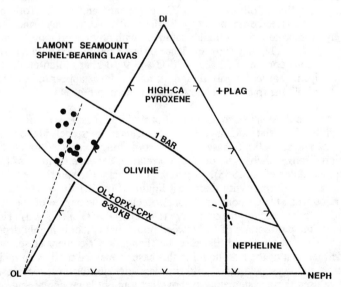

Figure 11. Compositions of spinel-bearing glasses from the Lamont seamont lavas plotted with the 1 atm. olivine + high-Ca pyroxene + plagioclase + spinel ($f_{O_2} \sim$ QFM) and high pressure olivine + orthopyroxene + high-Ca pyroxene cotectics for natural basic liquids. Liquidus boundaries and algorithms for computing proportions of nepheline (NEPH), diopside (DI), and olivine (OL) normative components are from Sack et al. (1987b). *Dashed line* radiating from OL vertex represents an average chord for liquids interrelated by olivine and plagioclase fractionation. From Allan et al. (1988).

silicate liquids is developed. In essence, this example demonstates that Cr in natural silicate liquids behaves as a perfect trace element: it obeys Henry's law, but its Henry's law coefficient is highly dependent on the composition of its matrix. The example drawn from HED representatives of the BAG meteorites is used to illustrate the importance of carefully examining composition data bearing on various exchange reactions before using the Cr- and Al-contents of pyroxenes or spinels to provide information about magmatic history. We show that it is not a safe practice to define liquid lines of descent or postulate different magma types based on such contents, because Cr- and Al-contents of pyroxenes or spinels are reset by Al-Cr exchange reactions during subsolidus re-equilibration. In our final example we compare composition data for natural spinels with calculated phase relations in the spinel prism for temperatures appropriate to the upper greenschist - lower amphibolite facies. We utilize the olivine-spinel Fe-Mg exchange geothermometer to demonstrate that some natural spinel pairs have tielines in general accord with those calculated for comparable Fe-Mg equipotential sections through the prism (i.e., isothermal sections for constant Fe/Mg ratio of coexisting Fe-Mg olivine). Using the same procedures we demonstrate that some other such coexisting spinels represent disequilibrium, or that some of the composition data for such spinels is appropriate to two-phase mixtures.

Chromian spinel in MORB-type lavas

Olivine- and plagioclase-phyric, MORB-type basalt glasses from the Lamont seamount chain adjacent to the East Pacific Rise near 10°N provide evidence that the Cr-content of spinel is a sensitive indicator of the degree of fractionation of the host liquids (Allan et al., 1988). Lavas from the Lamont seamount chain and other seamounts of near-ridge origin are systematically more primitive (higher $Mg/(Mg + Fe^{2+})$ = mg') than typical ridge-axis basalts (e.g., Walker et al., 1979; Batiza and Vanko, 1984; Langmuir et al., 1986; Allan et al., 1989). In terms of the systematics of Sack et al. (1987b) the Lamont basalts range from 'primitive' to 'evolved'. 'Primitive' basalts have ratios of nepheline/(nepheline + olivine + high-Ca pyroxene) and olivine/(olivine + high-Ca pyroxene) normative components (NEPH/(NEPH + OL + DI) and DI/(OL + DI), respectively) similar to those of liquids saturated with olivine, orthopyroxene and high-Ca pyroxene (e.g., Fig. 11); they show evidence of last equilibration with a peridotite mantle under near-solidus conditions, 1220-1230°C and 9-10 kbar (Allan et al., 1989). Within individual volcanic edifices 'evolved' and 'primitive' lavas appear to be simply related by plagioclase and olivine fractionation (e.g., dotted chord in Fig. 11).

Spinels in these lavas cover a range of molar $Cr/(Cr + Al)$ ratios of 0.20-0.54 for the entire suite; TiO_2 and calculated Fe_2O_3-contents in the spinels are low (0.16-0.85 and 5.5-9.2 wt %, respectively). In basalt samples containing spinels exhibiting only restricted compositional ranges there is no obvious correlation between Cr in glass (300-450 ppm) and the spinel (Allan et al., 1988), despite the fact that these spinels were in apparent Fe-Mg exchange equilibrium with their host liquids (Fig. 12). However, there is a strong correlation between the Cr-content of such spinels and the degree of fractionation of the host liquid, as measured by the normative ratio $DI/(DI + OL)$ (Fig. 13). This correlation indicates that the partitioning of Cr between liquid and spinel is highly dependent on the bulk composition of the liquid. Because the change in Cr-content of spinel is of greater magnitude than that of the host liquid in this case, it may be used as an indicator of melt composition and crystal-liquid equilibrium and disequilibrium processes. Allan et al. (1988) have used these systematics to show that Lamont lavas containing strongly zoned spinels or populations of spinels with broad compositional range record evidence of host melt interaction with other melts (magma mixing) or wall rock assimilation. Because MORB-type basalt glasses in which spinels exhibit only limited compositional ranges display features of apparent chemical equilibrium between basaltic liquids and spinels, chemical data from such assemblages will be useful in calibrating the Al-Cr exchange

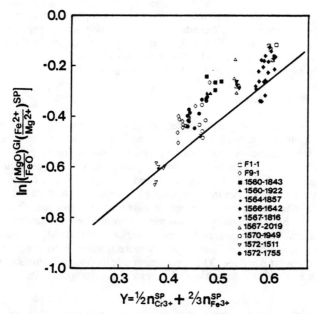

Figure 12. Glass-spinel partitioning data and spinel Cr-contents for Lamont seamount lavas with 'homogeneous' spinels compared with the empirical calibration of the spinel-basaltic liquid Fe-Mg exchange reaction of Allan et al. (1988). $n_{Cr^{3+}}$ and $n_{Fe^{3+}}$ indicate the numbers of chromium and ferric iron ions in a formula unit calculated assuming three cations and four oxygens. From Allan et al. (1988).

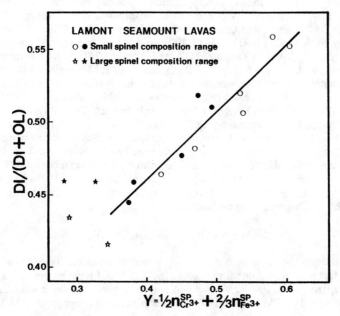

Figure 13. Normative DI/(DI + OL) ratios of Lamont seamount lavas represent in Figure 14 compared with Cr-content of their spinels. *Open* and *solid symbols* differentiate liquids which fall to the left and right of the *dotted chord* in Figure 11, respectively. From Allan et al. (1988).

reaction between basaltic liquids and spinels. The strong compositional dependency of Cr partitioning between spinel and basaltic liquid documented by the Lamont lavas underscores the difficulties associated with interpreting spinel zoning in the absence of a comprehensive model for the thermodynamic properties of chromium in natural silicate melts, particularly where the liquid phase has not been quenched.

Chromian spinels in HED meteorites

The compositions of 'unre-equilibrated' spinels and pyroxenes in olivine diogenites provide important clues regarding the origin of the layered crust of the parent body of meteorites of the howardite-eucrite-diogenite (HED) association (e.g., Mason, 1962; BVSP, 1981). Chromian spinels are a fairly ubiquitous accessory constituent of HED meteorites. Spinels in eucrites – basalts composed of pigeonite and calcic plagioclase (An_{80}-An_{93}) with accessory ilmenite, Cr-spinel, FeS, metal, and a silica polymorph (Fig. 14) – are typically aluminous chromites comparable with those in equilibrated ordinary chondrites (Table 1). Spinels in ordinary diogenites – 'cumulate' pyroxenites ($0.7 \leq mg' \leq 0.8$) composed of low-Ca pyroxene, minor calcic plagioclase (< 10%), and accessory FeS, metal, Cr-rich spinel, and tridymite – are typically more aluminous than spinels in eucrites, and have ranges in molar Al/(Al + Cr) ratios comparable to those of spinels in rare olivine diogenites (Fig. 15), the BAG equivalent of terrestrial harzburgites (Fig. 14; Sack et al., 1991).

Most spinels in ordinary diogenites and olivine diogenites have Mg/(Mg + Fe) ratios consistent with subsolidus re-equilibration to temperatures less than about 950°C (e.g., Fig. 15; Mukherjee et al., 1989; Sack et al., 1991), and thus record 'equilibration' temperatures comparable to, or slightly higher than, those obtained by olivine-spinel Fe-Mg exchange geothermometry for ordinary chondrites, irons, and pallasites (Fig. 6). However, spinels with Mg/(Mg + Fe) ratios more appropriate to solidus temperatures for assemblages composed of olivine + low-Ca pyroxene + calcic plagioclase (e.g., Stolper, 1977; Beckett and Stolper, 1987; Bartels and Grove, 1990) are found in one of the olivine diogenites examined by Sack et al. (1991), ALH 84001 (Fig. 15, Table 1), and several ordinary diogenites, Yamato 74013, Yamato 6902, and Yamato 74136, which may represent different portions of a single fall (e.g., Takeda et al., 1981; Mukherjee et al., 1989; Table 1). The Ca-contents of low-Ca pyroxenes in these ordinary diogenites are also appropriate to such high temperature processes (e.g., Takeda et al., 1981; Lindsley, 1983). In contrast, in ALH 84001 the range of (1) Fe/(Fe + Mg) ratios in spinels, (2) corresponding calculated Fe-Mg olivine-spinel temperatures (Fig. 15), and (3) Ca, Al, and Cr-contents of low-Ca pyroxenes (Fig. 16) indicates only partial preservation of such high-temperature, 'unre-equilibrated' mineral compositions. Most pyroxenes in ALH 84001 have higher Cr- and lower Al-contents then their counterparts in 'equilibrated' olivine diogenite ALH A77256 (Fig. 16). The lower Al-contents of such 'equilibrated' ALH 84001 pyroxenes undoubtedly reflects the lower activities of $MgAl_2O_4$ component in their coexisting spinels (Fig. 15; Table 6, Chapter 6) and the reaction

$$MgAl_2SiO_6 + Mg_2SiO_4 = MgAl_2O_4 + Mg_2Si_2O_6, \qquad (9)$$

which proceeds to the right with decreasing temperature (e.g., Gasparik and Newton, 1984). The fact that 'equilibrated' pyroxenes in ALH 84001 have higher Cr/Al ratios then their counterparts in ALH A77256 (Fig. 16) is a consequence of the higher Cr/(Cr + Al) ratios of the spinels in ALH 84001 (Fig. 15) and the spinel-pyroxene Al-Cr exchange reaction

$$MgCr(Al,Si)_2O_6 + \frac{1}{2}MgAl_2O_4 = \frac{1}{2}MgCr_2O_4 + MgAl_2SiO_6. \qquad (10)$$

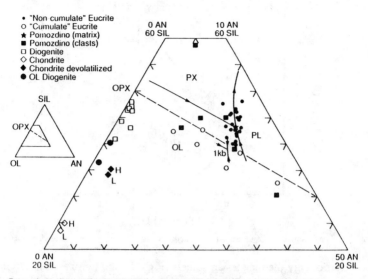

Figure 14. Comparison of a portion of the liquidus of the system silica-olivine-plagioclase with chemical analyses for olivine diogenites (Sack et al., 1991), ordinary diogenites (Fredriksson et al., 1976), 'non-cumulate' and 'cumulate' eucrites (Dymek et al., 1976; Ikeda and Takeda, 1985; Warren and Jerde, 1987; Warren et al., 1990; Warren, pers. comm.), and "average" ordinary H and L chondrites (Mason, 1971). Projection algorithms are from Stolper (1977), 1 atm and 1 kbar field boundaries (f_{O_2} close to metal saturation) are defined from the experimental data of Stolper (1977) and Bartells and Grove (1990), respectively, and abbreviations are as follows: OL, olivine; PX, low-Ca pyroxene; PL, plagioclase; SIL, silica; AN, anorthite; OPX, orthopyroxene. "Average", devolatilized H and L chondrites have all K_2O and sufficient Na_2O removed to result in a normative ratio of $CaAl_2Si_2O_8/(CaAl_2Si_2O_8 + NaAlSi_3O_8)$ components of 0.88. From Sack et al. (1991, Fig. 6).

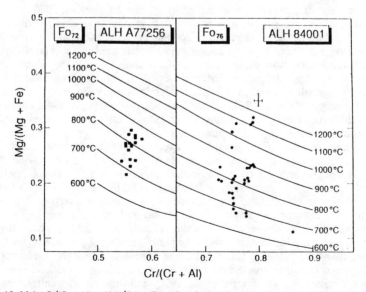

Figure 15. Molar $Cr/(Cr + Al)$ and $Mg/(Mg + Fe)$ ratios of spinels in olivine diogenites ALH A77256 (Fo 0.718 ± 0.009) and ALH 84001 (Fo 0.762±0.007) compared with isotherms calculated for Fo_{72} (left) and Fo_{76} (right) from the olivine-spinel geothermometer of Sack and Ghiorso (1991b) and a representative error bar (upper right) indicating maximum compositional uncertainty. From Sack et al. (1991, Fig. 5).

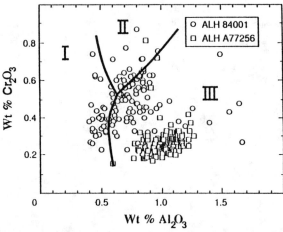

Figure 16. Comparison of the Cr_2O_3 and Al_2O_3 contents (wt %) of orthopyroxenes in olivine diogenites (Sack et al., 1991) with composition classes for pyroxenes in ordinary diogenites (Hewins, 1981).

In this scenario, only the magnesian spinels (Fig. 15) and pyroxenes in olivine diogenite ALH 84001 that have distinctly higher Al_2O_3-contents than average pyroxenes in ALH A77256 (Fig. 16) represent 'unre-equilibrated' mineral compositions representative of magmatic events on the parent body of meteorites of the HED association. The remaining compositions of pyroxenes and spinels in olivine diogenites ALH A77256 and ALH 84001 reflect the operation of the olivine-spinel Fe-Mg exchange reaction and Reactions (9) and (10) either during slow cooling or the intense granulation (impact metamorphism) that accompanied their extraction from the parent body.

It has been hypothesized that the Al- and Cr-contents of low-Ca pyroxenes in diogenites reflect differences in magma types (e.g., Fig. 16, Hewins, 1980, 1981; Hewins and Newsom, 1988) that may be used to differentiate various liquid lines of descent that interrelate members of the HED suite by fractional crystallization. In this view, and other variants of magma ocean-type models, eucrites are related to ordinary diogenites by pyroxene fractionation from parent magmas that initially crystallized olivine; individual crystals of olivine (e.g., Jérome and Michel-Lévy,1972; Desnoyers, 1982) and clasts of eucritic and diogenitic material in howardite polymicit breccias provide evidence for such genetic links (e.g., Mason, 1962; Ikeda and Takeda, 1985: Warren, 1985; Hewins, 1986). However, this type of evidence is equivocal. As shown above, differences in Al and Cr-contents of pyroxenes in diogenites undoubtably reflect variable degrees of operation of the Al-Cr exchange reaction (Eqn. 10) in response to differences in bulk composition and subsolidus annealing histories. Furthermore, the bulk compositions of olivine diogenites cannot be reconciled with a fractional crystallization model (e.g., Sack et al., 1991). Taken collectively the experimental results of Stolper (1977), Beckett and Stolper (1987), and Bartels and Grove (1990), and the composition data summarized in Figure 14 are consistent with the partial melting origin for the layered crust of the eucrite parent body (EPB) first advanced by Stolper (1977). In this scenario, main-stage eucrites (mg' ≤ 0.42) represent peritectic liquids formed early in the partial melting of a source region composed of olivine, low-Ca pyroxene, and plagioclase. These eucrites and their fractionated products formed the outer crust of the EPB. More magnesian 'cumulate' eucrites (e.g., Pomozdino eucrite, Fig. 14; Warren et al., 1990) and ordinary diogenites formed from later, more magnesian peritectic liquids by crystal fractionation in magma chambers beneath the eucritic crust. Olivine diogenites represent the residiuum of partial melting of a source region composed initially of devolatilized, ordinary H or L chondritic material and form the upper mantle of the EPB (Sack et al., 1991). When interpreted in light of the olivine-spinel

Fe-Mg exchange reaction and Reactions (9) and (10) the compositions of spinels and pyroxenes in olivine diogenites help refute arguements for a fractional crystallization origin of the layered crust of the EPB.

Miscibility gaps in the spinel prism

Phase relations in the spinel prism over the temperature range appropriate to the upper greenschist - lower amphibole facies are uncertain owing to the absence of credible experimental results at temperatures below about 600°C, problems in extrapolating results of higher temperature phase equilibrium studies to appropriate temperatures, and difficulties in interpretating phase assemblage data from rocks metamorphosed under these conditions. It is only well established that miscibility gaps between aluminate and ferrite spinels are extensive under these conditions and that miscibility gaps in the $Fe(Al,Cr)_2O_4$ and $Mg(Al,Cr)_2O_4$ binaries are restricted to temperatures below about 500°C (e.g., Fig. 1, Chapter 6; Petric and Jacob, 1982a; Sack, 1982; Oka et al., 1984; Carroll Webb and Wood, 1986). Conflicting inferences regarding the stabilities of Fe^{2+}-rich ferrichromite spinels below 600°C have been derived from petrological studies (e.g., Evans and Frost, 1975; Hoffman and Walker, 1978; Loferski and Lipin, 1983; Zakrzewski, 1989), and there are vitually no direct experimental constrains on the thermodynamic properties $Mg(Cr,Fe^{3+})_2O_4$ spinels at any temperature. In an effort to clarify the nature of these phase relations, calculated miscibility gaps for the sides and for several 500°, 550°, and 600°C Fe-Mg equipotential sections through the spinel prism (Sack and Ghiorso, 1991b) are compared with some of the relevant composition data from nature in Figures 17-19 and 21-22. In the bounding ternary spinel subsystems that form the sides of the prism (Figs. 17-19) Fe^{2+} - Mg partitioning relations between coexisting spinels are illustrated by contours for both temperature and the $Mg/(Mg + Fe^{2+})$ ratios of two-spinels pairs that would be in Fe-Mg exchange equilibrium with Fo_{60}, Fo_{80}, Fo_{90}, and Fo_{95} olivines. The ratios of trivalent cations and of $Mg/(Mg + Fe^{2+})$ in coexisting spinels are defined by the intersections of these contours.

Miscibility gaps in the bounding ternary spinel system $(Fe,Mg)(Al,Fe^{3+})_2O_4$ (Fig. 17) are well constrained by experimental studies. Gaps in the constituent aluminate - ferrite binaries $Fe(Al,Fe^{3+})_2O_4$ and $Mg(Al,Fe^{3+})_2O_4$ appear at temperatures below 875°-920°C (e.g., Fig. 1, Chapter 6; Turnock and Eugster, 1962; Sharma et al., 1973). These become wider in the $(Fe,Mg)(Al,Fe^{3+})_2O_4$ ternary and are most extensive where coexisting Fe^{3+}- and Al-rich spinels exhibit maximum differences in their $Mg/(Mg + Fe^{2+})$ ratios (e.g., Lehmann and Roux, 1986). The maximum in the critical curve is above 1000°C and mutual solubilities of ferrite components in aluminate spinels and *vice versa* are restricted to less than 10 mole percent below 600°C. Results of both high temperature phase equilibrium studies (Petric and Jacob, 1982; Oka et al., 1984; Carroll Webb and Wood, 1986) and theoretical considerations (e.g., Chapter 6; Sack and Ghiorso, 1991b) are consistent with the inference that miscibility gaps are not present in the constituent Mg and Fe^{2+} binaries of the $(Fe,Mg)(Al,Cr)_2O_4$ ternary spinel subsystem (Fig. 18), although results of the synthesis experiments of Cremer (1969) might be taken to indicate otherwise. Because of the nonidealities associated with the Fe-Mg substitution the critical curve for ternary $(Fe,Mg)(Al,Cr)_2O_4$ spinels extends to higher temperatures than those of the consolutes in these binaries. The thermal elevation of the maximum in the critical curve relative to these consolutes is probably comparable to that in $(Fe,Mg)(Al,Fe^{3+})_2O_4$ spinels, because comparable degrees of partitioning of Mg and Fe^{2+} betweeen coexisting spinels are exhibited at the respective temperatures associated with the miscibility gaps in the two ternary spinel subsystems. Finally, miscibilty gap features in the $(Fe,Mg)(Cr,Fe^{3+})_2O_4$ ternary spinel subsystem (Fig. 19) are almost certainly more complex than those associated with the other bounding ternary subsystems of the spinel prism, reflecting the more complex nature of ordering phenomena in these spinels.

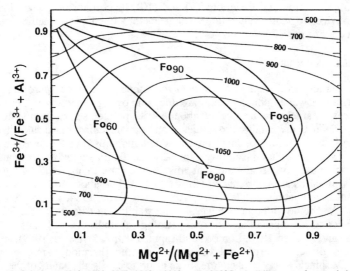

Figure 17. Calculated 500°, 700°, 800°, 900°, 1000°, and 1050°C miscibility gaps for coexisting (Fe,Mg)-(Al,Fe^{3+})$_2$O$_4$ spinels (Sack and Ghiorso, 1991b), contoured for Fe-Mg ratios of coexisting Fe-Mg olivines (Fo$_{60}$, Fo$_{80}$, Fo$_{90}$, and Fo$_{95}$).

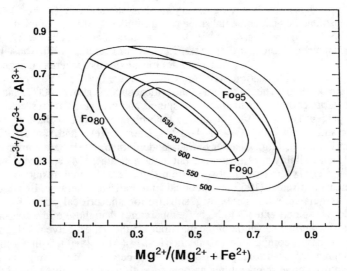

Figure 18. Calculated 500°, 550°, 600°, 620°, and 630°C miscibility gaps for coexisting (Fe,Mg)(Al,Cr)$_2$O$_4$ spinels (Sack and Ghiorso, 1991b), contoured for Fe-Mg ratios of coexisting Fe-Mg olivines (Fo$_{80}$, Fo$_{90}$, and Fo$_{95}$).

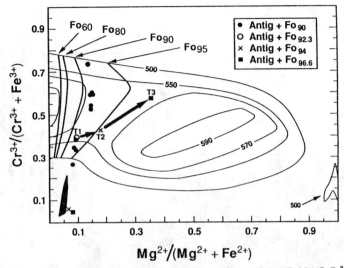

Figure 19. Calculated 500°, 550°, 570°, and 590°C miscibility gaps for coexisting $(Fe,Mg)(Cr,Fe^{3+})_2O_4$ spinels (Sack and Ghiorso, 1991b), contoured for Fe-Mg ratios of coexisting Fe-Mg olivines (Fo_{60}, Fo_{80}, Fo_{90}, and Fo_{95}). Also plotted are composition data reported by Evans and Frost (1975, their Figure 7) and Hoffman and Walker (1978) for serpentinites with olivine-antigorite (Antig) assemblages (upper greenschist facies). Labels in upper right corner indicate representative Mg/(Mg + Fe) ratios of olivines in these serpentinites (Fo_{90} - Evans and Frost; 1975; $Fo_{92.3-96.6}$ – Hoffman and Walker, 1978). *Arrows* connect representative analyses of ferrichromites (Hoffman and Walker, 1978, Table 1) in serpentines of textural types 1-3 (T1-T3). *Solid object* in lower left corner represents corresponding shaded area in Figure 7.

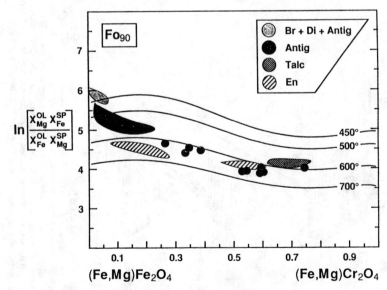

Figure 20. Figure 8 with additional composition data for low-temperature, fine-grained, ferrichromites. *Solid circles* represent isolated points plotted in Figure 7 of Evans and Frost (1975) which they inferred to correspond to homogeneous, single-phase ferrichromites with intermediate $Cr/(Cr + Fe^{3+})$ ratios.

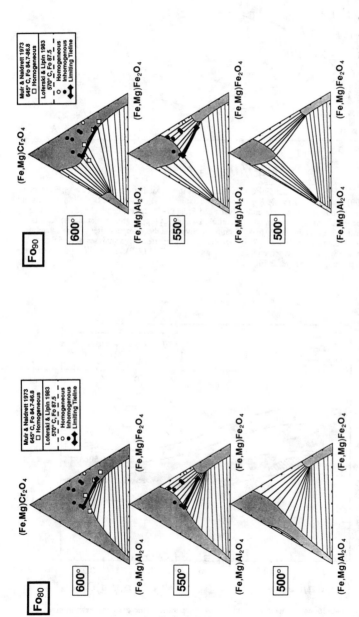

Figure 21 (left). Calculated 500°, 550°, and 600°C isothermal sections through the spinel prism (Sack and Ghiorso, 1991b) together with composition data from Muir and Naldrett (1973) and Loferski and Lipin (1983). Isothermal sections are calculated for Fe-Mg exchange potentials corresponding to those of Fo90 olivine. *Open squares* represent the most Cr-poor, homogeneous spinels with intermediate Al/(Al + Fe³⁺) ratios from the Giant Nickel Mine, Hope, British Columbia (Muir and Naldrett, 1973). Temperature and olivine composition given in associated label in upper right corner correspond to those calculated for less chromian spinel pairs from this deposit (see text). *Solid diamonds connected by thick line* represent the compositions of the most chromian, Red Lodge (MT), coexisting spinel pairs reported by Loferski and Lipin (1983). *Solid and open circles* indicate more chromian spinels that contain exsolution lamellae and are lamellae-free, respectively. Temperature and olivine composition given in associated label in upper right corner approximate those which yield optimal match between calculated curves and field constraints (see text).

Figure 22 (right). Calculated 500°, 550°, and 600°C isothermal sections through the spinel prism (Sack and Ghiorso, 1991b) together with composition data from Muir and Naldrett (1973) and Loferski and Lipin (1983). Isothermal sections are calculated for Fe-Mg exchange potentials corresponding to those of Fo90 olivine. Plotted points are identical to those in Figure 21.

Calculated miscibility gaps for $(Fe,Mg)(Al,Fe^{3+})_2O_4$ spinels (Fig. 17) are in at least qualitative agreement with the wealth of composition data from metamorphic rocks. Mutual solubilities of aluminate spinel in magnetite and *vice versa* are comparable to or less than those given in Figure 17 for estimated 'equilibration' temperatures. There have been no reports of coexisting $(Fe,Mg)(Al,Cr)_2O_4$ spinels from nature. The absence of such two-spinel pairs presumably reflects the instability of green spinels with respect to chlorite in hydrous assemblages (e.g., Evans and Frost, 1975) or extremely sluggish reaction rates associated with unmixing reactions at the relevant temperatures (e.g., Fig. 18). There is also no direct confirmation of the calculated miscibility gaps in $(Fe,Mg)(Cr,Fe^{3+})_2O_4$ spinels, although gaps similar to those displayed in Figure 19 have been reported from aluminous ferrichromite spinels in metamorphosed chromite ores (Loferski and Lipin, 1983; Zakrzewski, 1989) and will be discussed subsequently (cf. Figs. 21-22). In contrast, Evans and Frost (1975) and Hoffman and Walker (1978) have reached conflicting conclusions regarding the stability of such spinels on the basis of petrological studies of metamorphosed serpentinites with olivine + ferrichromite + antigorite-bearing assemblages (Fig. 19). For example, Evans and Frost (1975) concluded that solid solution in natural ferrichromites is complete down to temperatures of about 500°C, because some of their microprobe analyses for such spinels had intermediate $Cr/(Cr + Fe^{3+})$ ratios. However, their inference that such analyses correspond to equilibrium compositions of homogeneous single-phase aggregates may be problematic, as such ferrichromites are typically fine-grained, and are often complexly zoned on an exceedingly fine scale. The reported compositions are too magnesian to represent 'single-phase' ferrichromites in Fe-Mg exchange equilibrium with Fo_{90} olivines at appropriate temperatures (Fig. 20) and the distinctly bimodal distribution of $Cr/(Cr + Fe^{3+})$ ratios evident in their analyses (cf., their Fig. 1) is broadly consistent with the calculated miscibility gaps. Finally, the data summarized by Hoffman and Walker (1978) are also equivocal. They have observed that the thicknesses and compositions of fine-grained, concentrically zoned magnetite and ferrichromite overgrowths on chrome spinel are correlated to the extent of serpentinization of ultramafic rocks of the East Dover bodies (south-central Vermont) and have inferred that these spinels were produced simultaneously during metamorphism (upper greenschist - epidote amphibolite facies). With increasing degree of serpentiniztion olivine ($Fo_{92.3-96.6}$) and both the inner ferrichromite and outer magnetite-rich shells of opaque overgrowths become more magnesian and the relict chrome spinels cores become more magnesian and aluminous. Although these features are consistent with overall serpentinization reactions of the type proposed by Hoffman and Walker (1978), the products of such reactions do not represent equilibrium at least with respect to Fe-Mg exchange reactions. Both the chrome spinel cores and ferrichromite rims of the least altered serpentinites (Type 1) record temperatures of olivine-spinel Fe-Mg exchange 'equilibrium' above those of the calculated miscibility gaps displayed in Figures 18 and 19. Such temperatures decrease progressively for both chrome spinel cores and ferrichromite overgrowths with increasing extent of serpentinization (Types 2 and 3). Although the Fe-Mg exchange temperature calculated for ferrichromite in the most altered serpentine (Type 3, Table 1, Hoffman and Walker, 1978) is comparable to those calculated for olivine-magnetite pairs, 490°±20°C, the temperature calculated for the relict chrome spinel core, 412°C, clearly reflects either disequilibrium or that the reported analysis does not correspond to a single, homogeneous phase.

Despite the difficulties associated with deriving stability constraints from fine-grained mineral assemblages in low-grade metamorphic rocks, composition data from several studies of natural aluminous ferrichromite spinels annealed at higher temperatures (e.g., Muir and Naldrett, 1973; Loferski and Lipin, 1983) provide strong support for the inference that miscibility gaps in ferrichromites extend to temperatures above 550°C. In Figures 21 and 22 we compare their critical data with relevant calculated Fe-Mg equipotential sections through the spinel prism. Muir and Naldrett (1973) report composition data for both exsolved chromian aluminoferrite and chromian ferrialuminate spinels, and homogeneous aluminous ferrichromite spinels from the olivine-bearing ultramafic ores of the Giant Nickel Mine, Hope, British Columbia. Interpretation of their data is very straightforward. The compositions of exsolution pairs (e.g., Table 2) are consistent with calculated Fe-Mg spinel-spinel equilibration temperatures of 645°±24°C and local Fe-Mg exchange potentials corresponding to those of olivines with $Fo_{84.7-86.8}$. The least chromian aluminous ferrichromite spinels that do not display evidence of unmixing, outline, but are slightly more chromian than, the relevant gaps calculated for 600°C (Figs. 21 and 22). There is also excellent agreement between the calculated miscibilty gaps and composition data reported for similar spinels in the metamorphosed chromite ores of Red Lodge, Montana (Loferski and Lipin, 1983). These ores experienced peak metamorphic temperatures of at least 600°C, during which their coarse exsolution features presumably developed (cf. Loferski and Lipin, 1983). Interpretation of their data is less straighforward than for spinels from the Giant Nickel Mine, because many of these lamellae are inhomogeneous as a result, at least in part, of the development of exceedingly fine secondary lamellae. Nevertheless, the tielines corresponding to the average analyses obtained by Loferski and Lipin (1983) for coarse exsolution pairs are comparable to those calculated for temperatures between 550° and 600°C and Fe-Mg exchange potentials corresponding to those of olivines with molar $Mg/(Mg + Fe^{2+})$ ratios between 0.8 and 0.9. Miscibility gaps of comparable extent and with two-phase region widths and tieline slopes in optimal agreement with the constraints implied by the composition data for aluminous ferrichromites are calculated for a temperature of about 570°C and an olivine of about $Fo_{87.5}$ (cf. Figs. 21-22). Accurate spinel-spinel Fe-Mg exchange 'equilibration' temperatures cannot be obtained for many of the more Cr-poor spinel pairs, because there are large uncertainties associated with the MgO contents of the inhomogeneous, MgO-poor (<1 wt %) Fe^{3+}-rich spinels. However, olivine-spinel and spinel-spinel Fe-Mg exchange temperatures calculated for more magnesian spinels are typically between 550° and 600°C, or slightly higher. Furthermore, the compositions of olivine calculated to be in exchange equilibrium with the spinel-spinel pairs – Fo_{85} to Fo_{90} – are closely similar to those in actual olivines. It is also noteworthy that a tieline of similar length and slope as that displayed in Figures 21 and 22 is also defined by coexisting Cr^{3+} and Fe^{3+}-rich spinels in the gabbroic rocks and ores from the Ni-Cu mine of Kuså, Bergslagen, Sweden (Zakrzewski, 1989, his analyses 70-71; Table 2).

A PETROLOGICAL CONCERN (PASTICHE)

In this chapter we have summarized some of the petrological constraints that provide critical tests for thermochemical models of chromian spinels and coexisting ferromagnesian phases, and have provided several examples of the use of 'chromite' as a petrogenetic indicator. One of our purposes has been to demonstrate some of the petrological manifestations of Fe-Mg nonideality and cation ordering; another has been to illustrate how statements of exchange equilibria between several phases provide the multiple consistency tests required to examine the feasibilty of interpretations of natural phenomena. By virtue of their ubiquity and chemical diversity chromian spinels provide an ideal medium for constructing tests of this kind. It is our concern that such tests are seldom performed, either because the objectives of the analysis are limited or insufficient (i.e., 'representative') chemical analyses are reported with inadequate documentation of relationships between phase chemistry and local assemblage. It is our hope that future petrological studies employ only internally consistent exchange geothermometers/geobarometers in their analysis and/or provide sufficiently detailed characterization of phase assemblage chemistry to allow others to critically examine their hypotheses.

Table 2. Chemical analyses of chromian spinels in serpentinites (wt %).

	1	2	3	4	5	6	7	8	9	10	11	12
TiO_2	-----	0.08	-----	<0.04	0.1(1)	1.6(4)	1.4(2)	0.80	0.60	0.19	0.65	0.77
Al_2O_3	14.6	0.06	21.8	-----	38.0(48)	4.3(3)	14.8(9)	4.98	14.98	2.94	14.3	5.74
Cr_2O_3	52.4	26.1	46.0	3.43	20.8(22)	16.3(14)	31.7(10)	24.97	34.67	42.26	33.5	21.8
V_2O_3	0.16	0.12	0.21	<0.06	n.d.	n.d.	n.d.	n.d.	n.d.	n.d.	0.21	0.32
Fe_2O_3[a]	3.23	41.7	3.63	66.8	7.3(33)	44.2(26)	19.2(14)	36.91	14.88	23.99	26.51	28.47
FeO[a]	20.6	27.2	13.9	27.5	23.9(10)	31.8(10)	30.8(2)	31.90	31.25	29.81	18.55	39.04
MgO	8.33	1.58	12.8	1.48	9.1(10)	1.4(2)	3.2(2)	0.71	1.86	1.72	4.07	2.07
MnO	0.35	0.31	0.63	0.15	n.d.	n.d.	n.d.	n.d.	n.d.	n.d.	1.33	1.09
ZnO	0.60	0.43	1.39	0.29	n.d.	n.d.	n.d.	n.d.	n.d.	n.d.	0.70	0.29
NiO	-----	0.46	0.10	1.14	0.16(4)	0.60(5)	0.12(5)	0.56	0.36	0.12	n.d.	n.d.
Total	100.47	99.28	100.62	100.93	99.36	100.20	101.22	100.83	98.60	101.03	99.82	99.59

a) Recalculated electron microprobe analyses.

1-4. Relict chromian spinel core (type 1, CoO = 0.20 wt%), ferrichromite overgrowth (type 1, CoO = 0.24 wt%), relict spinel core (type 3, CoO = 0.16 wt%), and magnetite overgrowth (type 3, CoO = 0.14 wt%) from the East Dover ultramafic bodies, Vermont (Hoffman and Walker, 1978, p. 705, Table 1).

5-7. Coexisting chromian aluminoferrite (#5, ave. 3 analyses) and ferrialuminate (#6, ave. 3 analyses) spinels, and the bulk composition of a homogeneous aluminous ferrichromite spinel (#7, ave. 9 analyses) from the Giant Nickel Mine, Hope, B.C.(Muir and Naldrett, 1973, p. 934, Table 2).

8-10. Coexisting aluminous ferrichromite spinels (# HP-14) and the bulk composition of a ferrichromite exhibiting type C exsolution (# HP-9) from the chromite ores of Red Lodge, Montana (Loferski and Lipin, 1983, p. 783-784, Table 4).

11-12. Coexisting aluminous ferrichromite spinels from the Ni-Cu mine of Kusá, Bergslagen, Sweden (Zakrzewski, 1989, p. 452, Table 2, analyses 70-71).

REFERENCES

Agrell, S.O., Boyd, F.R., Bunch, T.E., Cameron, E.N., Dence, M.R., Douglas, J.A.V., Haggerty, S.E., James, O.B., Keil, K., Peckett, A., Plant, A.G., Prinz, M., Trail, R.J. (1970) Titanian chromite, aluminate and chromian ulvospinel from Apollo 11 rocks. In: Proc. Apollo 11 Lunar Sci. Conf. (Supl.1). Geochim. Cosmochim. 1, 81-86.

Aikawa, N., Kumazawa, M., and Tokonami, N. (1985) Temperature dependence of intersite distribution of Mg and Fe in olivine and the associated change of lattice parameters. Phys. Chem. Minerals 12, 1-8.

Allan, J.F., Sack, R.O., Batiza, R. (1988) Cr-rich spinels as petrogenetic indicators: MORB-type lavas from the Lamont seamount chain, Eastern Pacific. Am. Mineral. 73, 741-753.

Allan, J.F., Batiza, R., Perfit, M.R., Fornari, D.J., and Sack, R.O. (1989) Petrology of lavas from the Lamont seamount chain and adjacent East Pacific Rise, 10°N. J. Petrol. 30, 1245-1298.

Anovitz, L.M., Essene, E.J., and Dunham, W.R. (1988) Order-disorder experiments on orthopyroxenes: Implications for the orthopyroxene speedometer. Am. Mineral. 73, 1060-1073.

Banerjee, S.K., O'Reilly, W., Gibb, T.C., and Greenwood, N.N. (1967) The behavior of ferrous ions in iron-titanium spinels. J. Phys. Chem. Solids 28, 1323-1335.

Bartels, K.S., and Grove, T.L. (1990) High pressure experiments on magnesian eucrite magmas: constraints on magmatic processes in the eucrite parent body. Lunar Planet. Sci. XXI, 46-47 (abstract).

Batiza, R., and Vanko, D. (1984) Petrology of young Pacific seamounts. J. Geophys. Res. 89, 11235-11260.

Beckett, J.R. and Stolper, E.M. (1987) Constraints on the origin of eucrite melts; an experimental study. Lunar Planet. Sci. XVIII, 54-55 (abstract).

Berman, R.G. (1988) Internally-consistent thermodynamic data for minerals in the system Na_2O-K_2O-CaO-MgO-FeO-Fe_2O_3-Al_2O_3-SiO_2-TiO_2-H_2O-CO_2:. J. Petrol. 29, 445-522.

Bohlen, S.R., and Boettcher, A.L. (1981) Experimental investigations and geological applications of orthopyroxene geobarometry. Am Mineral. 66, 951-964.

Bohlen, S.R., Essene, E.J., and Boettcher, A.L. (1980) Reinvestigation and application of olivine-quartz-orthopyroxene barometry. Earth Planet. Sci. Lett. 47, 1-10.

Bunch, T.E., Keil, K. (1971) Chromite and ilmenite in non-chondritic meteorites. Am. Mineral. 56, 146-157.

Bunch, T.E., Keil, K., Olsen, E. (1970) Mineralogy and petrology of silicate inclusions in iron meteorites. Contrib. Mineral. Petrol. 25, 297-340.

Bunch, T.E., Keil, K., Snetsinger, K.G. (1967) Mineralogy and petrology of silicate inclusions in iron meteorites. Geochim. Cosmochim. Acta 31, 1569-1582.

BVSP (1981) Experimental petrology of basalts and their source rocks. In: Basaltic Volcanism of the Terrestrial Planets (New York: Pergamon Press), pp. 493-630.

Carroll Webb, S.A., and Wood, B.J. (1986) Spinel-pyroxene-garnet relationships and their dependence on Cr/Al ratio. Contrib. Mineral. Petrol. 92, 471-480.

Champness, P.E., Dunham, A.C., Gibb, F.G.F., Giles, H.N., MacKenzie, W.S., Stumpfl, E.F., and Zussman, J. (1971) Mineralogy and petrology of some Apollo 12 lunar samples. In: Proc. Apollo 11 Lunar Sci. Conf. (Supl.1). Geochim. Cosmochim. 1, 359-376.

Chatterjee, N.D., Leistner, H., Terhart, L., Abraham, K., and Klaska, R. (1982) Thermodynamic mixing properties of corundum-eskaolaiite, α-$(Al,Cr^{3+})_2O_3$, crystalline solutions at high temperatures and pressures. Am. Mineral. 67, 725-735.

Chatillon-Colinet, C., Newton, R.C., Perkins, D., III, and Kleppa, O.J. (1983) Thermochemistry of $(Fe^{2+}, Mg)SiO_3$ orthopyroxene. Geochim. Cosmochim. Acta 47, 1597-1603.

Cremer, V. (1969) Die Mischkristallbildung im System Chromit-Magnetit-Hercynit zwischen 1000°C und 500°C. Neues Jahrbuch für Mineralogie Abhandlungen 111, 184-205.

Davidson, P.M., and Lindsley, D.H. (1989) Thermodynamic analysis of pyroxene-olivine-quartz equilibria in the system CaO-MgO-FeO-SiO_2. Am. Mineral. 74, 18-30.

DeGrave, E., DeSitter, J., Vandenberghe, R. (1975) On the cation distributionin the spinel system Mg_2TiO_4 - $(1-y)MgFe_2O_4$. Appl. Phys. 7, 77-80.

Desnoyers, C. (1982) L'olivine dans les howardites: origine, et implications pour le corps parent de ces météorites achondritiques. Geochim. Cosmochim. Acta 46, 667-680.

Dick, H.J.B. (1977) Partial melting of the Josephine Peridotite, I. The effect of mineral composition and its consequence for geobarometry and geothermometry. Am. J. Sci. 277, 801-832.

Dunitz, J.D., and Orgel, L.E. (1957) Electronic properties of transition metal oxides II. J. Phys. Chem. Solids 3, 318-323.

Dymek, R.F., Albee, A.L., and Chodos, A.A., and Wasserburg, G.J. (1976) Petrography of isotopically-dated clasts in Kapeota howardite and petrological constraints on the evolution of its parent body. Geochim. Cosmochim. Acta 40, 1115-1130.

Ebel, D.S., and Sack, R.O. (1991) Arsenic-silver incompatibility in fahlore. Mineral. Mag. (in press).

El Goresy, A. (1976) Oxide minerals in lunar rocks. In D. Rumble III, Ed., Oxide Minerals , p. EG-1 - EG-45. Mineral. Soc. Am. Short Course Notes 3.

Engi, M. (1983) Equilibria involving Al-Cr spinel: Mg-Fe exchange with olivine. Experiments, thermodynamic analysis, and consequences of geothermometry. Am. J. Sci., 283A 29-71.

Epstein, D.J., and Frackiewicz, B. (1958) Some properties of quenched magnesium ferrites. J. Appl. Phys. 29, 376-377.

Evans, B.W., and Frost, B.R. (1975) Chrome-spinel in progressive metamorphism - a preliminary analysis. Geochim. Cosmochim. Acta 39, 959-972.

Evans, B.W., and Moore, R.G. (1968) Mineralogy as a function of depth in the prehistoric Makaopuhi tholeiitic lava lake. Contrib. Mineral. Petrol. 17, 85-115.

Evans, B.W., and Wright, T.L. (1972) Composition of liquidus chromite from the 1959 and 1965 eruptions of Kilauea Volcano Hawaii. Am. Mineral. 57, 217-230.

Fabriès, J. (1979) Spinel-olivine geothermometry in peridotites from ultramafic complexes. Contrib. Mineral. Petrol. 69, 329-336.

Floran, R.J., Prinz, M., Hlava, P.F., Keil, K., Nehru, C.E., and Hinthorne, J.R. (1978) The Chassigny meteorite; a cummulate dunite with hydrous amphibole-bearing melt inclusions. Geochim. Cosmochim. Acta 42, 1213-1229.

Fonarev, V.I., Korolkov, G.Y., Dokina, T.N. (1979) Laboratory data on the stability field of cummingtonite + olivine + quartz association. Geochem. Intl. 16, 21-32.

Francombe, M.H. (1957) Lattice changes in spinel-type iron chromites. J. Phys. Chem. Solids 3, 37-43.

Fredriksson, K., Noonan, A., Brenner, P., and Sudre, C. (1976) Bulk and major phase composition of eight hypersthene achondrites. Meteoritics 11, 278-280.

Frost, B.R. (1976) Limits to the assemblage forsterite-anorthite as inferred from peridotite hornfelses, Icicle Creek, Washington. Am. Mineral. 732-750.

Fujii, T. (1977) Fe-Mg partitioning between olivine and spinel. Carnegie Inst. Wash., Year Book 76, 563-569.

Gasparik, T., and Newton, R.C. (1984) The reversed alumina contents of orthopyroxene in equilibrium with spinel and forsterite in the system $MgO-Al_2O_3-SiO_2$. Contrib. Mineral. Petrol. 85 186-196.

Ghiorso, M.S. (1990) Thermodynamic properties of hematite-ilmenite-geikielite solid solutions. Contrib. Mineral. Petrol. 104, 645-667.

Ghiorso, M.S., and Sack, R.O. (1991) Fe-Ti oxide geothermometry: thermodynamic formulation and the estimation of intensive variables in silicic magmas. Contrib Mineral. Petrol. (in press).

Hafner, S.S., and Ghose, S. (1971) Iron and magnesium distribution in cummingtonites. $(Fe,Mg)_7Si_8O_{22}(OH)_2$. Zeit. Kristallogr. 133, 301-326.

Haggerty, S.E., and Meyer, H.O.A. (1970) Apollo 12: Opaque oxides. Earth Planet. Sci. Lett. 9, 379-387.

Hewins, R.H. (1980) Subdivision of diogenites into chemical classes. Lunar Planet. Sci. XI, 441-443 (abstract).

Hewins, R.H. (1981) Fractionation and equilibration in diogenites. Lunar Planet. Sci. XII, 445-447 (abstract).

Hewins, R.H. (1986) Serial melting or magma ocean for the HED achondrites? Meteoritics 21, 396-397 (abstract).

Hewins, R.H., and Newsom, H.E. (1988) Igneous activity in the early solar system. In: Meteorites and the Early Solar System (eds. J.F. Kerridge and M.S. Matthews), Chap. 3.2, pp. 73-101. Univ. Arizona.

Hill, R., and Roeder, P.L. (1974) The crystallization of spinel from basaltic liquids as a function of oxygen fugacity. J. Geol. 84, 709-709.

Hill, R.L., and Sack, R.O. (1987) Thermodynamic properties of Fe-Mg titaniferrous magnetite spinels. Canad. Mineral. 25, 443-464.

Hoffman, M.A., and Walker, D. (1978) Textural and chemical variations of olivine and chrome spinel in the East Dover ultramafic bodies, south-central Vermont. Geol. Soc. Amer. Bull. 89, 699-710.

Ikeda, Y., and Takeda, H. (1985) A model for the origin of basaltic achondrites based on the Yamato 7308 howardite. Proc. Lunar Planet. Sci. Conf. 15, J. Geophys. Res. Suppl. 90, C649-C663.

Irvine, T.N. (1965) Chromian spinel as a petrogenetic indicator. Part I. Theory. Can. J Earth Sci. 2, 648-672.

Irvine, T.N. (1967) Chromian spinel as a petrogenetic indicator. Part II. Petrological applications. Can. J Earth Sci. 4, 71-103.

Jackson, E.D. (1969) Chemical variations in co-existing chromite and olivine in chromite zones of the Stillwater complex. In H.D.B. Wilson, Ed. Magmatic Ore Deposits, p. 41-71, Econ. Geol. Monogr. 4.

Jamieson, H.E., and Roeder, P.L. (1984) The distribution of Mg and Fe^{2+} between olivine and spinel at 1300°C. Am. Mineral. 69, 283-291.

Jérome, D.Y., and Michel-Lévy, M.C. (1972) The Washougal meteorite. Meteor. 7, 449-461

Katsura, T., Wakihara, M., Shin-Ichi, H., and Sugihara, T. (1975) Some thermodynamic properties in spinel solid solutions with the Fe_3O_4 component. J.Solid State Chem. 13, 107-113.

Keil, K., and Fredriksson, K. (1964) The iron, magnesium and calcium distribution in coexisting olivines and rhombic pyroxenes of chondrites. J. Geophys. Res. 69, 3486-3515.
Kitayama, K., and Katsura, T. (1968) Activity measurements in orthosilicate and metasilicate solid solutions: I. Mg_2SiO_4-Fe_2SiO_4 and $MgSiO_3$-$FeSiO_3$ at 1204°C. Bull. Chem. Soc. Japan 41, 1146-1151.
Kretz, R. (1982) Transfer and exchange equilibria in a portion of the pyroxene quadrilateral as deduced from natural and exchange data. Geochim. Cosmochim. Acta 46, 411-421.
Kriessman, C.J., and Harrison, S.E. (1956) Cation distributions in ferrospinels. Magnesium-manganese ferrites. Phys. Rev. 103, 857-860.
Langmuir, C.H., Bender, J.F., and Batiza, R. (1986) Petrologic segmentation of the East Pacific Rise 5°N30' - 14°30'N. Nature 322, 422-429.
Lattard, D., and Evans, B.W. (1991) Experiments on the stability of grunerite. Eur. J. Mineral. (submitted).
Lehmann, J., and Roux, J. (1986) Experimental and theoretical study of $(Fe^{2+},Mg)(Al,Fe^{3+})_2O_4$ spinels: Activity-composition relationships, miscibility gaps, vacancy contents. Geochim. Cosmochim. Acta 50, 1765-1783.
Lindsley, D.H. (1983) Pyroxene thermometry. Am. Mineral., 68, 477-493.
Loferski, P.J., and Lipin, B.R. (1983) Exsolution in metamorphosed chromite from the Red Lodge district, Montana. Am. Mineral. 68, 777-789.
Mason, B. (1962) Meteorites, pp. 109-119. J. Wiley & Sons.
Mason, B. (1971) Handbook of Elemental Abundances in Meteorites. Gordon and Breach Science Publishers.
Mattioli, G.S., and Wood, B.J. (1988) Magnetite activities across the $MgAl_2O_4$-Fe_3O_4 spinel join, with application to thermobarometric estimates of upper mantle oxygen fugacity. Contrib. Mineral. Petrol. 98, 148-162.
Millard, R.L., Peterson, R.C., and Hunter, B.K. (1991) Temperature dependence of cation disorder in $MgAl_2O_4$ spinel using aluminum-27 and oxygen-17 magic-angle spinning nuclear magnetic resonance spectroscopy. Am. Mineral. (in press).
Mo, X., Carmichael, I.S.E., Rivers, M., and Stebbins, J. (1982) The partial molar volume of Fe_2O_3 in multicomponent silicate liquids and the pressure dependence of oxygen fugacity in magmas. Mineral. Mag. 45, 237-245.
Mozzi, R.L., and Paladino, A.E. (1963) Cation distributions in magnesium ferrites. J. Chem. Phys. 39, 435-439.
Muan, A., Hauck, J., and Lofall, T. (1972) Equilibrium studies with a bearing on lunar rocks. In E.A. King, Ed., Proceedings of the Third Lunar Science Conference (Supplement 3) Geochim. Cosmochim. Acta 1, 185-196.
Muir, J.E., and Naldrett, A.J. (1973) A natural occurrence of two-phase chromium-bearing spinels. Canadian Mineral. 11, 930-939.
Mukherjee, A.B., and Viswanath, T.A. (1987) Thermometry of diogenites. Mem. Natl. Inst. Polar Res., Spec. Issue 46, 205-215.
Mukherjee, A.B., Bulatov, V., and Kotelnikov, A. (1989) New high P-T experimental results on orthopyroxene-chrome spinel equilibria and a revised orthopyroxene-spinel cosmothermometer. Lunar Planet. Sci. XX, 731-732 (abstract).
Murck, B.W., and Campbell, I.H. (1986) The effects of temperature, oxygen fugacity and melt composition on the behavior of chromium in basic and ultrabasic melts. Geochim. Cosmochim. Acta 50, 1871-87.
Navrotsky, A., and Kleppa, O.J. (1967) The thermodynamics of cation distributions in simple spinels. J. Inorganic Nuclear Chem. 30, 479-498.
Oka, Y., Steinke, P., and Chatterjee, N.D. (1984) Thermodynamic mixing properties of $Mg(Al,Cr)_2O_4$ spinel crystalline solution at high temperatures and pressures. Contrib. Mineral. Petrol. 87, 197-204.
Olsen, E.J., and Bunch, T.E. (1984) Equilibration temperatures of the ordinary chondrites: A new evaluation. Geoch. Cosmoch. Acta 48, 1363-1365.
O'Neill, H. St. C. (1981) The transistion between spinel lherzolite and garnet lherzolite, and its use as a geobarometer. Contrib. Mineral. Petrol. 77, 185-194.
O'Neill, H. St. C., and Wall, V.J. (1987) The olivine-spinel oxygen geobarometer, the nickel precipitation curve and the oxygen fugacity of the upper mantle. J. Petrol., 28, 1169-1192.
Ottonello, G., Princivalle, F., and Guista, A.D. (1990) Temperature, composition, and f_{O_2} effects on intersite distribution and Mg and Fe^{2+} in olivines. Phys. Chem. Minerals 17, 301-312.
Paar, Von W.H., Chen, T.T., and Cunther, W. (1978) Extreme silberreicher Freibergit in Pb-Zn-Cu-Erzen des Bergbaues "Knappenstube", Hoctor, Salzburg. Carinthia II 168, 35-42.
Peterson, R.C., Lager, G.A., and Hitterman, I.P.N.S. (1991) A Time-Of-Flight neutron diffraction study of $MgAl_2O_4$ at temperatures up to 1273 K. Am. Mineral. (in press).
Petric, A., and Jacob, K.T. (1982a) Inter- and intra-crystalline ion-exchange equilibria in the system Fe-Cr-Al-O. Solid State Ionics 6, 47-56.
Petric, A., and Jacob, K.T. (1982b) Thermodynamic properties of Fe_3O_4-FeV_2O_4 and Fe_3O_4-$FeCr_2O_4$ spinel solid solutions. J. Am. Ceram. Soc. 65, 117-123.

Ramberg, H., and DeVore, G. (1951) The distribution of Fe^{2+} and Mg^{2+} in coexisting olivines and pyroxenes. J. Geol. 59, 193-216.
Robbins, M., Wertheim, G.K., Sherwood, R.C., and Buchanan, D.N.E. (1971) Magnetic properties and site distributions in the system $FeCr_2O_4\text{-}Fe_3O_4$ ($Fe^{2+}Cr_{2-x}Fe_x^{3+}O_4$). J. Phys. Chem. Solids 32, 717-729.
Roeder, P.L., Campbell, I.H., and Jamieson, H.E. (1979) A re-evaluation of the olivine-spinel geothermometer. Contrib. Mineral. Petrol. 68, 325-334.
Roth, W.L. (1964) Magnetic properties of normal spinels with only A-A interations. J. Phys. 25, 507-515.
Sack, R.O. (1982) Spinels as petrogenetic indicators: Activity-composition relations at low pressures. Contrib. Mineral. Petrol. 79, 169-182.
Sack, R.O. (1991) Thermochemistry of tetrahedrite-tennantite fahlores. In G.D. Price and N.L. Ross, Eds. The Stability of Minerals. Harper Collins Academic, London (in press).
Sack, R.O., and Ghiorso, M.S. (1989) Importance of considerations of mixing properties in establishing an internally consistent thermodynamic database: thermochemistry of minerals in the system Mg_2SiO_4-Fe_2SiO_4-SiO_2. Contrib. Mineral. Petrol. 102, 41-68.
Sack, R.O., and Ghiorso, M.S. (1991a) An internally consistent model for the thermodynamic properties of Fe-Mg- titanomagnetite - aluminate spinels. Contrib. Mineral. Petrol. 106. 474-505.
Sack, R.O., and Ghiorso, M.S. (1991b) Chromian spinels as petrogenetic indicators: thermodynamics and petrological applications. Am. Mineral. (in press).
Sack, R.O., Azeredo, W.L., and Lipschutz, M.E. (1991) Olivine diogenites: the mantle of the eucrite parent body. Geochim. Cosmochim. Acta (in press).
Sack, R.O., Ebel, D.S., and O'Leary, M.J. (1987a) Tennahedrite thermochemistry and metal zoning. In H.C. Helgeson, Ed. Chemical Transport in Metasomatic Processes, p. 701-731. D. Reidel, Dordrecht.
Sack, R.O., Walker, D., and Carmichael, I.S.E. (1987b) Experimental petrology of alkalic lavas: constraints on cotectics of multiple saturation in natural basic liquids. Contrib. Mineral. Petrol. 96, 1-23.
Saxena, S.L., and Ghose, S. (1971) Mg^{2+} - Fe^{2+} order-disorder and the thermodynamics of the orthopyroxene-crystalline solution. Am. Mineral. 56, 532-559.
Seifert, F. (1978) Equilibrium $Mg\text{-}Fe^{2+}$ cation distribution in anthophyllite. Am. J. Sci. 178, 1323-1333.
Sharma, K.C., Agawal, K.C., Kapoor, M.L. (1987) Determination of thermodynamic properties of (Fe,Mg)-pyroxenes at 1000 K by the EMF method. Earth Planet. Sci. Lett. 85, 302-310.
Sharma, K.K, Langer, K., and Sieffert, T. (1973) Some properties of spinel phases in the binary system $MgAl_2O_4$-$MgFe_2O_4$. Neues Jahrbuch für Mineralogie, Mitteilungen, 442-449.
Sigurdsson, H., and Schilling, J.-G. (1976) Spinels in Mid-Atlantic Ridge basalts: Chemistry and occurrence. Earth Planet. Sci. Lett. 29, 7-20.
Spiridonov, E.M. (1984) Species and varieties of fahlore (tetrahedrite-tennantite) minerals and their rational nomenclature. Dokl. Akad. Nauk SSSR, 279, 166-172.
Stolper, E.M. (1977) Experimental petrology of eucrite meteorites. Mem. Natl. Inst. Polar Res., Spec. Issue 8, 170-184.
Takeda, H. (1979) A layered crust model of a howardite parent body. Icarus 40, 455-470.
Takeda, H., Miyamoto, M., Yanai, K., and Haramura, H. (1978) A preliminary mineralogical examination of the Yamato-74 achondrites. Mem. Natl. Inst. Polar Res., Spec. Issue 8, 170-184.
Takeda, H., Mori, H., and Yanai, K. (1981) Mineralogy of the Yamato diogenites as possible pieces of a single fall. Mem. Natl. Inst. Polar Res., Spec. Issue 20, 81-99.
Taylor, L.A., Kullerud, G., and Bryan, W.B. (1971) Opaque mineralogy and textural features of Apollo 12 samples and a comparison with Apollo 11 rocks. Proc. Second Lunar Sci. Conf., 855.
Thompson, J.B., Jr (1967) Thermodynamics properties of simple solutions. In P.H. Abelson, Ed. Researches in Geochemistry, 2, p. 340-361. John Wiley, New York.
Trinel-Dufour, M.C., and Perrot, P. (1977) Etude thermodynamique des solution solides dans le systeme Fe-Mg-O. Ann. Chim. 2, 309-318.
Turnock, A.C., and Eugster, H.P. (1962) Fe-Al oxides; phase relationships below 1000°C. J. Petrol. 3, 533-565.
Walker, D., Shibata, T., and DeLong, S.E. (1979) Abyssal tholeiites from the Oceanographer Fracture Zone, II. Phase-equilibria and mixing. Contrib. Mineral. Petrol. 70, 111-125.
Warren, P.H. (1985) Origin of howardites, diogenites, and eucrites: A mass balance constraint. Geochim. Cosmochim. Acta 49, 577-586.
Warren, P.H., and Jerde, E.A. (1987) Composition and origin of Nuevo Laredo Trend eucrites. Geochim. Cosmochim. Acta 51, 713-725.
Warren, P.H., Jerde, E.A., Migdisova, L.F., and Yaroshevsky, A.A. (1990) Pomozdino: an anomalous, high-MgO/FeO, yet REE-rich eucrite. Proc. 20th Lunar Planet. Conf., 281-297.
Wechsler, B.A., and Navrotsky, A. (1984) Thermodynamics and structural chemistry of compounds in the system $MgO\text{-}TiO_2$. J. Solid State Chem. 55, 165-180.
Zakrzewski, M.A. (1989) Chromian spinels from Kuså, Bergslagen, Sweden. Am. Mineral. 74, 448-455.

Chapter 10
Stephen E. Haggerty
OXIDE MINERALOGY OF THE UPPER MANTLE

INTRODUCTION

Rock forming mineral oxides and oxide mineral structures are present throughout our planet: in the core, possibly as wüstite ($Fe_{1-x}O$); in the D" layer and the lower mantle as magnesiowüstite, ferropericlase, perovskite-based $MgSiO_3$, and ilmenite-based garnet; in the transition zone as Mg_2SiO_4, β-spinel; in the upper mantle as spinel and ilmenite (sensu stricto); and in the crust as spinel, ilmenite, rutile, pseudobrookite, perovskite and pyrochlore. Geophysical interpretation of Earth's interior is centered around the oxide mineral group. The oxide minerals are the very core of rock magnetism and are the most effective recorders of plate migration throughout time at the surface of the planet.

Although oxide minerals in upper mantle-derived rocks are generally a minor constituent (<1 to 3% by volume), wide ranges in mineral compositions and in mineral assemblages allow the oxides to be employed as useful petrogenetic indicators. Being sensitive to temperature, pressure, oxygen fugacity, bulk rock and fluid composition, the oxide minerals may record partial melt episodes, degrees of melt extraction, and events of subsequent metasomatic enrichment.

Lherzolite (olivine + orthopyroxene + clinopyroxene) is the dominant rock type in the asthenosphere and is the source of basalt. Aluminum is an essential component, and in lherzolites Al is largely present in plagioclase at low pressures (< 10 kbar), in spinel group minerals at intermediate pressures (10-20 kbar), and in garnet at high pressures (>20 kbar). Most of the vast basalt regions of Earth are considered to be derived by partial melting of spinel lherzolite. Garnet or spinel or plagioclase, and clinopyroxene undergo melting, and iron is extracted from olivine and orthopyroxene, leaving a basalt-depleted restite which is the upper mantle lithosphere. Dominant rock types are harzburgite (olivine + orthopyroxene) and dunite (>90% olivine), both are typically spinel-bearing, and bulk compositions are Mg and Cr-enriched and Si, Fe and Al-depleted. Magnesian chromites are characteristic of lithospherically-derived diamonds and are central to geochemical exploration programs for diamondiferous kimberlites and lamproites. Chromian and aluminian spinels have been used in the classification of ophiolites and in unravelling the origin of closely related podiform chromite deposits. Metasomatism of depleted upper mantle rocks, by volatile-rich melts, leads to the formation of amphibole and phlogopite and reaction of chromian spinels. Enrichment in large ion lithophile elements (LILE, such as Ba, K, Sr, Rb, Ca, Na, REE), and high field strength elements (HFSE, such as Ti, Zr, Nb, Ta) results in an exotic array of metasomatic titanates that include Nb-Cr rutile, Ca-Cr-(Zr,Nb) armalcolite, baddelyite, zirconolite, and members of the crichtonite, magnetoplumbite and hollandite structural groups. Recognized as potential repositories for LILE and HFSE in the upper mantle, these minerals may be central to the trace-element geochemical signatures recorded by alkali mafic rocks in arc, rift and cratonic settings.

This chapter will concentrate on recent advances in the application of the oxide mineral group as petrogenetic indicators (e.g., Irvine, 1965, 1967) and is structured within a framework of mineral chemistry. The experimental and theoretical basis is given in earlier chapters, and these reviews are complemented only where high pressure experiments are lacking and where natural examples may point to future and potentially important avenues of research. The source rocks are entrained, upper mantle-derived xenoliths in alkali basalts and related extrusives, xenoliths in kimberlites and related rocks such as lamproites and lamprophyres, ocean floor abyssal peridotites, continental emplaced ophiolites, and upper mantle orogenic diapirs. Wherever possible, primary mineralogies are distinguished from metamorphic or metasomatic effects, but this is not always possible from incomplete petrographic descriptions in the literature. The wide variations in mineral formula calculations and in data presentation have been reduced to the simplest possible presentation. A global coverage has been attempted, but for the purposes of illustrating principles, specific sites are emphasized and core references only are cited. These reflect the current status, but do not necessarily provide the most complete account of the distributions and compositions of all mineral oxides in the upper mantle. This is particularly the case for the exotic group of LILE and HFSE metasomatic oxides and ultra-deep (>300 km) oxide (magnesiowüstite-ferropericlase) inclusions in diamonds where we have much to learn and little attention has been given.

This review, like any other on xenoliths is strongly biased because it depends on accidental sampling of the upper mantle in specific tectonic environments, focused by plumes, continental rifts and oceanic fracture zones, and by alkali magmatism. Recognizing this as a major constraint, students of the upper mantle are also deeply aware of time and convection so that sampling by recent volcanics in ancient cratons is at best an integrated sampling of multiple thermal and metasomatic events. There is much debate on whether equilibrium is ever reached in the upper mantle in spite of ambient high temperatures and high pressures and rapid transport of xenoliths to the surface. Encapsulated inclusions in diamonds record ancient geochemical signatures, in contrast to arc basalts that have undergone continental and mid-ocean ridge extraction and rejuvenation by subduction; thus xenolith suites from continental and oceanic settings can only be viewed as chemistries in instances of time and tectonic setting. A totally integrated evolutionary view is at present not possible. Nonetheless, the spikes in time and tectonic setting are so widespread that some assessments can be made. Caution is encouraged, however, particularly where oxides are used in petrogenetic models; the data base of the upper mantle is far from complete.

SPINEL MINERAL GROUP

Spinel-bearing, upper mantle-derived peridotites are dominantly lherzolites in oceanic fracture zones and in continental rift settings. Emplacement in the former is possibly diapiric but remains uncertain (Dick and Bullen, 1984), entrainment in the latter is in alkali basalts and closely related rock types (Nixon, 1987). Spinels in ophiolites are either upper mantle-derived or are products of crystal fractionation in lithospheric cumulate sequences (Paktunc, 1990). Spinels in the deeper upper mantle are of multiple deviations and range from products of partial melting to metasomatically induced exsolution and silicate decomposition.

Regardless of tectonic setting or host rock entrainment, primary spinels are crystallochemically classified as normal spinels, where R^{2+} cations are nominally in tetrahedral (IV) coordination and R^{3+} cations are exclusively in octahedral (VI) coordination. This mineral group is typified by spinel, $MgAl_2O_4$, i.e., $^{IV}Mg[Al_2]^{VI}O_4$; but includes $^{IV}Fe^{2+}[Al^2]^{VI}O_4$, hercynite; $^{IV}Mg[Cr_2]^{VI}O_4$, picrochromite; and $^{IV}Fe^{2+}[Cr_2]^{IV}O_4$, chromite. These minerals are the coordinates for the base of the spinel prism illustrated in Figure 1. Spinels at the prism apex are inverse spinels in which either R^{2+} or R^{3+} ions are split between tetrahedral and octahedral sites as in, for example, either magnetite $^{IV}Fe^{3+}[Fe^{3+}Fe^{2+}]^{VI}O_4$, or ulvöspinel $^{IV}Fe^{2+}[Fe^{2+}Ti]^{VI}O_4$. Inverse spinels in upper mantle rocks are secondary.

Spinel-bearing xenoliths in alkali basalts

Representative data for spinels in lherzolites, dunites, harzburgites, websterites, pyroxenites, and olivine megacrysts from alkali basalts in China, Alaska, Kamchatca, Mexico, Saudi Arabia and Italy are presented in Table 1 and Figure 2a-c; a comprehensive review is given by Mattioli et al. (1989). Spinels in websterites and pyroxenites are the most aluminous, followed by spinels in lherzolites, dunites and harzburgites. The range in chrome number, $cr' = [100Cr/(Cr+Al)]$, between ~5 and a maximum of ~60 is typical of spinel-bearing xenoliths in alkali basalts from continental regions. Magnesium numbers, $mg' = [100Mg/(Mg+Fe^{2+})]$ are typically between ~65 and ~85, the more Fe-rich spinel compositions correlating with aluminum depletion. Examples of spinel-bearing xenoliths from plume-activated, ocean island alkali basalts are selected from Hawaii (Fig. 2d and e). Magnesium and chromium numbers for spinels in pyroxenites and lherzolites are very similar to those of the continental suite (Fig. 2a-c); however, in the depleted rocks, spinels in dunites and harzburgites are substantially more Cr-rich with cr' of 75. For the dunites in particular, there is a substantial enrichment in iron, and mg' as low as 20 is reached. Spinel-bearing xenoliths from other alkali host rocks (lamprophyres, nephelinites, monchiquites, alnoites, hawaiites and basanites) from the USA, Australia, India, New Guinea, and France show a similar spread in mg' and cr' for the more aluminous spinels but high cr' are attained (60-85) for the most alkalic suites (Fig. 2f, Table 2).

Alpine-type peridotites, abyssal peridotites and ophiolites

Spinels in these settings are grouped together because of the uncertainty as to whether Alpine-type peridotites are orogenically emplaced upper mantle diapirs or obducted slices of oceanic crust.

Dick and Fisher (1984) established the characteristic spinel compositional field for abyssal peridotites in terms of cr' and mg'. This was followed by Dick and Bullen (1984) in a proposal that ophlolites may be classified as ridge, arc or transitional types based on spinel compositional ranges (Table 3), specifically in terms of mg' and cr'. In an interesting application of the abyssal peridotite spinel data, Pober and Faupl (1988) used spinels from heavy-mineral stream concentrates from a large area in the Eastern Alps, and on the basis of spinels in lherzolites and harzburgites established that the greatest population of spinel data is in the abyssal harzburgite field (Fig. 3a), equivalent to Dick and Bullen's (1984) transitional ophiolite type, with cr' as high as 85.

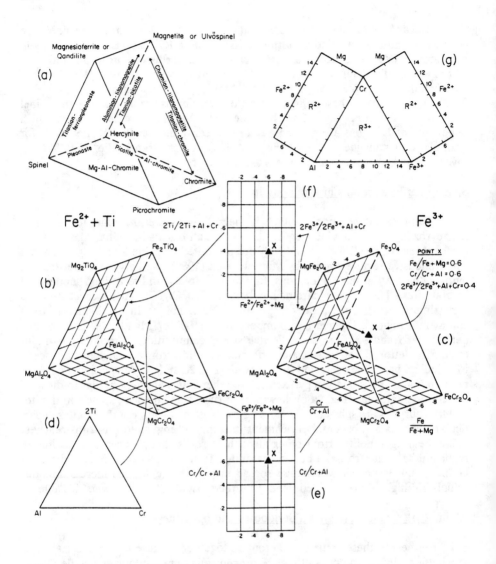

Figure 1. Nomenclature and plotting procedures for compositions in the multicomponent spinel prism. The bases of the prisms are defined by normal spinels and the apeces by inverse spinels. Inverse spinels are divided into Fe^{2+}, $Mg + Ti$ (b), and Fe^{2+}, $Mg + Fe^{3+}$ (c). The ternary on the left of each of these prisms has Mg as the common element, the ternary to the right has Fe^{2+} as the common element. For the divalent prism, Ti, Al and Cr are the variables (b); whereas the trivalent prism (c) has Fe^{3+}, Al and Cr as the variables. Because Fe^{2+} and Mg are common factors for the trivalent components of the respective ternaries, the ratios shown in (e) define variations along the prism base. Useful projections are shown in (d) and (f) but the most widely used projection is the spinel prism base (e). These prisms (b and c) are known as modified Johnson spinel prisms which differ slightly from the Pavlov prism in (g). In the Pavlov, Ti-free projection, an analysis is first plotted in the ternary using the trivalent components, and a second plot of the same analysis is transferred to the rectangle on the left if $Al > Fe^{3+}$, or to the rectangle on the right if $Fe^{3+} > Al$; this second plot is specific to the divalent ions. The scale values represent the number of cations in tetrahedral and octahedral coordination for the spinel formula based on 32 oxygens and 24 cations. An alternative plotting method that combines the Fe^{2+} and Fe^{3+} prisms (b and c) is shown in Figure 10 in a spinel trapezoid.

TABLE 1. Spinels in xenoliths in alkali basalts

	Web 1	Px 2	Lz 3	Lz 4	Lz 5	Lz 6	Hz 7	Hz 8	Hz 9	Hz 10	Dun 11
SiO_2	0.01	0.07	0.03	0.10	0.08	0.12	0.27	0.05	0.18	0.22	0.12
TiO_2	0.03	0.22	0.04	0.24	0.13		0.67	0.40	2.01	1.49	0.13
Al_2O_3	64.42	61.53	58.17	56.13	52.55	46.95	43.13	38.29	33.22	30.34	38.03
Cr_2O_3	2.30	2.29	9.14	11.21	14.75	20.67	21.29	28.66	24.73	34.99	30.27
Fe_2O_3	3.14	5.18	2.00	1.99	2.62	2.93	6.08	3.68	11.11	7.08	4.27
FeO	9.42	11.14	8.45	8.54	8.26	9.23	8.79	9.77	10.98	7.21	8.75
MnO	0.10	0.16	0.10	0.07	0.23	0.23	0.10	0.13	0.34	0.30	0.28
MgO	21.52	19.87	21.16	20.93	20.52	19.60	19.51	18.23	16.59	19.21	19.04
Total	100.94	100.46	99.09	99.21	99.14	99.73	99.84	99.21	99.16	100.84	100.89
cr'	2.3	2.4	9.5	11.8	15.9	22.8	24.9	33.4	33.3	43.6	34.8
mg'	80.3	76.1	81.7	81.4	81.6	79.1	79.8	76.9	72.9	82.6	79.5

	Lz 12	Lz 13	Lz 14	Lz 15	Lz 16	Weh 17	Lz 18	Hz 19	Lz 20	Lz 21	Lz 22
SiO_2									0.21	0.20	0.24
TiO_2	0.38	0.53	0.72	0.55	0.05	0.45	1.8	0.48	0.12	0.17	0.12
Al_2O_3	55.10	53.50	51.49	47.52	46.29	42.33	37.39	35.66	53.9	58.5	54.6
Cr_2O_3	8.21	9.77	11.12	16.76	20.68	20.79	12.56	31.26	12.8	7.45	12.4
Fe_2O_3	7.56	6.9	6.2	6.56	4.95	8.08	17.84	3.8	2.88	2.90	2.36
FeO	7.42	9.37	13.93	9.68	7.58	9.83	16.52	12.66	8.21	9.39	8.38
MnO	0.09	0.11	0.12	0.07	0.08	0.07	0.17	0.15	0.19	0.14	0.18
MgO	21.66	20.17	16.99	19.42	20.74	18.74	13.25	16.16	20.7	20.4	20.6
NiO	0.32	0.33	0.26	0.23	0.17	0.22	0.15	0.1	0.32	0.48	0.35
Total	100.74	100.68	100.83	100.79	100.54	100.51	99.68	100.27	99.33	99.63	99.23
cr'	12.0	14.0	16.0	24.0	29.0	29.0	19.0	44.0	13.4	7.70	12.9
mg'	75.0	68.0	55.0	67.0	73.0	66.0	45.0	56.0	81.8	79.5	81.4

	Px 23	Px 24	Px 25	Px 26	Lz 27	Lz 28	Lz 29	Lz 30	Lz 31	Lz 32	Lz 33
SiO_2	0.08	0.00	0.07	1.41	0.07	0.13			0.72	0.79	0.00
TiO_2	0.25	0.23	0.36	0.67	0.16	0.14	0.13	0.28	9.15	4.91	1.44
Al_2O_3	60.17	57.72	55.37	48.79	55.33	49.92	43.64	41.33	10.05	8.12	12.23
Cr_2O_3	2.65	3.85	6.54	10.56	9.07	15.57	20.96	20.52	16.39	20.45	37.81
Fe_2O_3	5.40	7.02	7.38	6.00	5.46	4.93	5.93	7.69	25.89	29.52	16.61
FeO	12.90	11.72	13.47	17.63	8.48	8.37	10.55	14.01	29.95	30.57	25.97
MnO	0.06		0.07		0.04	0.10	0.29	0.30			
MgO	18.95	19.20	18.35	16.24	21.12	20.45	18.22	15.82	8.25	4.95	5.70
CaO		0.11	0.01	0.07	0.03	0.32	0.20	0.22	0.00	0.00	0.33
Total	100.46	99.85	101.62	101.37	99.76	99.93	99.92	100.17	100.40	99.31	100.09
cr'	2.90	4.30	40.0	12.7	9.80	17.3	24.4	25.0	52.2	62.8	67.5
mg'	72.4	74.5	70.8	62.2	81.7	81.3	75.4	66.8	32.9	22.4	28.1

	Dun 34	Dun 35	Dun 36	Hz 37	Hz 38	Hz 39	Hz 40	Hz 41	Weh 42	Weh 43	Ol 44
TiO_2	3.52	3.51	3.06	0.02	0.82	0.10	0.17	0.68	2.79	2.86	2.64
Al_2O_3	19.5	22.3	17.5	24.2	19.1	17.3	14.5	11.1	15.1	11.8	16.2
Cr_2O_3	28.2	31.2	33.8	42.9	45.6	47.5	50.4	50.4	37.9	39.7	37.8
Fe_2O_3	18.4	12.4	16.1	5.86	5.00	7.67	7.79	8.86	13.6	16.0	13.1
FeO	17.8	16.5	14.7	9.86	11.4	13.1	13.6	16.3	17.7	18.1	17.3
MnO	0.28	0.27	0.4	0.24	0.24	0.30	0.29	0.36	0.29	0.28	0.27
MgO	13.2	14.3	14.6	16.8	15.2	13.9	13.5	11.3	12.1	11.8	12.5
Total	100.9	100.48	100.0	99.88	97.36	99.87	100.43	99.0	99.48	100.54	99.85
cr'	49.2	48.4	56.4	54.3	61.6	64.8	70.0	75.3	62.7	69.3	61.0
mg'	56.9	60.7	63.9	75.2	70.4	65.4	63.9	55.3	54.9	53.7	56.3

Analyses 1-11, Eastern China (Quicheng and Hooper, 1989); 12-19 Northern Italy (Siena and Coltorti, 1989); 20-22 Alaska (Swanson et al., 1987); 23-30 Hawaii (Sen, 1988); 31-33, Hawaii (Sen and Presnall, 1986); 34-34 Loihi, Hawaii (Clague, 1988). Px pyroxenite; Lz lherzolite; Hz harzburgite; Dun dunite; Weh wehrlite; Web websterite; Ol olivine megacryst.

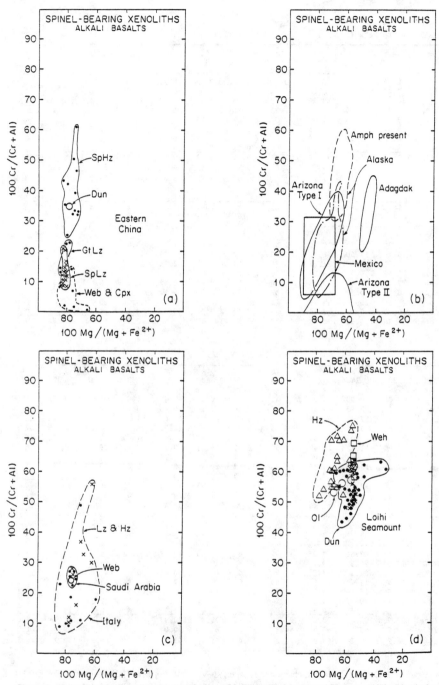

Figure 2. Spinel-bearing xenoliths in alkali basalts: (a) China (Quicheng and Hooper, 1989); (b) Alaska and Adagdak (Swanson et al., 1987), Mexico (Aranda-Gomez and Ortego-Gutierrez, 1987), Arizona (Kempton et al., 1987), Type I and Type II refer to Cr diopside and Al-Ti augite spinel lherzolites, respectively, (Wilshire and Shervais, 1975; Irving, 1980); (c) Italy (Siena and Coltorti, 1989; Baccaluva et al., 1989); Saudi Arabia (Lung-Chuan and Essene, 1986); (d) Loihi Seamount (Clague 1988).

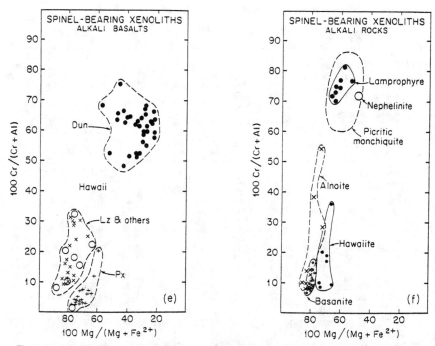

Figure 2. Spinel-bearing xenoliths in alkali basalts, cont.: (e) dunites, Hawaii (Sen and Presnall, 1986), remainder (Sen, 1988); (f) basanites, France (Fabries et al., 1987), hawaiites, Mexico (Gutmann, 1986), alnoites, Solomon Islands (Nixon and Boyd, 1979), picritic monchiquite, Australia (Jaques et al., 1989a), nephelinite, India (Mukherjee and Biswas, 1988), lamprophyre, Colorado (Ehrenberg, 1982).

TABLE 2. Spinels in xenoliths in alkali volcanics

	Lz 1	Lz 2	Lz 3	Lz 4	Lz 5	Lz 6	Hz 7	Lz 8	Hz 9	Hz 10	Hz 11
SiO_2	0.05	0.01		0.10	0.09	0.08	0.17	0.04	0.13	0.36	0.14
TiO_2		0.01	0.13	0.22	0.64	0.44	0.09	0.13	2.29	1.80	9.47
Al_2O_3	60.41	59.27	55.17	55.82	51.79	47.02	24.59	56.36	13.92	12.96	9.09
Cr_2O_3	7.43	7.63	13.73	7.64	7.72	18.32	43.39	9.11	47.28	47.44	42.57
Fe_2O_3	1.81	3.11	0.65				3.99	3.88	7.42	9.46	3.72
FeO	9.54	8.74	10.70	16.98	23.43	16.25	10.67	8.28	14.19	13.35	21.42
MnO		0.12		0.14	0.18	0.20	0.23	0.16	0.34	0.20	0.16
MgO	20.75	21.08	19.49	18.57	16.25	17.95	16.29	20.88	14.25	14.79	13.46
CaO		0.07					0.02	0.01			
NiO	0.39	0.25	0.18		0.25	0.28	0.24	0.43			
ZrO_2					0.24		0.13				
Total	100.38	100.29	100.05	99.71	100.35	100.67	99.68	99.28	99.82	100.36	100.03
cr'	7.60	8.00	14.3	8.40	9.10	20.7	54.2	9.80	69.5	71.0	75.9
mg'	79.5	81.1	76.4	72.7	64.6	72.3	73.2	81.8	64.2	66.3	52.8

Analyses 1-3 in basanites, Languedoc, France (Fabries et al., 1987); 4-6 in hawaiite, Mexico (Gutmann, 1986); 7-8 in alnoite, Solomon Islands (Nixon and Boyd, 1979); 9-11 in lamprophyre, Arizona-New Mexico (Ehrenberg, 1982). Abbreviations as in Table 1.

TABLE 3. Spinels in abyssal peridotites, Alpine peridotites and ophiolites

	Aby 1	Aby 2	Alp 3	Alp 4	Alp 5	Aby 6	Aby 7	Aby 8	Lz 9	Lz 10	Lz 11
SiO_2									0.04	0.05	0.01
TiO_2	0.04	0.05	0.09	0.06	0.10	0.01	0.04	0.13		0.01	0.02
Al_2O_3	48.76	25.08	58.7	27.73	13.23	38.12	29.40	23.22	60.06	58.69	55.51
Cr_2O_3	18.37	41.54	9.7	39.72	51.84	31.55	40.08	45.99	7.92	7.85	10.27
Fe_2O_3	2.67	3.68	0.97	3.00	5.67						
FeO	10.04	13.77	10.1	15.92	18.15	14.31	16.08	16.84	10.87	14.19	15.96
MnO	0.09	0.12	0.13	0.11	0.44	0.15	0.25	0.23	0.07		0.25
MgO	19.03	14.04	20.3	13.14	9.89	16.26	14.13	13.86	20.46	19.11	17.99
CaO									0.03		
NiO	0.33	0.18			0.09	0.18	0.09	0.13	0.19	0.33	0.18
Total	Av	Av	Av	Av	Av	100.58	100.07	100.40	99.66	100.23	100.19
cr'						35.7	47.8	57.1	8.10	8.20	8.40
mg'						68.9	62.9	63.1	78.8	74.1	75.9

	Hz 12	Hz 13	Weh 14	Weh 15	Weh 16	Lz 17	Lz 18	Lz 19	Dun 20	Dun 21	Dun 22
SiO_2	0.14	0.05	0.04	0.03	0.05	0.00	0.00	0.10	0.02	0.02	0.04
TiO_2	0.07		0.22	0.33	0.52	0.00	0.12	0.11	0.49	0.45	0.38
Al_2O_3	45.62	40.67	36.91	31.48	18.52	49.65	41.91	37.14	31.58	24.88	23.62
Cr_2O_3	22.07	23.56	27.32	32.45	44.27	15.95	21.51	25.25	35.30	37.83	42.95
V_2O_3			0.29	0.39	0.56				0.33	0.5	0.43
FeO	13.23	19.94	21.21	23.69	26.29	19.61	22.89	22.13	18.72	25.2	20.5
MnO		0.07	0.22	0.26	0.34	0.23	0.29	0.28	0.23	0.3	0.30
MgO	17.47	14.94	13.49	11.37	9.55	15.31	14.82	13.52	13.85	11.23	12.12
CaO	0.05	0.03									
NiO	0.34	0.15									
Total	98.99	99.41	99.7	100.0	100.10	100.75	101.54	98.53	100.49	100.43	100.33
cr'	24.5	28.0	33.2	38.8	61.6	17.7	25.6	31.3	42.9	50.5	55.0
mg'	72.0	63.6	58.3	50.7	45.2	62.3	61.6	59.1	60.3	51.1	55.3

	Dun 23	Dun 24	Hz 25	Pod 26	Pod 27	Pod 28	Pod 29	Pod 30	Pod 31	Pod 32	Pod 33
SiO_2	0.00	0.06	0.00	0.00	0.00	0.65	0.51	0.37	0.47	0.38	0.38
TiO_2	0.06	0.15	0.17	0.00	0.24	0.00	0.00	0.00	0.00	0.00	0.00
Al_2O_3	32.07	26.17	12.53	10.42	11.06	28.70	27.14	24.23	21.96	19.11	14.48
Cr_2O_3	31.96	42.61	56.27	58.09	60.19	37.09	41.74	45.51	48.04	54.09	58.39
Fe_2O_3	5.85	3.46	3.13	5.45	4.37	3.63	0.58	0.57	0.13	0.00	0.00
FeO	17.78	10.22	13.00	14.77	11.16	14.65	15.37	14.46	15.09	12.60	13.17
MnO	0.14	0.08	0.09	0.36	0.00	0.59	0.32	0.29	0.30	0.26	0.23
MgO	12.64	16.89	13.49	10.90	13.99	14.26	13.66	13.88	13.27	12.88	12.72
CaO	0.10	0.00	0.00	0.00	0.00	0.00	0.00	0.00	0.00	0.00	0.00
Ni,V*	0.14	0.08	0.00	0.13	0.11	0.29	0.30	0.24	0.33	0.18	0.19
Total	100.74	99.72	98.68	100.12	101.12	99.86	99.62	99.55	99.59	99.50	99.56
cr'	40.0	52.0	75.0	79.0	79.0	46.44	50.78	55.75	59.47	65.50	73.01
mg'	56.0	74.0	65.0	55.0	67.0	63.97	61.62	63.39	61.35	64.83	63.49

Analyses 1-5, averages of high Al (1) and low Al(2) abyssal peridotites; high Al (3, Rhonda, Spain, Type 1), low Al (4, Samail ophiolite, Type 2), and Twin Sisters (5, California, Type 3) Alpine peridotites - Dick and Bullen (1984); 6-8 abyssal peridotites - (Shibata and Thompson, 1986); 9-13 Caussou, Pyrenees (Fabries et al., 1989); 14-22 Newfoundland (Casey et al., 1985); 23-25 Oman (23 = base of cumulate series, 24 = top of cumulate series, 25 upper mantle series, Auge, 1987); 26-27 Greece (Alevizos and Sclar, 1988); 28-33 Turkey (Engin and Hirst, 1970) *Ni=NiO for analyses 23-27; V=V2O3 for 28-33. Lz - lherzolite.

Spinel compositions in lherzolites from alkali basalts in the Massif Central are compared with spinels in Fe-rich peridotites, lherzolites and harzburgites from Lherz, Freychinede, Cassou and Montferrier in the Pyrenees, and with spinels in peridotites and pyroxenites from Ronda in Spain and Beni Bouchera in Morocco (Table 3, Fig. 3b). Tight compositional envelopes exist for each data set with a maximum cr' of 60; the range of mg' from 44 to 86 is not unlike the compositional field for typical spinel peridotites in alkali basalts (Fig. 2).

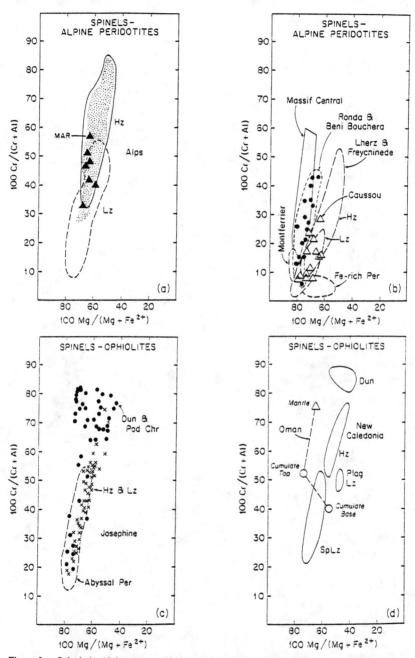

Figure 3. Spinels in Alpine-type peridotites and ophiolites: (a) Alps (Pober and Faupl, 1988), Mid-Atlantic Ridge (Shibata and Thompson, 1986); (b) Massif Central and Montferrier are xenoliths in alkali basalts used in a comparison with orogenic lherzolites from Caussou, Lherz and Freychinede (Fabries et al., 1989), Ronda and Beni Bouchera (also Beni Bousera)(Dick and Bullen, 1984; Cervilla and Leblanc, 1990); (c) Josephine (Dick and Bullen, 1984); (d) Oman (Auge, 1987), New Caledonia (Leblanc et al., 1980); (d) dunites, wherlites and lherzolites, Bay of Islands Complex (Casey et al., 1985), remaining data, Lewis Hills (Smith and Elthon, 1988).

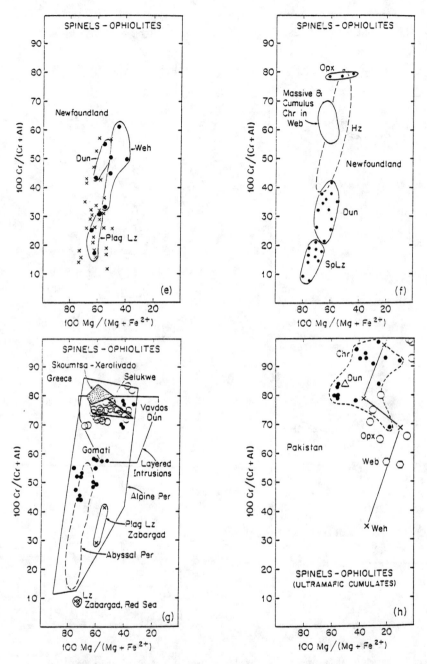

Figure 3. Spinels in Alpine-type peridotites and ophiolites, cont: (e) dashed field, Bay of Islands Complex (Malpas and Strong, 1975), remaining fields, White Hills (Talkington and Malpas, 1984); (g) Skoumtsa-Xenolivado (Alevizos and Sclar, 1988), Vardos and Gomati (Christodoulou and Hirst, 1985), Selukwe, Zimbabwe (Cotterill, 1979), Zabargad (Bonatti et al., 1986), abyssal peridotite field (Dick and Bullen, 1984), Alpine peridotite and layered intrusion fields (Irvine and Findlay, 1972); (h) Jan and Windley (1990).

A large spinel data set for the classic Josephine ophiolite in Oregon is given in Figure 3c, for spinels in harzburgites, lherzolites, dunites and podiform chromites. Most of the harzburgite and lherzolite data correspond to the compositional spinel field for spinels in abyssal peridotites, which typifies ridge-generated ophiolites in the Dick and Bullen (1984) classification. Spinels in dunites and podiform chromites, however, have an enormous range of cr' from 19 to 83 (Fig. 3c, Table 3).

Ophiolitic spinels from Oman and New Caledonia are compared in Figure 3d. These data provide important clues to the separation of spinel compositions as a function of rock type, and serve to distinguish between spinels that are upper mantle-derived from spinels that form in ultramafic cumulates by crystal fractionation (presumably at the base of magma chambers at the oceanic crust-upper mantle boundary). An excellent progression is established in New Caledonia for spinels with cr' and mg' that range from fertile, Al-enriched lherzolites, to intermediate harzburgites, and to depleted Cr-rich dunites. Most ophiolites are severely imbricated because of tectonic emplacement, but in Oman, Auge (1987) showed clearly that the upper mantle-derived, Cr-rich and depleted spinels differ from the more Al-rich spinels at the cumulate top, and are distinctive from the Al- and Fe-rich spinels at the ultramafic cumulate base (Fig. 3d, Table 3).

Newfoundland is another ophiolite classic that has received considerable attention. Spinel compositional ranges from the Lewis Hills, the North Arm and the Bay of Island Complexes are illustrated in Figures 3e and f. Spinels in plagioclase and spinel lherzolites are more fertile and Al-rich than spinels in harzburgites, dunites and cumulus chromites in websterites.

The final array of spinels in ophiolites is from podiform chromites in Greece, and for spinels in the Zabargad peridotite in the Red Sea (Fig. 3g). These data are compared with the Abyssal spinel peridotite field of Dick and Fisher (1984), and the Alpine peridotite and stratiform spinel compositional fields of Irvine and Findlay (1972) and Irvine (1967). Included in Figure 3g are spinels from Selukwe in Zimbabwe, which is considered possibly the oldest (3.5 Ga) ophiolite complex yet recognized (Cotterill, 1979). The Selukwe data are comparable to stratiform deposits and to three of the four sites in Greece, whereas spinels from the remaining site (Gomatie) are similar to spinels in abyssal peridotites; all of these spinel compositions, however, are very dissimilar to spinels in lherzolites from Zabargad which are uniformly more aluminous.

Spinel-bearing xenoliths in kimberlites and lamproites

Data for spinels are all from diamondiferous diatremes; thus the possible range in sampling is from about 40 to 200 km. Selected data from heavy mineral concentrates are included to broaden the geographical coverage, but only where geochemical patterns establish a derivation of spinels from xenoliths. For the voluminous literature on groundmass spinels, the interested reader is referred to Haggerty (1975), Mitchell (1986), and Tompkins and Haggerty (1985).

The most obvious compositional distinction between the spinel data set for xenoliths in kimberlites (Fig. 4a-f) and those presented previously (Figs. 2 and 3) is the extreme Al-depletion and Cr-enrichment of spinels from deep cratonic settings. Chromium numbers as high as 96 are achieved by spinels in diamond inclusions (Table 4, Fig. 4a); corresponding mg' range from 44 to 80. For diamond inclusions, there is only one known exception--the Zn-bearing chromites, which have cr' of 93 but mg' of only 6. There is a continuum from the diamond-inclusion spinel field to spinels in harzburgites and unspecified peridotites (mostly but unconfirmed dunites), as shown in Figure 4a, for data from southern Africa and the USSR.

Diamonds fall into two major classes of host rocks, based on inclusion suites: these are ultramafic suite diamonds from highly depleted spinel harzburgites, and fertile eclogitic suite diamonds that lack spinel but may contain rutile and rarely ilmenite; a third, and rare diamond suite that lacks mineral oxide inclusions is websteritic (Gurney, 1989; Meyer, 1987). Figure 4b compares a large data set of spinels from harzburgites entrained in the Jagersfontein kimberlite in South Africa, with lamproitic chromian spinels and spinels in a diamondiferous peridotite from Argyle. Apart from a single Fe-rich spinel in the peridotite, there is a substantial overlap among the data, with some cr' reaching values of 98. The characteristic signature for diamond inclusion spinels (Fig. 4a) is cr' >80 (i.e., greater than 60 wt % Cr_2O_3). Chromium numbers (70-92) from concentrate spinels in kimberlites from China (Fig. 4c) correspond closely to the diamond inclusion suite (Fig. 4a) and to the spinel harzburgites from Jagersfontein (Fig. 4b). Discrete spinels in kimberlites (possibly lamproites, Scott-Smith, 1989) and spinels in lherzolites from India are respectively more magnesian and more ferrous than the concentrate spinels from China (Fig. 4c).

At the other extreme of the depleted spinel suite in diamonds and spinel harzburgites are the aluminous spinel-bearing xenoliths known as alkremites. These are described from the USSR, the Moses Rock diatreme, Utah and in kimberlites from South Africa. A closely related suite of spinel xenoliths contain corundum (coined corgaspinite by Mazzone and Haggerty, 1989a), recognized so far only at Jagersfontein. Spinels in the corgaspinite suite are among the most aluminous yet recorded in the upper mantle (Table 4, Fig. 4d); alkremitic spinels are also aluminous but may contain up to 5 wt % Cr_2O_3. The origin of these xenoliths is unknown but a subducted sedimentary protolith is favored by Exley et al. (1983) and Mazzone and Haggerty (1989b).

Graphic or symplectic spinel intergrowths (fingerprint spinels in the terminology of Dawson and Smith, 1975) with olivine, orthopyroxene, and or clinopyroxene are widespread in peridotite xenoliths in kimberlites (Smith, 1977). These intergrowths appear to be products of reaction that involve olivine or garnet with clinopyroxene (Lock and Dawson, 1980). An alternative explanation is that these intergrowths represent equilibration from the garnet to the spinel lherzolite field in upwelling diapirs, or by xenoliths interrupted in passage from high to lower pressures in the diatreme (Ferguson et al., 1977; Dawson et al., 1978). Spinel symplectites are commonly associated with metasomatism but a definitive origin remains uncertain. A relatively narrow range in Fe/Mg is present for these spinels that range in cr' from 30 to 74 (Fig. 4d). Included in Figure 4d are spinel compositions in a rare eclogite, and in an equally rare orthopyroxene megacryst that has serially zoned inclusions composed of diopside, chromite, Cr pyrope, and olivine.

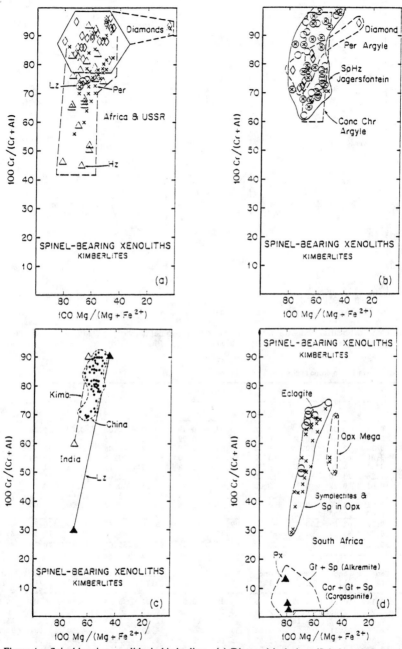

Figure 4. Spinel-bearing xenoliths in kimberlites. (a) Diamond inclusions (Sobolev, 1977; Meyer and Boyd, 1972; Nixon et al., 1987; Meyer, 1987; Daniels and Gurney, 1989; 1991), remaining xenolith data are from the review in Haggerty (1979); (b) Argyle concentrate and diamond peridotite data (Jaques et al., 1989a; 1990), Jagersfontein harzburgites (Field and Haggerty, unpublished); (c) China (Dong and Zhou, 1980), India (Nehru and Reddy, 1989); (d) eclogite (Shee and Gurney, 1979), orthopyroxene megacryst (Meyer et al., 1979), symplectites (crosses, Field and Haggerty, unpublished); open circles, Dawson and Smith (1975), pyroxenites (Kirkley et al., 1984); garnet + spinel (alkremite), and corundum + garnet + spinel (corgaspinite), Mazzone and Haggerty (1989a).

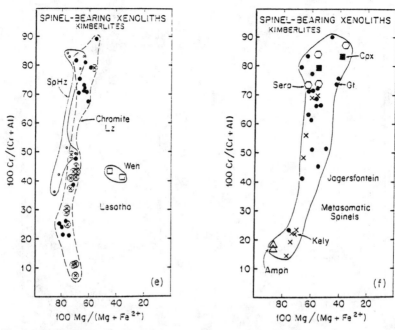

Figure 4. Spinel-bearing xenoliths in kimberlites, cont.: (e) spinel harzburgite data (Nixon et al., 1987), remaining data (Carswell et al., 1984); (f) Field and Haggerty, (unpublished).

TABLE 4. Spinels in xenoliths in kimberlites

	Diamond inclusions										
	1	2	3	4	5	6	7	8	9	10	11
SiO_2	0.26	0.08	0.13	0.26	0.23	0.29	0.00	0.00	0.00	0.00	0.00
TiO_2	0.05	0.55	0.32	0.05	0.03	0.09	0.00	1.36	0.12	0.00	0.00
Al_2O_3	6.74	5.59	2.24	6.74	5.94	3.23	9.48	2.08	7.77	12.98	7.49
Cr_2O_3	63.30	64.40	68.60	63.30	64.00	61.10	62.10	66.45	65.04	59.55	65.13
Fe_2O_3	3.49	4.00	1.83	3.74	3.75	2.68	1.19	3.19	2.55	1.57	2.44
FeO	11.26	13.70	17.70	11.00	11.70	28.80	12.63	12.48	10.00	11.06	10.26
MnO	0.01	0.27	0.36	0.00	0.00	0.41	0.21	0.21	0.20	0.23	0.23
MgO	14.40	13.10	9.77	14.40	13.80	0.54	13.32	13.45	15.06	14.74	14.61
CaO	0.01	0.00	0.00	0.01	0.00	0.26	0.00	0.00	0.00	0.00	0.00
NiO	0.00	0.00	0.00	0.00	0.00		0.07	0.10	0.00	0.12	0.00
Total	99.52	101.69	100.95	99.50	99.45	99.78*	99.00	99.32	100.74	100.25	100.16
cr'	86.30	88.54	95.36	86.30	87.85	92.70	81.5	95.6	84.9	75.5	85.4
mg'	69.52	63.29	50.11	70.00	67.77	4.57	65.3	65.8	72.8	70.3	71.7

	Conc 12	Conc 13	Conc 14	Conc 15	Hz 16	Hz 17	Hz 18	Hz 19	Hz 20	Lz 21	Hz 22
SiO_2	0.94	0.55	0.63	0.28	0.05	0.05	0.07	0.20	0.24	0.00	0.00
TiO_2	3.90	2.68	2.49	2.18	0.13	0.08	0.17	0.05	0.02	1.61	0.03
Al_2O_3	7.64	5.77	4.16	3.07	13.85	12.94	12.95	8.58	4.07	2.33	16.31
Cr_2O_3	52.16	56.31	58.21	60.44	53.70	54.83	54.12	62.20	66.90	66.02	53.36
Fe_2O_3	4.39	3.39	5.71	5.62	5.10	5.38	5.20	3.60	1.44	3.33	2.43
FeO	16.07	18.26	17.70	17.31	12.65	12.62	13.88	9.41	14.30	13.40	13.78
MnO	0.36	0.26	0.31	0.29	0.27	0.28	0.28	0.28	0.46	0.41	0.35
MgO	13.24	11.74	10.60	10.33	13.95	13.92	13.09	15.60	11.70	13.07	13.05
CaO	0.37	0.79	0.38	0.22	0.00	0.00	0.02	0.02	0.04	0.00	0.00
NiO	0.16	0.35	0.28	0.15	0.00	0.00	0.00	0.14	0.10	0.12	0.12
Total	99.23	100.10	100.47	99.89	99.70	100.10	99.78	100.08	99.27	100.29	99.43
cr'	82.09	86.76	90.37	92.06	72.23	73.98	73.71	82.94	91.69	95.00	68.70
mg'	61.29	56.25	54.05	54.25	66.52	66.54	62.98	74.91	59.85	63.5	62.8

TABLE 4. - Continued

	Per 23	Per 24	Per 25	Per 26	Per 27	Per 28	Per 29	Per 30	Per 31	Alk 32	Cgs 33
SiO$_2$	0.05	0.05	0.00	0.04	0.00	0.00	0.16	0.19	0.22	0.00	0.00
TiO$_2$	0.34	0.73	0.61	0.27	2.03	1.39	1.96	0.07	0.59	0.59	0.12
Al$_2$O$_3$	8.75	5.35	12.50	14.20	15.50	28.80	33.30	46.70	48.60	60.99	66.88
Cr$_2$O$_3$	63.10	59.90	55.90	51.70	47.80	38.40	31.40	21.90	18.70	4.16	0.17
Fe$_2$O$_3$	1.91	6.04	4.08	7.04	6.14	4.11	3.36	1.49	2.38	3.06	0.93
FeO	10.80	15.90	12.30	12.00	15.30	12.70	12.10	8.64	9.57	9.23	10.84
MnO	0.38	0.55	0.00	0.06	0.00	0.24	0.33	0.23	0.24	0.10	0.10
MgO	14.70	11.10	14.50	14.70	13.70	16.60	17.40	20.00	20.20	21.15	20.43
NiO	0.00	0.00	0.00	0.00	0.26	0.28	0.21	0.00	0.00	0.35	0.44
Total	100.03	99.62	99.89	100.01	100.73	102.52	100.22	99.22	100.5	99.63	99.91
cr'	82.90	88.30	25.60	70.90	67.40	47.20	38.80	23.90	20.60	4.37	0.15
mg'	70.80	55.30	76.70	68.50	61.50	69.90	72.10	80.40	79.00	80.33	77.07

	Ms 34	Ms 35	Ms 36	Ms 37	Ms 38	En 39	Ms 40	Sym 41	Sym 42	Sym 43	Sym 44
TiO$_2$	2.66	0.63	0.37	0.34	0.38	0.38	0.22	0.00	0.00	0.00	0.04
ZrO$_2$	0.17	0.11						0.11			
Al$_2$O$_3$	0.62	10.30	13.36	13.37	13.60	12.65	30.26	32.08	15.74	24.04	43.09
Cr$_2$O$_3$	55.60	56.19	52.74	53.05	52.47	52.14	36.99	36.28	52.70	44.40	27.56
Fe$_2$O$_3$	12.34	6.78	4.41	4.14	4.67	6.05	2.26	2.92	3.68	4.07	
FeO	15.03	10.99	16.48	16.85	16.14	15.45	17.32	10.19	14.65	9.81	10.65
MnO	0.47	0.42	0.61	0.62	0.60	0.64	0.74	0.29	0.24	0.17	
MgO	12.21	15.02	11.17	10.93	11.45	11.71	12.43	17.23	13.57	16.68	18.24
Total	99.10	100.44	99.14	99.30	99.31	99.02	100.22	99.10	100.58	99.17	99.58
cr'	98.37	78.53	92.62	72.70	72.13	73.43	45.06	43.14	69.20	55.30	30.10
mg'	59.13	70.88	54.71	53.62	55.83	57.50	56.13	75.10	59.20	68.80	75.30

Analyses 1-3 Sobolev (1977); 4-6 Meyer and Boyd (1972); 7-11 Daniels and Gurney (1991); 12-15 concentrate (Conc) spinels, China (Dong and Zhou, 1980); 16-18 Lesotho (Nixon and Boyd, 1973); 19-20 Premier (Danchin and Boyd, 1976); 21-22 and 42-44 Jagersfontein (Field and Haggerty, unpublished); 23-31 Pipe 200, Lesotho (Carswell et al., 1979); 32-33; Jagersfontein (Mazzone and Haggerty, 1989 a); 34-41 Bultfontein (Erlank et al., 1987). Per = peridotite; Alk = alkremite; Cgs = corgaspinite; Ms = metasomatized peridotite; En = lamellar spinel in orthopyroxene; Sym = symplectic spinel. Other abbreviations as in Table 1.
*Total includes 2.38 wt% ZnO.

Spinel-bearing xenoliths in kimberlites from Lesotho (Fig. 4e) show one moderately depleted population (Cr' 70-90) and another that is relatively fertile (Cr' 10-50). These data may be interpreted in the context of the spinel harzburgite data (Fig. 4e). Where recrystallization has taken place, it is not always simple to establish whether spinel is primary or secondary. The most complete data set on spinels derived from amphibole, garnet, clinopyroxene and olivine is from Jagersfontein (Fig. 4f). Amphiboles (edenite-pargasite) yield the most aluminous spinels, and garnets, clinopyroxene, and olivine the most chromian spinels (Table 4). Evidence for metasomatism is pervasive in the form of phlogopite and amphibole (Field et al., 1989; Winterburn et al., 1990), elsewhere on the Kaapvaal Craton, especially at Kimberley (Erlank et al., 1986). Spinel textures (Chapter 5) in decomposed silicates at both localities are very similar, but differ widely among the silicate groups, from wormy spinel around garnet to oriented spinel in orthopyroxene. Spinel in orthopyroxene marks the onset of metasomatism; although the spinel has an exsolution appearance, the preferred interpretation is decomposition of aluminous orthopyroxene to changing temperature (and possibly pressure) accompanying the influx of volatile-rich metasomatizing melts.

Discussion

From the wide range in spinel data presented in Figures 2-4, an interesting mineral oxide petrogenetic picture emerges for oceanic and continental upper mantles as a function of tectonic setting. In general, there are progressive increases in cr' as a function of host rock depletion. Spinels in websterites and clinopyroxenites in alkali basalts are the most aluminous and are matched only by the exotic peraluminous suite of alkremites and corgaspinites in kimberlites (Figs. 2a and 4d). Continental alkali-basalt-hosted lherzolites have spinels with cr' mostly in the range of ~10 to ~30. These are followed with increasing cr' to a maximum of ~60 by spinels in harzburgites and dunites. For ocean island hot-spot-generated alkali basalts, even higher cr' (75) are achieved in spinel xenoliths, and are matched by one suite of podiform chromites in ophiolites (e.g., Fig. 3c). Maximum cr' (up to 98) are reached by spinels in diamond inclusions and by spinels in highly depleted harzburgites in kimberlites (Figs. 4a and 4b).

In Dick and Bullen's (1984) spinel-based classification for ophiolites (Fig. 5a), type 1 is considered to represent the upper mantle at ocean ridge settings; the peridotites have spinel cr' of 10-60. Type 3 ophiolites have spinel cr' of 45-80 and are thought to be arc-related. Type 2 ophiolites encompass the spinel fields of types 1 and 3 and are interpreted as transitional or reworked. Spinels in ridge ophiolites are more aluminous than those of arc types and are related to passive-decompression partial melting in the source. Arc ophiolites, by contrast, represent higher degrees of partial melting by the lowering of the liquidus through the subduction of wet sediments and altered basalt; hence, spinels in the residuum are Cr-enriched. With even higher degrees of partial melting, or multiple episodes of melt extraction over long periods of time, as is the case in Archean cratonic root zones, the expectation is extreme depletion; this is recorded by spinels in diamond inclusions and in harzburgites. Higher degrees of partial melting are also recorded by hot spots with a distinct separation in spinel cr' between, for example, Hawaii (Fig. 2d) and continental rift zones (e.g., Fig. 2a).

Lateral, rather than vertical variations in depletion are also possible as proposed by Bonatti and Michael (1989). Their view is summarized in Figure 5b with spinel cr' plotted as a progressive function of the depletion expressed in xenoliths sampled by continental alkali basalts, pre-oceanic rift peridotites (Zarbargad in the Red Sea, Fig. 3g), passive ocean margin peridotites, oceanic (abyssal) fracture zone peridotites, and oceanic trench peridotites. There is considerable overlap in the spinel data of the last two settings, and a similar overlap exists in the transitional (type 2) ophiolite classification (Fig. 5a) of Dick and Bullen (1984). Some of these overlaps are possibly explained by secondary hydrothermal reaction of spinels as demonstrated by Kimball (1990). During hydrothermal alteration, Cr is preferentially retained in spinel, as Al is mobilized to form amphibole and chlorite. This reaction produces spinels of lower mg' and higher ferric iron. Data for secondary spinels are summarized in Figure 5c. The highest mg' are for metasomatic spinels in kimberlite xenoliths (Fig. 4f), and the lowest mg' are for ophiolitic spinels, fracture zone spinels, and for late stage overgrowth spinels in kimberlite groundmass.

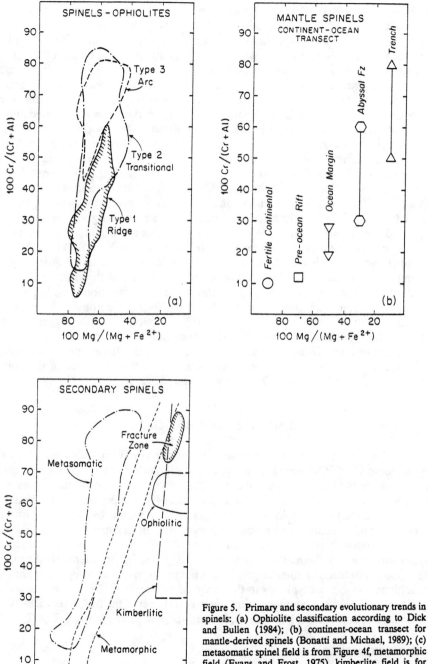

Figure 5. Primary and secondary evolutionary trends in spinels: (a) Ophiolite classification according to Dick and Bullen (1984); (b) continent-ocean transect for mantle-derived spinels (Bonatti and Michael, 1989); (c) metasomatic spinel field is from Figure 4f, metamorphic field (Evans and Frost, 1975), kimberlite field is for groundmass spinels (Haggerty, 1975, 1979); ophiolite and fracture zone fields (Kimball, 1990).

The origin of podiform chromites is a long-standing issue and until recently remained unresolved (Paktunc, 1990). Field relations in ophiolitic complexes are compounded by tectonic telescoping and the susceptibility of ultramafic rocks to serpentinization. Two compositional populations of podiform chromites are typically present (e.g., Fig. 3c and g), one highly refractory and Al-rich and the other Cr-rich. A common observation, expressed by Leblanc and Violette (1983) in their evaluation of chromites from the Phillipines, New Caledonia and the Troodos Massif, is that Cr-rich pods are located near the lherzolite-harzburgite transition whereas the Al-rich pods are associated with the harzburgite-dunite transition. On first inspection, this distribution is counterintuitive because the depleted Cr-rich pods should be associated with dunites and harzburgites, and the more fertile Al-rich horizons should have a lherzolitic affinity, as is the case for the spinel data sets in Figures 2-4. The stratigraphy of these settings and the compositions of coexisting olivine are key to the evident paradox. Malpas and Strong (1975) among others have shown that the olivine (Fo 91-94) compositions in ophiolitic harzburgites are more depleted than olivines (Fo 85-87) in closely associated dunites. The harzburgites are part of the depleted upper mantle sequence, whereas the dunites are cumulates from the fractional crystallization of an extracted tholeiitic melt (Malpas and Strong, 1975). Chromium-rich spinels are, therefore, upper mantle restites, whereas Al-rich spinels are magmatic cumulates. The distribution of chromites and chromite compositions in an ideal ophiolitic sequence (Fig. 6) shows that the transition zone between the cumulate sequence and the upper mantle is the focus of chromite concentration. Paktunc (1990) notes that the thickness of the transition zone, the degree of partial melting, the condition (i.e., fertility) of the upper mantle, and oceanic spreading rates are all factors that influence the accumulation and, hence, the economic potential of ophiolite-derived podiform chromite deposits. The ultimate source of Cr from Cr diopside in fertile lherzolite (Dickey and Yoder, 1972; Dickey, 1975).

ILMENITE MINERAL GROUP

This mineral group is composed of ilmenite ($FeTiO_3$), geikielite ($MgTiO_3$), pyrophanite ($MnTiO_3$), hematite (Fe_2O_3) and eskolaite (Cr_2O_3). Extensive mutual solid solution is present between ilmenite-geikielite, and between ilmenite-hematite in upper mantle xenoliths and in crystalline oxides from upper mantle-derived melts (Tables 5 and 6). The join $MgTiO_3$-Fe_2O_3 is interrupted by a large solvus that persists at high pressures and ambient temperatures in the upper mantle. Eskolaite in solid solution is rarely larger than about 10 mole %, is generally restricted to upper mantle assemblages, and is more typical of geikielite-rich members than ilmenite. Solid solution of hematite provides an accurate monitor of upper mantle fO_2 conditions and is most typically in the range of 1-15 mole % Fe_2O_3. Pyrophanite is a minor solid solution component in upper mantle xenoliths but is characteristic of carbonatites, carbonated kimberlites, lamproites, and lamprophyres (Fig. 7).

Ilmenite in alkali volcanics

The ilmenite mineral group is absent in the array of xenoliths typically sampled by alkali volcanics. Large (3-20 cms) discrete ilmenite xenoliths, although rare, are reported in basanites from the Ahagger Shield in Algeria (Leblanc et al., 1982).

OXIDE MINERALS IN OPHIOLITES

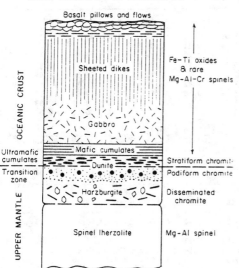

Figure 6. An idealized cross-section of the interface of oceanic crust and upper mantle illustrating the localization of stratiform chromites in the crustal segment and podiform chromites in the upper mantle component (adapted from Paktunc, 1990).

TABLE 5. Ilmenite-ilmenite and ilmenite-spinel xenoliths in kimberlites

	Ilm	Ilm	Ilm	Ilm	Ilm	Ilm	Ilm	Sp	Ilm	Ilm	Sp
	H	L	H	L	H	L	L	L	H	L	L
	1	2	3	4	5	6	7	8	9	10	11
TiO_2	54.56	42.32	48.77	40.86	29.88	42.29	46.23	7.10	36.68	50.39	14.58
Al_2O_3	0.00	1.20	0.32	2.91	0.35	0.10	0.01	0.76	0.44	0.01	0.60
Cr_2O_3	0.00	0.07	0.32	0.88	2.35	2.40	1.46	4.95	0.84	0.16	1.54
Fe_2O_3	3.68	26.64	14.42	26.18	41.80	22.08	12.25	49.76	32.39	7.80	38.96
FeO	30.76	20.93	25.15	19.08	21.19	27.90	33.03	36.12	25.44	35.11	42.23
MgO	10.10	9.45	10.39	9.77	3.34	5.58	2.29	0.41	4.10	5.55	1.04
MnO	0.19	0.23	0.16	0.23	0.11	0.29	4.47	1.16	0.16	0.19	0.76
CaO	0.07	0.04	0.01	0.01					0.05	0.08	0.03
Nb_2O_5					0.36	0.12	0.07	0.00			
Total	99.36	100.88	99.54	99.92	99.38	100.76	99.81	100.26	100.10	99.29	99.74
Ilm	61.04	42.05	50.15	39.53	46.05	58.02	70.08	47.08	13.54		
Geik	35.70	33.87	36.91	36.05	12.90	20.70	8.63	41.08	67.48		
Hem	3.26	24.08	12.94	24.42	40.84	20.66	11.66	11.84	18.98		
Pyr					0.21	0.62	9.63				

	Ilm	Sp	Sp	Ilm	Sp	Ilm	Sp	Ilm	Sp	Ilm	Sp
	12	13	14	15	16	17	18	19	20	21	22
TiO_2	36.38	1.44	11.58	47.39	9.50	55.05	26.03	48.66	15.05	49.30	8.42
Al_2O_3	0.42			0.44	4.69	0.11	5.21		0.80	1.16	10.41
Cr_2O_3	0.51	0.02	0.07	0.62	6.35	0.21	5.71	4.24	27.77	11.23	43.38
Fe_2O_3	33.25	66.12	46.40	15.67	40.88	6.92	11.72	11.08	14.87	5.43	3.57
FeO	25.47	32.25	41.67	25.89	34.80	22.12	43.14	25.36	33.64	19.48	18.71
MgO	3.97			9.22	3.76	15.18	8.41	10.13	7.49	13.81	14.64
MnO	0.13	0.06	0.02	0.27	0.27	0.32	0.32	0.33	0.43	0.19	0.28
CaO	0.02	0.02	0.04		0.01					0.03	0.02
Total	100.15	99.91	99.78	99.50	100.26	99.91	100.54	99.80	100.05	100.63	99.43
Ilm	40.74			52.42		42.32		52.38		41.87	
Geik	11.34			33.28		51.72		37.30		52.90	
Hem	47.92			14.30		5.96		10.32		5.23	

Representative analyses of exsolution (ilmenite-ilmenite), and reduction (ilmenite - spinel) pairs. Analyses 1-4 (Monastery) and 15-20 (Liberia) are from Haggerty and Tompkins (1984a,b); analyses 5-14 (Koidu) are from Tompkins and Haggerty (1984a,b); analyses 21-22, localities unspecified (Wyatt, 1979). Ilm = ilmenite; Sp = Spinel; H and L are host-lamellar pairs or intergrown asseblages.

TABLE 6. Ilmenite-bearing xenoliths in kimberlites

	1	2	3	4	5	6	7	8	9	10	11
TiO_2	52.53	45.86	52.18	44.93	47.09	48.68	53.84	52.38	47.18	51.33	50.93
ZrO_2			0.31	0.22	0.15	0.31	0.08				
Al_2O_3	0.00	1.02	0.15	0.48	0.13	0.22	0.56		0.37	0.65	0.66
Cr_2O_3	0.00	0.01	0.10	0.02	1.78	3.96	1.85	0.18	2.23	0.23	2.03
Fe_2O_3	7.69	11.97	8.38	20.25	13.38	10.95	6.12	6.84	15.80	11.44	10.82
FeO	31.05	33.36	27.62	22.48	29.26	25.82	22.70	33.44	25.13	24.44	20.60
MgO	8.89	8.08	10.40	9.65	7.29	10.15	14.19	7.24	9.58	11.99	13.97
MnO			1.04	1.01	0.33	0.20	0.31	0.74	0.20	0.33	0.26
CaO								0.00	0.00	0.00	0.01
Nb_2O_5			0.12	0.17	0.24	0.41	0.08				
NiO					0.10	0.20	0.23				
Total	100.16	100.30	100.30	99.21	99.75	100.90	99.96	100.82	100.49	100.41	99.28
Ilm	61.68		55.32		60.61	52.89	44.74	67.65	50.93	47.96	40.93
Geik	31.44		37.12		26.91	37.04	49.82	26.14	34.64	41.96	49.41
Hem	6.88		7.57		12.48	10.07	5.44	6.21	14.43	10.08	9.66

	12	13	14	15	16	17	18	19	20	21	22
TiO_2	48.40	48.49	48.86	50.65	54.39	56.40	45.60	51.70	48.20	52.00	53.70
ZrO_2	0.60	0.45	0.58								
Al_2O_3	0.02	0.03	0.01	0.54	0.51	0.33	0.60	0.70	0.39	0.61	0.66
Cr_2O_3	0.85	0.54	0.43	0.56	0.65	2.16	0.30	3.10	0.27	0.00	0.09
Fe_2O_3	10.55	13.13	12.59	12.30	5.57	6.60	18.80	8.91	16.40	9.22	9.29
FeO	33.70	30.01	27.47	24.63	25.78	10.39	27.49	21.68	25.70	25.90	21.50
MgO	5.75	7.56	9.58	11.62	12.86	22.63	7.50	13.80	9.33	11.70	15.00
MnO	0.25	0.33	0.34	0.18	0.25	0.64	0.20	0.20	0.31	0.26	0.39
CaO				0.01	0.04	0.04	0.02	0.02			
Nb_2O_5	0.35	0.01	0.57								
NiO	0.04	0.05	0.00				0.07	0.13			
Total	100.51	100.60	100.43	100.49	100.05	99.19	100.58	100.24	100.60	99.69	100.67
Ilm	69.20	60.79	54.74	48.42	50.36	19.32	55.78	43.10	51.69	50.88	41.05
Geik	21.07	27.25	33.97	40.69	44.76	75.13	27.07	48.95	33.47	40.99	50.99
Hem	9.73	11.96	11.29	10.89	4.88	5.55	17.15	7.95	14.84	8.13	7.96

	23	24	25	26	27	28	29	30	31	32	33
SiO_2	0.00	0.00	0.00	0.00	0.00	0.00	0.00	0.18	0.12	0.16	
TiO_2	45.87	49.41	48.98	48.75	51.71	50.36	54.55	50.50	51.90	51.40	51.73
Al_2O_3	0.53	0.44	0.76	0.59	0.28	0.97	0.41	0.22	0.21	0.36	0.40
Cr_2O_3	0.10	0.00	0.06	0.27	0.07	0.32	0.81	0.03	0.01	0.04	6.42
Fe_2O_3	18.23	13.58	13.10	14.97	11.38	14.65	9.30	3.11	1.92	3.03	7.48
FeO	27.19	27.07	25.98	22.70	22.00	17.48	15.03	44.60	45.88	45.47	16.45
MgO	7.79	9.64	10.04	11.67	13.54	15.30	18.75	0.11	0.14	0.14	16.24
MnO	0.16	0.16	0.16	0.28	0.30	0.48	0.43	0.73	0.64	0.68	0.53
CaO	0.00	0.00	0.00	0.03	0.03	0.03	0.12	0.07	0.03	0.01	0.34
Total	99.87	100.30	99.08	99.26	99.31	99.59	99.40	99.55	100.85	101.29	99.59
Ilm	55.19	53.74	52.21	45.17	42.87	34.09	28.57	96.58	97.66	96.58	33.72
Geik	28.15	34.14	35.96	41.40	47.12	53.08	63.49	0.41	0.51	0.51	59.36
Hem	16.66	12.12	11.83	13.43	10.01	12.83	7.94	3.01	1.83	2.91	6.93

	34	35	36	37	38	39	40	41	42	43	44
SiO_2	0.02	0.03	0.32	0.00	0.00	0.14	0.08	0.08	0.00	0.00	0.00
TiO_2	55.00	55.60	57.80	55.09	47.58	47.05	50.44	50.92	50.66	49.57	54.20
Al_2O_3	0.20	0.22	0.25	0.38	0.78	0.75	0.53	1.08	0.50	0.65	0.57
Cr_2O_3	0.13	0.23	0.04	1.92	0.07	0.05	0.14	1.24	1.07	1.17	0.33
Fe_2O_3	2.80	1.20	0.00	3.13	17.05	15.38	10.09	12.80	9.07	12.50	6.71
FeO	32.10	29.60	31.00	23.08	23.63	26.44	28.66	18.28	27.66	23.90	25.70
MgO	9.62	11.30	9.70	14.59	10.62	8.80	9.20	15.30	9.97	11.46	13.00
MnO	0.25	0.28	0.25	0.33	0.19	0.23	0.22	0.20	0.19	0.27	0.24
CaO	0.00	0.00	0.00		0.02	<0.05	0.10	0.07			
Nb_2O_5									0.16	0.03	
NiO				0.17		0.07			0.10		0.09
Total	100.12	98.46	99.36	98.69	99.94	98.91	99.46	99.97	99.38	99.55	100.84
Ilm	63.54	58.88	64.23	45.72	47.03	53.94	57.78	35.62	55.86	47.85	49.52
Geik	33.99	40.08	35.77	51.50	37.68	31.96	33.06	53.18	35.91	40.88	44.67
Hem	2.47	1.06	0.00	2.79	15.29	14.10	9.16	11.20	8.24	11.27	5.81

TABLE 6. - Continued

Analyses 1-2 (Leblanc et al., 1982) and 3-4 (Haggerty et al., 1985) are ilmenite (1 and 3) and bulk (ilmenite + spinel) chemical data for macrocrysts in basanites and melilitites, respectively; the remaining data in this Table are for ilmenites in kimberlites. Analyses 5-29 discrete ilmenite xenoliths; 30-37 ilmenite inclusions in diamonds; and 38-44 ilmenite - pyroxene intergrowths. Sources: 5-7, Orapa, Botswana (Tollo, 1982); 8-11, Liberia (Haggerty, unpublished); 12-15, Koidu, Sierra Leone (Tompkins and Haggerty, 1984a,b); 16-17 Green Mountain, Colorado and Wyoming (McCullum et al., 1975); 20-22, Kentucky (Agee et al., 1982); 23-29, Monastery (Haggerty, unpublished); 30-32, Brazil (Meyer and Svisero, 1975); 33-37, USSR (Sobolev, 1977); 38, Monastery (Haggerty et al., 1979); 39, Lesotho (Boyd and Nixon, 1973); 40-41 Monastery (Gurney et al., 1973); 42-43, Koidu (Tompkins and Haggerty, 1984); 44, Kentucky (Agee et al., 1982).

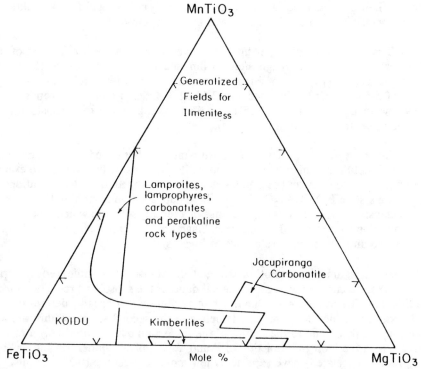

Figure 7. Upper mantle-derived ilmenite compositions as a function of pyrophanite ($MnTiO_3$), ilmenite ($FeTiO_3$) and geikielite ($MgTiO_3$). Summary from Tompkins and Haggerty (1985).

Ilmenite macrocrysts (1-2 cm in diameter) on the other hand are a common constituent of some melilitites from Namaqualand, South Africa (Haggerty et al., 1985). At both localities (Table 6), ilmenite has undergone subsolidus equilibration in which oriented spinels are formed along {0001} ilmenite planes (Chapter 5). Ilmenite reaction with the transporting melilitite results in the decomposition of ilmenite to titanomagnetite + perovskite ($CaTiO_3$). MgO in ilmenite varies between 6.5 and 10.5 wt % in the basanites and between 9 and 11 wt % in the melilitites; Fe_2O_3 contents are comparable (4-9 wt %), and Cr_2O_3 is low (0.01-0.1 wt %). The oriented lamellar spinels may be described as magnesian (4-8 wt % MgO) titanomagnetite. In addition to these iron-rich spinels, the basanitic ilmenite megacrysts also contain crystallographically oriented highly aluminous (61.8 wt % Al_2O_3) spinels. The formation of spinel in ilmenite is interpreted as subsolidus equilibration under reducing conditions (e.g., Buddington and Lindsley, 1964; Haggerty and Tompkins, 1984a), and appears to be widespread in upper mantle xenoliths (Tompkins and Haggerty, 1984a; McCallum, 1989).

Ilmenite xenoliths in kimberlites

Discrete ilmenite xenoliths (up to 15 cm in diameter) and ilmenite macrocrysts abound in type 1, non-micaceous kimberlites. These xenoliths are relatively robust in fluvial environments and are widely used in kimberlite exploration programs because their unusual mineral chemistries can readily be distinguished from ilmenites that are widespread in other igneous and metamorphic rocks.

The chemistries of upper-mantle derived ilmenites are a reflection of their ultramafic source rocks: magnesium in ilmenite is moderate to high (5-15 wt % MgO) and is at the expense of ferrous iron; chromium varies as a function of depletion (<0.5-5 wt % Cr_2O_3) or reaction (2-12 wt % Cr_2O_3); and ferric iron is moderately low in primary ilmenites (5-15 wt % hematite) but varies considerably in reacted ilmenites (5-30 wt % hematite).

In order to interpret the compositional distribution of primary, unreacted ilmenite xenoliths and macrocrysts in kimberlites, it is instructive initially to examine the effects of exsolution and subsolidus reduction as a function of phase relations and fO_2 in the system $FeTiO_3$-Fe_2O_3-TiO_2. These data are summarized in Figures 8 and 9. The most salient feature is the presence of a ternary solvus in which ilmenite solid solutions alone are replaced by the assemblage Ilm_{ss} + Sp_{ss} + Pb_{ss} (Figs. 8d and 9); this can be understood with reference to Figures 8b and c.

Ilmenite-ilmenite exsolution pairs and ilmenite-spinel ($\pm$ rutile) reduction pairs (Table 5) are compared with experimental data for the subsolidus reduction of kimberlitic ilmenites in Figure 9. One ilmenite exsolution pair broadly defines the nose of the decomposition loop, but the lamellae of the second set fall within the loop. This may partly be due to the nature of the projection but is more likely a result of variations in the loop as a function of P and T. All of the spinel-bearing ilmenites (i.e., reduction related) have expectedly low mole % hematite (0-5) contents. The compositions of spinels in ilmenites differ markedly from those discussed previously, specifically in Ti and Fe^{3+} contents. An alternative method of plotting spinel data, that accounts for eight rather than six end members, is shown in Figure 10 (Haggerty and Tompkins, 1984a). This alpha-numeric system permits a ready comparison between normal (alpha) and inverse (numeric) spinels and insights to modes of endmember substitution. The relative partitioning of Mg and Cr between ilmenite and spinel and the effect of coexisting rutile is sumarized in Figure 11a and b. Additional details are given in Haggerty and Tompkins (1984a,b).

Figure 12 is a global compilation of 1140 ilmenite analyses of discrete xenoliths and macrocrysts from kimberlites; representative analyses are listed in Table 6. A majority of the data are controlled by the ternary decomposition loops. The greatest population of data are relatively reduced and lie between about 25 and

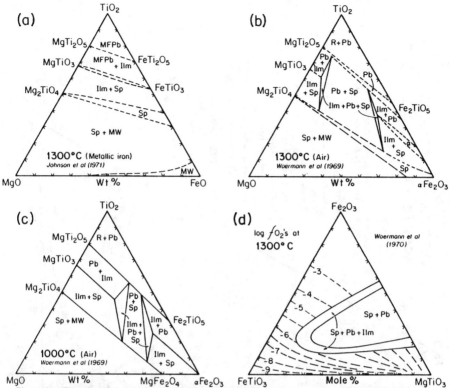

Figure 8. Experimental data for Fe-Mg-Ti-O relations between the extremes of metallic Fe and in contact with air. Solid solutions are complete in (a) but the series ilmenite-hematite shows a large miscibility gap at 1300° C in (b) and a somewhat smaller gap at 1000 C in (c). This is taken directly from Woermann et al. (1969). Relations in the lower right of (b) are ambiguous - the α-phase region should terminate at α-Fe_2O_3, and the α Fe_2O_3 corner should also include Fe_3O_4. The ilmenite plane between (a) and (b) is shown in (d) at 1300° C. The 2- and 3-phase fields define a decomposition loop that protrudes towards Ilm. The loop retracts with decreasing T and disappears at ~700° C. Oxygen fugacity isobars become more reducing away from the hematite apex towards ilmenite-geikielite, rotating about the geikielite apex.

60 mole % geikielite. The most oxidized ilmenites recorded are for the extremely rare exsolution pairs. Ilmenite xenolith data have been employed to estimate the fO_2 of the upper mantle (Haggerty, 1989a, 1990); the estimate is 10^{-6} to 10^{-7} bars at 1230-1300°C which is equivalent to WM-FMQ.

The high density of ternary ilmenite data at $Ilm_{55}Geik_{30}Hem_{15}$ is a good prospective indicator for kimberlites. An evaluation of the potential presence of diamonds, however, is better obtained from Mg-Cr relations (Figs. 13a and b). Each pipe or district appears to have a specific Mg-Cr signature (Schulze, 1987) but all are broadly parabolic (Haggerty, 1975); the empirical rule is that ilmenites on the right hand branch of the parabola are indicative of kimberlites with higher diamond contents than those on the left. The inset to Figure 13a shows possible equilibration paths for ilmenite on subsolidus reduction, with Cr strongly partitioned into spinel and with Mg to some extent into ilmenite (Fig. 11). The relationship of ilmenite to diamond is indirect because inclusions of ilmenite in diamond are extremely rare. High Mg and Cr and low Fe^{3+} contents in ilmenites are possibly more indicative of diamond survival than diamond genesis (Haggerty, 1986).

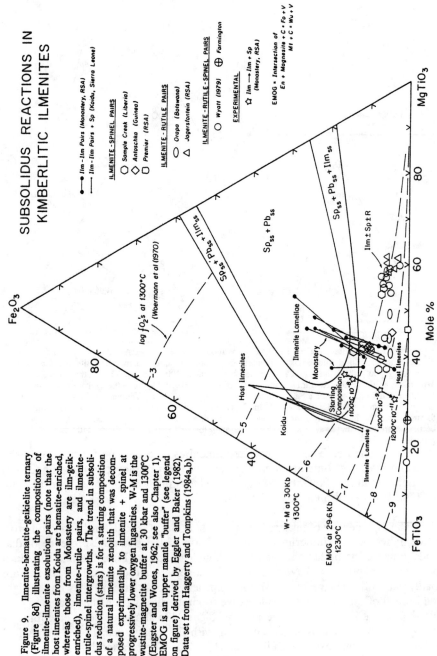

Figure 9. Ilmenite-hematite-geikielite ternary (Figure 8d) illustrating the compositions of ilmenite-ilmenite exsolution pairs (note that the host ilmenites from Koidu are hematite-enriched, whereas those from Monastery are ilm-geikenriched), ilmenite-rutile pairs, and ilmenite-rutile-spinel intergrowths. The trend in subsolidus reduction (stars) is for a starting composition of a natural ilmenite xenolith that was decomposed experimentally to ilmenite + spinel at progressively lower oxygen fugacities. W-M is the wustite-magnetite buffer at 30 kbar and 1300°C (Eugster and Wones, 1962; see also Chapter 1). EMOG is an upper mantle "buffer" (see legend on figure) derived by Eggler and Baker (1982). Data set from Haggerty and Tompkins (1984a,b).

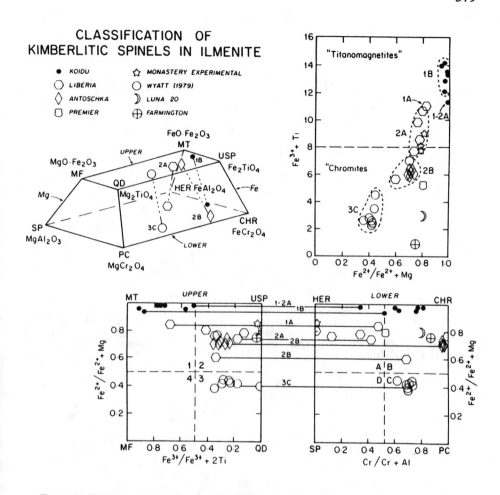

Figure 10. The base of the spinel trapezoid is the same as that conventionally used in the spinel prism (Fig. 1) but the upper plane is a section through the merged apeces of the oxidized (Fe^{3+}) and reduced (Fe^{2+}) prisms (Fig. 1b and c). Projections, with tie-lines, are shown in the lower two diagrams and each quadrant is dominated by an end member (e.g., Mt, Usp). Normal spinels are alphabetic (A-D) and inverse spinels are numeric (1-4); intermediate compositions are designated, therefore, as 1A, 2A, 2B, etc. A quantitative expression of octahedral site occupancy is given by Fe^{3+} + Ti, and hence represents the position of any composition in spinel space between the lower and upper planes of the trapezoid. Spinel data are for oriented spinels in ilmenite, generally considered and demonstrated experimentally to form by subsolidus reduction; the data set is from Haggerty and Tompkins (1984a,b).

The origin of the discrete megacryst suite, whether ilmenite, olivine, pyroxene or garnet remains uncertain (e.g., Moore, 1987; Schulze, 1987; Hops et al., 1989); proposed monomineralic horizons in the upper mantle or disaggregated mantle pegmatites are unverified, and possible megaphenocrysts in kimberlite or protokimberlitic magmas are inconclusive. A distinct possibility for ilmenite is an immiscible oxide liquid. There is circumstantial evidence for Fe-Ti-S-O liquids in the Koidu Kimberlite Complex where discrete ilmenite xenoliths are intimately intergrown with equilibrated immiscible monosulfide solid solutions (Tompkins and Haggerty, 1984a; Haggerty and Tompkins, 1982).

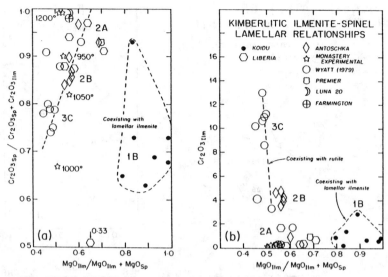

Figure 11. Cr-Mg relations for coexisting ilmenite-spinel ± rutile assemblages formed by subsolidus reduction. Experimental data for a Monastery ilmenite (Fig. 9) are also shown (Fig. 11a) along with the spinel trapezoid classification. Data from Koidu are for coupled exsolution (ilmenite-ilmenite) and subsolidus reduction assemblages (Tompkins and Haggerty, 1984a,b). The remaining are from Haggerty and Tompkins (1984a,b).

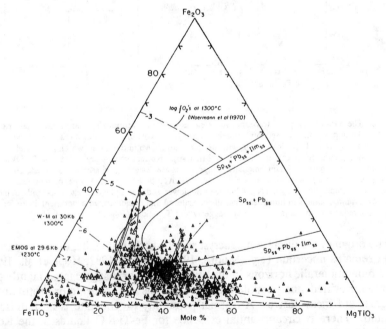

Figure 12. Upper mantle derived ilmenite compositions (1014 analyses) in kimberlites expressed as a function of ilmenite-hematite-geikielite. The decomposition loop is derived from Figure 8 and the fO_2 contours and buffers (W-M and EMOG) are from Figures 8 and 9. The data base is in part from the literature and in part unpublished.

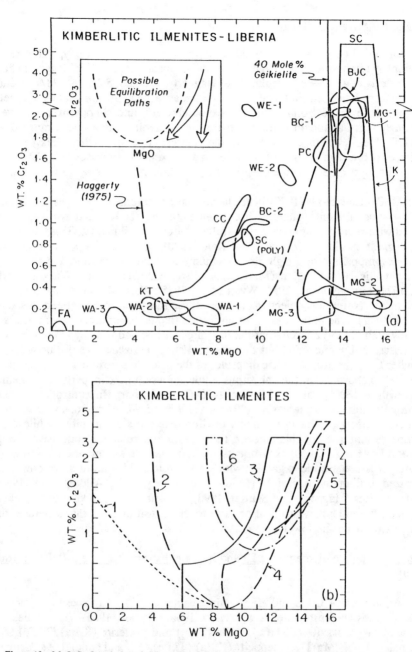

Figure 13. MgO-Cr$_2$O$_3$ relations for ilmenite xenoliths in kimberlites: (a) The parabolic trend (Haggerty, 1975) was established for ilmenites from west and southern Africa; the data from Liberia are for ilmenite xenoliths in dikes and small pipes from a large number of localities (Haggerty, unpublished). The inset shows the effect of subsolidus reduction on ilmenite, which, with the formation of spinel decreases both MgO and Cr$_2$O$_3$. FA is ilmenite associated with freudenbergite; and poly is polycrystalline ilmenite. (b) Hyperbolic and parabolic trends for ilmenite xenoliths in kimberlites from: 1, Iron Mountain (Schulze, 1987); 2, Koidu (Tompkins and Haggerty, 1984a, b); 3, Botswana (Moore, 1987); 4, Sloan and 5, Kampfersdam (Schulze, 1987); 7, Orapa (Shee and Gurney, 1979).

Ilmenite-pyroxene intergrowths

Graphic intergrowths of ilmenite and pyroxene (Chapt. 5) are typical of xenoliths in many kimberlites (Boyd and Nixon, 1973; Gurney et al., 1973; Haggerty et al., 1979; Pasteris et al., 1979; Tompkins and Haggerty, 1984a,b; Schulze, 1987). Xenoliths vary from 1 to 5 cms in length but some are as large as 20 cm. Ratios of ilmenite to pyroxene range from 1:1 to approximately 1:10, and clinopyroxene is overwhelmingly more abundant than orthopyroxene. Ilmenite compositions range from $Ilm_{35}Geik_{53}Hem_{12}$ to $Ilm_{64}Geik_{27}Hem_9$; the greatest population ($Ilm_{49}Geik_{42}Hem_9$) plot close to the high density of discrete ilmenite xenolith analyses (Fig. 12). Coexisting clinopyroxenes average $Wo_{35}En_{50}Fs_{15}$; orthopyroxenes average $Wo_5En_{90}Fs_5$.

McCallister et al. (1975) have shown that the intergrown phases are largely single crystals; ilmenite platelets are oriented along (0001); and pyroxenes are oriented with b crystallographic axes parallel to h0h0 of ilmenite. This coherence might suggest exsolution (Dawson and Reid, 1970), but other proposed origins include: decomposition of a high pressure titanium garnet (Ringwood and Lovering, 1970); eutectic crystallization (MacGregor, 1970; MacGregor and Wittkop, 1970; Boyd, 1971; Gurney et al., 1973; Wyatt et al., 1975; Wyatt, 1977); or metasomatic replacement. Numerous objections to all of these mechanisms have been raised by Mitchell (1977). The only areas of consensus are that equilibration temperatures are generally high (1175-1275° C), that the ilmenites correspond broadly in composition to the discrete ilmenite suite, but that the coexisting pyroxenes are distinctive from the higher temperature, subcalcic diopsides of the discrete pyroxene megacryst suite. Wyatt (1977) showed that the anhydrous solidus of a natural clinopyroxene-ilmenite intergrowth is 1300° C at 20 kbar and 1470° C at 47 kbar. In controlled cooling experiments, clinopyroxene cores (100 to several hundreds of microns in diameter) are surrounded by graphic intergrowths of clinopyroxene and ilmenite. While these experiments hint at a possible eutectic, the experiments remain inconclusive. Ringwood and Lovering's (1970) experiments at ~100 kbar are worthy of attention given that high pressure ultradeep garnets with pyroxene in solid solution have now been recognized in diamond inclusions (Moore and Gurney, 1985; Wilding et al., 1989), and in xenoliths (Haggerty and Sautter, 1990); pressures of ~130 kbar are invoked, equivalent to ~400 km, which is at or close to the transition zone that separates the upper mantle from the lower mantle.

Large ion lithophile element (LILE) and high field strength element (HFSE) oxide minerals

A plethora of LILE and HFSE minerals abound in alkali rocks that include oxides, silicates and carbonates (e.g., Vlasov, 1966; Mitchell, 1985). The oxides are dominated by perovskite ($CaTiO_3$), rutile (TiO_2) and priderite $(K,Ba)_2(Ti,Fe)_6O_{13}$, but armalcolite $(Fe,Mg)Ti_2O_5$, jeppeite $(K,Ba)_2(Ti,Fe)_6O_{13}$, and hollandite structured K and Ba titanates are also present (Mitchell and Haggerty, 1986; Mitchell and Meyer, 1989); other rare minerals include freudenbergite $Na_2Fe_2TiO_{16}$, baddeleyite (ZrO_2), zirconolite ($CaZrTi_2O_7$), zirkelite $(Ti,Ca,Zr)O_{2-x}$, tausonite ($SrTiO_3$), loparite $(Na,Ce,Ca)(TiNb)O_3$, latrappite $(Ca,Na)(NbTi,Fe)O_3$, leuchite $NaNbO_3$, kassite $CaTi_2O_4(OH)_2$, and cafetite $CaFe_2Ti_4O_{12}\cdot4H_2O$; other rock-specific mineral oxides are pyrochlore, complex U,Th,REE titanates, and zirconates (Vlasov, 1966).

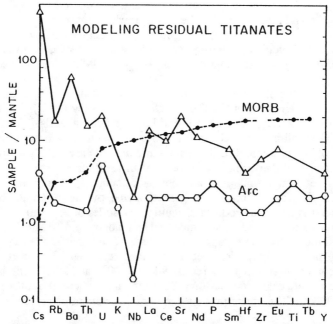

Figure 14. Spider diagram for incompatible elements, illustrating strongly negative Nb anomalies for arc settings coupled with either positive or weakly negative Ba anomalies, and anomalies for Sr, Hf and Ti. These are typical of the elemental distributions used in arguments for (e.g., Foley and Wheller, 1990) or against (e.g., Ryerson and Watson, 1987) residual titanates in the source regions of arc basalts. Selected data are from Ryerson and Watson (1987).

A major outstanding problem in upper mantle petrogenesis is whether the LILE and HFSE are grain-boundary controlled or are stored in high P, high T mineral oxides and silicates. The major constituents of olivine, pyroxene, garnet and spinel in lherzolites, harzburgites and dunites are unsuitable respositories. Potassium is essential in phlogopite and 3 to 4 wt % each of BaO and TiO_2 are possible in substitution. Richterite, edenite and pargasitic amphiboles are K- and Na-bearing. Although both phlogopite and amphibole are stable in the upper mantle, neither mineral is demonstrated to host Sr, Rb, Zr, Nb and REE in significant trace element concentrations. Alkali rocks in general and island arc basalts, andesites and boninites in particular, are characteristically enriched or depleted in certain groups of LILE and HFSE (Fig. 14). Petrogenetic models most commonly call upon residual titanates (rutile or titanite, i.e., "sphene") in the source region (e.g., Gill, 1981), as these minerals may in some cases contain elevated concentrations of Nb, Zr, Hf, Ta and REE (Vlasov, 1966).

It is indisputable that diamond-bearing kimberlites and lamproites are products of partial melting in the upper mantle. It is not, however, obvious that these CO_2 and water charged alkali magmas, enriched in LILE and HFSE, can be solely derived by partial melting of fertile, asthenospheric garnet lherzolite (e.g., Dawson, 1980; Nixon et al., 1981; Mitchell, 1986), unless a vanishingly small (<1%) degree of partial melting is invoked. While small-volume partial melts, which are necessary to concentrate large proportions of silicate incompatible elements, are theoretically possible (McKenzie, 1989), it has yet to be demonstrated that such melts can be mobilized and transported from depths of approximately 200 km through the overlying rigid lithosphere and into the crust. There is a growing consensus that upper mantle metasomatism plays a key role in the genesis of alkali melts (e.g., Menzies and Hawkesworth, 1987), and of kimberlites (e.g., Wyllie, 1987, 1989a,b), lamproites (e.g., Jaques et al., 1989b), and carbonatites (e.g., Haggerty, 1989a,b) in particular.

Focused research on upper mantle metasomatism over the past decade, prompted by the seminal contribution of Lloyd and Bailey (1975), has led to the recognition of an exotic group of LILE and HFSE oxide minerals. These minerals are titanates and fall into four classes: the crichtonite group; the magnetoplumbite group; armalcolite group; and rutile. Spinel and ilmenite are closely associated and figure prominently, either as nuclei for reaction or as products of reaction; ilmenite may also be directly metasomatic. Metasomatised xenoliths in kimberlites are garnet lherzolites, in which spinel forms from garnet, olivine, pyroxene and amphibole (Fig 4f); spinel harzburgites are products of multiple episodes of depletion. The assemblage of exotic oxide minerals is accompanied by phlogopite, K richterite, secondary diopside, and locally by apatite, zircon and calcite. The minerals are in veins (Jones et al., 1982) or are evenly dispersed throughout xenoliths (Erlank et al., 1987). Estimated temperatures and pressures are 900 to 1100°C and 20-30 kbar, placing the most advanced state of oxide metasomatism in the lithosphere at depths of approximately 75-100 km. The entire process is one of enrichment or replenishment.

Crichtonite mineral group

Members of the crichtonite mineral series are classified on the basis of the dominant large cation, $\underline{A}$ formula site in $\underline{AM_{21}O_{38}}$ (Grey et al., 1976). The large cation may be Sr (crichtonite), Ca (loveringite), Na (landauite), U + REE (davidite), Pb (senaite), or Re (unnamed), as defined by Grey and Lloyd (1976), Gatehouse et al. (1978, 1979), Grey and Gatehouse (1978), Kelly et al. (1979), Rouse and Peacor (1968) and Sarp et al. (1981). Barium (lindsleyite) and K (mathiasite) members ("LIMA") are new minerals and are specific to upper mantle xenoliths (Haggerty et al., 1983). The $\underline{M}$, small cation site, in all members of the series is mostly Ti, which accounts for approximately 60% of the $\underline{M}$ formula position; the remaining cations are Fe, Mg, Cr, Nb, Zr, Mn, Zn and V.

The crystal structure of crichtonite (Grey et al., 1976) is based on a framework of close-packed oxygens in a rhombohedral cell. In this structure (space group R3) there are eighteen octahedral cation sites (three nonequivalent sites, M_3, M_4, and M_5, each with a multiplicity of two) occupied primarily by Ti and Fe^{3+}. There are two tetrahedral sites (one non-equivalent site, M_2 with a multiplicity of two) and one additional octahedral site (M_1, with a multiplicity of one). One of the anion sites (M_o) is occupied by a large radius cation, coordinated by 12 oxygens. This gives a total of 22 cations per 38 oxygens, with Z = 1. Hexagonal cell parameters for lindsleyite are a = 1.037 nm, c = 2.052 nm, V = 1.911 nm^3, and for mathiasite are a = 1.035 nm, c = 2.058 nm, V = 1.909 nm^3. Lindsleyite and mathiasite are black with a metallic luster and no cleavage, are comparable to chromite in hardness, similar to chromite in reflection colors of pale gray, but weakly anisotropic in tones of pale tan, approaching ilmenite cut close to the basal section. A characteristic optical property of both minerals is curvilinear conchoidal cracks (Chapter 5), developed on decompression or inherited from precursor chromian spinel as a consequence of crystal volume expansion.

Lindsleyite is recorded in metasomatised spinel harzburgites from the DeBeers and Bultfontein kimberlites in Kimberley, South Africa (Haggerty, 1975; Jones et al., 1982; Haggerty et al., 1983). Mathiasite, although rare at Bultfontein (Haggerty et al., 1983), is relatively abundant in heavy mineral concentrates in the Jagersfontein (Haggerty, 1983a) and Kolonkwanen (Haggerty et al., 1983) kimberlites, South Africa, and in a kimberlite in Shandong Province, China (Zhou et al., 1984).

At all of these localities, whether in veined, unveined or in concentrate, Mg-Al chromite is an associated mineral. Armalcolite, Nb-Cr rutile and Cr ilmenite are associated minerals at Bultfontein and Jagersfontein. Chromium is an essential component of LIMA (Table 7) and reaction of spinel with metasomatising melts enriched in Ba,K,Ti,Zr, Nb and REE, results in either lindsleyite or mathiasite; excess Ti, Fe and Mg are subsequently precipitated in armalcolite, ilmenite and rutile. Lindsleyite is also present in zircon-bearing MARID (mica, amphibole, rutile, ilmenite, diopside-Dawson and Smith, 1977) xenoliths (Haggerty and Gurney, 1984), but these rocks are distinguished from classic MARID by the presence of either olivine or orthopyroxene, and thus bear a much closer affinity to harzburgites. In an unusual spinel harzburgite from Bultfontein, lindsleyite is also associated with hawthorneite (Ba magnetoplumbite), spinel, ilmenite and rutile (Haggerty et al., 1989).

A large-cation plot illustrating the distribution of LIMA in minerals relative to other crichtonite group minerals (Fig. 15a-b) shows that some lindsleyite members, although Ba-dominant and Sr-bearing (crichtonite), trend towards the Ca(loveringite) and REE (davidite) apex; whereas mathiasite, although K-dominant, is also Na (landauite), Ca and REE-bearing.

Small-cation variations are illustrated in Figures 16 and 17a-d. LIMA is distinguished from other members of the crichtonite series (Na-landauite is the exception) by lower Fe + Mg contents (Fig. 16) but is similar to an unidentified mineral from the Bushveld Complex (Cameron, 1978). The Fe + Mg versus Ti plot falls into four quadrants: a non-mantle crichtonite region; the upper mantle LIMA quadrant; and the two armalcolite quadrants. Ilmenite lies to lower values of TiO_2 and rutile to higher concentrations. Chromium-TiO_2 relations (Fig. 17a) are negatively correlated and are separated from other crichtonite minerals (with the exception of loveringite), all of which are Cr-depleted. The significant concentrations of Zr are emphasized in Figure 17b with both the Bushveld phase and loveringite falling into a common field with mathiasite, but not unrelated to lindsleyite. This distribution is repeated in Figure 17c for Ca-Zr. Niobium is variable in mathiasite, with a range of 0.1 to 1.0 wt % Nb_2O_5. These concentrations are relatively minor in comparison to armalcolite (Fig. 17d) or to rutile as discussed later, but the presence of Nb in LIMA distinguishes lindsleyite and mathiasite from other crichtonite series minerals.

Lindsleyite and mathiasite have 10^3 to 10^4 times chondritic abundances of LREE (Jones et al., 1982; Jones and Ekambaram, 1985; Haggerty, 1983a). They are depleted in the HREE, parallel loveringite in this respect, but are distinctive from REE-characteristic davidite which is both LREE- and HREE-enriched (Fig. 18).

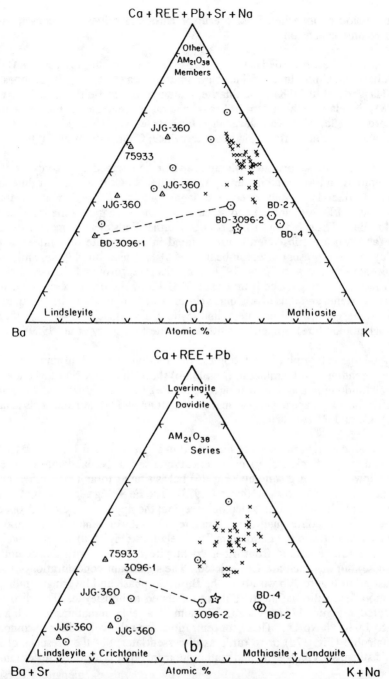

Figure 15. Large-cation plots for members of the $AM_{21}O_{38}$ crichtonite mineral series. Lindsleyite (Ba) and mathiasite (K) are specific to upper mantle metasomites. Approximately 70% of the endmembers are present (a) but significant amounts of Sr, Na, Ca and REE exist in solid solution (b). Only one sample (BD-3096) is recognized to contain both lindsleyite and mathiasite. Data are from Shandong Province, China (Zhou et al., 1984)--star; Bultfontein, South Africa (circles, Jones et al., 1982; triangles, Haggerty et al., 1983); Kolonkwanen (hexagons) and Jagersfontein (crosses) South Africa (Haggerty et al., 1983).

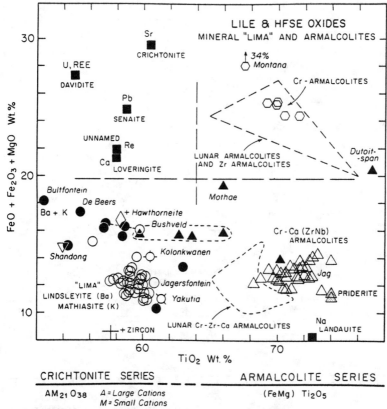

Figure 16. Titanium, iron and magnesium distributions for crichtonite series minerals and armalcolites that divide the diagram into four major quadrants: (1) upper left are non-kimberlitic crichtonites (see text for references) with one exception (landauite (Na) at 8 wt % FeO + Fe$_2$O$_3$ + MgO and 72.5 wt % TiO$_2$); (2) upper right are Cr-armalcolites from Jagersfontein (hexagons), armalcolite from Dutoitspan (Haggerty, 1975), armalcolite in a lamprophyre from Montana (Velde, 1975) and the general field of lunar armalcolites and Zr-armalcolites (Haggerty, 1973); (3) lower right are Cr-Ca (ZrNb) armalcolites from Jagersfontein (open triangles), the field of Cr-Zr-Ca armalcolites from the Moon (Haggerty, 1973), priderite and landauite; and (4) lower left are LIMA minerals in concentrate or metasomites, or in association with hawthorneite, zircon, and as a diamond inclusion (Yakutia). The Bushveld data (Cameron, 1978) and the Mothae kimberlitic point are ill-defined in this plot and may be either crichtonites or armalcolites.

Magnetoplumbite mineral group

Magnetoplumbite (PbFe$_{12}$O$_{19}$) is a very rare skarn mineral. Substitution for Pb and partial substitution of Fe in the structure has received considerable attention in materials science applications for permanent magnets and in lasers. A comprehensive review is given in Collongues et al (1990), with a recent structural refinement by Moore et al. (1989).

Upper mantle magnetoplumbites are new minerals. Yimengite, K(Cr,Ti,Fe,Mg)$_{12}$O$_{19}$ was first reported in heavy mineral concentrates from a kimberlite in Shandong Province, China, where it is associated with magnesian chromite (Dong et al., 1983); mathiasite (K-Cr crichtonite) is reported from the same locality. Yimengite is also described from a kimberlitic sill in the Guaniamo District of

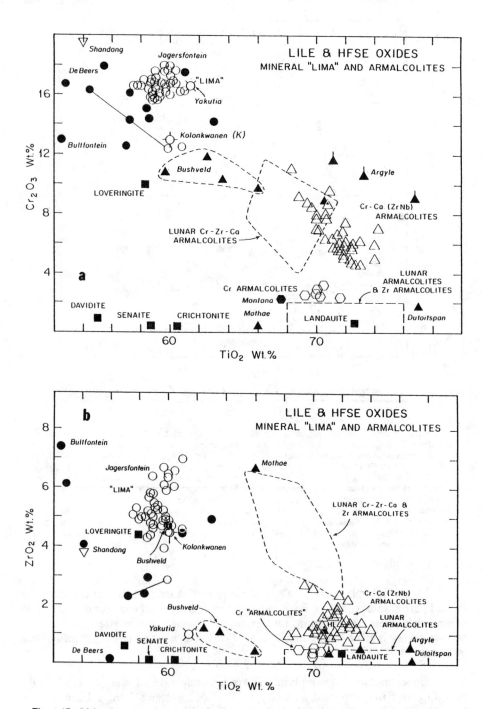

Figure 17. Major element variations for crichtonites and armalcolites distinguished on the basis of TiO$_2$ contents, CaO, Cr$_2$O$_3$, ZrO$_2$ and Nb$_2$O$_5$. Notable features are: (a) a broad negative correlation for TiO$_2$-Cr$_2$O$_3$ between the two mineral species; and (b) TiO$_2$-ZrO$_2$ correlations for kimberlitic crichtonites.

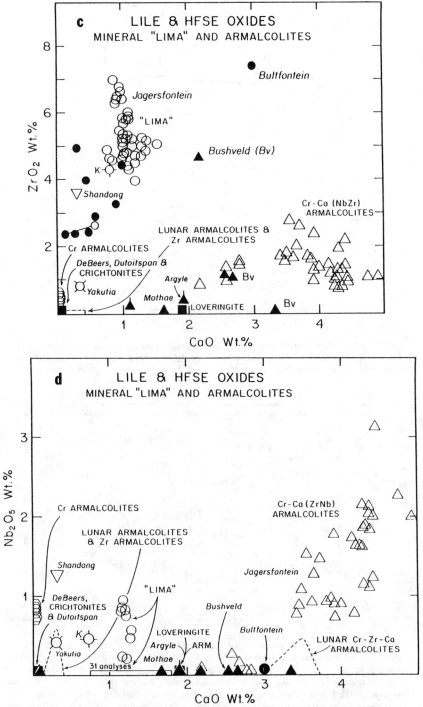

Figure 17, cont. Note (c) the $CaO-ZrO_2$ correlations for kimberlitic crichtonites; and (d) a very prominent $CaO-Nb_2O_5$ relation for Cr-Ca (Zr-Nb)-armalcolites. Symbols and references are similar to those in Figure 16.

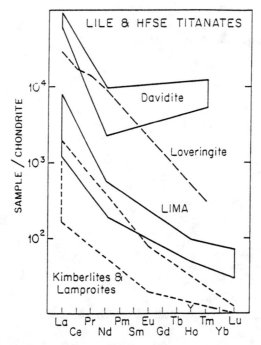

Figure 18. REE abundances and distribution patterns for crichtonite minerals (Jones, 1989), kimberlites and lamproites (Jaques et al., 1986) that emphasize LREE enrichment, and the reasonably close relationship between upper mantle metasomatic LIMA and the hybrid melts of kimberlites and lamproites.

TABLE 7. Lindsleyite, mathiasite, yimengite and hawthorneite in kimberlite xenoliths

	Lind 1	Lind 2	Lind 3	Lind 4	Math 5	Math 6	Yim 7	Yim 8	Haw 9	Lind 10	Magp 11
SiO_2		0.01	0.00	0.00	0.06		0.55	0.52	0.06	0.02	0.78
TiO_2	52.70	59.64	58.35	55.06	53.94	59.32	29.15	30.57	22.94	61.40	22.10
ZrO_2	6.10	3.41	0.26	4.78	3.79	5.35				0.96	
Al_2O_3		0.59	0.05	0.42	0.54	0.62	1.61	3.61		1.02	3.29
Cr_2O_3	16.80	15.10	12.40	14.96	19.83	16.71	37.06	39.37	34.05	16.60	48.80
V_2O_3		0.60			0.29					0.25	
Fe_2O_3				10.75	10.31			2.76	11.31		0.90
FeO	13.00	11.20	13.22	1.66		7.96	18.36	9.88	13.83	7.01	10.00
MgO	3.40	3.32	3.51	3.34	4.05	4.12	7.89	6.19	3.07	3.96	5.45
MnO		0.08	0.14		0.08	0.14	0.00		0.12	0.10	0.10
CaO	0.30	0.80	0.16	0.41	0.31	1.09	0.00			0.29	
SrO	2.40	1.18	1.25	2.37	0.65	0.66		0.36		0.50	4.00
BaO	5.40	2.56	5.91	2.35	1.42	0.96	1.61	2.31	12.52	4.74	0.80
Na_2O		0.10	0.01	0.03	0.01	0.09			0.10		
K_2O	0.20	0.55	0.12	0.19	0.96	1.22	3.75	4.50	0.55	0.43	2.46
Nb_2O_5	0.60	0.02	0.97	0.00	1.17	0.15	0.00		0.16	0.38	0.30
Ta_2O_5			0.01		0.16	0.08			0.18		
Y_2O_3			0.01	0.00		0.01					
La_2O_3			0.52	1.02	0.25	0.19			0.10		
Ce_2O_3			1.56	1.56	0.41	0.35			0.31		
*Other		0.03		0.15	0.31	0.02				0.04	0.07
Total	100.90	99.19	98.45	99.05	98.54	99.04	99.98	100.07	99.30	97.70	99.05

*REE in analyses 2,4,5,6; NiO in analyses 10 and 11. Analysis 1 Bulfontein (Haggerty et al., 1983); 2, average of six analyses, Bultfontein (Jones et al., 1982); 3, associated with hawthorneite, Bultfontein (Haggerty et al., 1986); 4, average of ten analyses, Sover, South Africa (Haggerty, unpublished); 5, associated with yimengite, Shandong Province (Dong et al., 1983); 6, Jagerfontein (Haggerty, 1983a); 7, Shandong Province (Dong et al., 1983); 8, Bolivar Province (Nixon and Condliffe, 1989); 9, Bultfontein (Haggerty et al., 1989); 10 and 11 (possible Sr magnetoplumbite inclusions in diamond (Sobolev et al., 1988); 11 is possibly Sr magnetoplumbite.

Lind = lindsleyite; Math = mathiasite; Yim = yimengite; Haw = hawthorneite; Magp = magnetoplumbite.

Bolivar Province, Venezuela (Nixon and Condliffe, 1989); here again the mineral is closely associated with Mg chromite. Hawthorneite, $Ba[Ti_3Cr_4Fe_4Mg]O_{19}$, the LILE analog of yimengite, is present in a strongly metasomatised harzburgite from the Bultfontein mine, in Kimberley, South Africa (Haggerty et al., 1989); associated minerals are phlogopite, K richterite, and secondary diopside, along with Mg chromite, lindsleyite (Ba-Cr crichtonite), Mg-Cr ilmenite and Nb-Cr rutile. Textural interpretation (Chapter 5), oxide mineral compositions, and X-ray data point to a reaction of hawthorneite formation from chromite + a metasomatizing melt enriched in LILE and HFSE; this reaction, subsequently or in parallel precipitates lindsleyite, ilmenite and rutile (Haggerty et al., 1986; 1989).

The crystal structure of hawthorneite (Grey et al., 1987) shows that it is isostructural with barium ferrite $BaFe_{12}O_{19}$ (Townes et al., 1967) which has the magnetoplumbite structure (Adelskold, 1938). Hawthorneite has hexagonal symmetry, the space group is $P6_3/mmc$, and cell parameters are a = 5.871 Å, c = 23.06 Å, V = 688.33 Å, and Z = 2. Corresponding unit cell dimensions for yimengite are a = 5.857 Å and c = 22.940 Å; and for barium ferrite are a = 5.892 Å, c = 23.183 Å. Hawthorneite is composed of slabs of a spinel-type structure, four anion layers thick, intergrown with perovskite-like BaO_3 close packed layers parallel to $[111]_{Sp}$. The Ba and Ti substitute, respectively, for tetrahedral Mg and octahedral Cr. Chromite and hawthorneite are crystallographically coherent at replacement contacts such that $[111]_{Sp}||[0001]_{Haw}$, and $(22\bar{4})_{Sp}||(30\bar{3}0)_{Haw}$. Hawthorneite and yimengite are black with metallic lusters, cleavage-free, and optical properties and hardnesses comparable to lindsleyite and mathiasite. Hawthorneite in particular is distinguished from LIMA by its greater bireflectance and weaker dispersion.

The compositions of yimengite and hawthorneite (YIHA) are compared in Table 7. Apart from the extraordinarily high BaO (13.4 wt %) contents of hawthorneite and the combined sum of 5 to 7 wt % K_2O, CaO, BaO and SrO in yimengite, hawthorneite is also distinguished from its K analog in small cations by lower Cr_2O_3 and TiO_2 and higher contents of FeO, ZrO_2, and Nb_2O_5.

The xenolith hosting hawthorneite is a complexly veined spinel harzburgite. Lindsleyite is in veins and the rock shows a mineral gradient with distance from the veins: the five-phase assemblage (spinel, hawthorneite, lindsleyite, rutile, ilmenite) is present in single crystals of Mg-Cr spinel at 1 to 2 cm from the veins; and at greater distances, Mg-Cr spinel is optically unaffected. Within the five-phase assemblage there is a systematic decrease in Cr_2O_3 (48 to 31 to 12 wt %) and an increase in TiO_2 (7 to 23 to 58 wt %) contents from spinel to hawthorneite to lindsleyite. This chemical progression, the absence of hawthorneite in veins, the crystallographic coherency between spinel and hawthorneite, and the very nature of the hawthorneite structure, demonstrate that spinel was at the nucleus of the metasomatic reaction that led initially to the formation of hawthorneite and subsequently to the growth of lindsleyite. From the common association of Mg-Cr spinel with yimengite and LIMA, the source of Mg and Cr must be from spinel, whereas Ti and Fe, Zr (in lindsleyite) Nb (in rutile), REE (in lindsleyite) and the LILE elements (Ba,K,Ca,Si,Na) are introduced by volatile-(phlogopite and K richterite) rich metasomatizing melts. Some unusual oxides, that have a common geochemical affinity to the oxides in metasomatised xenoliths are reported in diamond inclusions in kimberlites from Siberia (Sobolev et al., 1988). Lindsleyite was noted previously, but another mineral, approaching Sr magnetoplumbite in composition (Table 7) has equally important implications: (1) for the P-T stabilities of LIMA and YIHA; and (2) for the growth of diamond under metasomatizing conditions. These observations provide substantial support for the metastable growth of skeletal, fibrous mantles of diamond (on diamond); these coats are typically green and opaque with inclusions saturated in hydrous and carbonate minerals and high concentrations of LILE and HFSE (Navon et al., 1988).

Armalcolite mineral group

Armalcolite (FeMg)Ti_2O_5 is ideally the intermediate member of the ferropseudobrookite (FeTi_2O_5)-MgTi_2O_5 (previously named "karrooite") solid solution series, first recognized in lunar high-Ti basalts, subsequently in upper mantle xenoliths, and now more commonly in a wide range of rock types that include impact glasses and metal iron-bearing basalts but also lamprophyres and lamproites (Haggerty, 1987; Bowles, 1988).

Armalcolite is restricted to melt metasomatic assemblages in upper mantle xenoliths (Haggerty, 1983a, and is most typically present in overgrowths on LIMA, ilmenite and rutile or at silicate (olivine, orthopyroxene, zircon)-oxide (ilmenite, rutile) interfaces (Chapter 5). Rare euhedral crystals in macrocrysts of Nb-Cr rutile are recognized only at Jagersfontein; the other assemblages are recorded in Jagersfontein (Haggerty, 1983a), the Kimberley District (Haggerty et al., 1983; Shultze, 1990), and in Argyle (Jaques et al, 1989b; 1990).

Armalcolite compositions (Table 8) may be classified into three groups: type 1 are euhedral crystals of chromian (~2 wt % Cr_2O_3) armalcolite; type 2 armalcolites are in overgrowths or at mineral interfaces, have 5-12 wt % Cr_2O_3, 3-5 wt % CaO, 1-3 wt % Nb_2O_5, and are termed Cr-Ca-(Nb, Zr) armalcolites; type 3 armalcolite is recorded in a diamondiferous peridotite (Argyle) - it bears some compositional similarity to type 2 armalcolite but is distinctive in its high K_2O (1.3-2.1 wt %) content. These chemical variations are apparent in small-cation plots (Figs 16 and 17a-d) that include the fields of lunar armalcolites and selected chromian armalcolites in lamprophyres. An important function of these inter-element relationships is the establishment of a continuum in TiO_2, Cr_2O_3, ZrO_2, and CaO between and among LIMA and armalcolite group minerals. The relationship between Mg-Cr spinel and hawthorneite, and between hawthorneite and lindsleyite was previously established, so that these minerals are not only in contact assemblages but are also geochemically related.

Rutile

Rutile is the most commonly occurring opaque mineral oxide in eclogites (Smith and Dawson, 1975). Rutile is present in peridotites (McGetchin and Silver, 1970; Smith and Dawson, 1975), as inclusions in, or intergrowths with diamond (Sobolev, 1977; Meyer and Svizero, 1975; Prinz et al., 1975), as discrete grains in groundmass kimberlites (Haggerty, 1975; Elthon and Ridley, 1979; Boctor and Meyer, 1979), in MARID and zircon suite xenoliths (Dawson and Smith, 1977); Raber and

TABLE 8. Armalcolite in kimberlite xenoliths

	CrA 1	CCNZ 2	CCNZ 3	CrA 4	MnA 5	KCrA 6	KCrA 7	CrA 8
SiO_2	0.00	0.00	0.06	0.28	0.21	0.08	0.08	0.00
TiO_2	70.12	71.77	71.21	76.92	79.90	73.22	71.09	76.77
ZrO_2	0.44	1.93	0.50	0.00	0.00	0.28	0.27	0.51
Al_2O_3	0.02	0.16	0.76	0.02	1.60	0.89	0.44	0.88
Cr_2O_3	2.65	5.29	6.20	1.64	0.22	10.54	11.67	9.12
FeO	5.91	7.47	6.82	13.47	6.20	5.45	5.67	6.17
Fe_2O_3	6.17	0.00		0.00	0.00			
MgO	13.43	6.30	6.05	7.08	6.10	5.53	6.16	5.59
MnO	0.08	0.11	0.14	0.54	3.70	0.05	0.07	0.06
CaO	0.02	4.31	4.04	0.06	0.00	1.09	1.94	0.00
BaO	0.51	0.77		0.00	0.13	0.34	0.38	0.00
Nb_2O_5	0.85	1.14	0.82	0.00	0.00	0.07	0.04	0.21
Ta_2O_5	0.05	0.00		0.00	0.00			
La_2O_3	0.05	0.18		0.00	0.00			
Ce_2O_3	0.03	0.08		0.00	0.00			
SrO						0.17	0.22	0.00
Na_2O						0.07	0.10	
K_2O						2.10	1.32	0.00
Total	100.33	99.51	96.6	100.01	98.06	99.88	99.45	99.31

Analyses 1-2, Jagersfontein (Haggerty, 1983a); 3, Jagersfontein (Schulze, 1990); 4, DuToitspan (Haggerty, 1975); 5, Yakutia (Tatarintsev et al., 1983); 6-7, Argyle (Jaques et al., 1990); 8, Argyle concentrate (Jaques et al., 1989a). Cr, Mn, and K indicate respectively chromian, manganoan, and potassic armalcolites; CCNZ = chromium - calcium - niobium - zirconium armalcolite.

TABLE 9. Rutile in kimberlite xenoliths

	Di 1	Di 2	Di 3	Di 4	Kim 5	Kim 6	Conc 7	Brec 8	Brec 9	Zir 10	Zir 11
SiO_2	0.05	0.05		0.09					0.00		
TiO_2	98.53	97.85	99.0	99.60	98.10	93.30	92.19	96.09	98.05	94.62	93.87
Al_2O_3				0.27	0.20	0.10		0.06	0.03	0.09	0.00
Cr_2O_3		0.05	0.65	0.11		3.90	4.28	3.09	1.27	1.24	3.50
Fe_2O_3			0.35		0.50	0.90		0.40	0.37		
FeO	0.04	0.82		0.24			2.65			1.19	0.38
MgO				0.06	0.30	0.20		0.12	0.11	0.32	0.02
MnO				0.01	0.00	0.00			0.01	0.01	0.00
CaO	0.03	0.04		0.01					0.01		
ZrO_2	0.04	0.90								1.03	0.33
Nb_2O_5	1.77	0.26			1.20	1.80				1.35	1.45
Ta_2O_5										0.22	0.10
Total	100.46	99.97	100.00	100.39	100.30	100.20	99.12	99.77	99.84	100.07	99.65

	+Ilm 12	+Ilm 13	+Ilm 14	+Ilm 15	+Ilm 16	+Ilm 17	MRD 18	MRD 19	+LIMA 20	+ARM 21	+LIMA 22
SiO_2							0.10	0.01	0.00	0.00	
TiO_2	85.26	78.81	73.94	70.46	66.26	64.30	95.9	95.3	79.18	91.96	88.60
Al_2O_3	0.04	0.08	0.03	0.04	0.06	0.01	0.00	0.00	0.23	0.00	0.00
Cr_2O_3	5.97	6.76	7.75	7.29	7.78	8.21	0.70	3.16	7.40	2.15	5.92
Fe_2O_3							2.70	1.20			0.00
FeO	1.71	1.83	2.39	2.98	3.76	4.10			3.45	0.17	1.72
MgO	0.00	0.18	0.24	0.64	0.47	0.41	0.30	0.08	0.63	0.01	0.80
MnO				0.01	0.09	0.05		0.02	0.06	0.02	0.00
CaO								0.01		0.01	
ZrO_2	0.86	0.76	0.95	1.12	1.18	1.35			0.91	1.56	0.50
Nb_2O_5	6.94	10.46	13.38	15.74	18.28	20.90			7.65	2.24	2.44
Ta_2O_5	0.30	1.25	1.50	1.21	1.82	1.54				0.13	0.00
Total	101.08	100.13	100.18	99.49	99.70	100.87	99.70	99.78	99.51	98.25	99.98

Analyses 1-2, Argyle (Jaques et al., 1989a), 3, Zaire (Mvuemba et al., 1982); 4, Brazil (Meyer and Svisero, 1975); 5-6 Tunraq, Canada (Mitchell, 1979); 7, Kimberley (Clement et al., 1979); 8-9, Bultfontein and DeBeers (Wyat and Lawless, 1984); 10-11 Bultfontein (Haggerty, unpublished); 12-17 Orapa (Tollo and Haggerty, 1987); 18-19 Bultfontein (Dawson and Smith (1977); 20-21, Jagersfontein (Haggerty, 1983a); 22, Bultfontein (Haggerty, unpublished). Di = inclusions in diamond; Kim = xenocrysts in kimberlite; Conc = concentrate; Brec = polymict breccia; Zir = zircon-bearing MARID peridotite; +Ilm = rutile + oriented lamellar or symplecti ilmenite; MRD = MARID (mica, amphibole, rutile, ilmenite, diopside; +Arm = rutile + armalcolite; +LIM = rutile + lindsleyite-mathiasite.

Haggerty, 1979), as discrete xenoliths (Tollo, 1982; Tollo and Haggerty, 1987: Ottenburgs and Fieremens, 1979; Haggerty, 1983a), and in olivine-rutile symplectic intergrowths (Haggerty, 1983a; Schultze, 1990). The compositional characteristic of all mantle-derived rutiles is the rarity of stoichiometric TiO_2, and the abundances of Fe and Cr which may each be present in oxide concentrations of 2-5 wt % or greater (Table 9). A second distinctive feature is high concentrations of niobium, with values as large as 21 wt % Nb_2O_5 recorded in the Orapa kimberlite (Tollo and Haggerty, 1987).

Rutile is an essential mineral in MARID xenoliths; these rocks are considered by Dawson and Smith (1975), and Waters and Erlank (1989) to be the condensed melts responsible for upper mantle metasomatism, but the proposition remains uncertain. Rutile is also an essential mineral in one style of metasomatism, the type locality is the Matsoku kimberlite in Lesotho with the assemblage ilmenite, rutile, phlogopite, spinel and sulfides (IRPS, Harte et al., 1987). This assemblage (PRAIS in association with armalcolite) is also present at Bultfontein (Erlank et al., 1987); at both localities, the oxides, sulfides and phlogopite are in anastomizing veins.

Apart from primary rutile in eclogites and rutile in eclogitic suite diamonds (some crustal eclogites have rutile concentrations approaching ore deposit dimensions, e.g., Korneliussen and Foslie, 1985), all other rutiles in upper mantle-derived xenoliths are closely associated with phlogopite, or phlogopite and amphibole. The LIMA, YIHA and armalcolite mineral groups are also rutile-bearing, and the case for upper mantle metasomatism is irrefutable for the origin of these assemblages.

A ubiquitous texture in rutiles from the upper mantle, diamond inclusions being the exception, are sigmoidal lamellae of ilmenite (Chapter 5). This intergrowth may reflect either decomposition or replacement. Metasomatic rutiles may be classified into nine groups: (1) massive macrocrystic rutile xenoliths (up to 5 cm in length) having margins in graphic intergrowth with olivine; (2) macrocrystic rutile with coarse (100-200 micron) ilmenite lamellae in trellis networks; (3) macrocrystic ilmenite with sandwich laths of ilmenite; (4) rutile veins in discrete ilmenite xenoliths; (5) massive rutile with isolated inclusions of LIMA; (6) rutile + hawthorneite; (7) secondary rutile derived from LIMA, in association with spinel ilmenite and unidentified zirconates and titanates, (8) rutile + armalcolite; and (9) rutile + Nb titanite + latrappite (Nb perovskite).

The dominant substituting cations in upper mantle rutiles are Nb and Cr, but note that iron is also significant; niobium and tantalum are typical of rutiles in alkali rocks, specifically pegmatites (e.g., Cerny and Ercit, 1985; 1989), whereas rutiles in eclogites, diamond inclusions, lunar rocks and meteorites are generally depleted in Nb + Cr + Ta (Fig. 19). Niobium-chromium relations, as a function of texture, or LIMA, YIHA and armalcolite association (Fig. 20) show that the three major compositional populations are: (1) discrete Nb-Cr rutiles from the Orapa kimberlite (inset and positive correlation), (2) rutile macrocrysts with abundant, intergrown ilmenite (trellis, sandwich, irregular, sigmoidal); and (3) rutile-dominant, ilmenite-minor assemblages with LIMA, YIHA and armalcolite. The geochemical consanguinity among spinel, YIHA, LIMA and armalcolite shown previously can now be extended to chromian rutile with the distinction that Nb is strongly partitioned into TiO_2.

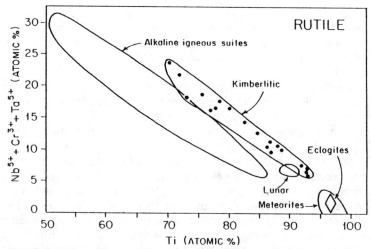

Figure 19. Niobium, Cr and Ta versus Ti to illustrate the distinctive compositional fields for rutiles. Upper mantle rutiles in kimberlite xenoliths are enriched in Nb + Cr in metasomites relative to rutiles in other alkalic rocks where high contents of Nb + Ta are typical.

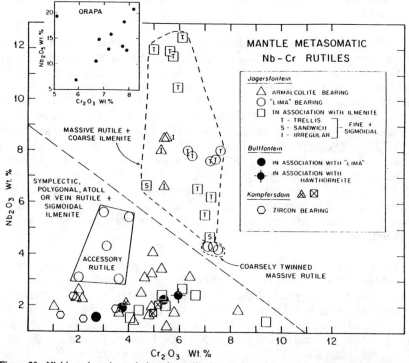

Figure 20. Niobium-chromium relations for rutiles in a variety of textural forms and associations. The plot is divided into three major fields: an upper field of rutile + ilmenite (Orapa); an intermediate field of massive rutile with coarse lamellar intergrowths of ilmenite in trellis, sandwich or irregular form (+ fine sigmoidal lenses); and a lower field dominated by armalcolite and ilmenite in association with rutile. The lower field also contains rutiles in symplectic intergrowths (and other textural forms) with silicates, and accessory rutile derived by decomposition of LIMA. Data sources are Haggerty (1983a), Tollo and Haggerty (1987), Haggerty et al. (1986) and Shulze (1990).

Among the outstanding problems related to the genesis and crystal chemistry of upper mantle rutiles are: (1) the origin of symplectic intergrowths of rutile and olivine; (2) the exsolution-like nature of ilmenite in rutile; and (3) the substitutional mechanism of Nb and Cr and the valence state of iron in rutile.

The graphic, rutile-olivine, and rarely rutile-orthopyroxene, intergrown xenoliths have not been identified in host rocks, and these assemblages, in common with ilmenite-pyroxene intergrowths remain enigmatic. Schulze (1983) has interpreted the olivine-rutile intergrowths as products of the decomposition of titanoclinohumite, $Mg_{16}Si_8O_{32}Mg(OH)_2TiO_2$, to $8\ Mg_2SiO_4$ (olivine) = TiO_2 (rutile) + $Mg(OH)_2$ (brucite). This suffers from the absence of brucite in all documented assemblages and the incorrect modal proportions of olivine:rutile. An alternative explanation is crystallization from a Ti-rich magma (Schulze, 1990). Melts generated in the upper mantle are inexcapably Ca- and Al-bearing, neither of which are present in the silicate-oxide assemblages. The experimental data (MacGregor, 1969) called upon to justify the intergrowths in this model show that a field of olivine + rutile alone does not exist. Bulk compositions appropriate to the assemblage are either liquid or armalcolite-bearing (strictly Mg_2TiO_5); however, armalcolite is mostly a reaction product between rutile and olivine (Chapter 5), and evidence for a discrete liquid is absent. A third possible origin is purely metasomatic (Haggerty, 1983a) but the issue remains open.

Trellis, sandwich and sigmoidal intergrowths of ilmenite in rutile bear some of the same problems of interpretation that ilmenite lamellae have to hosts of titanomagnetite (Chapter 5); notwithstanding crystallographic coherency, the latter are demonstrated to be products of oxidation and not exsolution (Buddington and Lindsley, 1964). Exsolution sensu stricto is unsatisfactory because of the limited solid solubility of FeO or Fe_2O_3 in TiO_2 (Chapter 3). Replacement of rutile by ilmenite is an interesting alternative (Pasteris, 1980), as is the decomposition of armalcolite, $(FeMg)Ti_2O_5$ to $(FeMg)TiO_3$ (ilmenite) + TiO_2 (rutile). The latter was tentatively proposed by Tollo and Haggerty (1987), but demonstrated to be inconclusive; because homogenization experiments (at P = 1 bar, 1000°C to 1400°C and fO_2 = 10^{-5} to 10^{-13} bars) of a coarse lamellar intergrowth of ilmenite in rutile, although producing armalcolite, persistently had rutile in excess. High pressures may be required but neither oxidation nor reduction can be invoked and replacement appears unlikely.

Even if the ilmenite constituent of these intergrowths is ignored, the concentrations of Cr and Nb (and Fe) are manifestly in excess of those nominally accommodated in the rutile structure. A possible alternative to normal rutile structures are crystallographic shear-based structures (CS). The type compound at high pressures is $\alpha\ PbO_2$ (Bursill et al., 1971). Crystallographic shear structures in the series M_nO_{2n-1} - M_nO_{3n-2} (where $\underline{M}$ = metal and n = 1,2,3) or $Cr_2Ti_{n-2}O_{2n-1}$ (where n = 3,4,5), described by Grey and Reid (1972) and Grey et al. (1973), are considered to be ideal candidates. For n = 4, the compound is $Cr_2Ti_2O_7$. In the system Cr_2O_3-TiO_2 (Gibb and Anderson, 1972) three distinct structural regions are present: (1) rutiles having 5 wt % or lower Cr_2O_3 contents are structurally homogeneous; (2) CS-based structures dominate in the range of 5-15 wt % Cr_2O_3; and (3) anionic deficient TiO_{2-x} Magneli phases exist with 15-17 wt % Cr_2O_3. The CS structures are demonstrated to persist with high ZrO_2 and Fe_2O_3 contents (Grey et al., 1973). Substitution of iron in $Cr_2Ti_2O_7$ to form $CrFeTi_2O_7$ is structurally related to αPbO_2 and may be considered to derive from it by crystallographic shear (Grey and Mumme, 1972). Niobium may also be accommodated leading to oxygen excess or metal deficiencies at high oxygen fugacities, but stoichiometric compounds at lower fO_2 values (Dirstine and Rosa, 1979a, b). Perhaps the most meaningful insight is provided by Bursill and Jun (1984) who conclude that $(TiCr)O_{2-x}$ is a "true non-stoichiometric compound"; this effectively extends the possible defect structures outlined by Gibb and Anderson (1972), but requires to be demonstrated in upper mantle rutile.

Apart from these unusual substitutions, Rossman and Smyth (1990) and Vlassopolus et al. (1990) show that up to 0.8 wt % H_2O is present in nominally anydrous rutile. These high concentrations are recorded in metasomatic rutiles from Jagersfontein, but significant concentrations of OH are also present in Nb-Cr rutiles, of probable metasomatic origin, in eclogites from the Roberts Victor kimberlite, South Africa (Rossman and Smyth, 1990).

<u>Ilmenite and spinel in LILE and HFSE mineral oxide assemblages</u>

Ilmenite and spinel display complex textures (Chapter 5) and have unusual chemistries (Figs. 21 and 22) in association with LIMA, YIHA, armalcolite and Nb-Cr rutile. Geikielite contents range from 34-44 mole % for the lamellar and sigmoidal ilmenites in rutile, 50-76 mole % in association with armalcolite, and 50-77 mole % in LIMA and hawthorneite assemblages. Chromium contents are extraordinary, and range from 2 to 8 wt % Cr_2O_3; these ilmenites are superseded only by a very unusual assemblage of ilmenite + lamellar spinel and lensoidal rutile in an unspecified kimberlite (Wyatt, 1979). Note that the Cr_2O_3-MgO relations shown in Figure 21 are markedly more enriched and show both positive and negative trends in comparison to the discrete ilmenite xenolith suite (Fig. 13a,b). Extensive substitution of Cr into ilmenite can evidently only be accomplished at elevated pressures (P > 20 kbar) and is temperature-dependent (Green and Sobolev, 1975; Wyatt, 1978). The marked negative correlations for Cr-Mg have yet to be resolved, but it appears that this relation and the parabolic or hyperbolic trends shown by the discrete xenolith suite depend on Mg-Cr partitioning with coexisting spinel. Partitioning is more complex than this simple substitution because Cr is also present in LIMA, hawthorneite, rutile, and armalcolite.

Typical compositions for discrete spinels are 52-62 wt % Cr_2O_3, 11-22 wt % FeO, 2-10 wt % Fe_2O_3 and 8-14 wt % MgO. Titanium in these spinels is low (0-3.5 wt %), intermediate in association with armalcolite (5-7 wt %), and relatively high (3-12 wt %) in YIHA and LIMA-bearing xenocrysts. The most aluminous are discrete spinels (max 17 wt % Al_2O_3), followed by Al-depleted LIMA and YIHA spinels (11 wt % max), and armalcolite spinels (4-6 wt %). These compositional variations are described in the alpha-numeric format in Figure 22. Discrete spinel xenocrysts are 1B and 4C. A single LIMA spinel is barely classified as 2B (approaching 3C) but the remainder are 3C. Thus, there is a marked distinction between the armalcolite and LIMA spinels, and the xenocrystic spinels. The Mg + Fe^{3+} component in discrete spinels is dominantly magnesioferrite, whereas armalcolite, YIHA and LIMA spinels are enriched in Mg_2TiO_4 (qandilite) with magnesioferrite as a minor com-

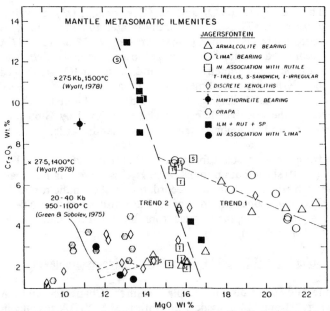

Figure 21. Cr_2O_3-MgO relations for metasomatic ilmenites in discrete xenoliths and other mineral associations. Full squares (Wyatt, 1979) are for ilmenites containing spinel + rutile lamellae, an assemblage that may have been derived from decomposed crichtonite. Full circles are from Jones et al. (1982) for ilmenite in association with lindsleyite and armalcolite. Experimental data are by Green and Sobolev (1975) and Wyatt (1978). Trend 1 is defined by ilmenites in association with LIMA and some armalcolites. Trend 2 is dominated by data from Wyatt (1979) but also includes discrete ilmenite xenolith data, ilmenites in association with rutile, and ilmenites coexisting with armalcolite (Haggerty, 1983a).

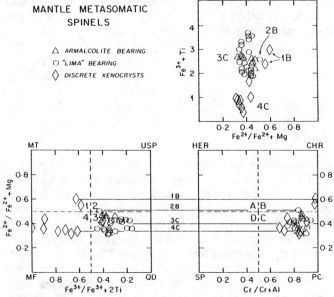

Figure 22. Metasomatic (residual or reacted) spinels plotted onto projections of the alpha-numeric, spinel trapezoid (Fig. 10). These plots enable a compositional distinction to be made between unreacted (1B and 4C) and reacted (3C) spinels (compare with Figure 10), and permit an evaluation to be made (without assumptions) of the dominant spinel endmembers present.

ponent. This distinction is indicative of somewhat lower fO_2 conditions for the exotic mineral assemblages than those of discrete spinel xenocrysts.

Priderite and related minerals

Although priderite $(K,Ba)(Ti_7Fe^{3+})O_{16}$ is relatively widespread and is typical of the groundmass mineralogy of leucite lamproites (e.g., Mitchell, 1985), priderite of upper mantle derivation is linked to metasomatism and is described in a peridotite in the Arkansas lamproite (Mitchell and Lewis, 1983), and in a discrete macrocryst with armalcolite from the Argyle lamproite, Australia (Jaques et al., 1989b). Compositionally, the priderites differ in Ba:K ratios, and in Cr_2O_3 (9% at Argyle and <0.1 wt % at Arkansas), and Fe^{3+} contents (Table 10).

Two phases that may be related to priderite are reported by Jones et al. (1982) as reaction rims on LIMA and ilmenite. Barium oxide is approximately equal to K_2O in one of these phases, but K_2O (9.4 wt %) is substantially larger than BaO (0.8 wt %) in the other; TiO_2 and FeO contents are less than priderite, but it is noteworthy that both minerals are Cr_2O_3-enriched (8-17 wt %). Complex and zoned reaction mantles around LIMA are also reported by Haggerty (1983a). Another, and as yet unidentified mineral occurs in reactions between zircon and rutile (Mothae, Lesotho), bearing some (e.g., BaO = 2.37 wt %; K_2O = 0.52 wt %; ZrO_2 = 6.71 wt %), but not all (e.g., Cr_2O_3 is only 0.1 wt %), of the chemical characteristics of LIMA and yet clearly deficient in large cations to qualify as priderite or a member of a possible priderite or jeppeite solid solution series. A phase with lower TiO_2 (51-54 wt %), but higher FeO (15-16 wt %) and BaO (16-17 wt %) contents is described by Scatena-Wachel and Jones (1984) in the Benfontein kimberlite sills in association with baddeleyite, Cr spinel and ilmenite. In all of the above cases, small grain sizes (20-50 microns) have precluded x-ray study and considerable uncertainty remains in mineral identification, stability or conditions of formation.

Baddeleyite

Baddeleyite (ZrO_2) is a geographically widespread secondary mineral in kimberlites (Kresten et al., 1975), and present as thin microcrystalline to cryptocrystalline coatings on zircon; it is assumed to result from the desilicification of zircon either through reaction with kimberlite, or in reaction with other minerals in the presence of carbonate. Discrete primary crystals (up to 1 cm) of baddeleyite in kimberlite are described from Mbuji-Mayi, Zaire (Fieremans and Ottenburgs, 1979), and as 50 micron grains in the Benfontein kimberlite sills, South Africa (Scatena-Wachel and Jones, 1984). Reaction between ilmenite and zircon from Monastery, South Africa and the Mothae kimberlite in Lesotho yields Ti-baddeleyite, diopside, calcite and zirconolite (Raber and Haggerty, 1979).

The compositions of baddeleyites in kimberlites, carbonatites, gabbros and lunar highland KREEP basalts are summarized by Haggerty (1987). Primary baddeleyite in kimberlite has a maximum of 0.8 wt % TiO_2, whereas in ilmenite-zircon reactions, the range is 3.7-6.1 wt % TiO_2 (Table 10). Hafnium is not commonly reported, and some of the low analytical totals may lack the 1-3 wt % HfO_2 that is characteristic of the mineral, particularly for baddeleyites in reaction contact with

TABLE 10. Priderite, baddeleyite, zirconolite, and freudenbergite in kimberlite xenoliths

	Prid 1	Prid 2	Prid 3	Bad 4	Bad 5	Bad 6	Zirc 7	Zirc 8	Zirc 9	Freu 10	Freu 11
SiO_2	0.27		0.00	0.00	0.00	0.04	0.00	0.00	0.00	0.00	0.01
TiO_2	74.54	72.20	72.66	3.68	6.10	0.32	29.16	16.70	40.48	80.10	74.61
ZrO_2	0.04		0.19	93.22	90.63	96.40	51.12	71.27	41.91	0.10	1.39
Al_2O_3			1.02	0.51	0.11	0.00	0.13	0.04	0.23	0.01	0.03
Cr_2O_3			9.06	0.18	0.11	0.00	1.10	0.06	0.09	0.08	1.02
FeO	8.59	11.30	3.32	0.68	0.63	0.43	6.94	2.38	5.15	9.20	7.27
Fe_2O_3											
MgO	0.07	1.00	0.96	0.19	0.09	0.06	1.20	0.31	0.49	0.75	2.07
MnO			0.00	0.18	0.05		0.11	0.15	0.11	0.02	0.01
CaO	0.24		0.00	0.00	0.09	0.65	0.31	10.11	9.09	11.10	0.42	0.01
BaO	8.67	6.70	0.93							0.59	0.70
Na_2O	0.04		0.03							8.75	8.91
K_2O	6.30	9.10	9.33							0.06	0.40
Nb_2O_5	0.68		0.17							0.01	2.16
Ta_2O_5			0.22								0.45
Ce_2O_3			0.29								0.12
HfO_2						1.72					
Total	99.44	100.30	98.18	98.73	98.37	99.28	99.87	100.00	99.56	100.09	99.16

Analysis 1, inclusion in diamond, Argyle (Jaques, et al., 1989 c); 2, peridotite Arkansas (Mitchell and Lewis (1983); 3, Argyle concentrate (Jaques et al., 1989b); 4, (Monastery), 5 (Mothae), Raber and Haggerty (1979); 6, Benfontein (Scatena-Wachel and Jones, 1984); 7-9, Bultfontein and Lesotho (Raber and Haggerty, 1979); 1 (Liberia), 11 (Bultfontein), Haggerty (1983b and unpublished).

zircon. Phlogopite and K richterite are commonly present and a metasomatic origin is implied.

The crystal symmetry of baddeleyite is temperature-dependent: it is monoclinic at 1,170° C, and tetragonal at 1,676° C, which is the upper limit of zircon stability at 1 bar (Butterman and Foster, 1967). Baddeleyite may also be cubic, and symmetry varies as a function of the Zr:O ratio (Rauh and Garg, 1980). It is stable at high pressures as evidenced from zircon decomposition in impactite glasses (El Goresy, 1965; Kleinman, 1969). The ocherous layers around kimberlitic zircons are monoclinic, but data are required on baddeleyite in reaction assemblages between zircon and ilmenite. These are also predictably monoclinic, but if shown to be otherwise would provide an important temperature indicator for such reactions.

Zirconolite

Confusion persists on whether zirconolite and zirkelite are the same mineral or distinct species despite the clear distinction by Vlasov (1966) and Bayliss et al. (1989) that zirconolite ($CaZrTi_2O_7$) differs from zirkelite $(Zr,Ca,Ti)O_{2-x}$ on the basis of cation:anion ratios and the characteristic presence of REE, Th and U. Substitution of the last named elements appears to be possible in zirconolite but limits remain to be established. Upper mantle-derived zirconolite is non-metamict.

In kimberlites, zirconolite has been identified only in zircon-ilmenite reactions (Raber and Haggerty, 1979) of the type illustrated in Chapter 5. At Monastery, reaction assemblages include baddeleyite and calcite, but phlogopite, diopside, olivine and orthopyroxene are also present. Kimberlitic zirconolites (Table 10) vary widely in the ratio of Ti:Zr (0.361 - 1.760), relative to other compositions that have ratios of 1.17 - 1.52. Most zirconolites contain FeO (2-9 wt %), although this is not an essential constituent of the mineral. Zirconolites in pyroxenites and in KREEP-rich basalts and lunar anorthosites, are characterized by high Nb_2O_5 contents, and these as well as zirconolites in gabbro are LREE-bearing.

The Ti:Zr ratio of zirconolite is temperature-dependent with low values (e.g., 1.0) stable at high synthesis temperatures (1550°C) and higher values (e.g., 2.0) at corresponding lower (1400°C) temperatures (Wark et al., 1973). The experimental data are for ideal compositions and yield unreasonably high (1450-1500°C) formation temperatures when applied to naturally occurring zirconolites. Reaction assemblages involving ilmenite and zircon in the presence of phlogopite and calcite is linked to metasomatism.

Freudenbergite

Freudenbergite ($Na_2Fe_2Ti_6O_{16}$) was first described in apatite-rich syenites from the Katzenbuckel alkali complex, Germany (Frenzel, 1961; Frenzel et al., 1971; McKie, 1963; and McKie and Long, 1970). The ferrous iron analogue ($Na_2FeTi_7O_{16}$) is recorded in lower crustal granulites from Liberia (Haggerty, 1983b) and in metasomatic zircon-bearing xenoliths from Bultfontein (Haggerty and Gurney, 1984). Examples illustrated in Chapter 5 show a serial sequence from cores of rutile to overgrowths of freudenbergite and ilmenite; in granulites this is followed by perovskite and titanite. Freudenbergite is accompanied by LIMA in mantle xenoliths in an assemblage of olivine, pyroxene, phlogopite, amphibole and calcite with 1 to 5 modal % zircon. Freudenbergite is important in the inventory of upper mantle LILE and HFSE minerals because it is Na and Nb-bearing (Table 10). With concentrations of 8 to 9 wt % Na_2O, freudenbergite is the most sodic of the metasomatic mineral suite (amphibole contains 3-4 wt % Na_2O, and LIMA has a maximum of about 0.5 wt % Na_2O); in the entire array of upper mantle minerals Na in freudenbergite is matched only by Na in eclogitic omphacite.

Freudenbergite (sensu stricto) is stable at high water pressures (1 kbar) and high fO_2 (MH) at T = 700-900°C (Flower, 1974). The upper mantle and lower crustal freudenbergites may have formed at comparable temperatures but fO_2 was substantially lower (<FMQ) because Fe^{3+} is extremely low or absent; substitution between $Na_2Fe^{3+}Ti_6O_{16}$ and $Na_2Fe^{2+}Ti_7O_{16}$ is accomplished by $2Fe^{3+}$ for Fe^{2+} + Ti.

Petrogenetic implications.

The upper pressure stabilities of armalcolite (Lindsley et al., 1974; Kesson and Lindsley, 1975; Friel et al., 1977) and K richterite (Kushiro and Erlank, 1970) demonstrate that these phases probably formed at 20-30 kbar or less, and at a maximum of 900-1100° C. Neither Nb nor Cr were present in the armalcolite experiments and both elements are likely to stabilize armalcolite to higher pressures. Lindsleyite and mathiasite have been synthesized at 1300° C and 20 kbar; and lindsleyite was also synthesized at 900° C and 22 kbar, but not at 1300° C and 35 kbar, or at 1 bar (Podpora and Lindsley, 1984). No experimental data are yet available on yimengite and hawthorneite but the close association of hawthorneite with LIMA suggests that stabilities of YIHA are probably not very different from LIMA.

These data place the most advanced form of metasomatism (LILE and HFSE oxides + phlogopite + K richterite + diopside) at depths of about 75 to 100 km in the depleted, subcontinental lithosphere. Haggerty (1989a, b) has argued for an extensive metasome horizon in cratonic lithospheres, and in McKenzie's (1989) view, it is a globally encompassing metasome. If these suggestions are correct, then the wide array of LILE and HFSE oxides identified from the upper mantle are likely to play an important role in the geochemical trace element signatures of alkali magmas. The development of metasomes is largely due to the liberation of the more volatile components of melts trapped in the lithosphere, and from melts that fail to exceed the CO_2-H_2O peridotite solidus, which has a large "shoulder" to it in the region of 20-30 kbar (Eggler, 1989; Wyllie, 1989a,b). The ultimate source of volatiles and LILE and HFSE is the deeper upper mantle, specifically the transition zone which is a region of subducted slab accumulation (e.g., Ringwood, 1990), or the lower mantle if plume activation is assumed (e.g., Anderson, 1989).

There is a general consensus for the link between mantle metasomatism and alkali magmatism (e.g., Menzies and Hawkesworth, 1987). In the context of this review it is relevant to link the high pressure upper mantle LILE and HFSE minerals in xenoliths with similarly exotic minerals in extrusive alkali rocks (Figs. 23a and b). The most significant features are the relative enrichments of Ba and K in the alkali suite relative to LIMA (Haggerty, 1989a,b; Jones, 1989). It is important to note that the minerals are all titanates. For BaO-TiO_2 relations (Fig. 23a), melt-derived, LILE and HFSE oxides in alkali rocks are negatively correlated and form a distinct field from LIMA. Note that the Cr priderites from Argyle and Bultfontein are displaced slightly towards LIMA and away from more typical priderite. A distinctive feature of LIMA, YIHA, armalcolite, rutile, and ilmenite is high Cr values for upper mantle metasomites. The behavior of chromium during fusion should be similar regardless of whether it is spinel or a compositionally exotic titanate. Deriving the alkali oxide suite from nominal LIMA, therefore, indeed appears possible.

Important experimental contributions to the question of residual titanates (or accessory phases) in the upper mantle during the extraction of partial melts have been made by Green (1981), Green and Pearson (1986a,b; 1987), Hellman and Green (1979), and Ryerson and Watson (1987). Rutile and titanite (i.e., "sphene") have received particular attention but spinel and ilmenite have also been considered. In these experiments the melts are commonly saturated with a titanate mineral. The TiO_2 contents of liquids (mafic, intermediate, and felsic, in the range 900-1100° C and P = 7.5 to 30 kbar for a range in fO_2) was shown by Green and Pearson (1986a) to decrease with increasing T, increasing P, and increasing SiO_2, alkali and REE contents; increasing fO_2 lowers the TiO_2 content of the liquid. In the study by Ryerson and Watson (1987), rutile saturation is achieved in basalt, andesite and dacite at 7-9 wt %, 5-7 wt %, and 1-3 wt % TiO_2, respectively, for a range of P, T and environmental conditions. These authors conclude that residual titanates cannot be responsible for the HFSE distribution patterns displayed by magmas generated at convergent plate boundaries (Fig. 14). Foley and Wheller (1990) on the other hand have evaluated five possible models for the generation of arc magmas, and conclude that rutile is indeed a candidate residual titanate in the upper mantle. Other candidate minerals, specifically LIMA, or LIMA in association with armalcolite, Nb-Cr rutile and Mg-Cr ilmenite should also be considered. HFSE depletion in arc

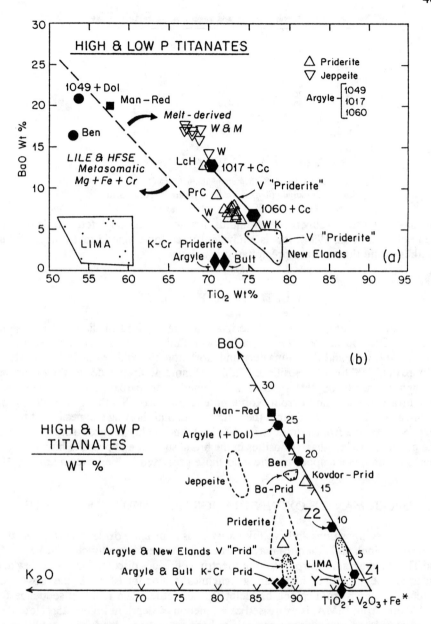

Figure 23. (a) Barium (large-cation) versus Ti (small-cation) plot for upper mantle-derived metasomatic LILE titanates in xenoliths (to the left of the dashed line), relative to the field of melt-derived LILE titanates in lamproites and kimberlites (to the right of the dashed line). **Man-Red** manardite-redledgeite series; **Bul**. Bultfontein, **Ben**. Benfontein, and **New Elands** are all in South Africa; **Argyle** is in Australia. Data base is from Haggerty (1989a). (b) K is essential in yimengite (Y) and mathiasite; Ba is essential in hawthorneite, lindsleyite, the manardite-redledgeite series and a host of other poorly characterized minerals (J,Ben,Z1,Z2) in mantle xenoliths; K + Ba are essential in priderite and jeppeite; and Ti is a major component in all of these minerals. It is from these plots (a and b) that a high pressure (upper mantle) to low pressure (melt emplaced into the crust) compositional link (see also Fig. 24) is proposed for the exotic suite of LILE and HFSE oxide minerals (Haggerty, 1989a,b). Kovdor is a carbonatite (USSR); other localities are given in (a).

TABLE 11. Ferropericlase and magnesiowustite in diamond

	1	2	3	4	5
SiO_2	0.08				
TiO_2					0.17
Al_2O_3		0.10		0.01	0.17
Cr_2O_3	0.49	0.84	0.57	0.16	
FeO	21.70	19.40	19.80	21.13	93.00
MgO	77.30	78.70	78.10	77.70	7.29
MnO	0.15	0.32	0.19	0.15	0.32
CaO		0.04			
Na_2O	0.29	0.07	0.30		
K_2O					
NiO			1.41	1.20	
Total	100.01	99.47	100.37	100.35	100.95

Analyses 1,3, and 5, Koffiefontein, South Africa (Moore et al., 1986); 2, Sloan, Colorado (ibid); 4, Ororoo, Australia (Scott - Smith et al., 1984).

magmas may also possibly result by liquids interacting with depleted mantle peridotite (Kelemen et al., 1990); this is equivalent to cratonic mantle metasomatism but whether exotic titanates are precipitated is unknown.

WÜSTITE AND PERICLASE

Although rare, wüstite and periclase are recognized in diamond inclusions from the Ororoo kimberlite, Australia (Scott-Smith et al., 1984), the Sloan kimberlite, Colorado, and the Monastery and Koffiefontein kimberlites in South Africa (Moore et al., 1986). Compositions (Table 11) more accurately define the inclusions as magnesiowüstite and ferropericlase. Syngenetic orthopyroxene and ferropericlase exist in a single diamond from Koffiefontein, and Scott-Smith et al. (1984), and Moore et al. (1986) conclude that the suite of diamonds hosting magnesiowüstite and ferropericlase are from ultradeep (>400 km), and highly reduced source regions. The minerals are less than curiosities in the broad spectrum of possible high pressure phases in the deeper upper mantle, and those predicted in the lower mantle.

UPPER MANTLE OXIDE STRATIGRAPHY AND REDOX STATE

The compositions and evolutionary trends of primary oxide minerals (wüstite, periclase, spinel, and ilmenite) may be traced in Figure 24 as a function of Fe + Mg and Ti. The diagram incorporates the spectrum of oxides discussed in the review and illustrates the potentially important geochemical region for LIMA at the crossroads to reaction trends from spinel, baddeleyite, and rutile. The reaction sequence for LIMA, and from LIMA, is expressed as a function of depth in Figure 25. Note that zircon-associated assemblages are calcite-bearing and are, therefore, stabilized at lower P and T.

The stratigraphic distribution of the major oxide minerals in the upper mantle is illustrated as a function of tectonic setting and xenolith sampling in Figure 26. The spinel mineral group can be traced throughout the lithosphere. Spinels are present in the upper asthenosphere in all settings except the asthenosphere below cratons where garnet lherzolite is stable rather than spinel lherzolite; the equivalent spinel lherzolite horizon in cratonic settings is lithospheric spinel harzburgite or spinel

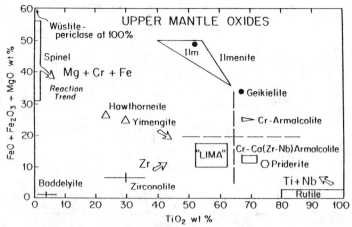

Figure 24. Compositional distribution of mineral oxides in the upper mantle as a function of total iron + MgO versus TiO$_2$. Arrows indicate trends of reaction: one trend is from spinel (in depleted harzburgites) which on reaction with metasomatizing melts enriched in LILE, TiO$_2$ (but also Nb) and ZrO$_2$ may form LIMA, YIHA, priderite and armalcolite; some Ti may also derive from ilmenite. Armalcolites may be expressed as rutile + ilmenite; LIMA and YIHA are equivalent to spinel + armalcolite; and priderite (or jeppeite) as LIMA or YIHA + rutile.

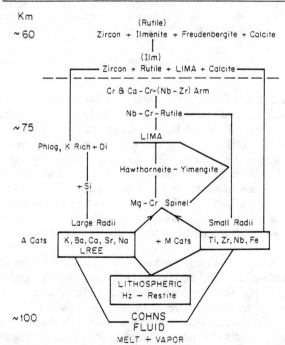

Figure 25. Detailed evolutionary paths for the development of metasomatic minerals from volatile-rich melts introduced into depleted lithospheric harzburgite at depths of 60-100 km. Replacement of Mg-Cr spinel (e.g., Fig. 24) by large radius (A) cations, and small (M) radius cations yield YIHA and LIMA, either sequentially or in parallel, followed by, or simultaneously with the precipitation of Nb-Cr rutile and armalcolite species. Zircon-bearing metasomites, zirconolite and baddelyite are typically accompanied by calcite and have lower P and T stabilities than LIMA and YIHA. The silicate track is also a large cation reaction trend producing phlogopite (K + Al), K richterite (K + Na), and diopside (Ca).

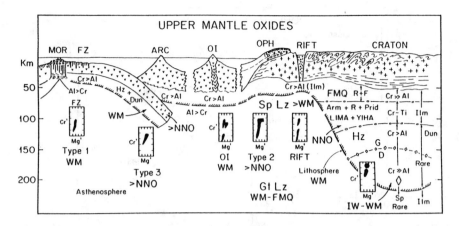

Figure 26. Regional tectonic setting for the sampling of upper mantle xenoliths and accompanying oxide minerals from mid-ocean ridge (MOR) and fracture zones (FZ), to arc, ocean islands (OI), ophiolite (OPH) complexes, continental rifts and stable cratons. Chrome-aluminum and cr'(Cr/[Cr + A]) and mg'(Mg/[Mg + Fe]) are generalized to reflect the degree of depletion or fertility as expressed by the compositions of spinels (Figs. 2-5) for the lithosphere (mantle) and the asthenosphere. The deep asthenosphere is garnet lherzolite (Gt Lz), shallow asthenosphere is spinel lherzolite (Sp Lz), and lithosphere is dominated by depleted harzburgite (Hz) and dunite (Dun). Types 1 and 3 refer to oceanic settings for MOR and arc generated ophiolites (Dick and Bullen, 1984); Type 2 is transitional and is of uncertain origin. Spinels are present in all settings; ilmenite is common in cratonic settings, rare in rift-generated xenoliths, and typically absent elsewhere; armalcolite, rutile and priderite, and YIHA + LIMA are identified only in cratonic, lithospheric xenoliths; however, priderite, armalcolite, jeppeite, ilmenite, rutile and perovskite are typical of alkalic melts (kimberlites, lamproites, lamprophyres, melilitites) generated from metasomatized upper mantle lithosphere in cratonic and rift settings. Redox states are expressed as IW (iron-wüstite); WM (wüstite-magnetite); FMQ (fayalite-magnetite-quartz).

dunite. The ilmenite mineral group, which is both primary or metasomatic, is prevalent in xenoliths from cratons, but is extremely rare in all other settings. The compositionally exotic LILE and HFSE oxides are recorded only from subcontinental, subcratonic metasome horizons (75-100 km). The deepest, the greatest diversity in mineral oxides, and possibly the most complete stratigraphic record of upper mantle mineral oxides are from xenoliths in kimberlites; alkali basalts provide a detailed cross-section of oxides in rifted but shallow upper mantles; and abyssal fracture zones, and ophiolites provide a critical record of oxides in the uppermost mantle. The oxide minerals reflect a broader geochemical view of the upper mantle: it is heterogeneous, both laterally and vertically, and has evolved with time.

This geochemical evolution has had profound effects on the redox state of the sub-continental sub-cratonic lithosphere and asthenosphere (Haggerty and Tompkins, 1983; Haggerty, 1990). Ranges in oxidation state are equal to or greater (Fig. 26) than those described for the shallower upper mantle (Chapter 11). This has resulted in part from the great antiquity of sub-cratonic lithospheric keels, in part from the growth of relatively oxidized metasomes, and in part from the highly reduced redox state that is inferred for diamond growth and preservation, at the base of the keel (Fig. 26). The keel is geochemically depleted in basaltic components (e.g., Ca,Al, Fe,Si) and in volatiles. Spinel is the major source of ferric iron and is in very low concentrations. By contrast, the metasomes contain spinels from the decomposition of a variety of silicate minerals (Fig. 4f); furthermore hydrous minerals such as phlogopite and amphibole, and also OH-bearing rutile are present. While there has been considerable debate on the redox state of the upper mantle, it is clear that neither a single nor an absolute value can or should be assigned. The range in upper mantle redox conditions varies between two log units below WM to two log units above FMQ (Haggerty, 1990). Among the more interesting questions to be addressed in future studies include: (1) What is the redox state of the transition zone? and (2) are plumes oxidized or reduced?

REFERENCES

Adelskold, V. (1938) X-ray studies on magneto-plumbite, $PbO.6Fe_2O_3$ and other substances resembling 'beta-alumina', $Na_2O.11Al_2O_3$. Arkiv for Kemi Mineralogioch Geologi Series A-12 no 29, 1-9.

Agee, J.J., Garrison, J.R. and Taylor, L.A. (1982) Petrogenesis of oxide minerals in Kimberlite, Elliott County, Kentucky. Am. Mineral. 67, 28-42.

Alevizos, A.A. and Sclar, C.B. (1988) The chemical composition of chromite from the Skoumtsa-Xerolivado chromite deposits, Vourinos ophiolitic complex, Greece, as a tectonic indicator. Proceedings of the Seventh Quadrennial IAGOD Symposium, 67-74, Schweizerbart'sche Verlagsbuchhandlung, Stuttgart.

Anderson, D.L. (1989) Theory of the Earth. 366p Blackwell Scientific, Oxford.

Aranda-Gomez, J.J. and Ortega-Gutierrez, F. (1987) Mantle xenoliths in Mexico. In: P.H. Nixon, ed., Mantle Xenoliths 75-84. John Wiley & Sons, New York.

Auge, T. (1987) Chromite deposits in the northern Oman ophiolite: Mineralogical constraints. Mineral. Deposita 22, 1-10.

Bayliss, P., Mazzi, F., Munno, R. and White, T.J. (1989) Mineral nomenclature: zirconolite. Mineral. Mag. 53, 565-569.

Beccaluva, L., Macciotta, G., Siena, F. and Zeda, O. (1989) Harzburgite-lherzolite xenoliths and clinopyroxene megacrysts of alkaline basic lavas from Sardinia. Chem. Geol. 77, 331-345.

Boctor, N.Z. and Meyer, H.O.A. (1979) Oxide and sulfide minerals in kimberlite from Green Mountain, Colorado. In: F.R. Boyd and H.O.A. Meyer, eds., Kimberlites, Diatremes and Diamonds: Their Geology, Petrology, and Geochemistry, 217-228. Am. Geophys. Union, Washington, D.C.

Bonatti, E. and Michael, P. (1989) Mantle peridotites from continental rifts to ocean basins to subduction zones. Earth Planet. Sci. Lett. 91, 297-311.

Bonatti, E., Ottonello, G. and Hamlyn, P.R. (1986) Peridotites from the island of Zabargad (St. John), Red Sea: Petrology and geochemistry. J. Geophys. Res. 91, 599-631.

Bowles, J.F.M. (1988) Definition and range of composition of naturally occurring minerals with the pseudobrookite structure. Am. Mineral. 73, 1377-1383.

Boyd, F.R. (1971) Enstatite-ilmenite and diopside-ilmenite intergrowths from the Monastery Mine. Carnegie Inst. Washington Yr. Bk. 70, 134-138.

Boyd, F.R. and Nixon, P.H. (1973) Origin of the ilmenite-silicate nodules in kimberlites from Lesotho and South Africa. In: P.H. Nixon, ed., Lesotho Kimberlites 254-268. Lesotho National Development Corporation, Maseru.

Buddington, A.F. and Lindsley, D.H. (1964) Iron-titanium oxide minerals and synthetic equivalents. J. Petrol. 5, 310-357.

Bursill, H.A., Hyde, B.G. and Philip, D.K. (1971) Crystal structure in the 132CS family of higher titanium oxides. Phil. Mag. 23, 1501-1511.

Bursill, L.A. and Jun, S.G. (1984) Temperature dependence of the solubility of chromia (Cr_2O_3) in titania (TiO_2). J. Solid State Chem. 51, 388-395.

Butterman, W.C. and Foster, W.R. (1967) Zircon stability and the ZrO_2-SiO_2 phase diagram. Am. Mineral. 52, 880-885.

Cameron E.M. (1978) An unusual titanium-rich oxide mineral from the Eastern Bushveld Complex. Am. Mineral. 63, 37-39.

Carswell, D.A., Griffin, W.L. and Kresten, P. (1984) Peridotite nodules from the Ngopetsoeu and Lipelaneng kimberlites, Lesotho: a crustal or mantle origin. In: J. Kornprobst, ed., Kimberlites: The Mantle and Crust-Mantle relations, 229-243, Elsevier, Amsterdam.

Carswell, D.A., Clarke, D.B. and Mitchell, R.H. (1979) The petrology and geochemistry of ultramafic nodules from Pipe 200, northern Lesotho. In: F.R.Boyd and H.O.A. Meyer, eds., The Mantle Sample: Inclusions in Kimberlites and other Volcanic Rocks, 127-155. Am. Geophys. Union, Washington, D.C.

Casey, J.F., Elthon, D.L., Siroky, F.X., Karson, J.A. and Sullivan, J. (1985) Geochemical and geological evidence bearing on the origin of the Bay of Islands and Coastal Ophiolites of Western Newfoundland. Tectonophysics 116, 1-40.

Cerny, P. and Ercit, T.S. (1985) Some recent advances in the mineralogy and geochemistry of Nb and Ta in rare-element granite pegmatites. Bull. Mineral. 108, 499-532.

Cerny, P. and Ercit, T.S. (1989) Mineralogy of niobium and tantalum: Crystal chemical relationships, paragenetic aspects and their economic implications. In: P. Moller, P. Cerny and F. Sallpe, eds., Lanthanides, Tantalum and Niobium, 27-79. Springer-Verlag. Berlin.

Cervilla, F. and Leblanc, M. (1990) Magmatic ores in high temperature Alpine-type lherzolite massifs (Rondo, Spain and Beni Bousera, Morocco). Econ. Geol. 85, 112-132.

Christodoulou, C. and Hirst, D.M. (1985) The Chemistry of chromite from two mafic-ultramafic complexes in northern Greece. Chem. Geol. 49, 415-428.

Clague, D.A. (1988) Petrology of ultramafic xenoliths from Loihi Seamount, Hawaii. J. Petrol. 29, 1161-1186.

Clement, C.R., Skinner, E.M., Hawthorne, J.B. and Kleinjan, L. (1975) Precambrian ultramafic dikes with kimberlite affinities in the kimberley area. In F.R. Boyd and H.O.A. Meyer, eds., Kimberlites, Diatremes and Diamonds: Their Geology, Petrology, and Geochemistry, 101-110. Am. Geophys. Union, Washington, D.C.

Collongues, R., Courier, D., Kahn-Harari, A., Lejus, A.M., Thery, J. and Vivien, D. (1990) Magnetoplumbite-related oxides. Ann. Rev. Mater. Sci. 20, 51-82.

Cotterill, P. (1979) The chromite deposits of Selukwe, Rhodesia. In: H.D.B. Wilson, ed., Magmatic Ore Deposits, Econ. Geol. Mono. 4, 154-186.

Danchin, R.V. and Boyd, F.R. (1976) Ultramafic nodules from the Premier kimberlite pipe, South Africa, Ann. Rep. Dir. Geophys. Lab., Carnegie Inst. Washington, Yr. Bk., 75, 531-535.

Daniels, L.R.M. and Gurney, J.J. (1989) The chemistry of the garnets, chromites and diamond inclusions of the Dokolowayo kimberlite, Kingdom of Swaziland. Geol. Soc. Australia, Spec. Pub. 14, 1012-1021.

Daniels, L.R.M. and Gurney, J.J. (1991) Oxygen fugacity constraints of the Southern African lithosphere as deduced from concentrate minerals, xenoliths and diamond inclusions from the Dokolwayo and other kimberlites from Southern Africa. Contrib. Mineral. Petrol. (in press).

Dawson, J.B. (1980) Kimberlites and Their Xenoliths. Springer-Verlag, Berlin.

Dawson, J.B. and Reid, A.M. (1970) A pyroxene-ilmenite intergrowth from the Monastery Mine, South Africa. Contr. Mineral. Petrol. 26, 296-301.

Dawson, J.B. and Smith, J.V. (1975) Chromite-silicate intergrowths in upper-mantle peridotites. Phy. Chem. Earth 9, 339-350.

Dawson, J.B. and Smith, J.V. (1977) The MARID (mica-amphibole-rutile-ilmenite-diopside) suite of xenoliths in kimberlite. Geochim. Cosmochim. Acta 41, 390-323.

Dawson, J.B., Smith, J.V. and Delaney, J.S. (1978) Multiple spinel-garnet peridotite transitions in the upper mantle: evidence from a harzburgite xenolith. Nature 273, 741-743.

Dick, H.J.B. and Bullen, T. (1984) Chromian spinel as a petrogenetic indicator in abyssal and Alpine-type peridotites and spatially associated lavas. Contrib. Mineral. Petrol 86, 54-76.

Dick, H.J.B. and Fisher, R.L. (1984) Mineralogical studies of the residues of mantle melting: Abyssal and alpine-type peridotites. In: J. Kornprobst, ed., Kimberlites II: The Mantle and Crust-Mantle Relationships. 295-308, Elsevier, Amsterdam.

Dickey, J.S. (1975) A hypothesis of origin for podiform chromite deposits. Geochim. Cosmochim. Acta 39, 1061.

Dickey, J.S. and Yoder, H.S. (1972) Partitioning of chromium and aluminum between clinopyroxene and spinel. Ann. Rep. Dir. Geophys. Lab Yr. Bk. 71, 384.

Dirstine, R.T. and Rosa C.J. (1979a) Defect structure and related thermodynamic properties of nonstoichiometric rutile (TiO_{2-x}(rutile)) and partial molar properties for oxygen solution at 1273°K. Zeit. Metallkde. 70, 322-329.

Dirstine, R.T. and Rosa, C.J. (1979b) Defect structure and related thermodynamic properties of nonstoichiometric rutile (TiO$_{2-x}$) and Nb$_2$O$_5$ doped rutile. Pt II The defect structure of TiO$_2$ solid solution containing 0.1 - 3.0 mole% Nb$_2$O$_5$ and partial molar properties for oxygen solution at 1273°K. Zeit. Metallkde. 70, 372-378.

Dong, Z. and Zhou, J. (1980) The typomorphic characteristics of chromites from kimberlites in China and the significance in exploration of diamond deposits. Acta Geol. Sinica 4, 284-298.

Dong, Z., Zhou, J., Lu, Q. and Peng, Z. (1983) Yimengite K (Cr, Ti, Fe, Mg)^{12}O^{19} - a new mineral. Kexue Tongbao (Bul. Sci) 15, 932-936. (Foreign language edition 29, 7, 920-923, Pub. Academy Sinica).

Eggler, D.H. (1989) Carbonatites: Primary melts and mantle dynamics. In: K. Bell, ed., Carbonatites: Genesis and Evolution. 561-579. Unwin Hyman, Boston.

Eggler, D.H. and Baker, D.R. (1982) Reduced volatiles in the system C-O-H: Implications to mantle melting, fluid formation and diamond genesis. In: S. Akimoto and M.H. Manghani, eds., High Pressure Research in Geophysics 237-250, Center Acad. Pub., Tokyo.

Ehrenberg, S.N. (1982) Petrogenesis of garnet lherzolite and megacrystalline nodules from the Thumb, Navajo volcanic field. J. Petrol. 23, 507-547.

El Goresy, A. (1965) Baddeleyite and its significance in impactite glasses. J. Geophys. Res. 70, 3453-3456.

Elthon, D. and Ridley, W.I. (1979) The oxide and silicate mineral chemistry of a kimberlite from the Premier Mine: Implications for the evolution of kimberlite magmas. In: R.R. Boyd and H.O.A. Meyer, eds., Kimberlites, Diatremes and Diamonds: Their Geology, Petrology and Geochemistry, 206-216. Am. Geophys. Union, Washington, D.C.

Engin, T. and Hirst, D.M. (1970) The Alpine chrome ores of the Andizlik--Zimparalik area, Fethiye, southwest Turkey. Trans Inst. Min. and Mett. Sec B, 79, 16-29.

Erlank, A.J., Waters, F.W., Hawkesworth, C.J., Haggerty, S.E., Allsopp, H.L., Rickard, R.S. and Menzies, M.A. (1987) Evidence for mantle metasomatism in peridotite nodules from the Kimberley pipes, South Africa. In: M.A. Menzies and C.J. Hawkesworth, eds., Mantle Metasomatism, 221-312. Academic Press Geology Series, New York.

Eugster, H.P. and Wones, D.R. (1962) Stability relations of the ferruginous biotite, annite. J. Petrol. 3, 81-125.

Evans, B.M. and Frost, B.R. (1975) Chrome-spinel in progressive metamorphism--a preliminary analysis. Geochim. Cosmochim. Acta 39, 959-972.

Exley, R.A., Smith, J.V. and Dawson, J.B. (1983) Alkremite, garnetite, and eclogite xenoliths from Bellsbank and Jagersfontein, South Africa. Am. Mineral. 68, 512-516.

Fabries, J. Bodinier, J.L., Dupuy, C., Lorand, J.P. and Benkerrou, C. (1989) Evidence for modal metasomatism in the orogenic spinel lherzolite body from Caussou (northeastern Pyrenees, France). J. Petrol. 30, 199-228.

Fabries, J., Figueroa, O. and Lorand, J.P. (1987) Petrology and thermal history of highly deformed mantle xenoliths from the Montferrier Basanites, Languedoc, southern France: A comparison with ultramafic complexes from the north Pyrenean Zone. J. Petrol 28, 887-919.

Ferguson, J., Ellis, D.F. and England, R.N. (1977) Unique spinel-garnet lherzolite inclusion in kimberlite from Australia. Geology 5, 278-280.

Field, S., Haggerty, S.E. and Erlank, A.J. (1989). Subcontinental metasomatism in the region of Jagersfontein, South Africa. Geol. Soc. Australia, Spec. Pub 14, 771-785.

Fieremans, M. and Ottenburgs, R. (1979). The occurrence of zircon and baddeleyite crystals in the kimberlite formations, at Mbuji-Mayi (Bakwanga, Zaire). Bull. Soc. Belge de Geol. 88, 25-31.

Flower, M.J. (1974). Phase relations of titan-acmite in the system Na$_2$O-Fe$_2$O$_3$-Al$_2$O$_3$-TiO$_2$-SiO$_2$ at 1,000 bars total water pressure. Am. Mineral. 59, 536-548.

Foley, S.F. and Wheller, G.E. (1990) Parallels in the origin of the geochemical signatures of island arc volcanics and continental potassic igneous rocks: The role of residual titanates. Chem. Geol. 85, 1-18.

Frenzel, G. (1961) Ein neues mineral: Freudenbergit (Na$_2$Fe$_2$Ti$_7$O$_{18}$). N. Jb. Mineral. Mh, 1961, 12-22.

Frenzel, G., Ottemann, J. and Nuber, B. (1971) Meue mikrosonden-untersuchun-gen an Freudenbergit. N. Jb. Mineral. Monatsh, 1971, 547-551.

Friel, J.J., Harker, R.I. and Ulmer, G.C. (1977) Armalcolite stability as a function of pressure and oxygen fugacites. Geochim. Cosmochim. Acta 41, 403-410.

Gatehouse, B.M., Grey, I.E., Campbell, I.H. and Kelly, P. (1978) The crystal structure of loveringite-a new member of the crichtonite group. Am. Mineral. 63, 28-36.

Gatehouse, B.M., Grey, I.E. and Kelly, P.R. (1979) The crystal structure of davidite from Arizona. Am. Mineral. 64, 1010-1017.

Gibb, R.M. and Anderson, J.S. (1972) The system TiO_2-Cr_2O_3: Electron microscopy of solid solutions and crystallographic shear structures. J. Solid State Chem. 1, 445-453.

Gill, J.B. (1981) Orogenic Andesites and Plate Tectonics. Springer, Berlin.

Green, D.H. and Sobolev, N.V. (1975) Coexisting garnets and ilmenites synthesized at high pressures from pyrolite and olivine basanite and their significance for kimberlitic assemblages. Contrib. Mineral. Petrol. 50, 217-229.

Green, T.H. (1981) Experimental evidence for the role of accessory phases in magma genesis. J. Volc. Geotherm. Res. 10, 405-422.

Green, T.H. and Pearson, N.J. (1986a) Ti-rich accessory phase saturation in hydrous mafic-felsic compositions at high P.T. Chem. Geol. 54, 185-201.

Green, T.H. and Pearson, N.J. (1986b) Rare-earth element partitioning between sphene and coexisting silicate liquid at high pressure and temperature. Chem. Geol. 55, 105-119.

Green, T.H. and Pearson, N.J. (1987) An experimental study of Nb and Ta partitioning between Ti-rich minerals and silicate liquids at high pressure and temperature. Geochim. Cosmochim. Acta 51, 55-62.

Grey, I.E. and Gatehouse, B.M. (1978) The crystal structure of landauite, Na $[MnZn_2 (Ti, Fe)_6 Ti_{12}]O_{38}$. Can. Mineral. 16, 63-68.

Grey, I.E. and Lloyd, D.J. (1976) The crystal structure of senaite. Acta Crystallogr. B32, 1509-1513.

Grey, I.E., Lloyd, D.J. and White, Jr., J.S. (1976) The crystal structure of crichtonite and its relationship to senaite. Am. Mineral. 61, 1203-1212.

Grey, I.E., Madsen, I.C. and Haggerty, S.E. (1987) Structure of a new upper-mantle, magnetoplumbite-type phase, $Ba[Ti_3Cr_4Fe_4Mg]O_{19}$. Am. Mineral. 72, 633-636.

Grey, I.E. and Mumme, W.G. (1972) The structure of $CrFeTi_2 O_7$. J. Solid State Chem. 5, 168-173.

Grey, I.E. and Reid, A.F. (1972) Shear structure compounds $(Cr, Fe)_2 Ti_{n-2} O_{2n-1}$ derived from the PbO_2 structural type. J. Solid State Chem. 4, 186-194.

Grey, I.E., Reid, A.F. and Alpress, J.G. (1973) Compounds in the system Cr_2O_3-Fe_2O_3-TiO_2-Zr_2 based on intergrowth of the αPbO_2 and V_3O_5 structural types. J. Solid State Chem. 8, 86-99.

Gurney, J.J. (1989) Diamonds. In: Kimberlites and Related Rocks. Geol. Soc. Australia Spec. Pub. 14, 935-965.

Gurney, J.J., Fesq, H.W. and Kable, E.J.D. (1973) Clinopyroxene-ilmenite intergrowths from kimberlite: a re-appraisal. In: P.H. Nixon, ed., Lesotho Kimberlites 238-253. Lesotho National Development Corporation, Maseru, Lesotho.

Gutmann, J.T. (1986) Origin of four- and five-phase ultramafic xenoliths from Sonora, Mexico. Am. Mineral. 71, 1076-1084.

Haggerty, S.E. (1973) Armalcolite and genetically associated opaque minerals in the lunar samples. Proc. 4th Lunar and Planet. Sci. Conf., Suppl. 4. Geochim. Cosmochim. Acta 1, 777-797.

Haggerty, S.E. (1975) The chemistry and genesis of opaque minerals in kimberlites. Phys. Chem. Earth 9, 295-307.

Haggerty, S.E. (1979) Spinels in high pressure regimes. In: Boyd, F.R. and Meyer, H.O. The Mantle Sample: Inclusions in kimberlites and other volcanic rocks, 183-196. Am. Geophys. Union, Washington, D.C.

Haggerty, S.E. (1983a) The mineral chemistry of new titanates from the Jagersfontein kimberlite, South Africa: Implications for metasomatism in the upper mantle. Geochim. Cosmochim. Acta 47, 1833-1854.

Haggerty, S.E. (1983b) A freudenbergite-related mineral in lower crustal granulites from Liberia. N. Jb. Mineral. Monatsh, 375-384.

Haggerty, S.E. (1986) Diamond genesis in a multiply constrained model. Nature 320, 34-38.

Haggerty, S.E. (1987) Metasomatic mineral titanates in upper mantle xenoliths. In: P.H. Nixon, ed., Mantle Xenoliths, 671-690. John Wiley & Sons, New York.

Haggerty, S.E. (1989a) Upper mantle opaque mineral stratigraphy and the genesis of metasomites and alkali-rich melts. Geol. Soc. Australia, Spec. Pub. 14, 687-699.

Haggerty, S.E. (1989b) Mantle metasomes and the kinship between carbonatites and kimberlites. In: K. Bell, ed., Carbonatites--Genesis and Evolution 546-560. Allen and Unwin, London.

Haggerty, S.E. (1990) Redox state of the continental lighosphere. In: M.A. Menzies, ed., Continental Mantle. 87-109. Claredon Press, Oxford.

Haggerty, S.E., Erlank, A.J. and Grey, I. (1986) Metasomatic mineral titanate complexing in the upper mantle. Nature 319, 761-763.

Haggerty, S.E., Grey, I.A., Madsen, I.C., Criddle, A.J., Stanley, C.J. and Erlank, A.J. (1989) Hawthorneite, $Ba[Ti_3Cr_4Fe_4Mg]O_{19}$: A new metasomatic magnetoplumbite-type mineral from the upper mantle. Am. Mineral. 74, 668-675.

Haggerty, S.E. and Gurney, J.J. (1984) Zircon-bearing nodules from the upper mantle, (Abs.) EOS Trans. Am. Geophys. Union 65, 301.

Haggerty, S.E., Hardie, R.B., McMahon, B.M. (1979) The mineral chemistry of ilmenite nodule associations from the Monastery diatreme. In: F.R. Boyd and H.O.A. Meyer, eds., The Mantle Sample: Inclusions in Kimberlites and other Volcanics, 249-256. Am. Geophys. Union, Washington, D.C.

Haggerty, S.E., Moore, A.E. and Erlank, A.J. (1985) Macrocryst Fe-Ti oxides in olivine melilitites from Namaqualand-Bushmanland, South Africa. Contrib. Mineral. Petrol. 91, 163-170.

Haggerty, S.E. and Sautter, V. (1990) Ultradeep (greater than 300 kilometers), ultramafic upper mantle xenoliths. Science 248, 993-996.

Haggerty, S.E., Smyth, J.R., Erlank, A.J. Rickard, R.S., Danchin, R.V. (1983) Lindsleyite (Ba) and mathiasite (K): Two new chromium-titanates in the crichtonite series from the upper manmtle. Am. Mineral. 68, 494-505.

Haggerty, S.E. and Tompkins. L.A. (1982) Sulfur solubilities in mantle-derived nodules from kimberlites. (Abs.) EOS Trans. Am. Geophys. Union 63, 463.

Haggerty, S.E. and Tompkins, L.A. (1983) Redox state of Earth's upper mantle from kimberlitic ilmenites. Nature 303, 295-300.

Haggerty, S.E. and Tompkins, L.A. (1984a) Subsolidus reactions in kimberlitic ilmenites: exsolution, reduction and the redox state of the mantle. In: J. Kornprobst, ed., Kimberlites 1 - Kimberlites and Related Rocks, 83-105. Elsevier Science Pub., Amsterdam.

Haggerty, S.E. and Tompkins, L.A. (1984b) Subsolidus reactions in kimberlitic ilmenites: exsolution, reduction and the redox state of the mantle--Appendix. In: J. Kornprobst, ed., Ann. Scientif. De L'Univ. De Clermont-Ferrand, Kimberlites III: Documents 74, 141-148.

Harte, B., Winterburn, P.A. and Gurney, J.J. (1987) Metasomatic and enrichment phenomena in garnet peridotite facies mantle xenoliths from the Matsoku kimberlite pipe, Lesotho. In: M.A. Menzies and C.J. Hawkesworth, Mantle Metasomatism, 145-220, Academic Press, London.

Hellman, P.L. and Green, T.H. (1979) The role of sphene as an accessory phase in the high-pressure partial melting of hydrous mafic compositions. Earth Planet. Sci. Lett. 42, 191-201.

Hops, J., Gurney, J.J., Harte, B. and Winterburn, P. (1989) Megacrysts and high temperature nodules from the Jagersfontein kimberlite pipe. Geol. Soc. Australia, Spec. Pub. 14, 759-770.

Irvine, T.N. (1965) Chromian spinel as a petrogenetic indicator. Part 1. Theory. Can. J. Earth Sci. 2, 648-672.

Irvine, T.N. (1967) Chromian spinel as a petrogenetic indicator. Part 2. Petrologic applications. Can. J. Earth Sci. 4, 71-103.

Irvine, T.N. and Findlay, T.C. (1972) Alpine-type peridotite with particular reference to the Bay of Islands Igneous Complex. Pub. Earth Phys. Branch, Dept. Energy Mines Resour. 42(3), 97-140.

Irving, A.J. (1980) Petrology and geochemistry of composite ultramafic xenoliths in alkalic basalts and implications for magmatic processes within the mantle. Am. J. Sci. 280-A, 389-426.

Jan, M.Q. and Windley, B.F. (1990) Chromian spinel-silicate chemistry in ultramafic rocks of the Tijal Complex, northwest Parkistan. J. Petrol. 31, 667-715.

Jaques, A.L., Kerr, I.D., Lucas, H., Sun, S-S. and Chapell, B.W. (1989a) Mineralogy and petrology of picritic monchiquites from Wandagee Carnavon Basin, Western Australia. Geol. Soc. Australia, Spec. Pub. 14, 120-139.

Jaques, A.L., Haggerty, S.E., Lucas, J. and Boxer, G.L. (1989b) Mineralogy and petrology of the Argyle (AK-1) lamproite pipe Western Australia. Geol. Soc. Australia, Spec. Pub. 14, 153-170.

Jaques, A.L., Hall, A.E., Sheraton, J.W., Smith, C.B., Sun, S-S., Drew, R.M., Foudoulis, C. and Ellingsen, K. (1989c) Composition of crystalline inclusions and C-isotopic composition of Argyle and Ellendale diamonds. Geol. Soc. Australia, Spec. Pub., 14, 966-989.

Jaques, A.L., Lewis, J.D. and Smith, C.B. (1986) The Kimberlites and Lamproites of Western Australia. Geol. Survey Western Australia, Bull 132.

Jaques, A.L., O'Neill, H.St.C., Smith, C.B., Moon, Jr. and Chapell, B.W. (1990) Diamondiferous peridotite xenoliths from the Argyle (AK1) lamproite pipe, Western Australia. Contrib. Mineral. Petrol. 104, 255-276.

Johnson, R.E., Woermann, E. and Muan, A. (1971). Equilibrium studies in the system MgO-"FeO"-TiO_2. Am. J. Sci. 271, 278-292.

Jones, A.P. (1989) Upper mantle enrichment by kimberlitic or carbonatitic magmatism. In: K. Bell, ed., Carbonatites: Genesis and Evolution, 448-463. Unwin Hyman, Boston.

Jones, A.P. and Ekambaram, V. (1985) New INAA analysis of a mantle-derived titanate mineral of the crichtonite series, with particular reference to the rare earth elements. Am. Mineral. 70, 414-418.

Jones, A.P., Smith, J.V. and Dawson, J.B. (1982) Mantle metasomatism in 14 veined peridotites from Bultfontein Mine, South Africa. J. Geol. 90, 435-453.

Kelemen, P.B., Johnson, K.T.M., Kingler, R.J. and Irving, A.J. (1990) High field strength element depletions in arc basalts due to mantle-magma interaction. Nature 345, 521-524.

Kelly, P.R., Campbell, I.H., Grey, I.E. and Gatehouse, B.M. (1979) Additional data on loveringite $(Ca,REE)(Ti,Fe,Cr)_{21}O_{38}$ and mohsite discredited, Can. Mineral. 17, 635-638.

Kempton, P.D., Dungan, M.A. and Blanchard, D.P. (1987) Petrology and geochemistry of xenolith-bearing alkalic basalts from the Geronimo volcanic field, Southwest Arizona. In: E.M. Morris and J.D. Pasteris, eds., Mantle Metasomatism and Alkaline Magmatism. Geol. Soc. Am. Spec. Paper 215, 347-371.

Kesson, S.E. and Lindsley, D.H. (1975) The effects of Al^{3+}, Cr^{3+}, and Ti^{3+} on the stability of armalcolite. Proc. Lunar Sci. Conf. 6th, Supp. 6, Geochim. Cosmochim. Acta 1, 911-920.

Kimball, K.L. (1990) Effects of hydrothermal alteration on the compositions of chromian spinels. Contrib. Mineral. Petrol. 105, 337-346.

Kirkley, M.E., McCallum, M.E. and Eggler, D.H. (1984) Coexisting garnet and spinel in upper mantle xenoliths from Colorado-Wyoming kimberlites. In: J. Kornprobst, ed., Kimberlites: The Mantle and Crust-Mantle Relationships, 85-96. Elsevier, Amsterdam.

Kleinman, B. (1969) The breakdown of zircon observed in the Libyan desert glass as evidence of its impact origin. Earth Planet. Sci. Lett. 5, 497-501.

Korneliussen, A. and Foslie, G. (1985) Rutile-bearing eclogites in the Sunnfjord region of Western Norway. Nor. Geol. Unders. 402, 65-71.

Kresten, P., Fels, P. and Berggen, G. (1975) Kimberlitic zircons--a possible aid in prospecting for kimberlite. Mineral. Deposita 10, 47-56.

Kushiro, I. and Erlank, A.J. (1970) Stability of potassic richterite. An. Rep. Dir. Geophys. Lab. Carnegie Inst. Washington Yr. Bk. 68, 231-233.

Leblanc, M., Dautria, J. and Girod, M. (1982) Magnesian ilmenite xenoliths in a basanite from Tahalra, Ahaggar (Southern Algeria). Contrib. Mineral. Petrol. 79, 347-354.

Leblanc, M., Dupuy, C. Cassard, D., Moutte, J., Nicolas, A., Pringhoffer, A., Rabinovitch, M. and Routhier, P. (1980). Essai sur la genese des corps podiformes de chromitite dans les peridotites ophiolitiques: Etude des chromites de Nouvelle-Caledonie et comparison avec celles de Mediterranee orientale. In: Ophiolites. Proc. Intl. Ophiolite Conf., Cyprus, 1979, 691-701, Cyprus Geol. Survey.

Leblanc, M. and Violette, J-F (1983) Distribution of aluminum-rich and chromium-rich chromite pods in ophiolite peridoties. Econ. Geol. 78, 293-301.

Lindsley, D.J., Kesson, S.E., Hartzman, M.J. and Cushman, M.K. (1974) The stability of armalcolite: experimental studies in the system MgO-Fe-Ti-O. Proc. Lunar Sci. Conf. 5th, Supp. 5, Geochim. Cosmochim. Acta 1, 521-534.

Lloyd, F.E. and Bailey, D.K. (1975) Light element metasomatism of the continental mantle: The evidence and the consequences. Phys. Chem. Earth 9, 389-416.

Lock, N.P. and Dawson, J.B. (1980) Garnet-olivine reaction in the upper mantle: evidence from peridotite xenoliths in the Letseng-la-Terae kimberlites, Lesotho. Trans. of the Royal Society of Edinburgh: Earth Sciences 71, 47-53.

Lung-Chuan Kuo and Essene, E.J. (1986) Petrology of spinel harzburgite xenoliths from the Kishb Plateau, Saudi Arabia. Contrib. Mineral. Petrol. 93, 335-346.

MacGregor, I.D. (1969) The system $MgO-SiO_2-TiO_2$ and its bearing on the distribution of TiO_2 in basalts. Am. J. Sci. 267-A, 342-363.

MacGregor, I.D. (1970) An hypothesis for the origin of kimberlite. In: B.A. Morgan, ed., Mineralogy and petrology of the upper mantle, sulfides, mineralogy and geochemistry of non-marine evaporites. Mineral. Soc. Am. Spec. Papers, 3, 51-62.

MacGregor, I.D. and Wittkop, R.W. (1970) Diopside-ilmenite xenoliths from the Monastery Mine, Orange Free State, South Africa. Geol. Soc. Am. Abs. Programs, 2, 113.

Malpas, J. and Strong, D.F. (1975) A comparison of chrome-spinels in ophiolites and mantle diapirs of Newfoundland. Geochim. Cosmochim. Acta 39, 1045.

Mattioli, G.S., Baker, M.B., Rutter, H.J. and Stolper, E.M. (1989) Upper mantle oxygen fugacity and its relationship to metasomatism. J. Geol. 97, 521-536.

Mazzone, P. and Haggerty, S.E. (1989a) Corganites and corgaspinites: Two new types of aluminous assemblages from the Jagersfontein pipe. Geol. Soc. Australia, Spec. Pub. 14, 795-808.

Mazzone, P. and Haggerty, S.E. (1989b) Peraluminous xenoliths in kimberlite: metamorphosed restites produced by partial melting of pelites. Geochem. Cosmochim. Acta 53, 1551-1561.

McCallister, R.H., Meyer, H.O.A. and Brooking, D.G. (1975) "Pyroxene"-ilmenite xenoliths from the Stockdale Pipe, Kansas: Chemistry, crystallography, and origin. Phys. Chem. Earth 9, 287-293.

McCallum, M.E. (1989) Oxide minerals in Chicken Park kimberlite, northern Colorado. Geol. Soc. Australia, Spec. Pub. 14, 241-263.

McCallum, M.E., Eggler, D.H. and Burns, L.K. (1975) Kimberlite diatremes in northern Colorado and southern Wyoming. Phys. Chem. Earth 9, 149-161.

McGetchin, T.R. and Silver, L.T. (1970) Compositional relations in minerals from kimberlite and related rocks in the Moses Rock dike, San Juan County, Utah. Am. Mineral. 55, 1738-1771.

McKenzie, D. (1989) Some remarks on the movement of small melt fractions in the mantle. Earth Planet. Sci. Lett. 95, 53-72.

McKie, D. (1963) The unit cell of freudenbergite. Z. Krist. 119, 157-160.

McKie, D. and Long, J.V.P. (1970) The unit cell contents of freudenbergite. Z. Krist. 132, 157-160.

Menzies, M.A. and Hawkesworth, C.J. (1987) Mantle Metasomatism. Academic Press, London.

Meyer, H.O.A. (1987) Inclusions in diamond. In: P.H. Nixon, ed., Mantle Xenoliths 501-522. John Wiley & Sons, N.Y.

Meyer, H.O.A. and Boyd, F.R. (1972) Composition and origin of crystalline inclusions in natural diamonds. Geochim. Cosmochim. Acta 36, 1255-1273.

Meyer, H.A.O. and Svisero, D.P. (1975) Mineral inclusions in Brazilian diamonds. Phys. Chem. Earth 9, 785-795.

Meyer, H.A.O., Tsai, H-m. and Gurney, J.J. (1979) A unique enstatite megacryst with coexisting Cr-poor and Cr-rich garnet, Wettevreden Floors, South Africa. In: F.R. Boyd and H.O.A. Meyer, eds., The Mantle Sample: Inclusions in kimberlites and other volcanics, 279-291. Am. Geophys. Union, Washington, D.C.

Mitchell, R.H. (1975) Mineralogy of the Tunraq kimberlite, Somerset Island, N.W.T., Canada. In: Boyd, F.R. and Meyer, H.O.A., eds., Kimberlites, Diatremes and Diamonds: Their Geology, Petrology and Geochemistry. 161-171. Am. Geophys. Union, Washington, D.C.

Mitchell, R.H. (1977) Geochemistry of magnesian ilmenites from kimberlites in South Africa and Lesotho. Lithos 10, 29-37.

Mitchell, R.H. (1985) A review of the mineralogy of lamproites. Trans. Geol. Soc. S. Africa 88, 411-437.

Mitchell, R.H. (1986) Kimberlites: Mineralogy, Geochemistry, Petrology. Plenum Press, N.Y.

Mitchell, R.H. and Haggerty, S.E. (1986) A new K-V-Ba titanate related to priderite from the New Elands Kimberlite, South Africa. N. Jb. Mineral. Monatsh., 376-384.

Mitchell, R.H. and Lewis, R.D. (1983) Priderite-bearing diopside-titanian-potassian richterite xenoliths from the Prairie Creek mica peridotite, Arkansas. Can. Mineral. 21, 59-64.

Mitchell, R.H. and Meyer, H.O.A. (1989) Niobian K-Ba-V titanates from micaceous kimberlite, Star Mine, Orange Free State, South Africa. Mineral. Mag. 53, 451-456.

Moore, A.E. (1987) A model for the origin of ilmenite in kimberlite and diamond: implications for the genesis of the discrete nodule (megacryst) suite. Contrib. Mineral. Petrol. 95, 245-253.

Moore, P.B., Sen Gupta, P.K. and Le Page, Y. (1990) Magnetoplumbite, $PbFe_{12}O_{19}$: Refinement and lone pair splitting. Am. Mineral. 74, 1186-1194.

Moore, R.O. and Gurney, J.J. (1985) Pyroxene solid solution in garnets included in diamond. Nature 318, 553-555.

Moore, R.O., Otter, M.L., Rickard, R.S., Harris, J.W. and Gurney, J.J. (1986) The occurrence of moissanite and ferro-periclase as inclusions in diamond. (Abs.) 4th Intl. Kimb. Conf. Perth, 409-411.

Mukherjee, A.B. and Biswas, S. (1988) Mantle-derived spinel lherzolite xenoliths from the Deccan volcanic province (India): Implications for the thermal structure of the lithosphere underlying the Deccan traps. J. Volc. Geotherm. Res. 35, 269-276.

Mvuemba, N.F., Moreau, J. and Meyer, H.O.A. (1982) Particularites des inclusion cristallines primaires des diamants du Kasai, Zaire. Can. Mineral. 20, 217-230.

Navon, O., Hutcheon, I.D., Rossman, G.R., and Wasserburg, G.J. (1988) Mantle-derived fluids in diamond micro-inclusions. Nature 335, 784-789.

Nehru, C.E. and Reddy, A.K. (1989) Ultramafic xenoliths from Vajrakarur kimberlites, India. Geol. Soc. Australia Spec. Pub. 14, 745-759.

Nixon, P.H. (1987) Mantle Xenoliths. P.H. Nixon, ed., John Wiley & Sons, N.Y.

Nixon, P.H. and Boyd, F.R. (1973) Petrogenesis of the granular and sheared ultrabasic nodule suite in kimberlites. In: P.H. Nixon, ed., Lesotho Kimberlites. 48-56. Lesotho National Development Corp., Maseru, Lesotho.

Nixon, P.H. and Condliffe, E. (1989) Yimengite of K-Ti metasomatic origin in kimberlitic rocks from Venezuela. Mineral. Mag. 53, 305-309.

Nixon, P.H., Rogers, N.W., Gibson, I.L. and Grey, A. (1981) Depleted and fertile mantle xenoliths from southern African kimberlites. Annu. Rev. Earth Planet. Sci. 9, 285-309.

Nixon, P.H., van Calsteren, P.M.C., Boyd, F.R. and Hawkesworth, C.J. (1987) Harzburgites with garnets of diamond facies from southern African kimberlites. In: P.H. Nixon, ed., Mantle Xenoliths, 523-533. John Wiley & Sons, N.Y.

Ottenburgs, R. and Fieremans, M. (1979) Rutile-silicate intergrowths from the kimberlite formations at Mguji-Mayi (Bakwanga, Zaire) Bull. Soc. Belge. Geol. 3, 197-203.

Paktunc, A.D. (1990) Origin of podiform chromite deposits by multistage melting, melt segregation and magma mixing in the upper mantle. Ore Geol. Rev. 5, 211-222.

Pasteris, J.D. (1980) The significance of groundmass ilmenite and megacryst ilmenite in kimberlites. Contrib. Mineral. Petrol. 15, 315-325.

Pasteris, J.D., Boyd, F.R. and Nixon, P.H. (1979) The ilmenite association at the Frank Smith Mine, R.S.A. In: F.R. Boyd and H.O.A. Meyer, eds., The Mantle Sample: Inclusions in kimberlites and other volcanics. 265-278. Am. Geophys. Union, Washington, D.C.

Pober, E. and Faupl, P. (1988) The chemistry of detrital chromian spinels and its implications for the geodynamic evolution of the Eastern Alps. Geologische Rundschau 77, 641-670.

Podpora, C. and Lindsley, D.H. (1984) Lindsleyite and mathiasite: Synthesis of chromium-titanates in the crichtonite $(AM_{21}O_{38})$ series (Abs.) EOS Trans. Am. Geophys. Union 65, 293.

Prinz, J., Manson, D.F., Hlava, P.F. and Keil, K. (1975) Inclusions in diamonds: garnet lherzolite and eclogite assamblages. Phys. Chem. Earth 9, 797-815.

Quincheng, F. and Hooper, P.R. (1989) The mineral chemistry of ultramafic xenoliths of Eastern China: implications for upper mantle composition and the paleogeotherms. J. Petrol. 30, 1117-1158.

Raber, E. and Haggerty, S.E. (1979) Zircon-oxide reactions in diamond-bearing kimberlites. In: F.R. Boyd and H.O.A. Meyer, eds., Kimberlites, Diatremes and Diamonds: Their Geology, Petrology and Geochemistry 229-240. Am. Geophys. Union, Washington, D.C.

Rauh, E.G. and Garg, S.P. (1980) The ZrO_{2-x} (cubic) - ZrO_{2-x} (cubic + tetragonal) phase boundary. Am. Ceram. Soc. 63, 239-240.

Ringwood, A.E. (1990) Slab-mantle interactions: 3. Petrogenesis of intraplate magmas and structure of the upper mantle. Chem. Geol. 82, 187-207.

Ringwood, A.E. and Lovering, J.F. (1970) Significance of pyroxene-ilmenite intergrowths among kimberlite xenoliths. Earth Planet. Sci. Lett. 7, 371-375.

Rossman, G.R. and Smyth, J.R. (1990) Hydroxyl contents of accessory minerals in mantle eclogites and related rocks. Am. Mineral. 75, 775-780.

Rouse, R.C. and Peacor, D.R. (1968) The relationship between sanaite, magnetoplumbite and davidite. Am. Mineral. 53, 869-879.

Ryerson, J.F. and Watson, E.B. (1987) Rutile saturation in magmas: implications for Ti-Nb-Ta depletion in island-arc basalts. Earth Planet. Sci. Lett. 86, 225-239.

Sarp, H., Bertrand, J., Deferne, J. and Liebich, B.W. (1981) A complex rhenium-rich titanium and iron oxide of the crichtonite-senaite group N. Jb. Miner. Mh. 10, 433-443.

Scatena-Wachel, D.E. and Jones, A.P. (1984) Primary baddeleyite (ZrO_2) in kimberlite from Benfontein, South Africa. Mineral. Mag. 48, pp. 257-261.

Schultze, D.J. (1983) Graphic rutile-olivine intergrowths from South African kimberlites. Ann. Rep. Dir. Geophys. Lab. Carnegie Inst. Washington, D.C., Yr. Bk. 82, 343-346.

Schulze, D.J. (1987) Megacrysts from alkalic volcanic rocks. In: P.H. Nixon, ed., Mantle Xenoliths, 433-451. John Wiley & Sons, New York.

Schulze, D.J. (1990) Silicate-bearing rutile-dominated nodules from South African kimberlites: Metasomatized cummulates. Am. Mineral. 75, 97-104.

Scott-Smith, B.H. (1989) Lamproites and kimberlites in India. Neus. Jahrb. Miner. Abh. 161, 193-225.

Scott-Smith, B.H., Danchin, R.V., and Stracke, K.J. (1984) Kimberlites near Orroroo, South Australia. In: J. Kornprobst, ed., Kimberlites: Kimberlites and Related Rocks, 121-142, Elsevier, Amsterdam.

Sen, G. (1988) Petrogenesis of spinel lherzolite and pyroxenite suite xenoliths from the Koolau shield, Oahu, Hawaii: Implications for petrology of the post-eruptive lithosphere beneath Oahu. Contrib. Mineral. Petrol. 100, 61-91.

Sen, G. and Presnall, D.C. (1986) Petrogenesis of dunite xenoliths from Koolau Volcano, Oahu, Hawaii: implications for Hawaiian volcanism. J. Petrol. 27, 197-217.

Shee, S.R. and Gurney, J.J. (1979) The mineralogy of xenoliths from Orapa, Botswana. In: F.R. Boyd and H.O.A. Meyer, eds., The Mantle Sample: Inclusions in Kimberlites and other volcanics, 37-49. Am. Geophys. Union, Washington, D.C.

Shibata, T. and Thompson, G. (1986) Peridotites from the Mid-Atlantic Ridge at 43°N and their petrogenetic relation to abyssal tholeiites. Contrib. Mineral. Petrol. 93, 144-159.

Siena, F. and Coltorti, M. (1989) Lithospheric mantle evolution: Evidences from ultramafic xenoliths in the Lessinian volcanics (northern Italy). Chem. Geol. 77, 347-364.

Smith, D. (1977) The origin and interpretation of spinel-pyroxene clusters in peridotite. J. Geol. 85, 476-482.

Smith, J.V. and Dawson, J.B. (1975) Chemistry of Ti-poor spinels, ilmenites and rutiles from peridotite and eclogite xenoliths. Phy. Chem. Earth, 9, 309-322.

Smith, S.E. and Elthon, D. (1988) Mineral compositions of plutonic rocks from the Lewis Hills Massif, Bay of Islands Ophiolite. J. Geophys. Res. 93, 3450-3468.

Sobolev, N.V. (1977) Deep Seated Inclusions in Kimberlites and the Problem of the Composition of the Upper Mantle. Am. Geophys. Union, Washington, D.C.

Sobolev, N.V., Yefimova, E.S., Kaminsky, F.V., Lavrentiev, Y.G. and Usova, L.F. (1988) Titanate of complex composition and phlogopite in the diamond stability field. (Abs.) Composition and Processes of Deep-Seated Zones of Continental Lithosphere, Novosibirsk, 185-186.

Swanson, S.E., Kay, S.M., Brearley, M. and Scarfe, C.M. (1987) Arc and back-arc xenoliths in Kurile-Kamchatka and western Alaska. In: P.H. Nixon, ed., Mantle Xenoliths, 303-318. John Wiley & Sons, N.Y.

Talkington, R.W. and Malpas, J.G. (1984) The formation of spinel phases of the White Hills peridotite, St. Anthony complex, Newfoundland. N. Jb. Mineral. Abh. 149, 65-90.

Tatarintsev, V.L., Tsymbal, S.N., Garanin, V.G., Kudryatseva, G.P. and Marshintsev, V.K. (1983) Quenched particles from kimberlites in Yakutia. Dokl. Earth Sci. Sect., 270, 144-148.

Tollo, R.P. (1982) Petrography and mineral chemistry of ultramafic and related inclusions from the Orapa A/K-1 kimberlite pipe, Botswana. Contrib. 39, Dept. Geol., Univ. of Massachusetts, Amherst, MA.

Tollo, R.P. and Haggerty, S.E. (1987) Nb-Cr rutile in the Orapa kimberlite, Botswana. Can. Mineral. 25, 251-264.

Tompkins, L.A. and Haggerty, S.E. (1984a) The Koidu kimberlite complex Sierra Leone: Geological setting, petrology and mineral chemistry. In: J. Kornprobst, ed., Kimberlites: Related Rocks. 83-105, Elsevier, Amsterdam.

Tompkins, L.A. and Haggerty, S.E. (1984b) The Koidu kimberlite complex Sierra Leone: geological setting, petrology and mineral chemistry--Appendix. In: J. Kornprobst, ed., Ann. Scientif. De L'Univ. De Clermont-Ferrand. Kimberlites III: Documents 74, 99-122.

Tompkins, L.A., Haggerty, S.E. (1985) Groundmass oxide minerals in the Koidu kimberlite dikes, Sierra Leone, West Africa. Contrib. Mineral. Petrol. 91, 245-263.

Townes, W.D., Fang, J.H., Perrotta, A.J. (1967) The crystal structure and refinement of ferrimagnetic barium ferrite. $BaFe_{12}O_{19}$. Zeit. Kristallogr. 125, 437-449.

Velde, D. (1975) Armalcolite-Ti-phlogopite-diopside-analcite-bearing lamproites from Smoky Butte, Garfield County, Montana. Am. Mineral. 60, 566-573.

Vlasov, K.A. (1966) Geochemistry and Mineralogy of Rare Elements and Genetic Types of Deposits. Israel Prog. Sci. Transl., Jerusalem.

Vlassopoulos, D., Rossman, G. and Haggerty, S.E. (1990) Hydrogen in natural and synthetic rutile (TiO_2): Distribution and possible controls on its incorporation. (Abs.) EOS Trans. Am. Geophys. Union 71, 626.

Wark, D.A., Reid, A.F., Lovering, J.F. and El Goresy, A. (1973) Zirconolite (versus sirkelite) in lunar rocks. In: Lunar Science IV (Abs. Vol), 764-766, Lunar Science Institute, Houston, Texas.

Waters. F.W. and Erlank, A.J. (1988) Assessment of the vertical extent and distribution of mantle metasomatism below Kimberley, South Africa. J. Petrol. Spec. Vol. Oceanic and Continental Lithosphere: Similarities and Differences. 185-204.

Wilding, M.C. Harte, B. and Harris, J.W. (1989) Evidence of asthenospheric source for diamonds from Brazil. (Abs) 28th Int. Geol. Congress, 3, 359-360. Washington, D.C.

Wilshire, H.G. and Shervais, J.M. (1975) Al-augite and Cr-diopside ultramafic xenoliths in basaltic rocks from western United States. Phys. Chem. Earth 9, 257-272.

Winterburn, P.A., Harte, B. and Gurney, J.J. (1990) Peridotite xenoliths from the Jagersfontein kimberlite pipe: 1. Primary and primary-metasomatic mineralogy. Geochem. Cosmochim. Acta 54, 329-341.

Woermann, E., Brezny, B. and Muan, A. (1969) Phase equilibrium in the system MgO-iron oxide-TiO in air. Am. J. Sci. 267A, 463-479.

Woermann, E., Hirschberg, A. and Lamprecht, A. (1970) Das system hematit-ilmenit-geikielith unter hohen temperaturen und hohen drucken. (Abs.) Fortschr. Mineral. 47, 79-80.

Wyatt, B.A. (1977) The melting and crystallization behavior of a natural clinopyroxene ilmenite intergrowth. Contrib. Mineral. Petrol. 61, 1-9.

Wyatt, B.A. (1978) Phase relationships in the system picroilmenite-clinopyroxene-Cr_2O_3: A preliminary assessment. In: Progress in experimental petrology. N.E.R.C., United Kingdom Pub D11, 181-185.

Wyatt, B.A. (1979) Kimberlitic chromian picroilmenites with intergrowths of titanian chromite and rutile. In: F.R. Boyd and H.O.A. Meyer, eds., The Mantle Sample: Inclusions in Kimberlites and Other Volcanics, 257-264. Am. Geophys. Union, Washington, D.C.

Wyatt, B.A. and Lawless, P.J. (1984) Ilmenite in polymict xenoliths from the Bultfontein and De Beers mines, South Africa. In J. Kornprobst, ed., Kimberlites 2: The Mantle and Crust-Mantle Relationships, 43-56, Elsevier, Amsterdam.

Wyatt, B.A., McAllister, R.H., Boyd, F.R. and Ohashi, Y. (1975) An experimentally produced clinopyroxene ilmenite intergrowth. Ann. Report Dir. Geophys. Lab. Carnegie Inst. Washington, Yr. Bk. 74, 536-539.

Wyllie, P.J. (1989a) The genesis of kimberlites and some low-SiO_2, high-alkali magmas. Geol. Soc. Australia, Spec. Pub. 14, 603-615.

Wyllie, P.J. (1989b) Origin of carbonatites: evidence from phase equilibrium studies. In: K. Bell, ed., Carbonatites: Genesis and Evolution. Unwin Hyman. 500-545.

Wyllie, P.J. (1987) Metasomatism and fluid generation in mantle xenoliths. In: P.H. Nixon, ed., Mantle Xenoliths, 609-621. John Wiley & Sons, New York.

Zhou, J., Yang, G. and Zhang, J. (1984) Mathiasite in a kimberlite from China. Acta Mineral. Sinica 9, 193-200.

Chapter 11 B. J. Wood
OXYGEN BAROMETRY OF SPINEL PERIDOTITES

INTRODUCTION

The current oxidation state of the upper mantle and how it evolves with time is a question of considerable petrologic importance. The composition of the fluid phase attending igneous and metamorphic processes is significantly affected by oxygen fugacity and this may, in turn, substantially affect the nature of primary melts (e.g., Eggler and Baker, 1982). Based on observations of fluid inclusions in peridotite xenoliths and of the compositions of volcanic gases (Andersen et al., 1984; Trial et al., 1984; Williams and McBirney, 1979, Arculus, 1985) it is generally accepted that mantle fluids may be reasonably approximated by the C-H-O system. In that case the fluid phase should be dominated by CO_2 and H_2O under relatively oxidising conditions (close to the fayalite-magnetite-quartz, FMQ, buffer) by H_2O and CH_4 at moderately reducing fO_2 (2-3 log units below FMQ) and by CH_4 and H_2 under strongly reducing (below iron-wüstite, IW) conditions (e.g., Wood et al., 1990).

Coupled to these generalizations are geologic observations that imply changes in oxidation state with time. Cycling of altered oceanic crust back to the mantle at subduction zones indicates that these are sources of secular hydration and oxidation. At mid-ocean ridges, on the other hand, the upper mantle partially melts to produce basalt, which leaves a residue in which the proportions of Fe^{3+}-bearing phases such as spinel and of hydrous phases such as amphibole and of carbonates are greatly diminished. Mid-ocean ridges appear, therefore, to be zones of dehydration, decarbonation, and reduction. More detailed evaluation of the processes involved requires the development of accurate means of determining the fugacities of fluid species in mantle and mantle-derived rocks. Oxygen fugacity is particularly important because of its influence on speciation.

There are at least three distinct methods which may be used to estimate the oxygen fugacities recorded by the upper mantle. The first of these is the oxygen-specific electrochemical measurement of the "intrinsic oxygen fugacity" described by Sato (1965, 1972). This method was used by Arculus and Delano (1981) and Arculus et al. (1984) to measure the intrinsic oxygen fugacities of spinels and spinel lherzolites from different xenolith suites. Overall, their results indicate considerable heterogeneity in upper mantle fO_2's, with, however, a marked clustering of the data into two groups. These are: a) type I spinel lherzolites (Cr-rich pyroxene suite) which give log fO_2-1/T values from slightly below IW to about 1 log unit above IW and b) type II spinel lherzolites (Al-augite suite) which cluster between FMQ and 1 log unit above FMQ (Arculus et al., 1984). The observation that the two groups of mineral samples appear to be separated by 3 to 4 log units in fO_2 is a surprising result in view of the fact that the contents of ferric iron in both type I and type II spinels from San Carlos, Arizona are similar. Alternatively, it has been suggested that the very reduced log fO_2-1/T trends are in error because of the ubiquitous presence of carbon in mantle xenoliths (Mathez et al., 1984; Virgo et al., 1988). In a study of whole rock and mineral separates from the Skaergaard intrusion, Sato and Valenza (1980) showed that carbon causes autoreduction of the sample at high temperatures in the electrochemical cell during intrinsic fO_2 measurements. It was assumed in their study, however, that log fO_2- 1/T values obtained at lower temperatures were representative of the conditions at the time of crystallization. More recently, Virgo et al. (1988) showed that C-bearing, ilmenite megacrysts from different kimberlite localities can be reduced to two phase assemblages (ilmenite$_{ss}$ + spinel$_{ss}$) at low temperatures in the electrochemical cell because of reaction between the ilmenite and carbon. Under these conditions, the initial reduced trends in the electrochemical cell are attributable to disequilibrium interactions between the decomposing sample and gas. With an increase in temperature, the more oxidized log fO_2- 1/T values (FMQ) correspond to the fO_2 of the ilmenite$_{ss}$ + spinel$_{ss}$ assemblage and not the intrinsic fO_2 of the mantle derived ilmenite. It seems probable, therefore, that intrinsic fO_2 data on mantle samples are unreliable.

A second, indirect, method of estimating upper mantle redox state is calibration of the Fe^{2+}/Fe^{3+} ratios of quenched silicate melts as functions of temperature and oxygen fugacity (e.g., Fudali, 1965; Christie et al., 1986). In one study (Christie et al., 1986) measurements of the redox states of a large number (78) of MORB pillow lavas have confirmed the importance of near-surface oxidation. It was found that the ferric/ferrous ratios of pillow cores are indicative of oxygen fugacities (at 1 bar) close to FMQ whereas Fe^{2+}/Fe^{3+} of rapidly quenched glasses from the pillow margins give oxygen fugacities 1 to 2 log units below FMQ (Fig. 1). These results suggest (Christie et al., 1986) that the MORB source regions are more reduced than FMQ and that lavas oxidize rapidly during cooling, possibly by hydrogen loss.

A third direct method is the one which will be described here. Coexistence of spinel, olivine and orthopyroxene in mantle peridotites enables calculation of a thermobarometer for upper mantle assemblages (O'Neill and Wall, 1987; Mattioli and Wood, 1988). The method is also potentially extendable to higher pressure garnet peridotites (Luth et al., 1990). In this review I briefly discuss the thermobarometric approach and

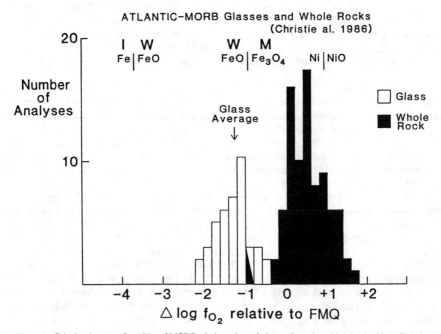

Figure 1. Calculated oxygen fugacities of MORB whole rocks and glasses from the mid-Atlantic ridge (Christie et al., 1986). The calculation method is based on homogeneous Fe^{2+}-Fe^{3+} equilibrium in the quenched melts. Note that rapidly quenched glassy margins are more reduced than whole rock pillow cores, possibly because of H_2 loss from the latter.

its recent application to peridotites from different tectonic environments. As will be seen, there are distinct correlations between average oxygen fugacity and geologic environment, and most of the relationships can be understood in terms of plate tectonic processes.

SPINEL PERIDOTITE OXYGEN BAROMETRY

Use of the peridotite assemblage for oxygen barometry relies on the coexistence of phases containing both ferric and ferrous iron. The assemblage olivine-orthopyroxene-spinel fulfills this requirement and enables fO_2 to be calculated in principle from the equilibrium (Mattioli and Wood, 1988):

$$6\,Fe_2SiO_4 + O_2 = 3\,Fe_2Si_2O_6 + 2\,Fe_3O_4 \qquad (1)$$
$$\text{olivine} \qquad \qquad \text{orthopyroxene} \quad \text{spinel}$$

The method involves measuring total Fe contents of all three phases by electron microprobe, determining the Fe_3O_4 content of spinel (as Fe^{3+}/Fe^{2+}) and applying end-member thermodynamic data for Equilibrium (1), together with appropriate activity-composition relations for the three phases. Mattioli and Wood (1988) obtained end-member data by combining results on the fayalite-magnetite-quartz (FMQ) buffer with measurements of fayalite-ferrosilite-quartz equilibrium. Olivine mixtures (Fe_2SiO_4-Mg_2SiO_4) were modeled (following Wood and Kleppa, 1981) as two-site regular solid solutions with an interaction parameter of 4.18 kJ/gm·atom while orthopyroxenes were treated as ideal two-site mixtures: $a^{opx}_{Fe_2Si_2O_6} = X^{M1}_{Fe} \cdot X^{M2}_{Fe}$. This yields, for oxygen fugacity relative to the FMQ buffer:

$$\log(fO_2)_{P,T} = \log fO_2(FMQ)_{P,T} + \frac{220}{T} + 0.35 - \frac{0.0369P}{T} - 12\log X^{ol}_{Fe}$$

$$- \frac{2620}{T} X^{ol\,2}_{Mg} + 3\log(X^{M1}_{Fe} \cdot X^{M2}_{Fe})^{opx} + 2\log a^{sp}_{Fe_3O_4} \qquad (2)$$

where X^{ol}_{Fe}, X^{ol}_{Mg} refer to mole fractions of Fe and Mg end-members in olivine, P is pressure in bars, T is absolute temperature and X^{M1}_{Fe}, X^{M2}_{Fe} in orthopyroxene refer to atomic fractions of Fe in the two orthopyroxene (opx) sites. The latter were calculated from the microprobe analysis with the assumption of equipartition of Mg and Fe after assignment of Al^{VI}, Cr and Ti to M1 and Ca, Na and Mn to M2. Uncertainties in fO_2 calculated from Equation (2) accrue from uncertainties in the end-member data (± 0.1 log fO_2 units), uncertainties in compositions of the silicate phases (± 0.05 log fO_2 units) and uncertainties in their mixing properties (± 0.5 log units approx.). The major uncertainty, however, is in the determination of the activity of Fe_3O_4 component ($a_{Fe_3O_4}$) in the spinel phase. This is discussed in more detail below.

A variant of this approach was described by O'Neill and Wall (1987) who preferred to use the thermodynamic properties of the FMQ buffer to calculate fO_2:

$$3\,Fe_2SiO_4 + O_2 = 2\,Fe_3O_4 + 3\,SiO_2 \qquad (3)$$
$$\text{olivine} \qquad \quad \text{spinel} \quad \text{quartz}$$

Since peridotites are SiO_2-undersaturated the activity of SiO_2, relative to a quartz standard state, was taken from the equilibrium:

$$Mg_2SiO_4 + SiO_2 = Mg_2Si_2O_6 \qquad (4)$$
$$\text{olivine} \quad \text{quartz} \quad \text{orthopyroxene}$$

O'Neill and Wall also assumed ideal two-site mixing in orthopyroxene and adopted a two-site regular model for olivine with an interaction parameter similar to that used by Mattioli and Wood, i.e., 5.0 kJ/gm·atom. (Note that a recent experimental study by Wiser and Wood (1991) give W^{ol} of 3.7±0.8 kJ/gm·atom.) O'Neill and Wall's expression yields for the activity of SiO_2 in the mantle assemblage:

$$\log a_{SiO_2} = -\frac{350}{T} + 0.016 - \frac{0.020P}{T} + \log a^{opx}_{Mg_2Si_2O_6} - \log a^{ol}_{Mg_2SiO_4} \quad (5)$$

The fugacity of oxygen is then given by:

$$\log fO_2 = \log(fO_2)_{FMQ} - 3 \log a^{ol}_{Fe_2SiO_4} + 3 \log a_{SiO_2} + 2 \log a^{sp}_{Fe_2Si_2O_6} \quad (6)$$

The main argument in favor of using (5) and (6) instead of (2) is that the former do not rely on estimating the activity of $Fe_2Si_2O_6$ component, which is very dilute in orthopyroxene. My analysis suggests, however, that the potential error is small and that uncertainties derived from uncertainties in Fe_3O_4 activity are much larger. We hence concentrated on trying to estimate $a^{sp}_{Fe_3O_4}$ as accurately as possible.

Mattioli and Wood (1988) made measurements of Fe_3O_4 activity in $MgAl_2O_4$ spinels by electrochemically determining the fO_2's recorded by mixtures of hematite and spinel solid solutions (see Chapter 7). When applied to mantle assemblages, however, their results show a marked dependence of apparent fO_2 on Cr content of the spinel (Wood, 1990; Ballhaus et al., 1990), implying that explicit provision must be made for the effects of $FeCr_2O_4$ component. In contrast to the experiments of Mattioli and Wood (1988), O'Neill and Wall (1987) used a complex spinel model to predict $a_{Fe_3O_4}$ at appropriate concentration levels. This incorporated order-disorder between octahedral and tetrahedral sites (O'Neill and Navrotsky 1983, 1984; Chapter 7) and available experimental data on macroscopic properties such as interphase partitioning and miscibility gaps (Chapter 7). As discussed by O'Neill and Wall (1987) and Wood (1990), this approach appears to compensate adequately for Cr substitution in spinel but tends slightly to underestimate the calculated fO_2. Nell and Wood (1991) have updated the O'Neill-Wall approach by incorporating the Mattioli and Wood (1988) results and new measurements on order-disorder in spinel. In the composition and temperature range of concern in the upper mantle (Fe_3O_4 mole fractions of 0.008 to 0.06 and temperatures of 800°-1400°C), the Nell-Wood model yields:

$$\log a^{sp}_{Fe_3O_4} = \log\left\{\frac{(Fe^{2+}) \cdot (Fe^{3+})^2}{4}\right\}$$

$$+ \frac{1}{T}[406(Al)^2 + 653(Mg)(Al) + 299(Cr)^2 + 199(Al)(Cr) + 346(Mg)(Cr)] \quad (7)$$

where terms in parentheses refer to total Mg, Fe^{2+}, Al, Cr and Fe^{3+} cations in the spinel structure on a four-oxygen basis.

Wood (1990) experimentally tested the approach embodied in Equations (2) and (7) by equilibrating olivine, orthopyroxene and spinel at known fO_2 at 1 atm pressure and 1200°C. The equilibrium Fe_3O_4 contents of spinel (as determined by microprobe -- see below) were approached from high and low concentrations (Fig. 2) and fO_2 calculated from the compositions of the coexisting phases. As can be seen from Figure 2, fO_2's

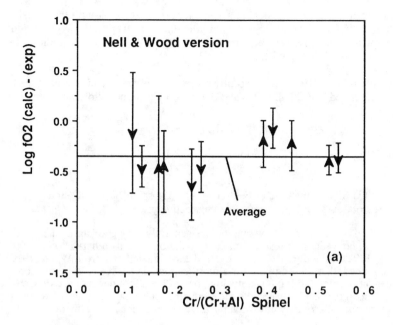

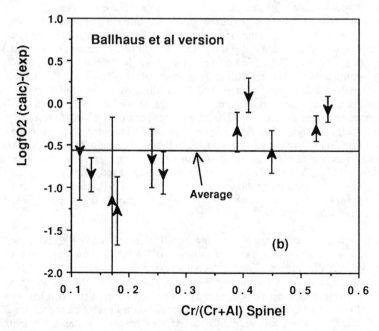

Figure 2. Comparison of two versions of the spinel peridotite oxygen barometer when tested against reversal data of Wood (1990). Note that both sets of calculations generate calculated fO_2 very close to the experimental value.

calculated from Equations (2) and (7) are in excellent agreement with the actual experimental values of fO_2. The average deviation is -0.35 log fO_2 units. We conclude that, given adequate analyses of all three phases, Equation (2) enables calculation of fO_2 to within, in general, ±0.5 log units.

Ballhaus et al. (1990) have recently made an empirical recalibration of the oxygen barometer described by O'Neill and Wall (1987). They give the following expression:

$$\log (fO_2)_{P,T} = \log fO_2(FMQ)_{P,T} + 0.27 + \frac{2505}{T} - \frac{0.04P}{T}$$
$$- 6 \log X_{Fe}^{ol} - \frac{3200}{T}(1-X_{Fe}^{ol})^2 + 2 \log (X_{Fe^{2+}}^{sp}) + 4 \log (X_{Fe^{3+}}^{sp}) + \frac{2630}{T}(X_{Al}^{sp})^2 \quad (8)$$

In Equation (8) $X_{Fe^{2+}}^{sp}$ is the ratio of $Fe^{2+}/(Fe^{2+}+Mg)$ in spinel, while X_{Al}^{sp} and $X_{Fe^{3+}}^{sp}$ refer to ratios of Al/total R^{3+} and Fe^{3+}/total R^{3+}, respectively.

Ballhaus et al. (1990) show that there is generally very good agreement (within 0.5 to 1 log units) between their results and the calculation described here (Wood, 1990, Nell and Wood, 1991). A comparison with reversed experimental data from Wood (1990) (Fig. 2) shows, indeed, that their equation is almost as good as the Nell-Wood version, giving fO_2 values on average 0.6 log fO_2 units low, but with a slight dependence on Cr/Al ratio. In view of the agreement between the different approaches taken by the different groups, oxygen thermobarometry may now be applied with confidence to mantle rocks.

Determination of Fe^{3+} contents of spinel

Given Equations (7) and (8), which incorporate relationships between $a_{Fe_3O_4}^{sp}$ and Fe^{2+} and Fe^{3+} concentrations, the major remaining problem is the accurate determination of the Fe_3O_4 content (or Fe^{3+}/Fe^{2+} ratio) of the spinel. Despite the general use of microprobe data to calculate Fe^{3+} contents from stoichiometry and despite various assertions to the contrary, microprobe data are usually too inaccurate for this purpose (Wood and Virgo, 1989; Dyar et al., 1989). The basic problem is to obtain low values (generally <0.05) 7 $X_{Fe_3O_4}$ to within ±0.002, corresponding to an uncertainty of ±0.2 log fO_2 units (Wood and Virgo, 1989). In the microprobe method this is done by calculating Fe^{3+}/Fe^{2+} using the assumption of 3 cations to 4 oxygens. Given the normal counting uncertainties this generally leads to two standard errors (20-25 analyses) of about ±0.0015 in $X_{Fe_3O_4}$, i.e., potentially precise enough. The main problem, however, is inaccuracy, since the microprobe correction factors are not exactly known, and spinels contain large concentrations of the light elements Mg and Al for which the uncertainties are greatest. As an example, Wood and Virgo (1989) showed that for one spinel using exactly the same standards and data, two commonly used correction schemes, ZAF and PAP, gave calculated Fe_3O_4 contents of 0.008 and 0.020, respectively. This is equivalent to nearly 2 log units difference in calculated fO_2.

In view of the inaccuracies it was suggested that the microprobe method could only be used when using spinel standards of known Fe^{3+} content which are very close in composition to the unknowns. The Fe^{3+} contents of a large number of mantle spinels, suitable for use as standards were measured by Wood and Virgo (1989) and Bryndzia and Wood (1990) using Mössbauer spectroscopy. They found that use of closely spaced standards did indeed enable generation of microprobe data in which accuracy in $X_{Fe_3O_4}$ approached ±0.002. Thus, although Mössbauer spectroscopy remains the method of choice for determination of Fe^{3+} contents, the microprobe may be used given extremely careful calibration and standardization. It must be emphasized, however, that most spinel analyses in the literature were not made with oxygen thermobarometry in mind, so that systematic errors in calculated fO_2 of around 2 log units are quite likely. I therefore consider the use of such literature data to support petrologic arguments (O'Neill and Wall, 1987; Ballhaus et al., 1990) as unacceptable. Accuracy and reproducibility of Fe^{3+} contents must be demonstrated first.

This review will now continue with the application of Equation (2) to rocks. Almost all of the spinel data which I shall discuss were obtained from Mössbauer determination of Fe^{3+}/Fe^{2+} with bulk composition being obtained by microprobe.

OXYGEN BAROMETRY OF CONTINENTAL SPINEL LHERZOLITE XENOLITHS

Continental spinel peridotite xenoliths are samples of the subcontinental lithosphere brought up from between 30 and 60 km depth by alkali basalt volcanism. They contain olivine, orthopyroxene, clinopyroxene and spinel. Figure 3 shows calculated oxygen fugacities of the 30 continental spinel lherzolites for which Mössbauer data were obtained by Wood and Virgo (1989). Temperatures recorded by these continental xenoliths, as obtained from the Wells (1977) two-pyroxene geothermometer, range from 914° to 1095°C. In order to facilitate comparison between data I have compared them all to the FMQ curve, because temperature errors of ±100°C only shift relative fO_2 by ±0.2 log units when it is normalized in this fashion. In the absence of an accurate geobarometer for spinel lherzolite assemblages I performed the calculation at a pressure of 15 kbar, essentially in the middle of the spinel peridotite field. As with temperature, fO_2 relative to FMQ is little affected by pressure errors within the stability field of spinel lherzolite.

Several important features may be observed from the data (Fig. 3). First, it is of interest to compare the results with oxygen barometry on rapidly quenched MORB glasses (Christie et al., 1986). In comparison to glass fO_2 data determined at 1 bar, almost all continental xenoliths lie outside the MORB range. The effect of pressure on ferric-ferrous equilibria in silicate liquids is poorly known, however, although some measurements

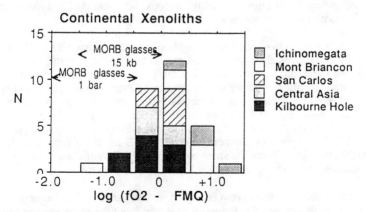

Figure 3. Histogram of oxygen fugacities (relative to FMQ) from continental spinel lherzolite xenoliths using the Nell-Wood version of the oxygen barometer. Regions of continental extension (San Carlos, Arizona; Kilborne Hole, New Mexico and Central Asia) show a range from MORB-like to more oxidized. Xenoliths from regions of current or recent subduction (Ichinomegata, Japan and Mont Briançon, France) are relatively oxidized.

indicate that it is negligible (Gudmundsson et al., 1988). If, in contrast, we use the volume data for Fe_2O_3 and FeO of Lange and Carmichael (1987), then the MORB liquids would lie at higher fO_2 (relative to FMQ) at high pressure than at 1 bar (Fig. 3). For purposes of comparison in Figure 3, I used a pressure of 15 kbar for both MORB and peridotites, although the actual pressure of MORB generation probably varies substantially. Even if this pressure correction, which is probably an upper bound, is applied, a large proportion of the continental xenoliths are more oxidized than the most oxidized MORB glass. Therefore, a large part of the continental lithosphere, although of similar composition, is distinct from the MORB source region.

A second important feature of Figure 3 is that almost all of the samples from Ichinomegata, Japan, and from Mont Briançon in the Massif Central of France lie above FMQ and are outside the range of MORB glasses even if a large pressure correction is applied to the latter. All of the Japanese samples contain amphibole or the remnants of amphibole undergoing decompression melting. Furthermore, this locality is close to the subducting Pacific plate and xenoliths containing H_2O-bearing fluid inclusions have been reported (Trial et al., 1984). These data strongly imply that the subduction environment is one where the mantle overlying the descending plate undergoes oxidation in addition to hydration. In support of a correlation between oxidation and hydration, data on five amphibole-bearing xenoliths from the Eifel region (currently a rift environment, but in an area of subduction in the past) of Germany also yield fO_2 values above FMQ. The Mont Briancon oxidized samples, although apparently anhydrous, are also from a region of subduction in the recent geologic past. Mattioli et al. (1989), using a large database of microprobe analyses, also noted that there was a correlation between hydrated and oxidized samples and, despite the reservations over the use of microprobe data, their observations have held up when the spinels were analyzed by Mössbauer spectroscopy. Further geochemical evidence for interaction of the mantle with a slab-derived fluid component is apparent open-system oxygen isotope behavior of the hydrous Eifel samples (Kempton et al., 1988) and the enrichment in oxidized samples of phlogopites in Ba and amphiboles in light rare earth elements (LREE), which are likely derived from the slab (K. E. Johnson, in prep.). It appears, therefore, that amphibole-bearing samples are oxidized, and enriched (commonly) in LREE and Ba, and that this process occurs above subduction zones. The agents of metasomatism are most probably hydrous melts and the fluids associated with melting and crystallization phenomena in the mantle wedge.

Continental xenoliths from Kilbourne Hole, San Carlos, and Central Asia are more reduced than those previously discussed and exhibit an fO_2 range from FMQ to about 1.5 log units below FMQ, that is, some are within the 1 bar MORB range and some are above it. Because these localities are in regions of continental extension where the asthenosphere is relatively close to the surface, it is not surprising that some samples have affinities with MORB. Recent trace element and isotopic data (Roden et al., 1988; Zindler and Jagoutz, 1988) have also shown that some of the San Carlos and Kilbourne Hole xenoliths have isotopic and LREE signatures characteristic of MORB-related mantle (that is, without LREE enrichment). Other xenoliths show evidence of LREE enrichment, but without any mineralogic manifestation of metasomatism.

In view of the correlation between high fO_2 values and trace element enrichments in the subduction environment, it is tempting to suggest that the samples enriched in trace elements in extensional environments also correspond to the more oxidized samples. This suggestion was tested by performing REE and other lithophile element analyses of clinopyroxene separates from Kilbourne Hole and San Carlos specimens (Johnson, 1990; Wood et al., 1990). Similar analyses were made on samples from Dish Hill, California and from the Anakie Hills of Southeast Australia. Surprisingly, there was no correlation between trace element enrichment and fO_2 in these anhydrous specimens. Thus, although some of the samples from extensional environments exhibit MORB-like fO_2's and some have MORB-like isotopic and trace element signatures, the sample groupings are not coincident.

In summary, the extensional continental environment is one of local fO_2, trace element and isotopic heterogeneity, caused by mixing of older continental lithosphere with younger asthenosphere. The lack of correlation between fO_2 and high trace element concentration is probably due to the ease with which fO_2 may be perturbed. Spinels constitute around 3% by volume of spinel lherzolites, and they contain about 1 wt % Fe_2O_3. The fO_2 recorded by these rocks is thus not well buffered, and it may be perturbed by the infiltration of relatively small volumes of oxidizing or reducing fluid. A straightforward calculation reveals that infiltration of C-H-O fluids can shift fO_2 in these lherzolites by 2 log units at fluid-rock ratios of 0.01 by volume. Thus, in an environment with a long and complex history such as that of continental extension, small amounts of fluid associated with magmatism can decouple fO_2 from trace element signatures. When one measures fO_2 in this environment one is looking at a mixture of old and new (magmatism-related) signatures. The subduction environment is different because the spinels are much more oxidized (up to 6% Fe_2O_3) and the rocks contain other phases rich in ferric iron, notably amphibole. Mössbauer experiments on amphibole indicate that this phase typically has a ferric/ferrous ratio about the same as that of the coexisting spinel and it is modally much more abundant. Thus, with a ratio of ferric to total iron between 0.3 and 0.4, these rocks have some buffering capacity and are much less readily altered than the more reduced samples from extensional environments. At high fO_2, therefore, correlations between LREE and other trace element enrichments and fO_2 are preserved, whereas at low fO_2 they are readily perturbed.

SUBOCEANIC PERIDOTITES

Abyssal peridotites have been dredged and drilled from a large number of fracture zones along the earth's mid-ocean ridge systems. These peridotites are generally hydrothermally altered, but by analysis of relict phases and determination of modes, primary rock and mineral compositions can be reconstructed (e.g., Dick et al., 1984). Furthermore, fresh unaltered spinel is quite common, enabling application of the spinel peridotite oxygen barometer. Dick et al. (1984) have provided convincing chemical evidence that these abyssal peridotites are the residues of partial melting and that they are thus the residuum of the mantle from which MORBs were formed.

Oxygen thermobarometry of abyssal peridotites from mid-Atlantic, central India, SW India, and America-Antarctica ocean ridge systems as well as the Cayman Trough (Bryndzia and Wood, 1990) reveals that these are substantially more reduced than the continental peridotites discussed earlier. Calculated values at an assumed pressure of 15 kbar give an average fO_2 of -0.9 log units relative to FMQ, about 1 log unit below the average for continental spinel peridotites (Fig. 4). This value is virtually coincident with the mean of the values for MORB glasses when the latter are corrected to 10 to 15 kbar. Even if we assume that the estimated pressure effect is too large, the MORB mean at 1 bar (-1.28 log units relative to FMQ) is essentially the same as that for the abyssal peridotites within uncertainty. The agreement between the two data sets solidifies the idea that there is a genetic link between the abyssal peridotites (residuum) and the MORB (melting products). Furthermore, the agreement with data for rapidly quenched MORB glasses from pillow margins suggests that the margins have not undergone significant degassing (hydrogen loss) during quenching. This is in contrast to the pillow cores, which are more oxidized (Christie et al., 1986).

Four of the six most reduced abyssal peridotites shown in Figure 4 came from the Islas Orcadas fracture zone, a region which produces "hotspot" or "plume" basalts and is close to the Bouvet Island hotspot of the southwest Indian Ridge (Dick et al., 1984). These four samples exhibit a range of fO_2 from -1.85 to -2.64 log units relative to FMQ, well below the average value of -0.9 log units. One possible interpretation is that mantle plumes are more reduced than normal asthenospheric mantle. An alternative explanation is that MORBs and their residues are carbon and fluid-saturated during the melting process (Bottinga and Javoy, 1989) and that the variations in fO_2 reflect an increasing extent of melting at hotspots relative to "normal" ridge. These relationships are discussed in more detail below.

In summary, abyssal peridotites are more reduced than continental spinel peridotite xenoliths and their range in fO_2 is consistent with the notion that they are residua from MORB generation. The association of the most reduced samples with a hotspot locality suggests either that low fO_2 is a characteristic of the hotspot environment or that it reflects a larger than "normal" extent of melting under carbon- and fluid-saturated conditions.

MASSIF PERIDOTITES

Peridotite massifs are large (km size) pieces of the upper mantle that become incorporated in the crust via diapiric upwelling and continental collision. As such they provide the opportunity for detailed study of the small scale variations in fO_2 and other geochemical parameters that may characterize the mantle. A preliminary study of the Ronda (Spain) and Beni Bousera (Morocco) massifs has been made by Woodland et al. (1989). Both of these massifs contain textural, mineralogical and geochemical evidence suggesting that they rose from depths in the garnet lherzolite field (Kornprobst, 1969; Suen and Frey, 1987), and that the garnet broke down to form spinel en route to the surface. In the case of Beni Bousera, the presence of pseudomorphs of graphite replacing diamond suggests that this massif had an origin at pressures of >50 kbar. Both massifs have veins and dikes indicating that they have undergone partial melting (Dickey, 1970). Initial measurements indicate that the bodies record two oxygen fugacities similar to those of abyssal peridotites, and unlike those of the continental xenoliths. Furthermore, a range of >2 log units in fO_2 is observed, and there is no systematic variation across the body. The spread in fO_2 values is similar to that in MORB glasses (Fig. 1), abyssal peridotites, and continental xenoliths (Fig. 3) and indicates that it represents real small scale variations in oxidation state of the upper mantle.

The Ronda and Beni Bousera massifs appear, therefore, to be two examples of peridotites that originated in the asthenosphere and that have the fO_2 ranges characteristic of suboceanic mantle. They did not have their oxygen fugacities modified during their short residence times in the lithosphere. Whether other peridotite massifs have similar ranges of fO_2 is not yet known.

Possible role of carbon in controlling fO_2

In the context of this study one of the most interesting results has been the reduced nature of the abyssal peridotites, particularly those from the Islas Orcadas fracture zone, which range from 1.67 to 2.32 log fO_2 units below FMQ. Many of these near-hotspot

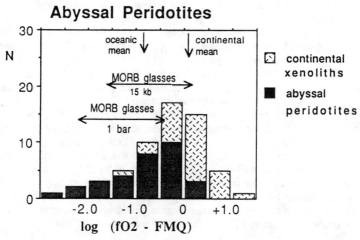

Figure 4. Comparison of oceanic abyssal peridotites with continental xenoliths. The former are more reduced and correspond closely to the fO_2 range of MORB glass.

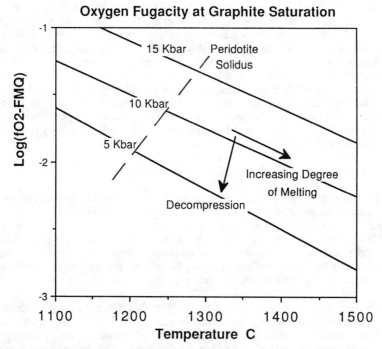

Figure 5. At graphite saturation in the presence of a C-O fluid (or C-H-O fluid with low X_{H_2O}), increased degrees of melting will produce relative reduction as illustrated. This may explain the relatively reduced nature of peridotites from the Islas Orcadas Fracture Zone.

samples are extremely refractory (Dick, 1989), suggesting that high degrees of partial melting influence fO_2. There are two possible explanations for the observation. One is that, during partial melting, Fe^{3+} is more incompatible than Fe^{2+} so that the restite becomes progressively depleted in Fe^{3+}, leading to a progressive decline in fO_2 in what would be an unbuffered system. The other, proposed several times before (e.g., French, 1966; Sato, 1978; Woermann and Rosenhauer, 1985, Ulmer et al., 1987) is that carbon exerts a controlling influence on fO_2. If, under conditions of partial melting the mantle were saturated in carbon and H_2O were not the dominant fluid species, as is suggested by spectroscopic data on MORB glasses (e.g., Eaby Dixon et al., 1988), the fO_2 would be fixed at a value near that defined by graphite saturation in the C-O system. This is very close to the reduced fO_2 values observed in many MORB glasses and abyssal peridotites (Fig. 5). Furthermore, progressive melting would lead to a decreasing fO_2 relative to FMQ because of the angle between FMQ and CCO buffers in log fO_2-T space (Fig. 5). Thus, the correlation between degree of melting and fO_2 and the apparently reduced nature of peridotites from near a hotspot could simply be due to carbon and fluid saturation. The total amount of carbon required in fertile mantle would be on the order of 700 ppmw, a value in the range indicated by $C/^3He$ values (Marty and Jambon, 1987).

Blundy et al. (1991) have taken these arguments a step further and shown that many of the measured fO_2 values for continental spinel peridotites are also close to those implied by C-saturation, albeit under lower temperature and higher pressure conditions where CCO and related equilibria lie close to FMQ. It becomes something of a "chicken and egg" argument, however, because it is clear that C (both native and CO_2) is virtually ubiquitous in mantle and mantle-derived samples and the Fe^{3+} and Fe^{2+} are also present in moderate concentrations. Therefore, because CCO and Fe^{2+}-Fe^{3+} solid equilibria are extremely close to one another at high pressure (e.g., Wood et al., 1990) there is an interplay between the two sets of equilibria. At the concentration levels inferred to be present in the mantle, up to 1000 ppm C and 1000-2000 ppm Fe_2O_3, neither will dominate under all conditions, and the result is a mixed signal with C perhaps being more important in the most fertile region of MORB generation. In depleted lithosphere, however, and in the oxidising environment above subduction zones, carbon is completely oxidised and substantial amounts of Fe^{3+} are produced, entering spinel, clinopyroxene and, at high f_{H_2O}, amphibole. Under these conditions as noted earlier, the ratio of Fe^{3+}/Fe^{2+} becomes sufficiently large that, with carbon exhausted, the assemblage is essentially Fe-buffered at 1-2 log units above FMQ.

SUMMARY

The intrinsic oxygen fugacity method can produce spuriously low results because of the presence of native carbon in many mantle samples. Oxide thermobarometry for peridotites is, on the other hand, now sufficiently well advanced for fO_2 values to be derived from eqn (2) to within, in general, ±0.5 log units. The most important parameter which requires careful measurement for accurate thermobarometry is the Fe^{3+} content of spinel. This may be determined accurately by Mössbauer spectroscopy or, with judicious choice of standards, by electron microprobe.

The fO_2 of the asthenosphere during basalt genesis is most likely represented by the oxidation states recorded in abyssal peridotites, about +0.5 to -2.5 log units relative to FMQ. The range arises from the interplay between Fe^{3+}, reduced C (and possibly sulfur), hydrous phases, and carbonates during mantle diapirism. The decompression results in oxidation of C by Fe^{3+} to generate CO_2, together with melting of any carbonate and hydrous phases. The result is to produce volatile-bearing basalts with a residuum of reduced abyssal spinel peridotites. Overall, this is a mechanism by which the mantle becomes more reduced.

The average oxygen fugacity of the mantle making up the continental lithosphere is 1 log unit higher than the asthenosphere, primarily because of oxidation and hydration associated with subduction processes. Oxidation is accompanied by large iron lithophile (LIL) element enrichment. Thus, although some continental xenoliths still show an asthenospheric or MORB-like signature, considerable heterogeneity and oxidation have occurred by secondary processes, and the subduction environment is clearly one of oxidation. In contrast to the xenoliths, large continental massifs (Ronda, Beni Bousera) that originated in the asthenosphere and have had only short residence times in the lithosphere exhibit predominantly MORB-like oxygen fugacities.

REFERENCES

Andersen, T., S.Y. O'Reilly and W.L. Griffin (1984) The trapped fluid phase in upper mantle xenoliths from Victoria, Australia. Implications for mantle metasomatism. Contrib. Mineral. Petrol. 88, 72-85.
Arculus, R.J., (1985) Oxidation status of the mantle: Past and present. Ann. Rev. Earth Planet. Sci. 13, 75-95.
Arculus, R.J. and J.W. Delano (1981) Intrinsic oxygen fugacity measurements: Techniques and results from upper mantle peridotites and megacryst assemblages. Geochim. Cosmochim. Acta 45, 899-913.
Arculus, R.J., J.B. Dawson, R.H. Mitchell, D.A. Gust and R.D. Holmes (1984) Oxidation states of the upper mantle recorded by megacryst ilmenite in kimberlite and type A and B spinel lherzolites. Contrib. Mineral. Petrol. 85, 85-94.
Ballhaus, C., R.F. Berry and D.H. Green (1990) Oxygen fugacity controls in the Earth's upper mantle. Nature 348, 437-440.
Blundy, J.D., J.P. Brodholt and B.J. Wood (1991) Carbon-fluid equilibria and the oxidation state of the upper mantle. Nature 349, 321-324.
Bottinga, Y. and M. Javoy (1989) MORB degassing: evolution of CO_2. Earth Planet. Sci. Lett. 95, 215-225.
Bryndzia, L.T. and B.J. Wood (1990) Oxygen thermobarometry of abyssal spinel peridotites: the redox state and C-O-H volatile composition of the Earth's suboceanic mantle. Am. J. Sci. 290, 1093-1116.
Christie, D.M., I.S.E. Carmichael and C.H. Langmuir (1986) Oxidation states of mid-ocean ridge basalt glasses. Earth Planet Sci. Lett. 79, 397-411.
Dick, H.J.B. (1989) Abyssal peridotites, very-slow spreading ridges and ocean ridge magmatism in Saunders, A.D. and M.J. Norry, Eds. Magmatism in the Ocean Basins: Geol. Soc. London Spec. Publ. 42, 71-105.
Dick, H.J.B., R.L. Fisher and W.B. Bryan (1984) Mineralogic variability of the uppermost mantle along mid-ocean ridges: Earth Planet. Sci. Lett. 69, 88-106.
Dickey, J.S. (1970) Partial fusion products in alpine-type peridotites: Serrania de la Ronda and other examples. Washington, D.C.: Mineral. Soc. Am. Spec. Paper 3, 33-49.
Dyar, M.D., A.V. McGuire and R.D. Ziegler (1989) Redox equilibria and crystal chemistry of coexisting minerals from spinel lherzolite mantle xenoliths: Am. Mineral. 74, 969-980.
Eaby Dixon, J., E. Stolper and J.R. Delaney (1988) Infrared spectroscopic measurements of CO_2 and H_2O in Juan de Fuca Ridge basaltic glasses: Earth Planet. Sci. Lett. 90, 87-104.
Eggler, D.H. and R.D. Baker (1982) Reduced volatiles in the system C-O-H: Implications to mantle melting, fluid formation, and diamond genesis. In: High Pressure Research in Geophysics, S. Akimoto and M. Manghnani, (eds.), 237-250. Center for Academic Pub., Tokyo, Japan.
French, B.M. (1966) Some geological implications of equilibrium between graphite and a C-H-O gas at high temperatures and pressures. Rev. Geophys. 4, 223-253.
Fudali, R.F. (1965) Oxygen fugacity of basaltic and andesitic magmas. Geochim. Cosmochim. Acta. 29, 1063-1075.
Gudmundsson, G., J.R. Holloway and I.S.E. Carmichael (1988) Pressure effect on the ferric/ferrous ratio in basaltic liquids: Am. Geophys. Union Trans. 69, 1511.

Johnson, K.E. (1990) An appraisal of mantle metasomatism based upon oxidation states, trace element and isotope geochemistry and fluid/rock ratios in spinel lherzolite xenoliths. Ph.D. dissertation Northwestern Univ., Evanston, Illinois.

Kempton, P.D., R.S. Harmon, H.G. Stosch, J. Hoefs and C.J. Hawkesworth (1988) Open system oxygen isotope behaviour and trace element enrichment in the sub-Eifel mantle. Earth Planet. Sci. Lett. 89, 273-287.

Kornprobst, J. (1969) Le massif ultrabasique de Beni Bousera. Etude des péridotites de haute température et de haute pression et des pyroxénites avec ou sans grenat qui leur sont associées. Contrib. Mineral. Petrol. 23, 283-322.

Lange, R.A. and I.S.E. Carmichael (1987) Densities of $Na_2O-K_2O-CaO-MgO-FeO-Fe_2O_3-Al_2O_3-TiO_2-SiO_2$ liquids: New measurements and derived partial molar properties. Geochim. Cosmochim. Acta. 51, 2931-2946.

Luth, R.W., D. Virgo, F.R. Boyd and B.J. Wood (1990) Ferric iron in mantle-derived garnets. Implications for thermobarometry and for the oxidation state of the mantle. Contrib. Mineral. Petrol. 104, 56-72.

Mathez, E.A., V.J. Dietrich and A.J. Irving (1984) The geochemistry of carbon in mantle peridotites. Geochim. Cosmochim. Acta. 48, 1849-1860.

Marty, B. and A. Jambon (1987) $C/^3He$ in volatile fluxes from the solid earth: implications for carbon geodynamics. Earth Planet. Sci. Lett. 83, 16-26.

Mattioli, G.S. and B.J Wood (1988) Magnetite activities across the $MgAl_2O_4 - Fe_3O_4$ join, with application to thermobarometric estimates of upper mantle oxygen fugacity. Contrib. Mineral. Petrol. 98, 148-162.

Mattioli, G.S., M.B. Baker, M.J. Rutter and E.M. Stolper (1989) Upper mantle oxygen fugacity and its relationship to metasomatism: J. Geol. 97, 521-536.

Nell, J. and B.J. Wood (1991) High temperature thermopower and electrical conductivity measurements on Cr-bearing spinels and development of a thermodynamic model. Am. Mineral. (in press).

O'Neill, H. St.C. and A. Navrotsky (1983) Simple spinels: Crystallographic parameters, cation radii, lattice energies and cation distributions. Am. Mineral. 68, 181-194.

O'Neill, H. St.C. and A. Navrotsky (1984) Cation distributions and thermodynamic properties of binary spinel solid solutions. Am. Mineral. 69, 733-753.

O'Neill, H. St.C. and V.J. Wall (1987) The olivine-spinel oxygen geobarometer, the nickel precipitation curve and the oxygen fugacity of the upper mantle. J. Petrol. 28, 1169-1192.

Roden, M.F., A.J. Irving and V. Rama Murthy (1988) Isotopic and trace element composition of the upper mantle beneath a young continental rift. Results from Kilbourne Hole, New Mexico. Geochim. Cosmochim. Acta. 52, 461-474.

Sato, M. (1965) Electrical thermometer: a possible new method of geothermometry with electroconductive minerals. Econ. Geol. 60, 812-818.

Sato, M. (1972) Intrinsic oxygen fugacities of iron-bearing oxide and silicate minerals under low total pressure. Geol. Soc. Am. Mem. 135, 289-307.

Sato, M. (1978) Oxygen fugacity of basaltic magmas and the role of gas-forming elements. Geophys. Res. Lett. 5, 447-449.

Sato, M. and M. Valenza (1980) Oxygen fugacities of the layered series of the Skaergaard intrusion, East Greenland. Am. J. Sci. 280-A, 134-158.

Suen, C.J. and F.A. Frey (1987) Origins of the mafic and ultramafic rocks in the Ronda peridotite. Earth Planet. Sci. Lett. 85, 183-202.

Trial, A.F., R.L. Rudnick, L.D. Ashwal, D.J. Henry and S.C. Bergman (1984) Fluid inclusions in mantle xenoliths from Ichinomegata, Japan: Evidence for subducted H_2O? (abstract). EOS Trans. Am. Geophys. Union 65, 306.

Ulmer, G.C., D.E. Grandstaff, D. Weiss, M.A. Moats, T.J. Buntin, D.P. Gold, C.J. Hatton, A. Kadik, R.A. Koseluk, M. Rosenhauer (1987) The mantle redox story: an unfinished story. In Mantle metasomatism and alkaline magmatism. Geol. Soc. Am. Spec. Paper 215, 5-23.

Virgo, D., R.W. Luth, M.A. Moats and G.C. Ulmer (1988) Constraints on the oxidation state of the mantle: an electrochemical and ^{57}Fe Mössbauer study of mantle-derived ilmenites. Geochim. Cosmochim. Act. 52, 1781-1794.

Wells, P.R.A. (1977) Pyroxene thermometry in simple and complex systems. Contrib. Mineral. Petrol. 62, 129-139.

Williams, H. and A.R. McBirney (1979) Volcanology. Freeman, Cooper and Co., San Francisco.

Wiser, N. and B.J. Wood (1991) Experimental determination of activities in Fe-Mg olivine at 1400 K. Contrib. Mineral. Petrol. (in press).

Woermann, E. and M. Rosenhauer (1985) Fluid phases and the redox state of the Earth's mantle. Fortschr. Mineral. 83, 263-349.

Wood, B.J. (1990) An experimental test of the spinel peridotite oxygen barometer. J. Geophys. Res. 95,15845-15851.
Wood, B.J. and O.J. Kleppa (1981) Thermochemistry of forsterite-fayalite olivine solutions. Geochim. Cosmochim. Acta. 45, 569-581.
Wood, B.J. and D. Virgo (1989) Upper mantle oxidation state: Ferric iron contents of lherzolite spinels by ^{57}Fe Mössbauer spectroscopy and resultant oxygen fugacities. Geochim. Cosmochim. Acta. 53, 1277-1291.
Wood, B.J., L.T. Bryndzia and K.E. Johnson (1990) Mantle oxidation state and its relationship to tectonic environment and fluid speciation. Science 248, 337-345.
Woodland, A., J. Kornprobst and B.J. Wood (1989) Oxygen fugacities of the peridotite massifs of Ronda (Spain) and Beni Bousera (Morocco). Geol. Soc. Am. Abstr. with program 21, A106.
Zindler, A. and E. Jagoutz (1988) Mantle cryptology. Geochim. Cosmochim. Acta 52, 319-334.

Chapter 12   B. R. Frost & D. H. Lindsley

OCCURRENCE OF IRON-TITANIUM OXIDES IN IGNEOUS ROCKS

INTRODUCTION

The composition of Fe-Ti oxides in an igneous rock provides petrologists with important information on the oxygen fugacity and temperature at which that rock crystallized (Buddington and Lindsley, 1964). When used in conjunction with the composition of coexisting ferromagnesian silicates, the oxide compositions can also provide the silica activity of the magma and the pressure at the time the assemblage crystallized. The oxide phases in igneous rocks are, of course, not composed solely of iron and titanium; they contain minor amounts of many other components such as Al, Cr, V, Mg, and Mn. The first part of this paper, therefore, is a discussion of the compositional range found in Fe-Ti oxides in typical igneous rocks. In the second part of this paper we discuss some of the exciting advances that have been made in understanding the silicate-oxide relations in igneous rocks and how they can be used to calculate the intensive parameters at which a rock crystallized.

THE COMPOSITION OF Fe-Ti OXIDES IN IGNEOUS ROCKS

Fe-Ti spinel in volcanic rocks

Although at magmatic temperatures there is nearly complete solid solution between the magnetite-ulvöspinel spinels and the chromate and aluminate spinels (Sack and Ghiorso, in press), in magmatic rocks the dominant spinel phase is usually either a chromite or a titanomagnetite. In some basaltic rocks both Cr-spinels and Fe-Ti spinels are present (see for example: Brown and Carmichael, 1971; Stormer, 1972; Thompson, 1973; Johnson and Arculus, 1978, Kyle, 1981; Perfit et al., 1982; Aurisicchio et al.,1983; Hasenaka and Carmichael, 1987; Fodor et al., 1989). In many occurrences the spinel is included in olivine (Brown and Carmichael, 1971; Stormer, 1972; Johnson and Arculus, 1978) and may be xenocrystic and not related to the surrounding phenocryst assemblage. In others, there is evidence that crystallization had proceeded from chromite to titanomagnetite during the evolution of the melt. This is particularly well displayed by the zoned spinels from the Snake River Plain (Thompson, 1973). It is also evident in the basalts and andesites of Michoacan-Guanajuato, where the spinels form a complete series from Mg-chromite to Ti-magnetite. Chromite is found in rocks that are relatively higher in Mg (and Cr), whereas titanomagnetite is found in those that are relatively poorer in Mg (Hasenaka and Carmichael, 1987).

There is a wide range in Ti content in spinels from volcanic rocks. Nearly pure ulvöspinel has been reported from highly reduced lavas from Greenland and Germany that have been intruded into coal (Medenbach and El Goresy, 1982). Spinels with more than 80% of the ulvöspinel component (in some instances nearly up to 100%) have been reported from reduced rhyolites (Ewart et al., 1977). Ulvöspinel from Greenland and Germany contains substantial amounts of Cr_2O_3 (up to

7.75%), Al_2O_3 (up to 5%), and MnO (up to 4%), whereas the ulvöspinel from rhyolites of S.E. Queensland contains Al_2O_3 (ca. 1%) as the only important impurity.

Tholeiitic rocks. Spinels from tholeiitic rocks tend to be enriched in Ti, with spinel from MORB rocks ranging from 50 - 80% Usp (Bence et al., 1975; Mazzulo and Bence, 1976; Perfit et al., 1982). Similar Ti-enriched spinels are also reported from tholeiitic rocks of Kilauea (Anderson and Wright, 1972) and Makaopuhi (Evans and Moore, 1968), and the flood basalts of the Parana Plateau (Bellieni et al., 1984). Spinels from these localities have a substantial amount of minor components. For example Al_2O_3 ranges from around 1 to 2 weight %; weight % Cr_2O_3, MnO, and V_2O_5 are usually between 0.5 and 1%. MgO content is variable with some spinels having as much as 2 or 3 %, whereas in others the MgO contents is well less than 0.1%. The extreme variability in MgO content of spinels, and to a lesser extent of the other oxides, is a reflection of the rapidity at which spinels tend to undergo cation-exchange with other minerals. A typical titanomagnetite in equilibrium with magnesian olivine at magmatic conditions should have a MgO content of 2% or more; those with low MgO contents will have lost Mg by cation exchange on cooling.

Calc-alkalic volcanics. Spinels from andesitic and dacitic lavas tend to be lower in Ti than those from tholeiites. For example, andesites from Cold Bay in the Aleutians have Ti-magnetite with Usp contents ranging from 24-44 mole % (Brophy, 1984, 1986), while a similar range (Usp_{20-40}) is shown at Colima (Carmichael and Luhr, 1980), the S. Cascades (Smith and Carmichael, 1968), and Sarigan island in the Marianas (Meijer and Reagan, 1981). Titanium content of the spinel is a bit higher in andesitic and dacitic rocks of Medicine Lake (Usp_{45-62}, Gerlach and Grove, 1982) and Santorini (Usp_{43-60}, Nicholls, 1971)) and a bit lower in rocks of Tonga (Usp_{18-26}, Ewart et al., 1973). Minor element contents of spinels from andesites and dacites are similar to those from tholeiitic rocks. Al_2O_3 and MgO range from 1 to 3%, while MnO and V_2O_5 are usually in the range of 0.5 to 1%.

Alkalic rocks. Spinels from alkalic basalts are also somewhat poorer in Ti than those from tholeiites. For example, alkali basalts from Annobon contain spinels with 52-62 mole % ulvöspinel (Cornen and Maury, 1980), a range which is similar to that of alkali basalts from the Central Pacific (Myers et al., 1975) and Hut Pt. Antarctica (Kyle, 1981). Alkali basalts from Banks peninsula (Usp_{66-71}, Price and Taylor, 1980), S. Aukland (Usp_{61-65}, Rafferty and Heming, 1979), and the S. Highlands of New South Wales (Usp_{42-70}, Wass, 1973) range to somewhat higher Ti-content, but still are lower in Ti than spinels from most tholeiites. Spinels in alkali basalts are richer in minor components than those from tholeiitic rocks or andesites. Aluminum and magnesium are markedly higher, with Al_2O_3 and MgO contents up to 6 and 5 %, respectively (Rafferty and Heming, 1979; Kyle, 1981; Cornen and Maury, 1980).

Spinels from other alkalic rocks tend to have moderate amounts of Ti (Usp_{30-70}), but there appears to be no consistent trend in Ti content as a function of rock type. The distinctive feature of spinels from alkalic rocks is their tendency toward high Al_2O_3 content, particularly in basanites and hawaiites. For instance a basanitic spinel from the Southern Highlands of New South Wales contains 7% Al_2O_3 (Wass, 1973) and a Ne-hawaiite from Hut Pt., Antarctica, contains a titanomagnetite with

10% Al_2O_3 (Kyle, 1981). Magnesium is correspondingly high, 5.8 % in the basanite from N.S.W. and 8.3% at Hut Pt.

The high Al content of spinels from undersaturated rocks reflects the fact that the aluminate content of spinels from a rock containing olivine, clinopyroxene, and plagioclase is a function of silica activity according to the equilibrium:

$$CaAlSi_2O_8 + Mg_2SiO_4 = MgAl_2O_4 + SiO_2 + CaMgSi_2O_6 \quad (1)$$
plagioclase olivine spinel in melt diopside

Decreasing silica activity will cause reaction (1) to move to the right thereby increasing the activity of aluminate in the spinel. Reaction (1) has a slope of 1.5 bar/degree. The aluminum content of spinel from an assemblage olivine-diopside-plagioclase-spinel may therefore be a geobarometer, provided one has an independent estimate of silica activity.

<u>Silicic volcanics</u>. Spinels from silicic volcanics show a complete range in titanium content from Usp_{100} in some fayalite rhyolites (Ewart et al. 1973) to around Usp_{10} in rhyolite from Los Chocoyos (Rose et al., 1979), and S. Nevada (Lipmann, 1971). Magnetite in dacite from Martinique reaches even lower values (Usp_3; d'Arco et al., 1981), but this low value is clearly product of subsolidus reequilibration. The Ti content of titanomagnetite is a function of the oxygen fugacity at which the magma crystallized. The spinels of relatively reduced fayalite rhyolites have high Ti (see list of occurrences in Frost et al., 1988), whereas those of the more oxidized rhyolites with biotite and hornblende have low Ti (see Carmichael, 1967a).

Minor element abundances in spinels from silicic volcanic rocks are variable. Al_2O_3 is in the range of 1 - 2 % in spinels from fayalite rhyolites (Carmichael, 1967a; Conrad, 1984; Novak and Mahood, 1986;) but it may be as high as 3% in peraluminous rhyolites (Bacon and Duffield, 1981). Similarly, MgO is less than 1% in the iron-rich fayalite rhyolites, but may be up to 2% in more magnesian rocks (Rose et al., 1979). MnO is usually present in abundances of less than 1%, although a few lavas contain spinel with MnO contents as much as 2 or even 3 % (Bacon and Duffield, 1981; Novak and Mahood, 1986). Hildreth (1977) showed that the MnO and MgO contents of magnetite from the Bishop Tuff are dependent on the apparent temperature of equilibration. With decreasing temperature (as determined by Fe-Ti oxide thermometry), the MnO in the magnetite increases from 0.5 to 1.0% whereas MgO decreases from more than 0.7 to less than 0.3%.

Fe-Ti spinel in plutonic rocks

The composition of spinels in plutonic rocks differs from that in volcanic rocks because the oxides tend to re-equilibrate readily during cooling. This behavior leads to a complex series of re-equilibration reactions. As outlined by Frost et al. (1988), at high temperatures these reactions will involve inter-grain exchange of ferric iron and titanium between titanomagnetite and ilmenite with the consequent enrichment of Fe_3O_4 in the spinel and of $FeTiO_3$ in the rhombohedral phase. At intermediate temperatures Ti-magnetite re-equilibrates by the process of "oxyexsolution" in which the Fe_2TiO_4 component of the titanomagnetite oxidizes to form granules of ilmenite within or around the spinel. At lower temperatures the process produces lamellae of ilmenite in (111) of the host spinel (see Chapter 5). In addition to the decrease of Ti in the spinel during cooling, both the spinel and the rhombohedral oxide will lose Mg through exchange with the surrounding silicates. As a result of these processes the host spinel in many plutonic rocks is nearly pure magnetite, with small amounts of Ti and only trace amounts of MgO and Al_2O_3 remaining in solid solution. An important exception to this general rule are the oxide-rich layers in some intrusions. Because they are isolated from silicates, they retain relatively high contents of MgO.

To use oxide compositions for geothermometry petrologists usually try to determine the original titanomagnetite composition by reconstituting the oxyexsolved titanomagnetites through various techniques (see Bohlen and Essene, 1977). In some instances the results produce ulvöspinel contents that are consistent with independent estimations of temperature. In other instances, it is evident that the resulting spinel composition reflects temperatures extending well below peak temperatures (Morse, 1980; Frost et al., 1988). In the following discussions of plutonic rocks we use the reconstituted magnetite compositions unless otherwise noted.

Mafic plutonic rocks. Magnetite occurs in the more differentiated portions of most layered mafic intrusions. For example it makes up about 15% in the Upper Zone of the Bushveld Intrusion (Willemese, 1969; Molyneux, 1972), it appears at the top of the Lower Zone of the Skaergaard intrusion and persists in abundances of 1 - 8 volume % throughout the rest of the crystallization history (Vincent and Phillips, 1954; Wager and Brown, 1968). It is also found in the upper portions of the Muskox (Irvine and Smith, 1969), Kiglapait (Morse, 1980), and Dufek intrusions (Himmelberg and Ford, 1977). Magnetite in these bodies is often markedly enriched in vanadium (1-3%), particularly in those horizons where it first appears as a cumulate phase.

Abundant magnetite is also found in mafic and ultramafic plutons associated with calc-alkalic plutonism. These include intrusions like the Duke Island (Irvine, 1974) and similar plutons throughout S.E. Alaska (Taylor and Noble, 1969), Bear Mountain Complex (Snoke et al., 1981), Somerset Dam (Mathison, 1975), and Lake Owen Complex (Patchen, 1984). Although the chrome content of magnetite from these complexes is highly variable, magnetite from these calc-alkalic mafic plutons does appear to be richer in Cr_2O_3 than that from tholeiitic layered mafic plutons. For example Cr_2O_3 ranges up to 3% in Somerset Dam (Mathison, 1975), up to 5% in the Bear Mountain Complex (Snoke et al., 1981), and up to 20% in the Lake Owen intrusion (Patchen, 1984). The few analyses of vanadium from magnetite in these plutons lie in the lower range of that found in tholeiitic mafic intrusions. For example V_2O_3 contents of magnetite from Somerset Dam are less than 0.5%.

Alkalic plutonic rocks. Magnetite or titanomagnetite is the dominant Fe-Ti oxide in alkalic plutonic rocks. Magnetite from carbonatites tends to be very poor in titanium (<2.0 % TiO_2, Prins, 1972; Mitchell, 1978; Gaspar and Wyllie, 1983), although amounts as high as 6.4% have been reported from the Sarfortôq carbonatite (Sechee and Larsen, 1980). Minor elements (Al_2O_3, MgO, MnO) in magnetite from carbonatites are generally present in abundances of less than 1%, but magnetite from Jacupiranga has MgO contents up to 4.0% (Mitchell, 1978).

Reconstructed titanomagnetite compositions from nepheline syenites show a wide range of compositions ranging from Usp_{20} in a nepheline-syenite inclusion from Tenerife (Ferguson, 1978) and in nepheline-bearing rocks from the Tugtutôq dike (Upton et al., 1985) to more than Usp_{70} in the Igdlerfigssalik (Powell, 1978) and Ilimaussaq (Larsen, 1976) intrusions. The most elevated minor element in these magnetites is MnO, which reaches 3.8% in Tenerife and 2.7% in Tugtutôq.

Felsic plutonic rocks. Magnetite in felsic plutonic rocks commonly approaches nearly pure Fe_3O_4, a fact usually attributed to extensive sobsolidus re-equilibration (Czamanske et al., 1977; Czamanske et al., 1981). For example the magnetites from the Finnmarka complex contain less than 0.50% TiO_2, 0.32% Al_2O_3, 0.75% MnO, and less than 0.06% MgO (Czamanske and Mihalik, 1972). Similar compositions are seen in magnetites from other granitic bodies (Barker et al., 1975; Stephenson and Hensel, 1978; Andersen, 1984). Even in instances where the reconstructed Ti content of magnetite is consistent with magmatic conditions, such as in some samples from Ballachulish (Weiss, 1986; Weiss and Troll, 1989), the minor elements Mg, Mn, and Al are far lower than the amounts found in magnetites from silicic volcanics. These compositional differences indicate that substantial re-equilibration occurred between magnetite and the matrix before the onset of or during oxyexsolution. This is consistent with the calculations of Frost et al. (1988), who found that reconstructed magnetite compositions in fayalite granites had oxide-silicate equilibration temperatures distinctly lower than those of most fayalite rhyolites. The extensive re-equilibration of magnetite in granitic rocks is likely to have been strongly enhanced by the presence of an aqueous fluid. Not only is the fluid likely to flux exsolution (Parsons, 1980), it is likely also to lower the blocking temperature of intergrain exchange.

Ilmenite-hematite in volcanic rocks

Tholeiitic volcanics. Ilmenite is absent from some sea-floor basaltic rocks (Bence et al., 1975; Myers et al., 1975) although it is present in others (Mazzulo and Bence, 1976). Where it is found in tholeiitic lavas, ilmenite usually has low to moderate Fe_2O_3 contents. Tholeiitic lavas from the E. Galapagos Rift (Perfit and Fornari, 1983), Garabaldi (Nicholls et al., 1982), the Gulf of California (Perfit et al., 1982), Kilauea (Anderson and Wright, 1972) and Makaopuhi (Evans and Moore, 1968) have between 10 and 15 mole% of the hematite component (up to 18% for Kilauea). In contrast, tholeiitic lavas of Thingmuli (Carmichael, 1967b) and the Nazca Plate (Mazzulo and Bence, 1976) contain only 4 - 10 mole % of the hematite component. The only significant minor element in ilmenites from tholeitic rocks is MgO, which is usually present in the range of 1 to 3%. Lower amounts of MgO (0.16 - 0.40%) are found in ilmenites from the Nazca Plate (Mazzulo and Bence, 1975) and higher amounts (4.8-7.2%) occur in ilmenites from Kilauea (Anderson and Wright, 1968). MnO is generally low in ilmenite from tholeiitic rocks, ranging from 0.5 to 1.0%.

Calc-alkalic lavas. Ilmenite is generally absent from high-Al basalts and basaltic andesites and is rare in andesitic lavas. Hemoilmenite and ilmenohematite have been reported from andesites of Cold Bay (Brophy, 1984), Colima (Carmichael and Luhr, 1980), Katmai (Hildreth, 1983), the Raton-Clayton volcanic field (Stormer,

1972) and Witu Island (Johnson and Arculus, 1978). The Fe_2O_3 content of ilmenite from andesites is variable but generally higher than that of ilmenite from tholeiites. Ilmenite with the lowest hematite content comes from Witu Island (Hem_{10}) and Cold Bay (where the rims of the ilmenite grains are Hem_8). In contrast the cores of the ilmenite grains from Cold Bay are Hem_{24}, which is compositionally more similar to the ilmenite from Colima (Hem_{29}) and Katmai (Hem_{30}). The most hematite-rich rhombohedral oxides come from the Sierra Grande andesitic lavas of the Raton-Clayton volcanic field, which have compositions from Hem_{54} to Hem_{94}. Minor element contents of ilmenite from andesitic rocks are similar to those from tholeiites, with MgO between 1 and 4% and MnO from 0.5 to 1.0%.

<u>Ilmenite in alkalic lavas.</u> Some alkali basalts, such as those from Banks Peninsula (Price and Taylor, 1980) lack ilmenite; titanomagnetite is the sole Fe-Ti oxide. Most alkali basalts do contain ilmenite, with the Fe_2O_3 component of the ilmenite ranging between 3 to 8 mole % (Bizouard et al., 1980; Maury et al., 1980; Kyle, 1981; Nicholls et al., 1982). The most abundant minor element in ilmenite from these rocks is Mg, which is usually present in abundances of 1 - 5 wt % MgO (although ilmenite from alkalic basalts of Hut Point has between 6 and 7% MgO, Kyle, 1981). MnO contents for ilmenite from alkali basalts are similar to those from tholeiites, specifically 0.5 to 1%.

Ilmenite is absent from most feldspathoid-bearing lavas, although it has been reported to occur with nephelene in mafic lavas from Hut Point (Kyle, 1981), Itcha (Nicholls et al., 1982) and in trachyte from Tejeda (Crisp and Spera, 1987). Ilmenite from such rocks is low in the hematite component (Hem_3 - Hem_{12}), and has high MgO (3.5 - 5.2 %) and moderate MnO (0.58-0.72%). Ilmenite is also reported to occur in lavas with leucite from New South Wales (Cundari, 1973) and Cosgrove (Birch, 1978). Ilmenite from these rocks has only 1 to 3 mole % of the hematite component, 1-3% MgO, and 0.7-1.2% MnO. Ilmenite of an unusual composition (33 mole % of the hematite component, more than 10% MgO, and 1.3% MnO) is reported to occur with hauyne from the Raton-Clayton volcanic field (Stormer, 1972).

<u>Ilmenite in silicic lavas.</u> Ilmenite from silicic volcanic rocks shows a wide range of hematite content, an observation first made by Carmichael (1967a). Ilmenite from fayalite-bearing rhyolites has the lowest hematite contents, ranging from Hem_1 to Hem_7 (see compilation of Frost et al., 1988). Because of the evolved nature of such rocks, the MgO content of ilmenite from fayalite rhyolites is low, generally well under 1%, and the MnO content is correspondingly high, usually in the range of 1 to 2 %. Fe_2O_3 in ilmenite from orthopyroxene-bearing silicic lavas ranges from Hem_1 to Hem_{29}, generally higher than in fayalite-bearing lavas. Ilmenite with the lowest hematite contents comes from assemblages that lack either quartz or magnetite, whereas that with the highest hematite content is from rocks having the assemblage q-opx-mt-ilm with moderately magnesian opx (Carmichael, 1967a; Ewart, 1981). MgO content of ilmenite from orthopyroxene-bearing silicic lavas (slightly less than 1% to nearly 3%) is generally higher than that of MnO (usually less than 1%). As with magnetite, the minor element content of ilmenite from the Bishop Tuff is strongly dependent on the apparent equilibration temperature (Hildreth, 1977).

Ilmenites from tuffs with biotite or hornblende have hematite contents that range from 6 mole % to more than 30 mole %. The ilmenite with the lowest ferric iron (Hem_{6-9}) is from the Tajao ignimbrite on Tenerife (Wolf and Storey, 1983). In other occurrences, such as Crater Lake (Druitt and Bacon, 1989), Fish Canyon (Whitney and Stormer, 1985), Infiernite (Henry et al., 1988), and S.W. Nevada (Lipman, 1971) the ilmenite is much richer in hematite (Hem_{15-30}). Ilmenite from these rocks may show ranges in both MnO and MgO from less than 1% to more than 4%.

Ilmenite from plutonic rocks

Understanding the compositional range of ilmenite from plutonic rocks is complicated by the fact that in plutonic rocks ilmenite may form in two ways. It may be a product of direct crystallization, or it may be produced by oxyexsolution from titanomagnetite (Buddington and Lindsley, 1964). Oxyexsolved ilmenite is easy to recognize texturally when it remains as the familiar (111) lamallae, but ilmenite formed by oxyexsolution at high temperatures tends to migrate to grain boundaries (this is the granule-oxyexsolution of Buddington and Lindsley, 1964) where it may form separate grains or overgrowths on primary ilmenite. In some occurrences ilmenite formed by oxyexsolution can be distinguished from primary ilmenite by its higher MnO content (Himmelberg and Ford, 1977; Fuhrman et al., 1988). In yet other occurrences there is no consistent relation between the composition of oxyexsolved ilmenite and discrete ilmenite (Pasteris, 1985).

Ilmenite from mafic plutons. In some mafic plutons, such as Bear Mountain Complex (Snoke et al., 1981), Duke Island (Irvine, 1974), McIntosh Intrusion (Mathison and Hamlyn, 1987) and perhaps the Bushveld (Willemse, 1969; Molyneux, 1972) primary ilmenite is absent. In most other mafic intrusions primary ilmenite is present, and in some it is the dominant Fe-Ti oxide. Ilmenite has also been found in gabbros dredged from the sea floor (Prinz et al., 1976). In all instances the ilmenite is poor in ferric iron, the hematite component being less than 20 mole %, and most commonly less than 10 mole %. MgO in ilmenite from these rocks is variable (0.1 to 2.3%), as is MnO (0.1 to 1.4%). The MgO values are significantly lower than those found in ilmenite from volcanic rocks, indicating that the composition of ilmenite in mafic plutons has been significantly altered by ion exchange with the silicates during cooling.

Ilmenite from alkalic plutons. Ilmenite is very rare in alkalic plutons. It is reported from carbonatites (Mitchell, 1978; Sechee and Larsen, 1980) but the ilmenite from the Jacupiranga carbonatite is unusually rich in MgO (up to 23%) and MnO (up to 8%) (Mitchell, 1978). Ilmenite that is very poor in ferric iron may be present in the mafic and ultramafic portions of alkalic plutons, such as the Gardiner Complex (Nielsen, 1981) and in the Tugtutôq Older Giant Dyke (Upton et al., 1985). It is usually absent in feldspathoidal rocks. One exception is in a nepheline-syenite inclusion from Tenerife (Ferguson, 1978), which contains an "ilmenite" with 29% MnO. An unusual oxidized alkali gabbro from Hawaii (Johnston and Stout, 1984) contains hemoilmenite with X_{hem} = 0.65 that also has more than 6% MgO.

Ilmenite in felsic plutons. Ilmenite from felsic plutons has a compositional trend similar to that seen in silicic volcanic rocks. In plutons containing fayalite, ilmenite is poor in ferric iron, generally Hem_6 or less (see Frost et al., 1988). Ilmenite that is poor in ferric iron also occurs in biotite and hornblende-bearing granites where magnetite is absent (Czamanske et al., 1981; Whalen and Chappell, 1988). Ilmenite from magnetite-bearing granitoids tends to be moderately rich in ferric iron. For example, hematite component reaches 16 mole % in quartz diorites of Ballachulish (Weiss and Troll, 1989), 13 mole % in the Finnmarka complex (Czamanske and Mihalik, 1972) and 22 mole % in granites from S. W. Japan (Czamanske et al. 1981).

The MnO content of ilmenite from felsic plutonic rocks shows a similar variation. It is low (less than 3% MnO) in fairly unevolved granitoids, such as the pyroxene-bearing diorites and quartz diorites of the Ballachulish intrusion (Weiss and Troll, 1989) and the pyroxene-bearing Sande monzonites and quartz monzonites (Andersen, 1984). The MnO content of ilmenite is more variable in evolved fayalite-bearing rocks wherein it ranges from less than 1.0% in the Sybille Monzosyenite (Fuhrman et al., 1988) to as high as 11% in fayalite granites from Pikes Peak (Barker et al., 1975). Ilmenite from granites may be very rich in MnO. For example ilmenite from the Finnmarka complex has MnO contents of 30% (Czamanske and Mihalik, 1972) and similar high contents are reported from an adamellite in the Sierra Nevada (Snetsinger, 1969) and from S.W. Japan (Czamanske et al., 1977).

Pseudobrookite in igneous rocks.

Minerals of the pseudobrookite-ferropseudobrookite solid solution series are found in two terrestrial environments: (1) as an oxidation product of primary ilmenite, and (2) as a primary crystallization phase in some, generally highly oxidized, igneous rocks. It is likely that most of the occurrences of pseudobrookite summarized by Ottemann and Frenzel (1965) and Bowles (1988) formed as the result of oxidation of ilmenite. In many instances, when textural criteria are described (Frenzel, 1971; Negendank, 1972) it is evident that pseudobrookite has replaced ilmenite in lamellar ilmenite-magnetite intergrowths. Where this process has gone to completion, the tabular ilmenite lamellae within an origial titanomagnetite have altered to pseudobrookite whereas the magnetite host has oxidized to hematite (Frenzel, 1971).

There are, however, some occurrences of pseudobrookite, that, because of the extremely oxidizing nature of the rock in which they occur, arguably represent part of an equilibrium assemblage. Lufkin (1976) reported the occurrence of pseudobrookite in association with hematite and bixbyite from miarolitic cavities in rhyolite. Similarly, Johnston and Stout (1984) reported pseudobrookite (which they called kennedyite) in association with magnesioferrite-rich spinel and hematite in an oxidized alkali gabbro. Other occurrences of pseudobrookite in volcanic rocks, such as Gough Island (Le Roex, 1985), Kilauea (Anderson and Wright, 1972) and in ultrapotassic lavas of the Sierra Nevada (van Kooten, 1980), are more difficult to interpret. In these occurrences there is no textural evidence to suggest that pseudobrookite is a secondary mineral; indeed usually the textures strongly indicate that the pseudobrookite is a magmatic phase. Since these rocks all contain magnetite rather than hematite and are not markedly reduced, it is difficult to reconcile the textural observations with the experimental evidence that under magmatic conditions ilmenite is more stable than the pseudobrookite and magnetite (MacChesney and Muan, 1959;

Taylor, 1964). It is possible that pseudobrookite in volcanic rocks may be a metastable phase (Rice et al., 1971) or an early-crystallizing phase that would have reacted to ilmenite had it not been quenched by solidification of the melt (Anderson and Wright, 1972). Alternatively, since pseudobrookite from both Gough Island and from the high-K rocks of the Sierra contains around 5% MgO, it is possible that this amount of additional components is sufficient to stabilize pseudobrookite in the presence of magnetite. In other cases the pseudobrookite may have been deposited as a late phase from the vapor.

Armalcolite in terrestrial rocks

Armalcolite, an Fe-Ti oxide first reported from lunar samples, is found in two terrestrial environments: (1) as a primary phase in highly reduced igneous rocks, and (2) in high-T, Mg-rich hornfelses. Armalcolite is found in the native iron-bearing lavas of Disko, where the assemblage armalcolite-rutile-iron is present instead of ilmenite and ulvöspinel (Pedersen, 1981). It is also found in lamprolites from Smoky Butte, Montana, where it is the sole oxide phase (Velde, 1975). Armalcolite also occurs in magnesian hornfelses (Smith 1965; Pedersen, 1979), where it may be stabilized by its high Al_2O_3 content, which may range up to 11%.

OXIDE-SILICATE REACTIONS IN MAGMATIC SYSTEMS

The composition and occurrence of Fe-Ti oxides in igneous rocks as outlined above is not only a function of the composition of the melt from which the oxides crystallized, but is also determined by reactions between the oxides and the silicates in the rock. As early as 1935 Bowen and Schairer noted how increased oxidation would drive Fe into the oxide phase and enrich the silicates in Mg. Kennedy (1955), Osborn (1959), and Carmichael (1967a) also published important papers describing the relations between oxide and silicate minerals in igneous rocks. Recent papers by Frost et al. (1988), Lindsley et al. (1990), Ghiorso and Sack (in press), and Lindsley and Frost (in press) have begun a detailed discussion of this problem.

The interaction between oxides and silicates in magmatic systems is best understood by first considering relations in the simple system Fe-Ti-Si-O and then enlarging that system to include the effects of Mg and then Ca. Reactions in various hydrous systems will then be discussed.

System Fe-Ti-Si-O

In the system Fe-Si-O the important petrologic reactions are the FMQ and HM buffers (see Chapter 1). The presence of Ti affects both of these reactions but the effect is most obvious in the HM equilibrium. The HM buffer relates the stability of spinels and rhombohedral oxides, and in the system Fe-Si-O, where the composition of both phases is fixed, the buffer is isobarically univariant. Because titanium preferentially substitutes into the rhombohedral oxide rather than the spinel, increasing the activity of the $TiFe^{2+}Fe^{3+}_{-2}$ exchange vector will stabilize the rhombohedral oxide over spinel and displace the HM equilibrium to lower oxygen fugacities.

As a result, a spinel phase and a rhombohedral oxide can occur over most of the temperatures and oxygen fugacities found in crustal rocks. The composition of the coexisting phases is a monitor of both the temperature and oxygen fugacity at which they equilibrated (Fig. 1A). This is of course the Fe-Ti oxide thermometer and oxygen barometer (Buddington and Lindsley, 1964; see also Chapter 6).

The FMQ equilibrium is isobarically divariant in Ti-bearing systems because aFe_3O_4 in magnetite decreases as the Ti content increases. When magnetite coexists with ilmenite, however, its Ti content is fixed and the FMQ reaction becomes:

$$SiO_2 + 2\ Fe_2TiO_4 = 2\ FeTiO_3 + Fe_2SiO_4 \quad \text{(QUIlF)}$$
quartz   ulvöspinel   ilmenite   fayalite

which was termed QUIlF by Frost et al. (1988). As written, the QUIlF reaction is independent of oxygen fugacity. However, in all but the most reduced conditions, the compositions of coexisting ulvöspinel (or titanomagnetite) and ilmenite are dependent on oxygen fugacity (this equilibrium is, of course, the displaced HM buffer). At low temperatures, where titanomagnetite coexisting with ilmenite can dissolve only small amounts of titanium, the QUIlF equilibrium becomes coincident with the FMQ buffer (Fig. 1A). With increasing temperature magnetite coexisting with ilmenite becomes progressively enriched in Ti, displacing the QUIlF surface to lower oxygen fugacity. At temperatures above 1000°C the QUIlF curve becomes nearly independent of oxygen fugacity and terminates at the IQF buffer (Fig. 1A). At 1 bar this intersection lies at 1058 ± 3°C and at a log fO_2 of -14.7. With increasing amounts of Mg in olivine (i.e. with decreasing activity of fayalite) the QUIlF surface moves to higher T and oxygen fugacity, whereas with decreasing activity of silica the QUIlF surface shifts to lower temperatures and oxygen fugacities (Fig. 1B).

Topologic relations in orthopyroxene-bearing systems

To illustrate how the topology of the Mg-bearing systems containing orthopyroxene relates to the topology of QUIlF in the iron-rich system, we first address the topology of the system Fe-Mg-Si-O (Speidel and Nafziger, 1968), from which it will be easier to understand the topology of the titanium-bearing system. The many assemblages that are possible among olivine, pyroxenes, Fe-Ti oxides, and quartz and the abbreviations used by Lindsley and Frost (in press) are given in Table 1.

System Fe-Mg-Si-O. The best way to illustrate this system is to consider isobaric-isothermal log fO_2 - X_{Fe} sections. Because we are interested in the composition of orthopyroxene, which is not present throughout the section, instead of X_{Fe} we plot $\mu MgFe_{-1}$ as the compositional variable (Fig. 2). The equilibria involved in this system are listed in Table 2. In the pure Fe-system at low pressures, there is but a single reaction, that of the FMQ buffer. The substitution of Mg into olivine changes the FMQ buffer into the OMQ (olivine-magnetite-quartz) assemblage and displaces it to higher oxygen fugacities. At some critical value of $\mu MgFe_{-1}$, olivine + quartz will be no longer stable and orthopyroxene appears instead (the QOOp equilibrium). The intersection of this reaction with the displaced OMQ surface generates two other equilibria, OpMQ (orthopyroxene-magnetite-quartz) and OMOp (olivine-magnetite-orthopyroxene).

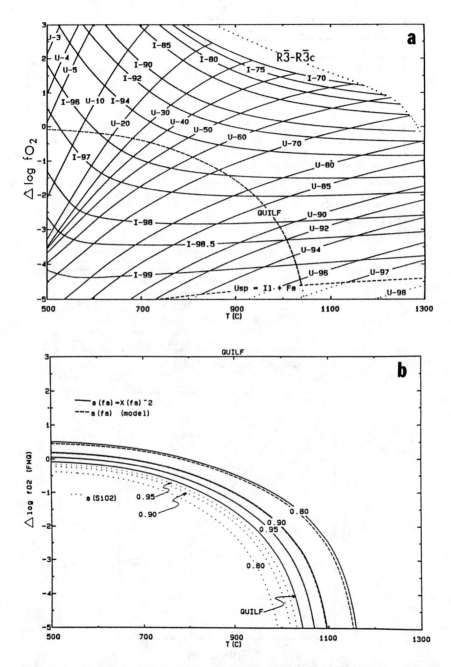

Figure 1. A. $\Delta \log fO_2$ - T diagram for the system $Fe-O-TiO_2-SiO_2$, showing the location of the QUIlF surface and isopleths for Ilm and Usp. B. $\Delta \log fO_2$ - T diagram showing how changes in silica activity and olivine composition affect the location of the QUIIF surface (Frost et al., 1988), assuming that the other variable is kept fixed at 1.0. In these and the following diagrams, $\Delta \log fO_2$ is with respect to the FMQ buffer.

TABLE 1
LOW-VARIANCE ASSEMBLAGES INVOLVING OLIVINE, PYROXENES, OXIDES, AND QUARTZ

	Ca-free	Ca-bearing
1. Oxide-free	QOOp: qtz-ol-opx	QOOpA: qtz-ol-opx-aug QOOpP: qtz-ol-opx-pig QOAP: qtz-ol-aug-pig
2. Ti-Free		
2a: High Fe	FMQ: fay-mt-qtz OMQ: ol-mt-qtz	OMQA: ol-mt-qtz-pig
2b: High Mg, quartz-saturated	OpMQ: opx-mt-qtz	OpAMQ: opx-aug-mt-qtz OpPMQ: opx-pig-mt-qtz APMQ: aug-pig-mt-qtz
2c: High Mg, oliv.-saturated	OMOp: ol-mt-opx	OMOpA: ol-mt-opx-aug OMOpP: ol-mt-opx-pig OMAP: ol-mt-opx-pig
3. Ti-bearing		
3a. High Fe	QUIIF: qtz-Timt-ilm-fay QUIIO: qtz-Timt-ilm-ol	QUIIOA: qtz-Timt-ilm-ol-aug
3b. High Mg, q-saturated	QUIIOp: qtz-Timt-ilm-opx	QUIIOpA: qtz-Timt-ilm-opx-aug QUIIOpP: qtz-Timt-ilm-opx-pig QUIIAP: qtz-Timt-ilm-aug-pig
3c. High Mg, ol-saturated	OpUIIO: opx-Timt-ilm-ol	OpAUIIO: opx-aug-Timt-ilm-ol OpPUIIO: opx-pig-Timt-ilm-ol APUIIO: aug-pig-Timt-ilm-ol

abbreviations: augite = A,aug; fayalite = F,fay; olivine = O,ol; orthopyroxene = Op,opx; pigeonite = P,pig; quartz = Q,qtz; ilmenite = Il,ilm; Ti-free magnetite = M,mt; titanomagnetite = U,Timt;

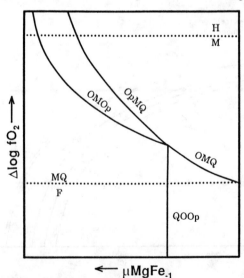

Figure 2. Schematic $\Delta \log fO_2$ - $\mu MgFe_{-1}$ diagram showing relations in the system Fe-O-MgO-SiO$_2$. OMQ = olivine-magnetite-quartz, QOOp = quartz-olivine-orthopyroxene, OpMQ = orthopyroxene-magnetite-quartz, OMOp = olivine-magnetite-quartz. The values for the FMQ and MH buffers are shown for reference. Those equilibria become displaced to higher fO$_2$ with increasing $\mu MgFe_{-1}$, as Mg preferentially enters olivine over magnetite, and magnetite over hematite. Above the displaced MH equilibrium, OMQ and OMOp become OHQ (olivine-hematite-quartz) and OHOp (olivine-hematite-orthopyroxene) respectively.

Table 2

Critical Equilibria in the System Fe-Mg-Si-O

SiO_2 + Fe_2SiO_4	=	$Fe_2Si_2O_6$		(QOOp)
quartz olivine		orthopyroxene		
3 Fe_2SiO_4 + O_2	=	Fe_3O_4 + 3 SiO_2		(FMQ,OMQ)
olivine		magnetite quartz		
3 $Fe_2Si_2O_6$ + O_2	=	2 Fe_3O_4 + 6 SiO_2		(OpMQ)
orthopyroxene		magnetite quartz		
6 Fe_2SiO_4 + O_2	=	2 Fe_3O_4 + 3 $Fe_2Si_2O_6$		(OMOp)
olivine		magnetite orthopyroxene		

With increasing $\mu MgFe_{-1}$ both of these equilibria will move to higher oxygen fugacity, with the OpMQ equilibrium lying at higher oxygen fugacity than the OMOp equilibrium. Although it is not important to the present discussion, at very high $\mu MgFe_{-1}$, both the OpMQ and the OMOp surfaces intersect the HM equilibrium, which will have been displaced to higher fO_2 as Mg enters magnetite. There are a few natural assemblages that record oxygen fugacity conditions near or on these intersections and indicate how magnesian the rocks must be for these intersections to occur. For example, the assemblage orthopyroxene (X_{Fe}^{opx} = 0.23)-hematite-magnetite-quartz occurs in metamorphosed iron-formations (Butler, 1969), and the assemblage olivine-orthopyroxene (X_{Fe}^{opx} = 0.06)-magnesioferrite-ilmenohematite has been reported from oxidized alkali gabbros by Johnston and Stout (1984).

System Fe-Mg-Ti-Si-O. The phase relations in the Fe-Mg-Ti-Si-O system are determined by the manner in which the oxidation-reduction equilibria in the system Fe-Mg-Si-O are affected by the substitution of Ti. Because Ti preferentially substitutes into oxides relative to the silicates and into hematite relative to magnetite in the Ti-bearing system, the OMQ, OpMQ, and OMOp curves, which are isobarically, isothermally univariant in the Fe-Mg-Si-O system, become divariant surfaces that curve toward lower oxygen fugacities with increasing activity of the $TiFe^{2+}Fe^{3+}_{-2}$ exchange vector. At sufficiently high values of $\mu TiFe^{2+}Fe^{3+}_{-2}$ the displaced HM buffer intersects the OMQ, OpMQ, and OMOp surfaces, forming the family of QUIlF equilibria (Table 3). When discussing these pyroxene-bearing equilibria we will refer to them as pyroxene-QUIlF to emphasize the relation between them and the initial QUIlF equilibrium described by Frost et al. (1988). Along with the QUIlF equilibrium noted above these reactions include QUIlO, QUIlOp, and OpUIlO (Table 1).

Topologically the relation between QUIlOp and OpUIlO is similar to that between OMQ and OMOp (Fig. 3). Both equilibria move to higher oxygen fugacities with increasing $\mu MgFe_{-1}$, with QUIlOp lying at higher fO_2 than OpUIlO for any given value of $\mu MgFe_{-1}$. However, unlike the OMOp and OpMQ surfaces, the pyroxene-QUIlF surfaces cannot be displaced to oxygen fugacities above those of the HM buffer, since they are defined by the intersection of the <u>displaced</u> HM equilibrium with the OMOp or OpMQ surfaces.

The relationship between the isobarically, isothermally univariant equilibria and the isobarically, isothermally divariant surfaces that lie between them is shown

Table 3

Critical Equilibria in the System Fe-Mg-Ti-Si-O

SiO_2 + 2 Fe_2TiO_4 = 2 $FeTiO_3$ + Fe_2SiO_4 (QUIIF,QUIIO)
quartz ulvöspinel   ilmenite   olivine

2 SiO_2 + 2 Fe_2TiO_4 = 2 $FeTiO_3$ + $Fe_2Si_2O_6$ (QUIIOp)
quartz ulvöspinel   ilmenite   orthopyroxene

$Fe_2Si_2O_6$ + 2 Fe_2TiO_4 = 2 $FeTiO_3$ + 2 Fe_2SiO_4 (OpUIIO)
orthopyroxene ulvöspinel   ilmenite   olivine

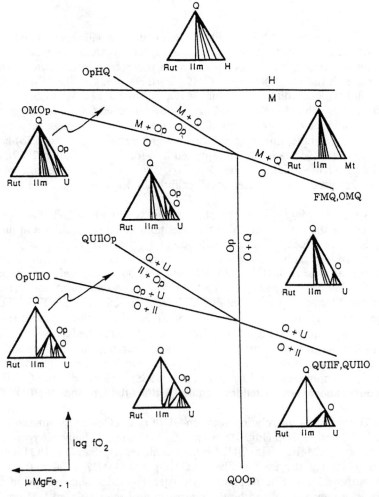

Figure 3. Schematic Δ log fO_2 - μ $MgFe_{-1}$ diagram showing relations in the system Fe-O-MgO-SiO_2-TiO_2 (modified from Lindsley et al., 1990). The upper part of this diagram is the same topology as is shown in Figure 2. Abbreviations not used in Figure 2 are: QUIIO = quartz-ulvöspinel-ilmenite-olivine, QUIIOp = quartz-ulvöspinel-ilmenite-orthopyroxene, OpUIIO = orthopyroxene-ulvöspinel-ilmenite-olivine. Chemographic diagrams are projections from O and Mg onto the Fe-Ti-Si plane; U in the chemographic diagrams stands for magnetite. Abbreviations for the univariant curves are given in Table 1.

in Figure 3 by use of chemographic projections from oxygen and magnesium onto the Si-Fe-Ti plane. In these chemographies the assemblages listed in Table 1 are shown as three-phase assemblages. It is important to note that each three-phase field shown in the chemographic triangles on Figure 3 is valid only for a fixed oxygen fugacity (as well as fixed T and P); the positions of the tie lines shift to higher Ti in ilmenite and magnetite with decreasing oxygen fugacity. With adequate thermodynamic data, one could use any of these assemblages as a monitor of oxygen fugacity, provided of course the assemblage happens to be stable. The assemblage ilm-mag ($\pm$ q, ol, or opx) is, of course the ilmenite-magnetite oxygen barometer. The manner in which the composition of the two Fe-Ti oxides varies as a function of T and fO_2 (at constant pressure) has been calculated by Andersen and Lindsley (1988) and is reproduced in Figure 1. How the other buffering surfaces behave in isobaric, isothermal $\mu MgFe_{-1}$ - $\mu TiFe^{2+}Fe^{3+}_{-2}$ space is described by Lindsley et al. (1990), and Lindsley and Frost (in press).

In addition to acting as monitors of oxygen fugacity, some of the equilibria described above are also important indicators of silica activity. Silica activity is, of course fixed at unity along all quartz-bearing curves (FMQ, OMQ, QUIlF, and QUIlOp). It is also buffered in any assemblage containing olivine and orthopyroxene (displaced OMOp, OpUIlO, or displaced OHOp), with $aSiO_2$ decreasing as $\mu MgFe_{-1}$ increases. Finally, silica activity and oxygen fugacity are co-variants in assemblages that contain two oxides and only one ferromagnesian silicate, such as UIlO and UIlOp (QUIlO and QUIlOp equilibria each displaced to $aSiO_2 < 1.0$).

<u>System Ca-Fe-Mg-Ti-Si-O.</u> The addition of Ca to the system Fe-Mg-Ti-Si-O has two important effects. It introduces pigeonite as a low-Ca pyroxene and Ca substitutes into orthopyroxene, thus changing its activity. Consequently each orthopyroxene-bearing reaction in the Ca-free sustem is represented by three reactions in the Ca-bearing system. In one reaction, the Ca-content of orthopyroxene is saturated by the presence of augite, in another it is saturated by the presence of pigeonite, and in the third the activity of ferrosilite is determined by the coexistence of pigeonite and augite (see Table 1). A $\Delta \log fO_2$ - X_{Fe}^{opx} diagram of this system is shown for 1000°C and 3 kbar (Lindsley and Frost, in press) in Figure 4. Note that the $\mu MgFe_{-1}$ variable of Figure 3 has been replaced by X_{Fe}^{opx} [= Fs/(Fs + En)]. This variable can be determined even for opx-free assemblages by calculating X_{Fe} for the fictive opx that would be in exchange equilibrium (but not phase equilibrium) with other phases such as olivine. Because the topology of this system is essentially a variant of the topology of the Ca-free system (Fig. 3) and because the pigeonite-bearing reactions are displaced from Ca-free opx reactions by 0.5 log units of oxygen fugacity or less (and from Ca-saturated opx reactions by 0.25 log units or less), in the ensuing discussions we will consider all low-Ca pyroxenes as if they were Ca-free orthopyroxene and use the Ca-free topologies to discuss relations in natural systems. Should a more detailed discussion of the system Ca-Fe-Mg-Ti-Si-O be desired, readers are referred to Lindsley and Frost (in press) for their reading enjoyment.

<u>QUIlF-type equilibria with biotite and hornblende.</u> In addition to pyroxene and olivine, oxide-silicate equilibria can also involve biotite and hornblende. Because these minerals are hydrous and difficult to characterize thermodynamically, little has

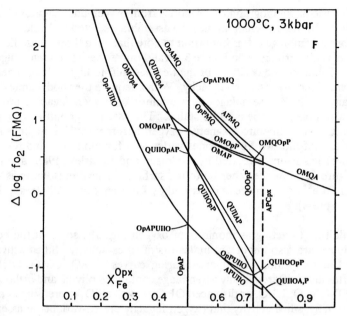

Figure 4. Δ log fO_2 - X_{Fe}^{opx} diagram (where X_{Fe}^{Opx} replaces $\mu MgFe_{-1}$ in Figures 2 and 3) showing relations in the system Fe-O-MgO-CaO-TiO$_2$-SiO$_2$ at 1000°C and 3 kbar (from Lindsley and Frost, in press). Abbreviations for the univariant curves are given in Table 1.

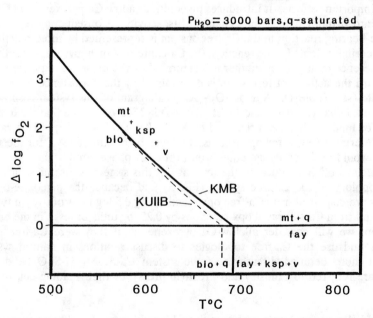

Figure 5. Δ log fO_2-T diagram for the K-feldspar-magnetite-biotite (KMB; heavy line) and KUIIB (light dashed line) assemblages. Note how both curves move to higher relative oxygen fugacity with decreasing temperature, indicating that a melt crystallizing biotite at constant water activity will tend to undergo relative oxidation at it cools.

been done to evaluate these reactions quantitatively (but see Diamond, 1989; Frost, 1990). The oxygen-dependent equilibria for these phases can be written as:

$$2\ KFe_3AlSi_3O_{10}(OH)_2 + O_2 = 2\ KAlSi_3O_8 + 2\ Fe_3O_4 + H_2O \quad (KMB)$$
in biotite   in feldspar   magnetite   fluid

$$2\ Ca_2Fe_5Si_8O_{22}(OH)_2 + O2$$
in hornblende
$$= 4\ CaFeSi_2O_6 + 2\ Fe_3O_4 + 8\ SiO_2 + 2\ H_2 \quad (HMAQ)$$
in augite   magnetite   quartz   fluid

The corresponding ilmenite-bearing reactions are:

$$KFe_3AlSi_3O_{10}(OH)_2 + 3\ FeTiO_3 = KAlSi_3O_8 + 3\ Fe_2TiO_4 + H_2O \quad (KUIlB)$$
in biotite   ilmenite   in feldspar in magnetite   fluid

$$Ca_2Fe_5Si_8O_{22}(OH)_2 + 3\ FeTiO_3 =$$
in hornblende   ilmenite
$$4\ CaFeSi_2O_6 + 2\ Fe_2TiO_4 + 8\ SiO_2 + 2H_2O \quad (QUAHIl)$$
in pyroxene   in magnetite   quartz   fluid

The KMB equilibrium was evaluated experimentally by Eugster and Wones (1962), Rutherford (1969), and Hewitt and Wones (1984). Combination of their expressions with the Fe-Ti oxide thermometer of Andersen and Lindsley (1988) allows one to calculate the KUIlB equilibrium (Fig. 5). There are insufficient thermodynamic data available to calculate either the HMAQ or QUAHIl equilibria but much can be inferred about them just from their stoichiometries.

The important feature of the KMB and HMAQ equilibria that distinguishes them from the other QUIlF-like equilibria discussed above is the presence of H_2O. Inspection of these reactions indicates that increases in the activity of H_2O during the evolution of a magma system will drive these equilibria to the left, consuming magnetite plus K-feldspar (or magnetite + augite + quartz), producing biotite (or hornblende), and liberating oxygen. Consequently the production of hornblende or biotite during the evolution of a magma should increase oxygen fugacity (Frost, 1990). This is probably part of the reason why hornblende- or biotite-bearing volcanic rocks are more oxidizing than rocks of similar composition that lack hydrous silicates (Carmichael, 1967a; Ghiorso and Sack, in press).

Petrologic significance

The oxide-silicate relations discussed above are important because, whereas oxides have a tendency to reset readily on cooling, the oxide-silicate reactions have higher blocking temperatures and are thus more likely to record magmatic conditions. Using QUIlF-like equilibria the petrologist has far more tools with which to evaluate the history of a rock than are available using oxide and silicate thermometry independently. This concept was discussed in detail by Frost et al. (1988), Lindsley et al. (1990), and Lindsley and Frost (in press) and was used to summarize the crystallization conditions of many igneous rocks by Frost and Lindsley (in press).

A good example of its utility involves the crystallization conditions of the Bishop Tuff. In his work on the Bishop Tuff, Hildreth (1977, 1979) found a consistent decrease in the equilibration temperatures of the Fe-Ti oxides with increasing stratigraphic depth. He interpreted this as an indication of the temperature of the original magma body and that the Bishop Tuff records a thermally zoned magma body. This interpretation carries the assumption that the T and fO_2 recorded by the oxides was locked in at the time of eruption. In the absence of contrary data, this is a not an unreasonable assumption. With recent advances in thermodynamic modelling, however, it has become evident that the orthopyroxene in the Bishop Tuff is not in equilibrium with the two Fe-Ti oxides and quartz (Lindsley et al., 1990; Ghiorso and Sack, in press). This fact raises questions as to whether the present composition of the oxides reflects pre-eruption equilibrium.

One might dismiss the Bishop Tuff data as merely reflecting a disequilibrium assemblage. However, if we assume that the oxides lost Mg during cooling and that the ilmenite compositions re-set during cooling, a different picture emerges, for we can use the silicate-oxide equilibria to obtain an estimate of the equilibrium conditions at the time the magma was erupted. A key to this assumption is the comment in Hildreth (1977) that magnetite is one to two orders of magnitude more abundant in the Bishop Tuff than is ilmenite. As a result, closed-system interoxide cooling should follow an Usp isopleth, with magnetite composition remaining constant as the composition of the ilmenite changes to accommodate the decreasing temperature (Frost et al., 1988, Fig. 4; Chapter 14, Fig. 2). Indeed, this is what seems to have happened in the Bishop Tuff, for magnetite has a constant composition of around Usp_{26}, whereas ilmenite varies from Hem_7 to Hem_{14} (Hildreth, 1977).

Because much of the Bishop Tuff contains the assemblage orthopyroxene (opx)-augite(aug)-titanomagnetite(Ti-mag)-ilmenite(ilm)-quartz(q), we can obtain an independent estimate of temperature by using the two-pyroxene thermometer. Because the pyroxene compositions show negligible variation, they yield tightly clustered temperatures, and both opx and augite temperatures are virtually identical. We can then calculate the oxygen fugacity and pressure from which a given assemblage was erupted by determining the conditions at which quartz and the pyroxenes in each sample would have been in equilibrium with titanomagnetite of Usp_{26} (i.e. the assemblage QUIlOpA). Because of our assumptions that the Mg contents of both oxides and the Fe_2O_3 contents of the ilmenite have reset, during the calculations those values are allowed to vary. Because the pyroxenes throughout the Bishop Tuff have relatively uniform composition, this technique produces a very restricted range of T, P and fO_2 conditions (Fig. 6). The 21 samples of Hildreth (1977) that contain two pyroxenes, and that pass the Bacon and Hirschmann (1988) test for Fe-Ti oxides, equilibrated at $824 \pm 10°C$, 2700 ± 600 bars, and at oxygen fugacities of 1.39 ± 0.05 log units above FMQ. None of these variables shows any consistent trend with respect to stratigraphic location, indicating that the range in values is a function of error inherent in the determination.

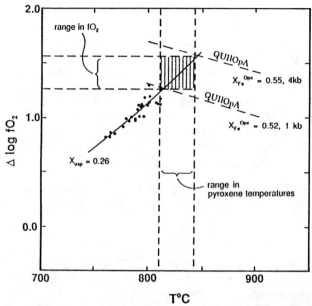

Figure 6. Δ log fO_2 - T diagram showing inferred pre-eruption conditions for pyroxene-bearing portions of the Bishop Tuff. Points give T-fO_2 conditions as calculated from Fe-Ti oxides; note that they plot close to the isopleth for Usp_{26} and that they are not in equilibrium with the quartz-titanomagnetite-ilmenite-orthopyroxene-augite (QUIlOpA) surfaces for the composition of pyroxenes found in the Bishop Tuff. Intersection of the Usp_{26} isopleth with the QUIlOpA surface lies within the temperature range from the opx-aug thermometer for pressures of 1 to 4 kbar (after Frost and Lindsley, in press).

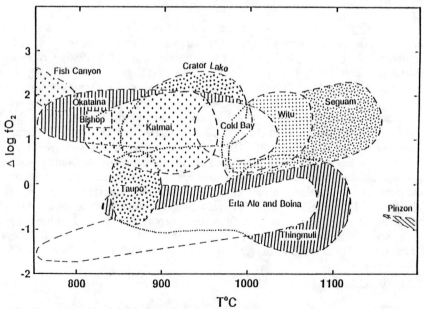

Figure 7. Δ log fO_2 - T diagram showing the crystallization trends for some tholeiitic (Thingmuli and Pinzon) and alkalic suites (Erta Ale) compared to those of calc-alkalic suites (Bishop, Cold Bay, Fish Canyon, Katmai, Mazama, Okataina, Seguam, and Witu). Modified after Frost and Lindsley (in press) with data from Ghiorso and Sack (in press) for Fish Canyon.

CRYSTALLIZATION CONDITIONS OF COMMON IGNEOUS ROCKS

Igneous rocks crystallize over a temperature interval. In the discussion below it is assumed (unless stated otherwise) that the inferred temperatures, pressures, and oxygen fugacities are those at which the mineral assemblage was in equilibrium within that interval. Trends for rock suites thus represent snapshots of equilibration temperatures for the various rock types of the suites. In all cases the numeric value of oxygen fugacity decreases with decreasing temperature. However, the log fO_2 values in the figures are relative to the FMQ buffer, the oxygen fugacity of which also decreases with decreasing temperature. In the discussion below, then, reported changes in fO_2 are with respect to those of the FMQ buffer.

The crystallization conditions of some major volcanic suites are shown in Figures 7 after the compilation of Frost and Lindsley (in press).

Volcanic rocks

Tholeiitic and alkalic rocks. The more evolved tholeiitic rocks, such as those of Thingmuli (Carmichael, 1967b) and Pinzon Island (Baitis and Lindstrom, 1980), lie at distinctly lower oxygen fugacities (ca. 1 log unit below FMQ) than do lavas of the calc-alkalic suite. Like the rocks of Thingmuli, the alkalic rocks of Erta Alta and Boina (Bizouard et al., 1980) are markedly more reduced than calc-alkalic volcanic rocks. In addition, there seems to be a distinct decrease in relative fO_2 in this suite with increasing differentiation. The most evolved rocks, which contain fayalite-hedenbergite-magnetite-ilmenite, crystallized at oxygen fugacities below those of the QUIlF surface.

Calc-alkalic rocks. The consistent range of oxygen fugacities shown by rock suites from calc-alkalic magmatic arcs is remarkable. Pyroxene-bearing rocks from the Bishop Tuff (Hildreth, 1977, 1979), Cold Bay (Brophy, 1984, 1986), Crater Lake (Druitt and Bacon, 1989), Katmai (Hildreth, 1983) and Seguam (Singer, 1990; Singer et al., in press), and Witu Islands (Johnson and Arculus, 1978), all crystallized at oxygen fugacities of 1 ($\pm$1) log units above FMQ, even though rocks in these suites range from basaltic andesite to rhyolite. The role of hydrous minerals in the evolution of silicic melts (Carmichael, 1967a) is also evident in Figure 7. The Fish Canyon Tuff (Whitney and Stormer, 1985; Ghiorso and Sack, in press), which has no orthopyroxene, lies at higher oxygen fugacities than orthopyroxene-bearing volcanic rocks listed above, around 2.5 log units above FMQ. In addition, the oxygen fugacity of the lavas of New Zealand (Ewart et al., 1975) is distinctly related to the presence or absence of hornblende. Rhyolites containing hornblende and pyroxenes (the rhyolites of Okataina) lie at oxygen fugacities around 1.5 log units above FMQ, whereas those that lack hornblende (such as those of Taupo) seem to follow a reducing trend that extends to oxygen fugacities as low as 0.5 log units below FMQ.

Mafic plutonic rocks

It is clear from the discussion of the Bishop Tuff that Fe-Ti oxides are prone to recrystallization even in volcanic rocks and that it may be difficult to apply these thermometers to plutonic rocks. We have shown, however, that with careful analyses one might be able to reconstruct the composition of the phases in volcanic rocks and to determine the crystallization conditions. There is no reason why this cannot be done with plutonic rocks as well, although the task is daunting indeed. Because few of the references in the literature have provide sufficient data to allow reconstruction of the crystallization conditions, we have taken a different approach in our discussion of plutonic rocks. Rather than calculate the pressure, temperature, and fO_2 for a suite of rocks, we have opted to assume a P and T and to use a variation of Figure 4 as a petrogenetic net from which to infer the changes of oxygen fugacity within and between suites.

In obtaining a petrogenetic net for mafic plutons we have assumed that the crystallization temperature (or, more precisely, the mineral equilibration temperature) of the pluton is roughly a function of bulk composition. We assume that a hypothetical Fe-free melt crystallizes at 1200°C; an intermediate melt crystallizes orthopyroxene with $X_{Fe}^{opx} = 0.5$ at 1100°C; and the Mg-free differentiate of this melt freezes at 1000°C. We then project a simplified pyroxene-bearing topology onto a corresponding polythermal section at 3 kbar (Figs. 8-10). The topology was simplified by removing those curves that contain pigeonite. Since there is only minor difference between the pigeonite-bearing and opx-bearing surfaces in Figure 4, such a change simplifies the topology without doing undue harm to the information that can be derived from the projection. For felsic plutonic rocks an isothermal, isobaric projection at 800°C and 3 kbar was used.

Because the pressure and temperature of these diagrams has been arbitrarily fixed, the oxygen fugacity values on these figures should be considered as reference only. What is important is not the absolute values of the oxygen fugacity but the different buffering assemblages found in the various bodies. These assemblages indicate important relative <u>differences</u> in oxygen fugacity, even though differences in P and T may cause the absolute difference in oxygen fugacity between different suites to be uncertain.

<u>Tholeiitic plutons</u>. Tholeiitic and weakly alkalic mafic plutons are marked by a distinct iron-enrichment trend that in many instances produces final assemblages with nearly pure fayalite (Figs. 8A,B). Magnetite appears in these plutons when the fictive X_{Fe}^{Opx} is in the range of 0.3 to 0.5, and in most plutons is it accompanied by ilmenite. The Bushveld Intrusion is distinctive in this regard since primary ilmenite is apparently lacking (Molyneux, 1972). It is interesting to note that each of the plutons shown in Figure 8A and 8B follows its own fO_2 path with differentiation. To some extent the trend followed is a function of the silica activity of the melt. Melts that have a sufficiently high silica activity that olivine is not stable, such as the Dufek intrusion (Himmelberg and Ford, 1976, 1977), intersect a QUIIF-like curve (QUIIOA) only at the late stages, when low-Ca pyroxene has become unstable. Others, with lower silica activity, such as the Fongen-Hyllingen ((Wilson et al., 1981; Thy, 1982) or the Skaergaard (Vincent and Phillips, 1954; Brown, 1957; Brown and Vincent, 1963) follow a QUIIF buffering surface (OpAUIIO or QUILOA) over much of their differentiation history. Those plutons with such low silica activities that low-Ca pyroxene is not stable, such as Kiglapait (Morse, 1979, 1980) and Klokken (Parsons, 1981), are more closely related to alkali basalts than to tholeiites. These follow crystallization trends that lie entirely below the QUIIOA and OpAUIIO

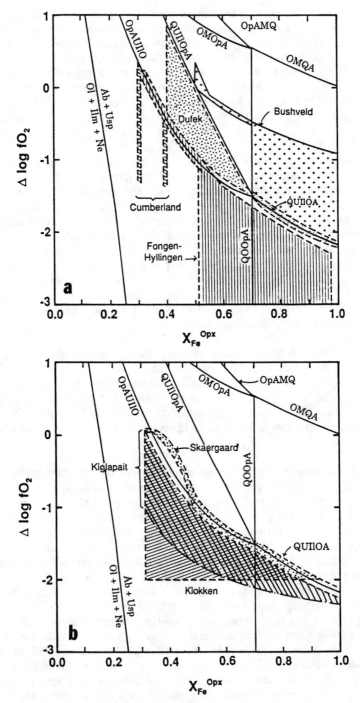

Figure 8 A,B. Polythermal $\Delta \log fO_2$ - X_{Fe}^{Opx} diagram showing inferred crystallization trends of some mafic plutons of tholeiitic and alkalic affinities (after Frost and Lindsley, in press).

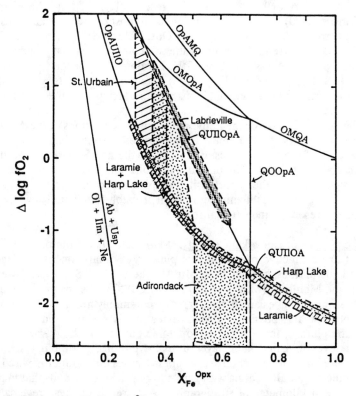

Figure 9. Polythermal $\Delta \log fO_2$ - X_{Fe}^{Opx} diagram showing inferred crystallization conditions of some anorthositic complexes (after Frost and Lindsley, in press).

surfaces (although in its final stages the Klokken becomes quartz-bearing and has the QUIIOA assemblage).

Anorthositic complexes. In many respects anorthositic plutons behave much like the tholeiitic intrusions mentioned above. Most show a distinct iron-enrichment trend and attain fairly reducing conditions (Fig. 9). We consider them here in a separate category here because there has been some question as to whether oxygen fugacity is an important variable by which to distinguish anorthosite plutons.

Although the Fe-enriched potassic rocks associated with anorthosite complexes all show similar oxygen fugacities, the anorthositic portions of these complexes appear to have crystallized over a wide range of oxygen fugacity. It is likely that this range in oxidation state reflects a range in silica activity. For example the anorthosite complexes with hemoilmenite lack olivine and must have crystallized at oxygen fugacities above the OpAUIlO surface. Indeed, Labrieville (Anderson, 1966) contains minor to trace amounts of quartz and is likely to have crystallized on or close to the QUIlOpA surface. St. Urbain (Dymek and Gromet, 1984) contains neither quartz nor olivine but has relatively magnesian pyroxenes, restricting the oxygen fugacity of the rocks to lie above that of the FMQ buffer.

The Laramie Anorthosite Complex (Frost et al., 1991) and the Harp Lake Complex (Emslie, 1980) contain olivine throughout most of the suite and thus follow oxygen fugacity trends that are parallel to the OpUIlO and QUIlF surfaces. The only difference between the trends followed by the two complexes is that magnetite is absent from the iron-rich potassic plutons of the Laramie Anorthosite Complex (Fuhrman et al., 1988; Kolker and Lindsley, 1990), meaning that the oxygen fugacity of these rocks must have evolved to conditions more reducing than the QUIlF surface.

In view of the lack of olivine in the more primitive portions of the Adirondack anorthosite (Ashwal, 1982), they must have followed a trend that is intermediate between the path of the anorthosites of the Labrieville and St. Urbain and those of the Laramie and Harp Lake. The path of the Adirondack anorthosite, however, is less certain than that of other anorthosite complexes because growth of garnet in response to subsequent granulite metamorphism probably eliminated any olivine that may have been present in the original rock.

Mafic plutons of calc-alkalic affinity. Mafic calc-alkalic plutons have several features that are distinctive from tholeiitic plutons. Primary ilmenite is lacking in many calc-alkalic plutons, such as Bear Mtn (Snoke et al., 1981), Lake Owen (Patchen, 1987), and McIntosh (Mathison and Hamlyn, 1987); and magnetite appears in rocks that are quite magnesian (fictive X_{Fe}^{Opx} is commonly around 0.2; Fig. 10A,B). A more important and possibly correlative feature is the fact that, unlike most tholeiitic mafic plutons, there is not a monotonic increase in X_{Fe}^{Opx} with increasing stratigraphic level in calc-alkalic layered mafic plutons. Although some iron-enrichment is found, more common are abrupt increases in Mg number with increasing stratigraphic height alternating with gentle Mg-enrichment trends that may occur over as much as a kilometer of stratigraphy. As a result of these reversals, calc-alkalic layered mafic plutons do not reach the extreme iron-enrichment that is characteristic of most layered mafic plutons of tholeiitic affinity.

In addition to plutons that are indisputably calc-alkaline, we include the Sudbury irruptive in Figure 10B. Geochemically, one could hardly consider Sudbury to be calc-alkalic. It is associated with one of the largest nickel deposits in the world and calc-alkalic rocks are notably poor in nickel. However, the Sudbury irruptive (Naldrett et al., 1970) has many aspects that are similar to the calc-alkalic mafic plutons described above. It is considerably more oxidized than typical tholeiitic plutons and it displays distinct excursions to higher relative fO_2 and higher Mg number. Frost and Lindsley (in press) concluded that the behavior of the Sudbury irruptive is probably a reflection of the intense contamination of the magma.

Alkalic plutons

At very low silica activities an additional QUIlF-type reaction takes place, one involving nepheline:

$$NaAlSi_3O_8 + 4\,Fe_2TiO_4 = 2\,Fe_2SiO_4 + 4\,FeTiO_3 + NaAlSiO_4$$
$$\text{plagioclase} \quad \text{ulvöspinel} \quad \text{olivine} \quad \text{ilmenite} \quad \text{nepheline}$$

As shown in Figures 8 through 10 this reaction occurs at petrologically accessible oxygen fugacities only in magnesian rocks. However, because all alkalic plutons that

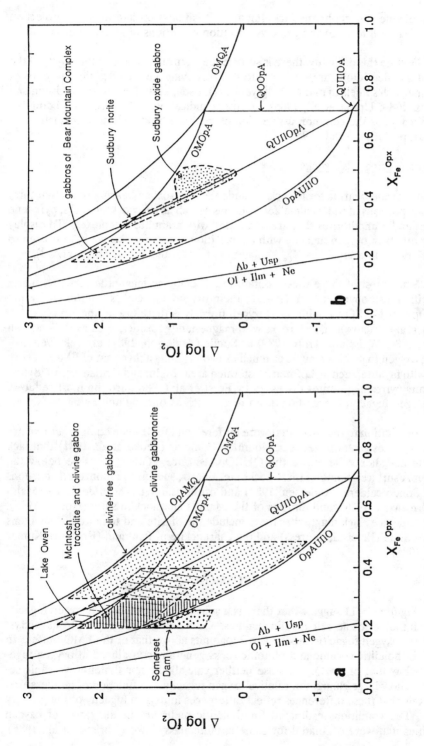

Figure 10 A,B. Polythermal $\Delta \log fO_2$ - X_{Fe}^{Opx} diagram showing inferred crystallization conditions of some calc-alkalic mafic plutons (after Frost and Lindsley, in press).

contain olivine and nepheline lack ilmenite, it is inappropriate to use a projection such as Figure 8 to constrain the crystallization conditions of alkalic plutons.

Despite this difficulty, there is sufficient evidence to conclude that many of the evolved alkalic plutons are quite reduced. This evidence includes the presence of ulvöspinel-rich spinels in many alkali syenites (Larsen, 1976; Powell, 1978; Upton and Thomas, 1980; Upton et al., 1985). Another indication of the reducing nature of alkali syenites is the common association of methane fluid inclusions in such rocks (Konnerup-Madsen et al., 1985).

Pyroxene-bearing felsic rocks

Pyroxene-bearing felsic rocks include those pyroxene-bearing rocks that form the mafic portions of some calc-alkalic plutons as well as charnockitic rocks, pyroxene granites and granodiorites that are associated with granulite complexes. To display the variations in oxygen fugacity with composition in these rocks, we have chosen to plot relations on a log fO_2 - X_{Fe}^{Opx} diagram at 800C and 3 kbar (Figs. 11A,B).

Since most of these rocks contain the assemblage magnetite-ilmenite-quartz with either orthopyroxene or fayalite, their oxygen fugacity is controlled by the QUIIOpA or QUIIOA equilibria. Oxygen fugacity is therefore a function of X_{Fe}^{Opx} (either real or fictive). Those rocks with magnesian pyroxenes, such as Ballachulish (Weiss, 1986; Weiss and Troll, 1989) or Sonde (Andersen, 1984) are relatively oxidizing; oxygen fugacities may be as much as two log units above that of FMQ. Those rocks with fayalite, such as Lofoten-Vesterålen area (Malm and Ormaasen, 1978) and the many pyroxene-granites discussed by Frost et al. (1988), are the most reduced, with oxygen fugacities extending down to as much as one log unit below FMQ.

In addition to the rocks that were buffered on QUIIF-type equilibria there are also rock suites such as the Bear Mountain Complex (Snoke et al., 1981) that lack ilmenite and, hence lie above the QUIIOpA surface. Other suites lie below the QUIIOpA surface. Some of these rocks lack quartz, such as the monzonitic portions of the Sonde caldera (Andersen, 1984) and of the Lofoten-Vesterålen area (Malm and Ormaasen, 1978), and diorites of the Madras charnockite suite (Howie, 1955). Other sequences lack magnetite; these include the diorite and tonalite of Des Liens Igneous Suite (Percival, in press) and the charnockite of Ponmudi (Ravindra Kumar et al., 1985).

DISCUSSION

Figures 7 - 11 suggest that there is a substantial difference in oxygen fugacity between tholeiitic and calc-alkalic magmas. Calc-alkalic rocks appear to have crystallized at oxygen fugacities about 1 to 2 log units above that of the FMQ buffer. In contrast, tholeiitic vocanic and plutonic rocks generally crystallized at oxygen fugacities below that of FMQ. Because neither calc-alkalic nor tholeiitic volcanic sequences show large changes in relative oxygen fugacity as a function of temperature, it is likely that these differences reflect differences in oxygen fugacity of the primary melts. The conditions estimated for tholeiitic melts are in the range of oxygen fugacities that were estimated for some basaltic glassy lavas (Christie et al., 1986).

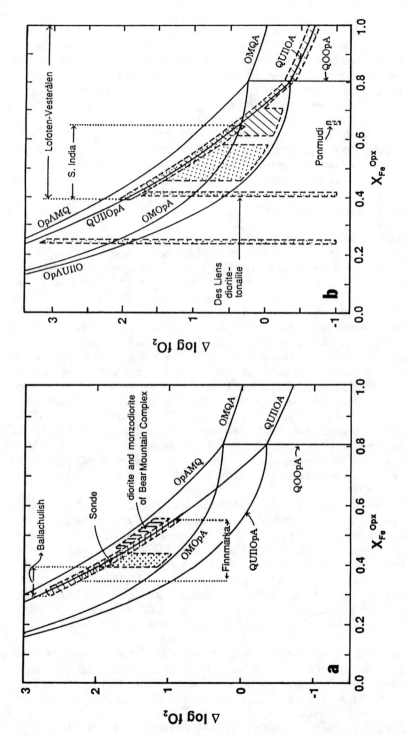

Figure 11 A,B. Isothermal Δ log fO_2 - X_{Fe}^{Opx} diagram taken at 800°C and 3000 bars showing crystallization trends of pyroxene-bearing granitoids (after Frost and Lindsley, in press).

Another important point evident from Figure 7 is that the relative oxygen fugacity of tholeiitic or calc alkalic volcanic rocks does not seem to change much during differentiation. Frost and Lindsley (in press) concluded that this is because iron-enrichment in the silicates is accompanied by decreases in relative oxygen fugacity only for low-variance assemblages such as OpAUIlO or QUIlOpA. Frost and Lindsley (in press) also noted that in only a few of the volcanic rocks having low-variance assemblages such as OpAUIlO were all the phases in equilbrium. Thus the differentiation of most volcanic suites is governed by assemblages of higher variance. In these higher variance assemblages, such as orthopyroxene-clinopyroxene-ilmenite-titanomagnetite, the iron-enrichment produced in the silicates by differentiation is apparently accomodated by increases in silica activity rather than decreases in oxygen fugacity (Frost and Lindsley, in press).

Crystallization trends for tholeiitic rocks

Simplified paths for the differentiation of tholeiitic and calc-alkalic melts are shown in Figure 12. The paths are predicated upon the assumption that oxygen fugacities estimated for the volcanic rocks (Fig. 7) and for mafic plutons (Figs. 8, 10) represent the actual oxygen fugacities of their parental melts. Much of our model for tholeiitic melts (Fig. 12, path A to F) is based upon Thingmuli (Carmichael, 1967b), since that is the best-studied tholeiitic system available. The initial crystallizing assemblage in most tholeiites is olivine-augite-chromite, although chromite is relict in many, occurring as inclusions in olivine or rimmed by titanomagnetite (Thompson, 1973; Bence et al., 1975). Because this is a low-variance assemblage, crystallization of it need not lead to any change in oxygen fugacity (path A, Fig. 12). Ilmenite and pigeonite appear at about the same time as titanomagnetite. If there is substantial iron-enrichment in the silicates as the assemblage olivine-pigeonite-augite-titanomagnetite-ilmenite (OpAUIlO) crystallizes, there should be a distinct decrease in relative oxygen fugacity for the system (path B'-C'-D'-F, Fig. 12). This does not seem to have happened in Thingmuli since pigeonite appears late in the rocks that have olivine and olivine is absent from many of the tholeiites where pigeonite shows a substantial crystallization history (Carmichael, 1967b).

After the disappearance of olivine, the crystallizing assemblage in tholeiites becomes low-Ca pyroxene-augite-titanomagnetite-ilmenite and increases in Fe/Mg ratio in silicates are apparently accomodated by increases in silica activity rather than decreases in oxygen fugacity (paths C and C', Fig. 12). When the magma becomes saturated with quartz (i.e. it reaches the QUIlOpA or QUIlAP surface) the differentiation path shows distinct relative reduction (paths D and D', Figure 12) down to the QUIlF surface. At this point, further crystallization may follow the QUIlF surface (path E), as seen at Thingmuli, or it may cross the QUIlF surface (path F) with the subsequent loss of magnetite, as is seen at Sierra Primavera (Mahood, 1981).

Differentiation of tholeiitic magmas will thus follow a band of oxygen fugacities that for most part lie below the oxygen fugacities of the FMQ buffer. This band will be roughly defined by the OpAUIlO and QUIlOA curves between X_{Fe}^{Opx} = 0.3 and 1.0 (Fig. 8). For those sequences that lose olivine, the upper limits of the band will be the QUIlOpA surface, which for X_{Fe}^{Opx} = 0.5 or more lies less than one log unit of oxygen fugacity above OpAUIlO. The lower limits of this band are

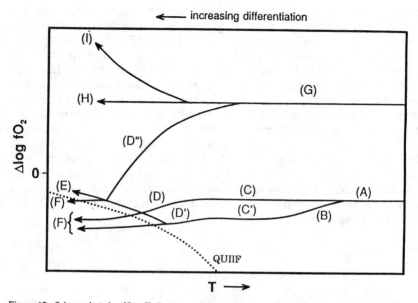

Figure 12. Schematic $\Delta \log fO_2$ - T diagram showing inferred crystallization paths for tholeiitic and calc-alkalic rocks. Path (A) is inferred path followed for rocks with the assemblage olivine-augite-titanomagnetite ($\pm$ ilmenite). Path (B) is inferred path for tholeiitic melts with the assemblage olivine-two pyroxenes-two oxides. Paths C and C' show the inferred path followed by tholeiitic rocks with the assemblage two pyroxenes-two oxides. Paths D and D' show paths for rocks of tholeiitic parentage with the assemblage quartz-two pyroxenes-two oxides. Path E is the path governed by the QUIlF assemblage. Paths labelled F are for evolved rocks with the assemblage quartz-fayalite-titanomagnetite. Path (G) is for calc alkalic rocks with the assemblage olivine-augite-titanomagnetite, olivine-two pyroxenes-titanomagnetite, two pyroxenes - titanomagnetite, and two pyroxenes - two oxides. Path (D") is for calc-alkalic rocks with the assemblage quartz-two pyroxenes-two oxides. Path (H) is for lavas of calc-alkalic affinity that have the assemblage quartz-two pyroxenes - biotite or hornblende - two oxides. Path (I) is for lavas of calc-alkalic affinity that contain hornblende or biotite without orthopyroxene.

difficult to define, for there is no low-variance equilibrium (such as QUIlF) that can be used as a limit. However, considering that tholeiitic rocks have oxygen fugacities lying on or near the critical plane of silica saturation, it is unlikely that the oxygen fugacity will be more than a log unit below the OpUIlO surface.

Differentiation trends in calc-alkalic magmas

In calc-alkalic lavas the early crystallizing phases are olivine-augite-titanomagnetite. Ilmenite appears late in the history of the magma, usually after olivine has ceased to be a crystallizing phase. Because the OpAUIlO assemblage does not appear, changes in $\mu MgFe_{-1}$ can be accommodated either by changes in silica activity or activity of Ti. Consequently, differentiation apparently takes place at a nearly fixed relative oxygen fugacity (path G, Fig. 12). This is probably the main reason for the consistent oxygen fugacity found in calc-alkalic suites (Fig. 9). It is also true that calc-alkalic rocks lack the marked iron-enrichment seen in tholeiitic suites and that this lack would limit the change in oxygen fugacity one may see with differentiation. However, that cannot be the sole reason for the consistent oxygen fugacity observed in calc-alkalic volcanics because, at low X_{Fe}^{Opx}, the OpAUIlO surface is extremely sensitive to changes in silicate composition, so that even an iron enrichment from X_{Fe}^{Opx} = 0.2 to 0.3 would produce a noticeable decrease in oxygen fugacity (Fig. 10A,B). The appearance of quartz in calc-alkalic rocks along with two pyroxenes and two oxides is expected to lead to relative reduction in a manner similar to that postulated for evolved tholeiitic melts (path D", Fig. 12). This is perhaps the path followed by Taupo (Fig. 7). In many calc-alkalic lavas, hornblende and biotite appear at the same stage in the evolution of the melt as does quartz. Because the effects of the KUILB and the QUAHII equilibria lead to relative oxidation on cooling, the presence of these hydrous minerals may counter any tendencies toward reduction (path H, Figure 12). This is likely the path followed by rhyolites from Okataina (Fig. 7). Indeed, if the H_2O activity of the melt is sufficiently high, those equilibria could lead to marked relative oxidation (path I, Fig. 12), with the loss of orthopyroxene and, perhaps, an increase in μ MgFe$_{-1}$. This is probably the path followed by oxidized felsic rocks such as the Fish Canyon tuff and the Finnmarka Complex.

ACKNOWLEDGMENTS

We thank David J. Andersen for writing the computer program QUILF that made this paper possible, and Tony Morse for providing, in a remarkable short time, a review that materially improved it.

REFERENCES

Andersen, D. J., and Lindsley, D. H. (1988) Internally consistent solution models for Fe-Mg-Mn-Ti oxides: Fe-Ti oxides. Am. Mineral., 73, 714-726.

Andersen, D. J., Lindsley, D. H., and Davidson, P. M. QUIIF: A PASCAL program to assess equilibria among Fe-Mg-Ti oxides, pyroxenes, olivine, and quartz. Computers in Geosciences, in press.

Andersen, T. (1984) Crystallization history of a Permian composite monzonite - alkali syenite pluton in the Sande Cauldron, Oslo rift, southern Norway. Lithos, 17, 153-170.

Anderson, A. T. (1966) Mineralogy of the Labrieville Anorthosite, Quebec. Am. Mineral., 51, 1671-1711.

Anderson, A. T., and Wright, T. L. (1972) Phenocrysts and glass inclusions and their bearing on oxidation and mixing of basaltic magmas, Kilauea Volcano, Hawaii. Am. Mineral., 57, 188-216.

Anderson, J. L. (1980) Mineral equilibria and crystallization conditions in the Late Precambrian Wolf River Rapakivi Massif, Wisconsin. Am. J. Sci., 280, 298-332.

Ashwal, L. D. (1982) Mineralogy of mafic and Fe-Ti oxide-rich differentiates of the Marcy anorthosite massif, Adirondacks, New York. Am. Mineral., 67, 14-27.

Aurisicchio, C., Brotzu, P., Morbidelli, L., and and Traversa, G. (1983) Basanite to peralkaline phonolite suite: Quantitative crystal fractionation model (Nyambeni Range, East Kenya Plateau). Neues Jahrbuch Mineral. Abhandlung, 148, 130-140.

Bacon, C. R., and Duffield, W. A. (1981) Late Cenozoic rhyolites from the Kern Plateau, southern Sierra Nevada, California. Am. J. Sci., 281, 1-34.

Bacon, C. R., and Hirschmann, M. M. (1988) Mg/Mn partitioning as a test for equilibrium between coexisting Fe-Ti oxides. Am. Mineral., 73, 57-61.

Baitis, H. W., and Lindstrom, M. M. (1980) Geology, Petrography, and Petrology of Pinzon Island, Galapagos Archipelago. Contrib. Mineral. Petrol., 72, 367-386.

Barker, F., Wones, D. R., Sharp, W. N., and Desborough, G. A. (1975) The Pikes Peak Batholith, Colorado Front Range, and a model for the origin of the gabbro-anorthosite-syenite-potassic granite suite. Precambrian Research, 2, 97-160.

Bellieni, G., Brotzu, P., Comin-Chiaramonti, P., Ernesto, M., Melfi, A., Pacca, I. G., and Piccirillo, E. M. (1984) Flood basalt to rhyolite suites in the southern Parana Plateau (Brazil): Paleomagnetism, petrogenesis, and geodynamic implications. J. Petrol. 25, 579-618.

Bence, A. E., Papike, J. J., and Ayuso, R. A. (1975) Petrology of submarine basalts from the cenrtal Caribbean: DSDP Leg 15. J. Geophys. Res., 80, 4775-4804.

Birch, W. D. (1978) Mineralogy and geochemistry of the leucitite at Cosgrove, Victoria. J. Geol. Soc. Australia, 25, 379-385.

Bizouard, H., Barberi, F., and Varet, J. (1980) Mineralogy and petrology of Erta Ale and Boina volcanic series, Afar Rift, Ethiopia. J. Petrol., 21, 401-436.

Bohlen, S. R., and Essene, E. J. (1977) Feldspar and oxide thermometry of granulites in the Adirondack Highlands. Contrib. Mineral. Petrol., 62, 153-169.

Bowen, N. L., and Schairer, J. F. (1935) The system $MgO-FeO-SiO_2$. Am. J. Sci., 29, 151-217.

Bowles, J. F. W. (1988) Definition and range of composition of naturally occurring minerals with the pseudobrookite structure. Am. Mineral., 73, 1377-1383.

Brophy, J. G. (1984) The chemistry and physics of Aleutian arc volcanism. The Cold Bay volcanic center, southwestern Alaska. Ph.D. Dissertation, Johns Hopkins University, 422p.

Brophy, J. G. (1986) The Cold Bay volcanic center, Aleutian volcanic arc I: Implications for the origin of high-aluminum arc basalt. Contrib. Mineral. Petrol., 93, 368-380.

Brown, F. H., and Carmichael, I. S. E. (1971) Quaternary volcanos of the Lake Rudolf Region: II The lavas of North Island, South Island and the Barrier. Lithos, 4, 305-323.

Brown, G. M. (1957) Pyroxenes from the early and middle stages of fractionation of the Skaergaard intrusion, east Greenland. Mineral. Mag., 31, 511-543.

Brown, G. M. and Vincent, E. A. (1963) Pyroxenes from the late stages of fractionation of the Skaergaard Intrusion, East Greenland. J. Petrol.. 4, 175-197.

Buddington, A. F., and Lindsley, D. H. (1964) Iron-titanium oxide minerals and synthetic equivalents. J. Petrol., 5, 310-357.

Butler Jr., P. (1969) Mineral compositions and equilibrium in the metamorphosed iron formation of the Gagnon Region, Quebec, Canada. J. Petrol., 10, 56-101.

Carmichael, I. S. E. (1967a) The iron-titanium oxides of salic volcanic rocks and their associated ferromagnesian silicates. Contrib. Mineral. Petrol., 14, 36-64.

Carmichael, I. S. E. (1967b) The mineralogy of Thingmuli, a Tertiary volcano in eastern Iceland. Am. Mineral., 52, 1815-1841.

Carmichael, I. S. E., and Luhr, J. F. (1980) The Colima Volcanic Complex, Mexico. I. Post-caldera andesites from Volcan Colima. Contrib. Mineral. Petrol., 71, 343-372.

Christie, D. M., Carmichael, I. S. E., and Langmuir, C. H. (1986) Oxidation states of mid-ocean ridge basalt glasses. Earth Planetary Sci. Letters, 79, 397-411.

Conrad, W. K. (1984) The mineralogy and petrology of compositionally zoned ash flow tuffs and related silicic volcanic rocks, from the McDermitt Caldera, Nevada-Oregon. J. Geophys. Res., 89, 8639-8664.

Cornen, G., and Maury, R. C. (1980) Petrology of the volcanic island of Annobon, Gulf of Guinea. Marine Geology, 36, 253-267.

Crisp, J. A. , and Spera, J. A. (1987) Pyroclastic flows and lavas of the Mogan and Fataga formations, Tejeda Volcano, Gran Canaria, Canary Islands: Mineral chemistry, intensive parameters, and magma chamber evolution. Contrib. Mineral. Petrol., 96, 503-518.

Cundari, A. (1973) Petrology of the leucite-bearing lavas in New South Wales. J. Geol. Soc. Australia, 20, 465-491.

Czamanske, G. K., Ishihars, S., and Atkin, S. A. (1981) Chemistry of rock-forming minerals of the Cretaceous - Paleocene batholith in Southwestern Japan and implications for magma genesis. J. Geophys. Res., 86, 10431-10469.

Czamanske, G. K., and Mihalik, P. (1972) Oxidation during magmatic differentiation, Finnmarka Complex, Oslo area, Norway: Part I, The Opaque Oxides. J. Petrol., 13, 493-509.

Czamanske, J. G., Wones, D. R., and Eichelberger, J. C. (1977) Mineralogy and petrology of the intrusive complex of the Pliny Range, New Hampshire. Am. J. Sci., 277, 1073-1123.

d'Arco, P., Maury, R. C., and Westercomp, D. (1981) Geothermometry and geobarometry of a cummingtonite-bearing dacite from Martinique, Lesser Antilles. Contrib. Mineral. Petrol., 77, 177-184.

Diamond, J. L. (1989) Mineralogical controls on rock magnetism in the Sierra San Pedro Martir Pluton, Baja California, Mexico. M.S. Thesis, University of Wyoming

Druitt, T. H., and Bacon, C. R. (1989) Petrology of the zoned calcalkaline magma chamber of Mount Mazama, Crater Lake, Oregon. Contrib. Mineral. Petrol., 101, 245-259.

Dymek, R. F., and Gromet, L. P. (1984) Nature and origin of orthopyroxene megacrysts from the St. Urbain Anorthosite Massif, Quebec. Can. Mineral., 22, 297-326.

Emslie, R. F. (1980) Geology and petrology of the Harp Lake Complex, central Labrador: an example of Elsonian magmatism. Geol. Survey Canada, Bull. 293.

Eugster, H. P., and Wones, D. R. (1962) Stability relations of the ferruginous biotite, annite. J. Petrol., 3, 82-125.

Evans, B. W., and Moore, J. G. (1968) Mineralogy as a function of depth in the prehistoric Makaopuhi lava lake. Contrib. Mineral. Petrol., 17, 85-115.

Ewart, A. (1981) The mineralogy and chemistry of the anorogenic Tertiary silicic volcanics of S. E. Queensland and N. E. New South Wales, Australia. J. Geophys. Res., 86, 10242-10256.

Ewart, A., Bryan, W. B., and Gill, J. B. (1973) Mineralogy and geochemistry of the younger islands of Tonga, S. W. Pacific. J. Petrol., 14, 429-465.

Ewart, A., Hildreth, W., and Carmichael, I. S. E. (1975) Quaternary acid magmas in New Zealand. Contrib. Mineral. Petrol., 51, 1-27.

Ewart, A., Oversby, V. M., and Mateen, A. (1977) Petrology and isotope geochemistry of Tertiary lavas from the northern flank of the Tweed volcano, southeastern Queensland. J. Petrol., 18, 73-113.

Ferguson, A. K. (1978) The occurrence of Ramsayite, Titan-lavenite and a fluorine-rich eucolite in a nepheline-syenite inclusion from Tenerife, Canary Islands. Contrib. Mineral. Petrol., 66, 15-20.

Fodor, R. V., Mukasa, S. B., Gomes, C. B.,, and Cordani, U. G. (1989) Ti-rich Eocene basaltic rocks, Abrolhos Platform, offshore Brazil, 18° S: Petrology with respect to South Atlantic magmatism. J. Petrol., 30, 763-786.

Frenzel, G. (1971) Die Mineralparagense der Albersweiler Lamprophyre (Südpfalz). Neues Jahrbuch Mineral. Abhandlungen. 115, 164-191, 1971.

Frost, B. R. (1990) Biotite crystallization as an oxidation agent in granitic rocks. Geol. Soc. Am., Abs. w. Prog., 22, A301.

Frost, B. R., Frost, C. D., Lindsley, D. H., Scoates, J. S., and Mitchell, J. N. (1991) The Laramie Anorthosite Complex and the Sherman Batholith: Geology, evolution, and theories for origin. in Snoke, A. W. and Steidtmann, J. R., editors. Geology of Wyoming, Geological Survey of Wyoming, Laramie, Wyoming.

Frost, B. R., and Lindsley, D. H. Equilibria among Fe-Ti oxides, pyroxenes, olivine, and quartz: 2. Applications. Am. Mineral., in press,

Frost, B. R., Lindsley, D. H., and Andersen, D. J. (1988) Fe-Ti oxide-silicate equilibria: Assemblages with fayalitic olivine. Am. Mineral., 73, 727-740.

Fuhrman, M. L., Frost, B. R., and Lindsley, D. H. (1988) Crystallization conditions of the Sybille Monzosyenite, Laramie Anorthosite Complex, Wyoming. J. Petrol., 29, 699-729.

Gaspar, J. C., and Wyllie, P. J. (1983) Magnetite in the carbonatites from the Jacupiranga Complex, Brazil. Am. Mineral., 68, 195-213.

Gerlach, D. C., and Grove, T. L. (1982) Petrogenesis of Medicine Lake Highland volcanics: Characterization of endmembers of magma mixing. Contrib. Mineral. Petrol., 80, 147-159.

Ghiorso, M. S., and Sack, R. O. Fe-Ti oxide geothermometry: Thermodynamic formulation and the estimation of intensive variables in silicic magmas. Contrib. Mineral. Petrol., in press,

Hasenaka, T., and Carmichael, I. S. E. (1987) The cinder cones of Michoacan-Guanajuato, Central Mexico: Petrology and chemistry. J. Petrol., 28, 241-269.

Henry, C. D., Price, J. G., and Smyth, R. C. (1988) Chemical and thermal zonation in a mildly alkaline magma system Infiernito caldera, trans-Pecos Texas. Contrib. Mineral. Petrol., 98, 194-211.

Hewitt D. A., and Wones D. R. (1984) Experimental phase relations of the micas. In Bailey, S.W., Ed., Rev. Mineral., v. 13.

Hildreth, W. (1977) The magma chamber of the Bishop Tuff: Gradients in pressure, temperature, and composition. Ph.D. Dissertation, University of California, 328 p.

Hildreth, W. (1979) The Bishop Tuff: Evidence for the origin of compositional zonation in silicic magma chambers. Geol. Soc. Am. Special Paper, 180, 45-75.

Hildreth, W. (1983) The compositionally zoned eruption of 1912 in the Valley of Ten Thousand Smokes, Katmai National Park, Alaska. J. Volcanol. Geotherm. Res., 18, 1-56.

Himmelberg, G. R., and Ford, A. B. (1976) Pyroxenes of the Dufek Intrusion. J. Petrol., 17, 219-243.

Himmelberg, G. R., and Ford, A. B. (1977) Iron-titanium oxides of the Dufek Intrusion, Antarctica. Am. Mineral., 62, 623-633.

Howie, R. A. (1955) The geochemistry of the charnockite series of Madras, India. Trans of the Roy. Society of Edinburgh, 62, 725-768.
Irvine, T. N. (1974) Petrology of the Duke Island ultramafic complex, southeastern Alaska. Geol. Soc. Am. Memoir 138.
Irvine, T. N., and Smith, C. H. (1969) Primary oxide minerals in the layered series of the Muskox Intrusion. Econ. Geol., Mono. 4, 76-94.
Johnson, R. W., and Arculus, R. J. (1978) Volcanic rocks of the Witu Islands, Papua New Guinea: The origin of magmas above the deepest part of the New Britain Benioff Zone. Bull. Volcanology, 41, 609-655.
Johnston, A. D., and Stout, J. H. (1984) A highly oxidized ferrian salite-, kennedyite-, forsterite- and rhonite-bearing alkali gabbro from Kauai, Hawaii, and its mantle xenoliths. Am. Mineral., 69, 57-68.
Kennedy G. C. (1955) Some aspects of the role of water in rock melts. pp. 489-504 in Arie Poldervaart, Ed. Crust of the Earth, Geological Soc. Am. Special Paper 62, New York.
Kolker, A., and Lindsley, D. H. (1989) Geochemical evolution of the Maloin Ranch Pluton, Laramie Anorthosite Complex, Wyoming: Petrology and mixing relations. Am. Mineral., 74, 307-324.
Konnerup-Madsen, J. Dubessy, J., Rose-Hansen, J. (1985) Combines Raman microprobe spectrometry and microthermometry of fluid inclusions in minerals from igneous rocks of the Gadar province (south Greenland). Lithos, 18, 271-208.
Kyle, P. R. (1981) Mineralogy and geochemistry of a basanite to phonolite sequence at Hut Point Peninsula, Antarctica based on core from Dry Valley Drilling Project drill holes 1, 2, and 3. J. Petrol., 22, 451-500.
Larsen, N. M. (1976) Clinopyroxenes and coexisting mafic minerals from the alkaline Ilimaussaq intrusion, South Greenland. J. Petrol., 17, 258-290.
Le Roex, A. P. (1985) Geochemistry, mineralogy and magmatic evolution of basaltic and trachytic lavas from Gough Island, South Atlantic. J. Petrol., 26, 149-186.
Lindsley, D. H., and Frost, B. R. Equilibria among Fe-Ti oxides, pyroxenes, olivine, and quartz: 1. Theory. Am. Mineral., in press,
Lindsley D. H., Frost B. R., Andersen D. J., and Davidson P. M. (1990) Fe-Ti oxide - silicate equilibria: Assemblages with orthopyroxene. in R. J. Spencer and I.-M. Chou, Eds. Fluid-Mineral Interactions: A Tribute to H. P. Eugster. The Geochemical Society, Special Publication No. 2.
Lipman, P. W. (1971) Iron-titanium oxide phenocrysts in compositionally zoned ash-flow sheets from Southern Nevada. J. Geol., 79, 438-456.
Lufkin, J. L. (1976) Oxide minerals in miarolitic rhyolite, Black Range, New Mexico. Am. Mineral., 61, 425-430.
MacChesney, J. B., and Muan, A. (1959) Studies in the system iron oxide - titanium oxide. Am. Mineral., 44, 926-945.
Mahood, G. A. (1981) Chemical evolution of a Pleistocene rhyolite center: Sierra la Primavera, Jalisco, Mexico. Contrib. Mineral. Petrol., 77, 129-149.
Malm, O. A., and Ormaasen, D. E. (1978) Mangarite-charnockite intrusives in the Lofoten-Vesterålen area, North Norway: Petrography, Chemistry and Petrology. Norges Geol. Unders., 338, 83-114.
Mathison, C. I. (1975) Magnetite and ilmenite in the Somerset Dam layered basic intrusion, southeastern Queensland. Lithos, 8, 93-111.
Mathison, C. I., and Hamlyn, P. R. (1987) The McIntosh layered troctolite-olivine gabbro intrusion, East Kimberley, Western Australia. J. Petrol., 28, 211-234.
Maury, R. C., Brousse, R., Villemant, B., Joron, J.-L., Jaffrezic, H., and Treuil, M. (1980) Cristallisation fractionnée d'un magma basaltique alcalin: série de la chaîne des Puys (Massif Central, France). Bull. Minéral., 103, 250-266.
Mazzulo, L. J., and Bence, A. E. (1976) Abyssal tholeiites from DSDP leg 34: The Nazca Plate. J. Geophys. Res., 81, 4327-4351.
Medenbach, O., and El Goresy, A. (1982) Ulvöspinel in native iron-bearing assemblages and the origin of these assemblages in basalts from Ovifak, Greenaldn and Buhl, Federal Republic of Germany. Contrib. Mineral. Petrol., 80, 358-366.
Meijer, A., and Reagan, M. (1981) Petrology and geochemistry of the island of Sarigan in the Mariana Arc: Calc-alkaline volcanism in an oceanic setting. Contrib. Mineral. Petrol., 77, 337-354.
Mitchell, R. H. (1978) Manganoan magnesian ilmenite and titanian clinohumite from the Jacupiranga carbonatite, Sao Paolo, Brazil. Am. Mineral., 63, 544-547.

Molyneux, T. G. (1972) X-ray data and chemical analyses of some titanmagnetite and ilmenite samples from the Bushveld Complex, South Africa. Mineral. Mag., 38, 863-871.
Morse, S. A. (1979) Kiglapait geochemistry. II. Petrography. J. Petrol., 20, 394-410.
Morse, S. A. (1980) Kiglapait Mineralogy. II. Fe-Ti oxide minerals and the activities of oxygen and silica. J. Petrol., 21, 685-719.
Myers, C. W., Bence, A. E., Papike, J. J., and Ayuso, R. A. (1975) Petrology of an alkali-olivine basalt sill from site 169 of DSDP Leg 17: The central Pacific basin. J. Geophys. Res., 80, 807-822.
Naldrett, A. J., Bray, J. G., Gasparrini, E. L., Podolsky, T., and Rucklidge, J. C. (1970) Cryptic variation and the petrology of the Sudbury nickel irruptive. Econ. Geol., 65, 122-155.
Negendank, J. F. W. (1972) Volcanics of the Valley of Mexico. Part II: The opaque mineralogy. Neues Jahrbuch Mineral., Abhandlungen, 117, 183-195.
Nielsen, T. F. D. (1981) The ultramafic cumulate series, Gardiner Complex, east Greenland. Cumulates in a shallow level magma chamber of a nephelinite volcano. Contrib. Mineral. Petrol.,. 76, 60-72.
Nicholls, I. A. (1971) Petrology of Santorini Volcano, Cyclades, Greece. J. Petrol., 12, 67-119.
Nicholls, J., Stout, M. Z., and Fiensinger, D. W. (1982) Petrologic variations in Quaternary volcanic rocks, British Columbia, and the nature of the underlying upper mantle. Contrib. Mineral. Petrol., 79, 201-218.
Novak, S. W., and Mahood, G. A. (1986) Rise and fall of a basalt-trachyte-rhyolite magma system at Kane Springs Wash Caldera, Nevada. Contrib. Mineral. Petrol., 94, 352-373.
Osborn, E. F. (1959) Role of oxygen pressure in the crystallization and differentiation of basaltic magma. Am. J. Sci., 257, 609-647.
Ottemann, J., and Frenzel, G. (1965) Der Chemismus der Pseudobrookite von Vulkaniten. Schweizerische Mineral. Petrogr. Mitt., 45, 819-836.
Parsons, I. (1980) Alkali-feldspar and Fe-Ti oxide exsolution textures as indicators of the distribution and subsolidus effects of magmatic 'water' in the Klokken layered syenite intrusion, South Greenland. Trans. of the Roy. Soc. Edinburgh, Earth Sciences, 71, 1-12.
Parsons, I. (1981) The Klokken Gabbro-Syenite Complex, South Greenland: Quantitative interpretation of mineral chemistry. J. Petrol., 22, 233-260.
Pasteris, J. D. (1985) Relationships between temperature and oxygen fugacity among Fe-Ti oxides in two regions of the Duluth Complex. Can. Mineral., 23, 111-127.
Patchen, A. D. (1984) Petrology and stratigraphy of the Lake Owen layered mafic complex, S.E. Wyoming. M.S. Thesis, University of Wyoming, 163 p.
Pedersen, A. K. (1979) A shale buchite xenolith with Al-armalcolite and native iron in a lava from Asuk, Disko, Central West Greenland. Contrib. Mineral. Petrol., 69, 83-94.
Pederson, A. K. (1981) Armalcolite-bearing Fe-Ti oxide assemblages in graphite-equilibrated volcanic rocks with native iron from Disko, Central West Greenland. Contrib. Mineral. Petrol., 77, 307-324.
Percival, J. A. Orthopyroxene-poikilitic tonalites of the Desliens igneous suite, Ashuanipi granulite complex, Labrador - Quebec, Canada. Can. J. Earth Sci., in press,
Perfit, M. R., and Fornari, D. J. (1983) Geochemical studies of abyssal lavas recovered by DSRV Alvin from Eastern Galapagos Rift, Inca Transform, and Ecuador Rift 2. Phase chemistry and crystallization history. J. Geophys. Res., 88, 10530-10550.
Perfit, M. R., Saunders, A. D., and Fornari, D. J. (1982) Phase chemistry, fractional crystallization, and magma mixing in basalts from the Gulf of California, Deep Sea Drilling Project Leg 64. Initial Reports of the Deep Sea Drilling Project, 64, 649-666.
Powell, M. (1978) The crystallization history of the Igdlerfigssalik Nepheline Syenite Intrusion, Greenland. Lithos, 11, 99-120.
Price, R. C., and Taylor, S. R. (1980) Petrology and geochemistry of the Banks Peninsula volcanos, South Island, New Zealand. Contrib. Mineral. Petrol., 72, 1-8.
Prins, P. (1972) Composition of magnetite from carbonatites. Lithos, 3, 227-240.
Prinz, M., Keil, K., Green, J. A., Reid, A. M., Bonatti, E., and Honnorez, J. (1976) Ultramafic and mafic dredge samples from the equatorial Mid-Atlantic Ridge and fracture zones. J. Geophys. Res., 81, 4087-4103.
Rafferty, W. J., and Heming, R. F. (1979) Quaternary alkaline and sub-alkaline volcanism in South Aukland, New Zealand. Contrib. Mineral. Petrol., 71, 139-150.
Ravindra Kumar, G. R., Srikantappa, C., and Hansen, E. (1985) Charnockite formation at Ponmudi in southern India. Nature, 313, 207-209.

Rice, J. M., Dickey Jr., J. S., and Lyons, J. B. (1971) Skeletal crystallization of pseudobrookite. Am. Mineral., 56, 158-162.

Rose Jr. W. I., Grant N. K., and Easter J. (1979) Geochemistry of the Los Chocoyos Ash, Quezaltenango Valley, Guatemala. In Chapin, C. E. and Elstron, W. E., Eds. Ash Flow Tuffs, Geol. Soc. Am., Special Paper 180, 87-99.

Rutherford, M. J. (1969) An experimental determination of iron biotite-alkali feldspar equilibria. J. Petrol., 10, 381-408.

Sack, R. O., and Ghiorso, M. S. Chromian spinels as petrogenetic indicators: Thermodynamics and petrological applications. Am. Mineral., in press,

Sechee, K. and Larsen, L. M. (1980) Geology and mineralogy of the Sarfortôq carbonatite complex, southern west Greenland. Lithos, 13, 199-212.

Singer, B. S. (1990) Petrology and geochemistry of mid-Pleistocene lavas from Seguam Island, Central Aleutian Island, Alaska: Implications for the chemical and physical evolution of oceanic island arc magmatic centers. Ph.D. Dissertation, University of Wyoming, 188 p.

Singer, B. S., Myers, J. D., and Frost, C. D. Mid-Pleistocene lavas from the Seguam volcanic center, central Aleutian arc: Closed-system fractional crystallization of a basalt to rhyodacity eruptive suite. Contrib. Mineral. Petrol., in press,

Smith, A. L., and Carmichael, I. S. E. (1968) Quaternary lavas from the southern Cascades, western U.S.A. Contrib. Mineral. Petrol., 19, 212-238.

Smith, D. G. W. (1965) The chemistry and mineralogy of some emery-like rocks from Sithean Slauigh, Strachur, Argyllshire. Am. Mineral., 50, 1982-2022.

Snetsinger, K. G. (1969) Manganoan ilmenite from a Sierrian adamelite. Am. Mineral., 54, 431-436.

Snoke, A. W., Quick, J. E., and Bowman, H. R. (1981) Bear Mountain igneous complex, Klamath Mountains, California: An ultrafasic to silicic calc-alkalic suite. J. Petrol., 22, 501-552.

Speidel, D. H., and Nafziger, R. H. (1968) P-T-fO_2 relations in the system Fe-O-MgO-SiO$_2$. Am. J. Sci., 266, 361-379.

Stephenson, N. C. N., and Hensel, H. D. (1978) A Precambrian fayalite granite from the south coast of Western Australia. Lithos, 11, 209-218.

Stormer, J. C. (1972) Mineralogy and petrology of the Raton-Clayton volcanic field, northeastern New Mexico. Geol. Soc. Am., Bull., 83, 3299-3322.

Taylor, H. P., and Noble, J. A. (1969) Origin of magnetite in the zoned ultramafic complexes of Southeastern Alaska. Econ. Geol., Mono. 4, 209-230.

Taylor, R. W. (1964) Phase equilibria in the system FeO-Fe_2O_3-TiO_2 at 1300°C. Am. Mineral., 49, 1016-1030.

Thompson, R. N. (1973) Titanian chromite and chromian titanomagnetite from a Snake River Plain basalt, a terrestrial analogue to lunar spinels. Am. Mineral., 58, 826-830.

Thy, P. (1982) Titanmagnetite and ilmenite in the Fongen-Hullingen Complex, Norway. Lithos, 15, 1-16.

Upton, B. J. G. and Thomas, J. E. (1980) The Tugtutôq younger dyke comples, South Greenland: Fractional crystallization of transitional olivine basalt magma. J. Petrol., 21, 167-198.

Upton, B. J. G., Stephenson, D., and Martin, A. R. (1985) The Tututôq older Dyke Complex: Mineralogy and geochemistry of an alkali gabbro-augite syenite-foyaite association in the Gadar Province of South Greenland. Mineral. Mag., 49, 623-642.

van Kooten, G. K. (1980) Mineralogy, petrology and geochemistry of an ultrapotassic basaltic suite, Central Sierra Nevada, California, USA. J. Petrol., 21, 651-684.

Velde, D. (1975) Armalcolite-Ti-phlogopite-diopside-analcite-bearing lamproites from Smoky Butte, Garfield County, Montana. Am. Mineral., 60, 566-573.

Vincent, E. A., and Phillips, R. (1954) Iron-titanium oxide minerals in layered gabbros of the Skaergaard Intrusion, East Greenland. Geochim. Cosmochim. Acta, 6, 1-26.

Wager, L. R., and Brown, G. M. (1968) Layered Igneous Rocks. W. H. Freeman and Company, San Francisco

Whalen, J. B., and Chappell, B. W. (1988) Opaque mineralogy and mafic mineral chemistry of I- and S-type granites of the Lachlan fold belt, southeast Australia. Am. Mineral., 73, 281-296.

Wass, S. (1973) Oxides of low pressure origin from alkali basaltic rocks, Southern Hignlands N.S.W., and their bearing on the petrogenesis of alkali basaltic magmas. J. Geol. Soc. Australia, 20, 427-447.

Weiss, S. (1986) Petrogenese des intrusivkomplexes von Ballachulish, Westschottland: Kristallisationsverlauf in einem zonierten Kaledonischen Pluton. Ph.D. Dissertation, Univ. of Munich, 228 p.

Weiss, S., and Troll, G. (1989) The Ballachulish Igneous Complex, Scotland: Petrography, mineral chemistry, and order of crystallization in the monzodiorite-quartz diorite suite and in the granite. J. Petrol., 30, 1069-1115.

Whitney, J. A., and Stormer, J. C. (1985) Mineralogy, petrology, and magmatic conditions from the Fish Canyon tuff, Central San Juan volcanic field, Colorado. J. Petrol., 26, 726-762.

Willemse, J. (1969) The vanadiferous magnetic iron ore of the Bushveld Complex. Econ. Geol., Mono. 4, 187-208.

Wilson, J. R., Ebensen, K. H., and Thy, P. (1981) Igneous petrology of the synorogenic Fongen-Hyllingen layered basic complex, South-Central Scandinavian Caledonides. J. Petrol., 22, 584-627.

Wolf, J. A., and Storey, M. (1983) The volatile component of some pumice-forming alkaline magmas from the Azores and Canary Islands. Contrib. Mineral. Petrol., 82, 66-74.

Chapter 13

B. Ronald Frost

STABILITY OF OXIDE MINERALS IN METAMORPHIC ROCKS

INTRODUCTION

When discussing the occurrence of oxides in metamorphic rocks one can consider the oxides in two distinct groups: the non-magnetic oxides, corundum, chromite, the "green" spinels (i.e. the aluminates); and the Fe-Ti oxides, magnetite, ilmenite, hematite and rutile. Although there is complete solid solution between magnetite and the other common spinels, magnetite can be considered separate from the other two spinels because it has a distinctly different mode of occurrence; the coexistence of magnetite with either chromite or hercynite is not common and will be noted in the appropriate areas below. Another reason for considering the Fe-Ti oxides separately from the non-magnetic oxides comes from the fact that the two types of oxides provide the petrologist with different types of information. The Fe-Ti oxides are important indicators of oxygen fugacity and, hence are monitors of the metamorphic fluid. In addition, of course, understanding the controls of their occurrence and abundance is important to understanding the way rock magnetism varies in the crust (see the chapter on magnetic petrology in this volume). The non-magnetic oxides are more important as indicators of temperature or pressure and, although they may monitor oxygen fugacity because they contain dilute amounts of ferric iron-bearing species, in most assemblages they have limited application to determination of oxygen fugacity.

NON-MAGNETIC OXIDES

Corundum

The presence of corundum in a rock is an indication of silica undersaturation. Although quartz and corundum are known to occur together in some high-grade rocks (Santosh, 1987) and may even be in direct contact (Bohlen, 1986) this coexistence is considered to be metastable (Shulters and Bohlen, 1989). In metamorphic environments corundum is found in metamorphosed laterites (Feenstra, 1985), in partially melted pelitic rocks (Harris, 1981; Grant and Frost, 1990), and in high-Mg and high-Al schists that have probably formed after hydrothermally altered mafic rocks (Schumacher and Robinson, 1987; Kerrich et al., 1987). Corundum from crustal rocks is usually fairly pure, with Cr_2O_3 and Fe_2O_3 as the main impurities. Corundum from eclogites however may contain substantial amounts of Cr_2O_3.

"Green spinel"

The term green spinel describes aluminate spinels with various proportions of Fe and Mg and Zn as the divalent cations, that are green in thin section. These spinels occur in four major protoliths: (1) metamorphosed impure dolomitic marbles, (2) metamorphosed and partially melted pelitic rocks, (3) high temperature metabasites, and (4) metamorphosed ultramafic rocks. In all rock types spinel is an indicator of high temperatures of metamorphism. Because spinels in metaperidotites form a series that includes magnetite and chromite, as well as green spinel, the occurrence of spinel in metaperidotites will be discussed separately.

Impure dolomitic marbles. In impure dolomitic marbles spinel forms as the result of the breakdown of chlorite at temperatures equivalent to those of middle amphibolite facies (Rice, 1977). Compositionally spinel from metamorphosed dolomitic marbles can be nearly pure $MgAl_2O_4$. Aluminum is usually by far the dominant trivalent ion and X_{Mg} commonly exceeds 0.95 and often is higher than 0.99 (see Fig. 1). Zinc is usually very low in spinels from amphibolite facies marbles, but it can be considerable in spinels from granulite facies where spinels with $X_{Zn} > 0.15$ have been reported (Tracy et al., 1978; Böhme et al., 1989).

Metapelitic rocks. This discussion of spinel from metapelitic rocks includes spinel from cordierite-anthophyllite rocks, which are usually considered not to be true pelitic rocks but to have formed from other protoliths, such as hydrothermally altered mafic rocks (Andrew, 1984; Schumacher and Robinson, 1987; Bernier, 1990). Although spinels from such rocks tend to be richer in zinc, and locally chrome and vanadium as well, than those from pelitic rocks, they are included with pelitic rocks in this discussion because the mineral equilibria that control spinel composition are the same in both rock types, regardless of the origin of these high Mg and high Al rocks.

The green spinel found in metapelites is highly aluminous. Cr_2O_3 is usually present in only minor amounts, but in some rocks spinel can contain up to 10% Cr_2O_3 (Atkin, 1978) and spinel with more than 1% Cr_2O_3 is not at all uncommon in metapelitic rocks (Stoddard, 1979; Harley, 1986; Bingen et al., 1988; Percheck et al., 1989). Calculated ferric iron never makes up more than 10% of the trivalent ions. Even spinel coexisting with magnetite or hematite is relatively low in ferric iron (Sandiford et al., 1987; Santosh, 1987; Currie and Gittins, 1988; Kamineni and Rao, 1988), with ferric iron contents no higher than those spinels that do not coexist with magnetite. These low values of ferric iron clearly do not reflect the primary composition of the spinel, since under granulite conditions green spinel should be able to accomodate well more than 10% of the ferric iron component (see for example Turnock and Eugster, 1962). This indicates that aluminate spinels are as prone to re-equilibration on cooling as the Fe-Ti oxides (Frost et al., 1988) and that the spinel compositions reported from such rocks are artifacts of the cooling process.

Spinel in pelitic rocks is generally rich in hercynite, although Zn-rich (gahnitic) spinel and Mg-rich spinels occur as well (Fig. 2). The composition of the spinel is strongly dependent on the associated mineral assemblage. Spinel coexisting with quartz may be moderately zinc rich, and is generally richer in iron than in magnesium. Spinel coexisting with corundum, in contrast, tends to be richer in Fe than Mg.

Under most conditions hercynitic spinel is incompatible with quartz, the limiting reactions being:

3 hercynite + 5 quartz = almandine + 2 sillimanite   (1)

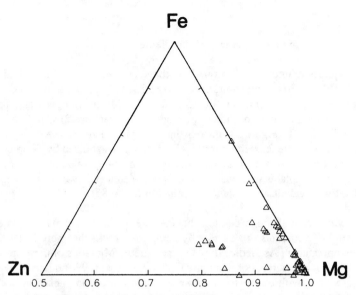

Figure 1. Fe-Mg-Zn diagram showing composition of spinels from metamorphosed impure dolomitic marbles. Data from Rice (1977), Tracy et al. (1978), Bucher-Nurminen (1981, 1982), Lieberman and Rice (1986), Spry (1987), and Böhme et al. (1989).

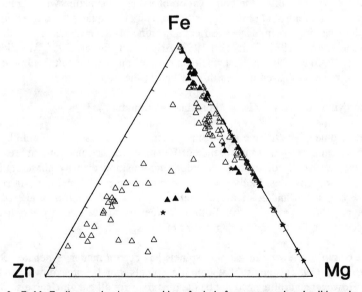

Figure 2. Fe-Mg-Zn diagram showing composition of spinels from metamorphosed pelitic rocks and from cordierite-anthophyllite rocks. Open triangles = quartz-bearing assemblages; stars = corundum-bearing assemblages, filled triangles = assemblages with neither quartz nor corundum. Data from Shannon (1923), Frost (1973a), Berg (1977), Atkin (1978), Stoddard (1979), Dietvorst (1980), Ellis et al. (1980), Kars et al. (1980), Harris (1981), Spry (1982), Horrocks (1983), Vielzeuf (1983), Arima and Barnett (1984), Droop and Bucher-Nurminen (1984), Windley et al. (1984), Hicks et al. (1985), Waters and Moore (1985), Montel (1985), Harley (1986), Spry and Scott (1986), Waters (1986), Ackermand et al. (1987), Lal et al. (1987), Sandiford et al. (1987), Santosh (1987), Schumacher and Robinson (1987), Schumacher et al. (1987), Spry (1987), Bingen et al. (1988), Currie and Gittins (1988), Kamineni and Rao (1988), Droop (1989), Percheck et al. (1989), Stüwe and Powell (1989), Bernier (1990), Grant and Frost (1990).

and

$$2 \text{ hercynite} + 5 \text{ quartz} = \text{Fe-cordierite} \tag{2}$$

These two equilibria intersect at around 3 kbars and 800°C (Bohlen et al., 1986), indicating that the quartz + hercynite assemblage is likely to occur only at fairly low pressures and extreme temperatures (Fig. 3). Because Mg is strongly fractionated into cordierite over spinel, in rocks of increasing Mg content reaction (2) will be displaced to higher P and T (Fig. 3). Figure 3 indicates that the spinel + quartz assemblage is likely to be restricted to extreme temperatures unless the spinel is stabilized by a component that is not easily accomodated in garnet or cordierite. In many spinels this role is played by zinc (Fig. 2), which explains why quartz coexists with gahnite-rich spinels at temperatures far lower than with gahnite-poor spinels (Montel et al., 1986).

In accordance with the discussion above, those spinels that coexist with quartz and do not contain large amounts of zinc usually come from rocks that were metamorphosed at extreme temperatures. As predicted by Figure 3 the Fe/Mg ratios of these spinels seem to be a function of pressure. For example, Grant and Frost (1990) describe nearly pure hercynite coexisting with quartz from a contact aureole metamorphosed at 3 kbars. In contrast, the relatively magnesian spinels coexsiting with quartz reported by Currie and Gittins (1988) come from rocks metamorphosed at pressures of more than 10 kbars.

The factors controlling the composition of spinels coexisting with corundum are not as clear as those controlling the composition of spinels in quartz-bearing assemblages. The corundum-bearing assemblages also contain minerals such as garnet, cordierite, or orthopyroxene (Harris, 1981; Horrocks, 1983; Arima and Barnett, 1984; Windley et al., 1984; Waters, 1986) and, thus are not stable unless the spinel-sillimanite tie line is broken (Fig. 4). The reaction that eliminates this tie line in the iron system is:

$$2 \text{ hercynite} + 2 \text{ sillimanite} = \text{almandine} + \text{corundum} \tag{3}$$

Shulters and Bohlen (1989) have shown that reaction (3) is easily accessible by crustal metamorphism, indicating that the absence of Fe-spinel with corundum is more likely to be caused by the paucity of rocks with proper composition than by phase equilibrium controls. Corundum is most likely found in restites from partial melting (Harris, 1981; Grant and Frost, 1990) and Al-rich rocks inferred to be hydrothermally altered basalts (Windley et al., 1984; Waters, 1986). Both rock types tend to be strongly enriched in Mg over Fe, accounting for the magnesian nature of the spinel.

The origin of the zinc in gahnitic spinels has been a matter of debate. Stoddard (1979) and Spry (1982) suggest that gahnitic spinels can be breakdown products of staurolite. However staurolite need not be the precursor for all gahnitic spinels. Spry and Scott (1986) argued that gahnitic spinel can form by desulfurization of sphalerite and noted that Zn-bearing spinels are found in many rock types in association with metamorphosed ore deposits. This argument was further elaborated by Spry (1987), who showed how the composition of Zn-bearing spinels in the assemblage spinel-garnet-quartz-sphalerite-pyrite (or pyrrhotite) is an indicator of changes in sulfur and oxygen fugacities. With decreasing sulfur fugacity spinel in this assemblage should become enriched in Zn.

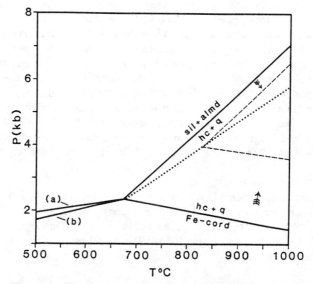

Figure 3. Phase diagram showing stability of assemblages with quartz and spinel. Reaction (a): Fe-cord = almd + q + sil, reaction (b): Fe-cord + hc = almd + sil, almd = almandine, Fe-cord = Fe-cordierite, hc = hercynite, q = quartz, sil = sillimanite. Arrows show the displacement of reactions caused by the presence of Mg. Because Mg is strongly favored in cordierite and there is only a small difference in Mg preference between spinel and garnet, the cordierite-bearing reaction is more strongly displaced in Mg-bearing systems than is the cordierite-absent curve. Dashed lines show the inferred location of the curves at some intermediate, but unspecified, value of X_{Fe} in spinel. Dotted line gives the inferred trace of the invariant point with progressive enrichment of the system in Mg.

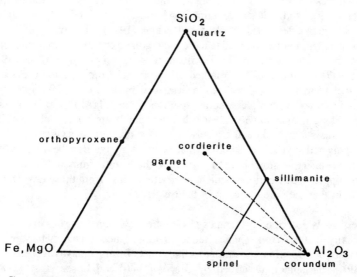

Figure 4. Chemographic diagram showing the topologic relations in the system $(Fe,MgO)-SiO_2-Al_2O_3$ between minerals commonly found occurring with spinel in high-grade pelitic rocks. Note that to get corundum in garnet- or cordierite-bearing pelitic rocks (dashed lines) the tie-lie spinel-sillimanite (solid line) must be broken. See text for discussion.

Metabasites. Green spinel is a minor phase in gabbroic rocks metamorphosed under high-pressure granulite facies. Spinels from these rocks are dominantly aluminates. Chrome is usually present in only minor amounts, but in some metabasites spinel can have significant Cr_2O_3, with amounts up to 10% having been reported (Francis, 1976). Ferric iron is low; even those spinels that coexist with magnetite have less than 5% of the trivalent ions occupied by ferric iron (van Lamoen, 1979; Ellis and Green, 1985). As in spinels from pelitic rocks, this indicates that aluminates are particularly effective in eliminating ferric iron from their structure during cooling. The divalent ions are dominantly iron and magnesium; zinc makes up less than 5% of the divalent ions. Magnesium is usually more abundant than iron (Fig. 5), which is probably a reflection of bulk composition.

The occurrences of spinel in metabasites are reported from rocks metamorphosed in granulite facies, either from exposed granulite terranes (e.g., Johnson and Essene, 1982; Ellis and Green, 1985; Mall and Sharma, 1988)) or in xenoliths (Francis, 1976; Meyer and Brookins, 1976; Selverstone and Stern, 1983). The model reaction that produces spinel in these assemblages is:

forsterite + anorthite = enstatite + diopside + spinel

In the pure system this reaction occurs at moderate pressures; 6-8 kbars (Kushiro and Yoder, 1966). The presence of chrome and iron in natural systems stablizes spinel down to lower pressures, with the result that, in aluminous ultramafic hornfelses the assemblage olivine-anorthite-orthopyroxene-clinopyroxene-spinel can be found at pressures as low as 3 or 4 kbars (Frost, 1976).

Other protoliths. Aluminate spinels are found in a variety of other rare rock types, including buchites (Pedersen, 1978; Grapes, 1986; Foit et al., 1987), tactites (Shedlock and Essene, 1979), rodingites (Lieberman and Rice, 1986), and an unusual iron and aluminum-rich gneiss described by Berg and Wiebe (1985). The most unusual spinels come from buchites, pelitic rocks that have undergone extreme thermal metamorphism at low pressure. In the occurrences described by Pedersen (1978), Grapes (1986), and Foit et al. (1987) the spinel differs markedly from spinel from plutonic rocks described above. Spinel from the coal-fire buchite described by Foit et al.(1987) coexists with hematite and shows a complete solid solution from $MgAl_2O_4$ to magnesioferrite ($MgFe_2O_4$). There is also some evidence that the spinels are non-stoichiometric in that they contain excess ferric iron and are deficient in the divalent ions (Fig. 6). The spinels from the highly reduced shale xenoliths that occur in native-iron bearing andesites on Disko Island are also non-stoichiometric. Because these come from highly reduced rocks, they have an opposite sense of non-stoichoimetry than those from those described by Foit et al. (1987) in that they are deficient in the trivalent ions.

The spinels described by Grapes (1986) are stoichiometric, but even so they are compositionally distinct from spinels found in most igneous and metamorphic rocks, though they are compositionally similar to the spinels from paralavas reported by Cosca et al. (1989). The spinels described by Grapes (1986) and Cosca et al. (1989) show extensive solution between magnetite and hercynite; some of the aluminous spinels have up to 20% of the magnetite end member. The difference between the spinels from these buchites and those described from other metamorphic rocks probably reflects the fact that

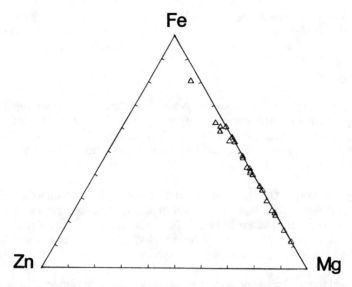

Figure 5. Fe-Mg-Zn diagram showing the composition of spinels commonly found in high-grade metabasites. Data from Francis (1976), Meyer and Brookins (1976), van Lamoen (1979), Johnson and Essene (1982), Stephenson and Hensel (1982), Selverstone and Stern (1983), Ellis and Green (1985), Mall and Sharma (1988), Carswell et al. (1989),

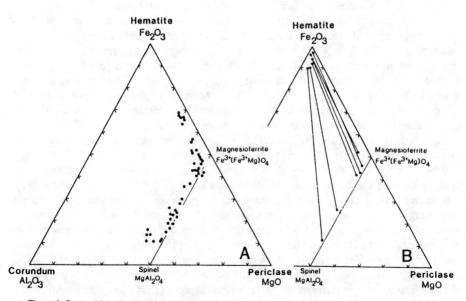

Figure 6. Composition of spinels from a coal-fire buchite. (A) composition of spinels showing extent of non-stoichiometry, particularly in magnesionferrites. (B) composition of coexisting spinels and hematite from same rocks (modified from Foit et al., 1987).

buchites form at extreme temperatures and are rapidly chilled. Consequently the high-temperature compositions are likely to have been preserved. In more slowly cooled metamorphic rocks the spinels evidently adjust compositions on cooling.

Chrome spinels

In metamorphic rocks chrome spinels are found mainly in metaperidotites, where their composition is strongly determined by metamorphic grade (Evans and Frost, 1975). Chrome-spinels do occur in metasedimentary rocks, such as at Otokumpu, Finland (Trelor, 1987) and in the Naire deposit, Australia (Graham, 1978), where they can assume unusual compositions.

Metaperidotite. Generally spinels from metaperidotites contain Cr, Al, Fe, and Mg as the major cations. ZnO, MnO, NiO, V_2O_3, and TiO_2 are usually present in abundances of less than 1% (Evans and Frost, 1975). However, substantial amounts of ZnO (up to 20.0%, Wylie et al., 1987) and MnO (up to 14.4%, Paraskevopoulos and Economou, 1981) have been reported from spinels in serpentinites.

Spinels from metaperidotites have a wide range in composition that reflects changes in metamorphic conditions (see Evans and Frost, 1975) (Fig. 7). In serpentinite from zeolite and prehnite-pumpellyite facies there is usually two types of spinel. One is fine-grained magnetite dust that is present throughout the rock and the other is a chromite that has has been inherited from the peridotitic protolith. Commonly the chromite is rimmed by ferritchromite (a chrome-spinel that is rich in ferric iron) (Beeson and Jackson, 1969; Frisch, 1971; Onyeagocha, 1974; Bliss and Maclean, 1975). Traditionally it has been assumed that these spinels are not an equilibrium assemblage and that the chromite core is relict from the parent peridotite. However, recent calculations by Sack and Ghiorso (in press) show that an extensive solubility gap is present in chrome spinels at low temperature and that chromite and ferritchromite could be in equilibrium at these temperatures. The spinels from greenschist-grade serpentinites commonly have a wide range in chrome content (Fig. 7, triangle A). The coarser-grained spinels that formed from primary chromite are ferritchromite and have a higher Cr content than the fine-grained matrix magnetite (Shive et al., 1988). At metamorphic grades above those of antigorite stability the spinel in metaperidotites is ferritchromite, which becomes progressively enriched in Cr with increasing grade (Fig. 7, triangles C and D). In some localities there is evidence of a solubility break between aluminous and chromiferous spinels and ferric spinels (Muir and Naldrett, 1973, Loferski and Lipin, 1983; Zakrzewski, 1989). In other localities apparently homogenous spinels occur that have compositions lying within the solvus outlined by Loferski and Lipin (1983) (Evans and Frost, 1975). The spinels having compositions that lie within the proposed solvus come from prograde metamorphic rocks with the assemblage forsterite - anthophyllite, which forms at temperatures around 600°C. It is possible that the crest of the proposed solvus lies below this temperature and that exsolution phenomena will be observed only in rocks that have had prolonged cooling histories. With the breakdown of chlorite at the upper limits of amphibolite facies, spinels in metaperidotites become more aluminous (Fig. 7 triangle E). At this grade the composition of the spinel is dependent on the composition of the protolith. Most metaperidotites will have Cr-rich spinels but those that were originally fairly aluminous, or are associated with metamorphosed blackwall rocks, may contain aluminous spinels that are green in thin section.

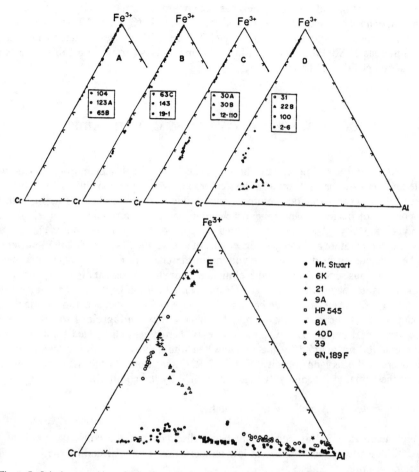

Figure 7. Spinel compositions from metaperidotites from the following assemblages: A. diopside-olivine-antigorite, B. tremolite-olivine-antigorite, C. talc-olivine, D. anthophyllite-olivine, E. enstatite-olivine. Samples 19-1, 12-110, and 2-6 are from Frost (1973b) others are from the Swiss and Italian Alps locations for these are given in Evans and Frost (1975). Figure is modified after Evans and Frost (1975).

Other protoliths. Chrome spinels occur in metasedimentary rocks at Otokumpu, Finland. Protoliths include skarns, quartzite, schist, and cordierite-anthophyllite rocks. Spinels from these rocks show a range in Cr content from pure aluminate to pure chromate (Weiser, 1967; Trelor, 1987). Zinc is a common component in the spinels and is often the dominant divalent ion. Trelor (1987) shows that there is a distinct negative correlation between Zn and Cr in spinels from some horizons, suggesting that there is a coupled substitution of $FeCr_2O_4$ and $ZnAl_2O_4$. Spinels from Otokumpu are also moderately enriched in manganese, with up to 6% MnO being reported (Trelor, 1987). Minor amounts of V_2O_3, CoO, and CuO are also reported (Trelor, 1987). Chrome spinels have also been reported from skarns in the Helmo area of Ontario (Pan and Fleet, 1989). Like the spinels from Otokumpu they contain considerable ZnO (up to 6%) and 1-2% MnO.

Another unusual suite of spinels has been reported from the metamorphosed pyritic sedimentary horizons in the Nairne area of South Australia (Graham, 1978). These spinels are vanadiferous chromites that contain up to 24% Mn. They are associated with unusual concentrations of Sb, Pd, as well as Mn and V and their occurrence probably reflects the unusual composition of their host rock.

OCCURRENCE AND CHEMISTRY OF Fe-Ti OXIDES

Magnetite and ilmenite are the major Fe-Ti oxides in metamorphic rocks, with rutile and hematite being somewhat less common. Magnetite and ilmenite coexist over most of the oxygen fugacities commonly accessible in metamorphism (Fig. 8), with the composition of magnetite and ilmenite changing in response to changes in temperature or oxygen fugacity. The manner in which the composition of these phases change with respect to temperature and oxygen fugacity is, of course, the Fe-Ti oxide thermometer and has been discussed in detail in this volume in the chapters on phase equilibria and on oxides in igneous rocks. Magnetite is eliminated from this system at high oxygen fugacity by the hematite-magnetite buffer (see Chapter 1), which marks the oxygen fugacities at which magnetite reacts to Ti-free hematite. Because hematite can accomodate significant amounts of Ti, Ti-bearing hematite may be stable to oxygen fugacities somewhat below those of the MH buffer. At low temperatures Ti-hematite is eliminated from the system at the equilibrium marked by the ilmenite-hematite solvus. At higher temperatures it disappears at the oxygen fugacity of the R3-R3c transition. In the magnetite-saturated system the hematite-ilmenite solvus is represented by the equilibrium:

$$\text{ilmenite}_{ss} + \text{magnetite} + O_2 = \text{hematite}_{ss} \tag{4}$$

At high temperatures this equilibrium curve disappears at the consolute point of the hematite-ilmenite solvus whereas at low temperatures it is truncated by the isobarically invariant reaction:

$$\text{ilmenite}_{ss} + \text{hematite}_{ss} = \text{magnetite} + \text{rutile}.$$

Two reactions radiate from this point. One is the reaction:

$$\text{magnetite} + \text{rutile} = \text{ilmenite}_{ss} + O_2 \tag{5}$$

the other is the reaction:

$$\text{hematite}_{ss} = \text{magnetite} + \text{rutile} + O_2 \tag{6}$$

Because of stoichiometry, reaction of magnetite + rutile to form a hematite-ilmenite solid solution will involve oxidation if the solution has more than 50% hematite and it will involve reduction if the solution has less than 50% hematite.

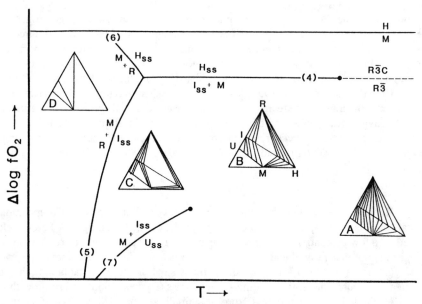

Figure 8. Δ log fO_2 - T diagram for the low-temperature portion of the system Fe-Ti-O. Δlog fO_2 refers to the deviation of oxygen fugacity from the FMQ buffer. H = hematite, I = ilmenite, M = magnetite, R = rutile, U = ulvöspinel. Subscript ss = solid solution. Chemographies A through D show how the phase relations change as a function of temperature.

Another curve at low oxygen fugacities and low temperatures marks the magnetite-ulvöspinel solvus. On this diagram it is represented by the reaction:

$$\text{magnetite} + \text{ilmenite} = \text{ulvöspinel}_{ss} + O_2 \tag{7}$$

This curve is truncated at "high" temperatures by the consolute point of the magnetite-ulvöspinel solvus.

Although the general shape of this diagram can be inferred from mineral relations, an exact calibration is not possible at present. Ghiorso and Sack (in press) calculate the hematite-ilmenite solvus to be about 3 to 4 log units above the FMQ buffer and about 1 to 2 log units below the MH buffer. The consolute point of the ilmenite-hematite solvus is estimated to be around 680°C by Burton (1985) and around 750°C by Ghiorso (1990). The consolute point of the magnetite-ulvospinel solvus is below 500°C and at oxygen fugacities more than 3 log units below those of the FMQ buffer (Andersen and Lindsley, 1988). The location of the reaction magnetite + rutile = ilmenite$_{ss}$ + hematite$_{ss}$ is poorly known. Lindsley and Lindh (1974) maintain that at 1 kbar it lies below 550°C, a conclusion that is consistent with petrologic observations indicating that it occurs in upper greenschist facies (Braun and Raith, 1985; Feenstra, 1985).

Mineral relations in this system are shown for a few temperatures through use of an FeO-TiO$_2$-Fe$_2$O$_3$ chemographic triangle (Buddington and Lindsley, 1964). In this chemography oxygen fugacity increases from left to right parallel to the base of the triangle. At high temperatures (above ca. 680°C) chemography A is stable and there are two simple assemblages, the common assemblage involving spinel and rhombic oxides and the Ti-rich assemblage of a rhombic oxide and rutile. In both assemblages the composition of the rhombic oxide becomes more Ti-rich with lower oxygen fugacity. In the temperature range of 550 to 680°C (chemography B) there are two three-phase assemblages: rutile-hemoilmenite-ilmenohematite and magnetite-hemoilmenite-ilmenohematite. At temperatures below those of reaction (4) (500°C or below) chemography C is stable, with the three-phase assemblages being magnetite-rutile-ilmenohematite and magnetite-rutile-hemoilmenite. Chemography D show relations at a still lower temperature where magnetite + rutile coexist with either a hematite or an ilmenite having nearly end-member compositions. Relationships similar to those show in chemography C have been reported from pelitic rocks that show a wide range in oxygen fugacity by Williams and Grambling (1990).

Rutile

Rutile is most commonly found in pelitic rocks and metabasites. Compositionally crustal rutile is nearly pure, with the major impurities, Cr_2O_3, SiO_2, and Fe_2O_3, seldom exceeding a few percent. Niobium, which may be a major constituent of rutile from some eclogites is not generally abundant in rutile from low-pressure environments. However, rutile with as much as 6.0% Nb_2O_5 (and perhaps small amounts of tantalum) has been reported from cordierite-anthophylliite-garnet gneiss (Dymek, 1983) and rutile with more than 4% Nb_2O_5 has been reported from amphibolite-grade pelitic schist (Ghent, pers. comm.).

Figure 8 indicates that rutile is stable in equilibrium with magnetite at low temperature. This is consistent with relations reported by Itaya and Banno (1980), who found rutile to be the stable Ti phase in weakly metamorphosed pelitic rocks whereas ilmenite was stable at higher grades. Figure 8 also indicates that rutile is stable over a whole range of temperatures in rocks metamorphosed at oxygen fugacities above those of the HM buffer. This is also consistent with natural assemblages, for rutile and hematite are found in oxidizing, Mn-rich rocks (Kawachi et al., 1983; Reinecke, 1986a,b).

In the absence of magnetite, rutile may occur in a range of rock types. It is found at high pressures in both pelitic rocks and metabasites. The reaction controlling its occurrence in pelitic rocks, known as GRAIL (Bohlen et al., 1983) is:

$$Fe_3Al_2Si_3O_{12} + 3\ TiO_2 = 3\ FeTiO_3 + Al_2SiO_5 + 2\ SiO_2 \qquad (8)$$
almandine   rutile   ilmenite   kyanite   quartz

An analogous reaction occurs in metabasites (Bohlen and Liotta, 1986):

$$CaFe_2Al_2Si_3O_{12} + 2\ TiO_2 = 2\ FeTiO_3 + CaAl_2Si_2O_8 + SiO_2 \qquad (9)$$
garnet   rutile   ilmenite   anorthite   quartz

In their end-members both equilibria occur in pressure ranges of 12-15 kbars at temperatures of medium to high-grade metamorphism. However, because of solid solutions, particularly of Mg in garnet, the high-pressure, rutile-bearing assemblages limited by reactions (8) and (9) can be encountered in rocks of the appropriate bulk composition metamorphosed at pressures as low as 6 kbars (see the occurrences tabulated by Bohlen et al., 1983 and Bohlen and Liotta, 1986).

Rutile is the major Ti mineral in eclogites (see Chapter 10) and it is also common in blueschists (Itaya et al., 1985). However, the presence of rutile in metamorphosed pelitic and mafic rocks need not indicate high pressures. In rocks without magnetite it may be found in a range of pressures (not usually in equilibrium with garnet, though). In pelitic rocks with high sulfur fugacity rutile may form by the reaction:

$$2\ FeTiO_3 + S_2 = 2\ FeS + 2\ TiO_2 + O_2$$
ilmenite   pyrrhotite   rutile

A particularly good example of this reaction was discussed by Tracy and Robinson (1988).

Hematite

Hematite is found in the following environments: metamorphised iron-formations, metabasites metamorphosed at low to medium grades, oxidized pelitic rocks, and metamorphosed manganiferous rocks. Hematite from iron-formations can be very pure, with negligible amounts of MgO and TiO_2 (Floran and Papike, 1978). Hematite found in oxidizing Mn-rich metasediments may have 1 to 4 % Mn_2O_3 (Abraham and Schreyer, 1975; Kramm, 1979; Reinecke, 1986a). Titanohematite may also occur in any rock type that has been subjected to slow cooling or retrogressive metamorphism. Because most ilmenite isopleths intersect the hematite-ilmenite solvus at low temperature one would expect exsolution of ilmenite to ilmenite + titanohematite to occur in rocks that initially were rather oxidized (i.e., initially had ilmenite with a moderately high X_{hem}) if these rocks were cooled slowly enough or if they were subjected to low-temperature deformation and metamorphism. Such exsolution relations were described in metabasites by Braun and Raith (1985) and in metamorphosed granitoids by Diamond (1989).

Ilmenite

Ilmenite is most commonly found in metapelitic and metabasic rocks, although it may also occur in metaperidotites (Trommsdorff and Evans, 1980) and more rarely in metamorphosed carbonates (Wise, 1959). In all rock types ilmenite tends to be absent in the lowest grade rocks. In metaperidotites the low-temperature equivalent of ilmenite is titanoclinohumite or perovskite (Trommsdorff and Evans, 1980). In weakly metamorphosed pelitic rocks, rutile is usually the stable Ti-oxide. For example, Ferry (1984) reports that the biotite-in reaction can be modelled by the reaction muscovite + ankerite + rutile + pyrite + graphite + siderite = biotite + plagioclase + ilmenite. In weakly metamorphosed metabasites the stable low-grade assemblage is titanite +/- hematite or magnetite (Laird, 1980; Cassidy and Groves, 1988). Ilmenite does not appear in metabasites until mid-greenschist (Cassidy and Groves, 1988) or lower amphibolite facies (Peacock and Norris, 1989). As noted above, the stability field for ilmenite in most rock types is truncated at high pressures by reactions leading to the formation of rutile.

Along with variable amounts of iron and titanium, indicative of solution between hematite and ilmenite, ilmenite from natural systems can also contain substantial amounts of MgO and MnO. In both pelitic rocks and metabasites MnO is more abundant in ilmenite than is MgO. The abundance of MnO in ilmenite tends to decrease markedly with increasing metamorphic grade in both metabasites and metapelites (Cassidy and Groves, 1988). In metapelites the MnO content of ilmenite is strongly dependent on the

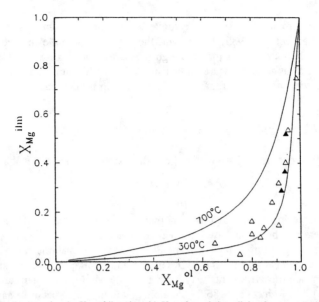

Figure 9. Graph comparing the X_{Mg} of ilmenite with X_{Mg} of coexisting olivine from metaperidotites. Curves show the calculated distribution using the olivine-ilmenite thermometer of Andersen and Lindsley (1981). Solid symbols = mineral pairs from forsterite-tremolite - grade serpentinites (Trommsdorff and Evans, 1980); open symbols = mineral pairs from high-grade metaperidotites (Frost, 1973b; Desmarais, 1981; Dymek et al., 1988; Windley et al., 1989).

abundance of garnet in the rocks. Mn strongly favors garnet over ilmenite with K_d ranging from 6 to 10 (Tracy, 1982) thus ilmenite from rocks rich in garnet tends to be poorer in MnO than ilmenite from garnet-absent rocks.

The most magnesian ilmenite is found in metaperidotites (Trommsdorff and Evans, 1980; Windley et al., 1989) and in marbles (Wise, 1959; Giere, 1987), where it may be sufficiently enriched in Mg to be geikielite ($X_{Mg} > 0.5$). In metaperidotites Trommsdorff and Evans (1980) found ilmenite (or geikielite) to appear as a consequence of the breakdown of titanoclinohumite. This occurs in antigorite schists at approximately the first appearance of the assemblage olivine + tremolite. Figure 9 shows that X_{Mg} in ilmenite from metaperidotites is strongly correlated with X_{Mg} of the coexisting olivine. Geikielite is found in rocks containing olivine that is more magnesian than Fo_{95}. This correlation is to be expected if there is an equilibrium distribution of Fe and Mg between olivine and ilmenite. Comparison with the olivine-ilmenite thermometer of Andersen and Lindsley (1981) however, shows that the olivine-ilmenite pairs from metaperidotites lie on distribution curves that indicate very low temperatures of equilibration. This shows that, like many other ion-exchange equilibria involving oxides, the ilmenite-olivine system re-equilibrates down to very low temperatures.

Magnetite

In most metamorphic rocks magnetite is nearly pure Fe_3O_4. As noted above, magnetite in metaperidotites may accomodate considerable Cr_2O_3, which could serve to lower their Curie temperature significantly (Shive et al., 1988). In high-grade rocks magnetite in equilibrium with ilmenite may also contain considerable Ti. However, most magnetite in metamorphic rocks has expelled Ti during re-equilibration on cooling, leaving Ti as a minor element.

Magnetite is found in a wide variety of rock types, including metaperidotites, metabasites, iron-formations, many quartzofeldspathic gneisses and some metapelites. Because it is the major source of magnetism in the crust, the factors that control the occurrence of magnetite in various protoliths are of interest to geophysicists as well as to petrologists. For this reason we will consider this subject in detail in chapter 14.

REFERENCES

Abraham, K. and Schreyer, W. (1975) Minerals of the viridine hornfels from Darmstadt, Germany. Contrib. Mineral. Petrol. 49, 1-20.
Ackermand, D., Herd, R.K., Reinhardt, M., and Windley, B.F. (1987) Sapphirine parageneses from the Caraiba complex, Bahia, Brazil: the influence of Fe^{2+} - Fe^{3+} distribution of the stability of sapphirine in natural assemblages. J. Met. Geol. 5, 323-339.
Andersen, D.J. and Lindsley, D.H. (1981) A valid Margules formulation for an asymmetric ternary solution: Revision of the olivine-ilmenite thermometer. Geochim. Cosmochim. Acta 45, 847-852.
Andersen, D.J. and Lindsley, D.H. (1988) Internally consistent solution models for Fe-Mg-Mn-Ti oxides: Fe-Ti oxides. Am. Mineral. 73, 714-726.
Andrew, A.S. (1984) P-T-X(CO_2) conditions in mafic and calc-silicate hornfelses from Oberon, New South Wales, Australia. J. Met. Geol. 2, 143-163.
Arima, M. and Barnett, R.L. (1984) Sapphirine bearing granulites from the Sipiwesk Lake area of the late Archean Pikwitonei terrain, Manitoba, Canada. Contrib. Mineral. Petrol. 88, 102-112.
Atkin, B.P. (1978) Hercynite as a breakdown product of staurolite from within the aureole of the Ardara Pluton, Co. Donegal, Eire. Mineral. Mag. 42, 237-239.
Beeson, M.H. and Jackson, E.D. (1969) Chemical composition of altered chromites from the Stillwater Complex, Montana. Am. Mineral. 54, 1084-1100.
Berg, J.H. (1977) Dry granulite mineral assemblages in the contact aureoles of the Nain Complex, Labrador. Contrib. Mineral. Petrol. 64, 33-52.
Berg, J.H. and Wiebe, R.A. (1985) Petrology of a xenolith of ferro-aluminous gneiss from the Nain complex. Contrib. Mineral. Petrol. 90, 226-235.
Bernier, L.R. (1990) Vanadiferous zincian chromian hercynite in a metamorphosed basalt-hosted alteration zone, Atik Lake, Manitoba. Can. Mineral. 28, 37-50.
Bingen, B., Demaiff, D., and Delhal, J. (1988) Aluminous granulites of the Archean craton of Kasai (Zaire): Petrology and P-T conditions. J. Petrol. 29, 899-919.
Bliss, N.W. and MacLean, W.H. (1975) The paragenesis of zoned chromite from central Manitoba. Geochim. Cosmochim. Acta 39, 973-990.
Böhme, V.O., Wand, U., and Ullrich, B. (1989) Kristallchemischs Untersuchungen an Spinellen granulitfazieller Marmore des zentralen Königin-Maud-Landes Ostantarktika. N. Jb. Miner. Mh. 289-299.
Bohlen, S.R. (1986) Instability of the assemblage corundum + quartz. Trans. Am. Geophys. Union 67, 1280.
Bohlen, S.R., Dollase, W.A., and Wall, V.J. (1986) Calibration and applications of spinel equilibria in the system $FeO-Al_2O_3-SiO_2$. J. Petrol. 27, 1143-1156.
Bohlen, S.R. and Liotta, J.J. (1986) Barometer for garnet amphibolites and garnet granulites. J. Petrol. 27, 1025-1034.
Bohlen, S.R., Wall, V.J., and Boettcher, A.L. (1983) Experimental investigations and geological applications of equilibria in the system $FeO-TiO_2-Al_2O_3-SiO_2-H_2O$. Am. Mineral. 83, 52-61.
Braun, E. and Raith, M. (1985) Fe-Ti oxides in metamorphic basites from the Eastern Alps, Austria: a contribution to the formation of solid solutions of natural Fe-Ti oxide assemblages. Contrib. Mineral. Petrol. 90, 199-213.
Bucher-Nurminen, K. (1981) Petrology of chlorite-spinel marbles from NW Spitsbergen (Svalbard). Lithos 14, 203-213.

Bucher-Nurminen, K. (1982) On the mechanism of contact aureole formation in dolomitic country rock by the Adamello intrusion (northern Italy). Am. Mineral. 67, 1101-1117.

Buddington, A.F. and Lindsley, D.H. (1964) Iron-titanium oxide minerals and synthetic equivalents. J. Petrol. 5, 310-357.

Burton, B.P. (1985) Theoretical analysis of chemical and magnetic ordering in the system Fe_2O_3-$FeTiO_3$. Am. Mineral. 70, 1027-1035.

Carswell, D.A., Möller, C., and O'Brien, P.J. (1989) Origin of sapphirine-plagioclase symplectites in metabasites from Mitterbachgraben, Dunkelsteinerwald granulite complex, Lower Austria. Eur. J. Mineral. 1, 455-466.

Cassidy, K.F. and Groves, D.I. (1988) Manganoan ilmenite formed during regional metamorphism of Archean mafic and ultramafic rocks from Western Australia. Can. Mineral. 26, 999-1012.

Cosca, M.A., Essene, E.J., Geissman, J.W., Simmons, W.B. and Caotes, D.A. (1989) Pyrometamorphic rocks associated with naturally burned coal beds, Powder River Basin, Wyoming. Am. Mineral. 74, 85-100.

Currie, K.L. and Gittins, J. (1988) Contrasting sapphirine parageneses from Wilson Lake, Labrador and their tectonic implications. J. Met. Geol. 6, 603-622.

Desmarais, N.R. (1981) Metamorphosed Precambrian ultramafic rocks in the Ruby Range, Montana. Precamb. Res. 16, 67-101.

Diamond, J.L. (1989) Mineralogical controls on rock magnetism in the Sierra San Pedro Martir Pluton, Baja California, Mexico. M.S. Thesis, University of Wyoming.

Dietvorst, E.J.L. (1980) Biotite breakdown and the formation of gahnite in metapelitic rocks from Kemiö, Southwest Finland. Contrib. Mineral. Petrol. 75, 327-337.

Droop, G.T.R. (1989) Reaction history of garnet-sapphirine granulites and conditions of Archean high-pressure granulite-facies metamorphism in the Central Limpopo Mobile Belt, Zimbabwe. J. Met. Geol. 7, 383-403.

Droop, G.T.R. and Bucher-Nurminen, B. (1984) Reaction textures and metamorphic evolution of sapphirine-bearing granulites from the Gruf Complex, Italian Central Alps. J. Petrol. 25, 766-803.

Dymek, R.F. (1983) Fe-Ti oxides in the Malene supracrustals and the occurrence of Nb-rich rutile. Rapp. Grølands geol. Unders. 112, 83-94.

Dymek, R.F., Brothers, S.C., and Schiffries, C.M. (1988) Petrogenesis of ultramafic metamorphic rocks from the 3800 Ma Isua supracrustal belt, West Greenland. J. Petrol. 29, 1353-1397.

Ellis, D.J. and Green, D.H. (1985) Garnet-forming reactions in mafic granulites from Enderby Land, Antarctica - Implications for geothermometry and geobarometry. J. Petrol. 26, 633-662.

Ellis, D.J., Sheraton, J.W., England, R.N., and Dallwitz, W.B. (1980) Osumilite-sapphirine-quartz granulites from Enderby Land Antarctica - Mineral assemblages and reactions. Contrib. Mineral. Petrol. 72, 123-143.

Evans, B.W. and Frost, B.R. (1975) Chrome-spinel in progressive metamorphism - a preliminary analysis. Geochim Cosmochim Acta 39, 959-972.

Feenstra, A. (1985) Metamorphism of bauxites on Noxos, Greece. Ph.D. Dissertation, Universtiy of Utrecht.

Ferry, J.M. (1984) A biotite isograd in south-central Maine, U.S.A.: Mineral reactions, fluid transfer, and heat transfer. J. Petrol. 25, 871-893.

Floran, R.J. and Papike, J.J. (1978) Mineralogy and petrology of the Gunflint Iron Formation, Minnesota-Ontario: Correlation of compositional and assemblage variations at low to moderate grade. J. Petrol. 19, 215-288.

Foit, F.F. Jr., Hooper, R.L., and Rosenberg, P.E. (1987) An unusual pyroxene, melilite, and iron-oxide mineral assemblage in a coal-fire buchite from Buffalo, Wyoming. Am. Mineral. 72, 137-147.

Francis, D.M. (1976) Corona-bearing pyroxene granulite xenoliths and the lower crust beneath Nunivak Island, Alaska. Can. Mineral. 14, 291-298.

Frisch, T. (1971) Alteration of chrome spinel in a dunite nodule from Lanzarote, Canary Islands. Lithos 4, 83-91.

Frost, B.R. (1973a) Ferroan gahnite from quartz-biotite-almandine schist, Wind River Mountains, Wyoming. Am. Mineral. 58, 831-834.

Frost, B.R. (1973b) Contact metamorphism of the Ingalls ultramafic complex at Paddy-Go-Easy Pass, Central Cascades, Washington. Ph.D. Dissertation, University of Washington.

Frost, B.R. (1976) Limits to the assemblage forsterite-anorthite as inferred from peridotite hornfelses, Icicle Creek, Washington. Am. Mineral. 61, 732-750.

Frost, B.R., Lindsley, D.H., and Andersen, D.J. (1988) Fe-Ti oxide - silicate equilibria: Assemblages with

fayalitic olivine. Am. Mineral. 73, 727-740.
Ghiorso, M.S. (1990) Thermodynamic properties of hematite-ilmenite-geikielite solid solutions. Contrib. Mineral. Petrol. 104, 645-667.
Giere, R. (1987) Titanian clinohumite and geikielite in marbles from the Bergell contact aureole. Contrib. Mineral. Petrol. 96, 496-502.
Graham, J. (1978) Manganochromite, palladium antimonide, and some unusual mineral associations at the Nairne pyrite deposit, South Australia. Am. Mineral. 63, 1166-1174.
Grant, J.A. and Frost, B.R. (1990) Contact metamorphism and partial melting of pelitic rocks in the aureole of the Laramie Anorthosite Complex, Morton Pass, Wyoming. Am. J. Sci. 290, 425-472.
Grapes, R.H. (1986) Melting and thermal reconstitution of pelitic xenoliths Wehr volcano, East Eifel, West Germany. J. Petrol. 27,343-396.
Harley, S.L. (1986) A sapphirine-cordierite-garnet-sillimanite granulite from Enderby Land, Antarctica: implications for FMAS petrogenetic grids in the granulite facies. Contrib. Mineral. Petrol. 94, 452-460.
Harris, N. (1981) The application of spinel-bearing metapelites to P/T determinations: An example from South India. Contrib. Mineral. Petrol. 76, 229-233.
Hicks, J.A., Moore, J.M., and Reid, A.M. (1985) The co-occurrence of green and blue gahnite in the Namaqualand metamorphic complex, South Africa. Can. Mineral. 23, 535-542.
Horrocks, P.C. (1983) A corundum and sapphirine paragenesis from the Limpopo Mobile Belt, southern Africa. J. Met. Geol. 1, 13-23.
Itaya, T. and Banno, S. (1980) Paragenesis of titanium-bearing accessorite in pelitic schists of the Sanbagawa metamorphic belt, Central Shikoku, Japan. Contrib. Mineral. Petrol. 73, 267-276.
Itaya, T., Brothers, R.N. and Black, P.M. (1985) Sulfides, oxides, and sphene in high-pressure schists from New Caledonia. Contrib. Mineral. Petrol. 91, 151-162.
Johnson, C.A. and Essene, E.J. (1982) The formation of garnet in olivine-bearing metagabbros from the Adirondacks. Contrib. Mineral. Petrol. 81, 240-251.
Kamineni, D.C. and Rao, A.T. (1988) Sapphirine granulites from the Kakanuru area, Eastern Ghats, India, Am. Mineral. 73, 692-700.
Kars, H., Jansen, B.H., Tobi, A.C., and Poorter, R.P.E. (1980) The metapelitic rocks of the polymetamorphic Precambrian of Rogaland, SW Norway. Contrib. Mineral. Petrol. 74, 235-244.
Kawachi, Y., Grapes, R.H., Coombs, D.S., and Dowse, M. (1983) Mineralogy and petrology of a piemontite-bearing schist, western Otago, New Zealand. J. Met. Geol. 1, 353-372.
Kerrich, R., Fyfe, W.S., Barnett, R.L., Blair, B.B. ,and Willmore, L.M. (1987) Corundum, Cr-muscovite rocks at O'Briens, Zimbabwe: the conjunction of hydrothermal desilicification and LIL-element enrichment - geochemical and isotopic evidence. Contrib. Mineral. Petrol. 95, 481-498.
Kramm. U. (1979) Kanonaite-rich viridines from the Venn-Stavelot Massif, Belgian Ardennes. Contrib. Mineral. Petrol. 69, 387-395.
Kushiro, I. and Yoder, H.S. (1966) Anorthite-forsterite and anorthite-enstatite reactions and their bearing on the basalt-eclogite transformation. J. Petrol. 7, 337-362.
Laird, J. (1980) Phase equilibria in mafic schists from Vermont. J. Petrol. 21, 1-37.
Lal, R.K., Ackermand, D., and Upadhyay, H. (1987) P-T-X relationships deduced from corona textures in sapphirine-spinel-quartz assemblages from Paderu, southern India. J. Petrol. 28, 1139-1168.
Lieberman, J.E. and Rice, J.M. (1986) Petrology of marble and peridotite in the Seiad ultramafic complex, northern California, USA. J. Met. Geol., 4, 179-199.
Lindsley, D.H. and Lindh, A. (1974) A hydrothermal investigation of the system $FeO-Fe_2O_3$ $-TiO_2$: A discussion with new data. Lithos 7, 65-68.
Loferski, P.J. and Lipin, B.R. (1983) Exsolution in metamorphosed chromite from the Red Lodge district, Montana. Am. Mineral. 68, 777-789.
Mall, A.P. and Sharma, R.S. (1988) Coronas in olivine metagabbros from the Proterozoic Chotanagpur terrain at Mathurapur, Bihar, India. Lithos, 21, 291-230.
Meyer, H.O.A. and Brookins, D.C. (1976) Sapphirine, sillimanite, and garnet in granulite xenoliths from Stockdale kimberlite, Kansas. Am. Mineral. 61, 1194-1202.
Montel, J-M. (1985) Xenolities peralumineux dans les dolerites du Peyron en Velay (Massif Central, France). Indications sur l'evolution de la croule profonde tardihercynienne. C. R. Acad. Sci. 301, 615-620.
Montel, J-M., Weber, C., and Pichavant, M. (1986) Biotite-sillimanite-spinel assemblages in high-grade metamorphic rocks: Occurrences, chemographic analysis and thermobarometric interest. Bull.

Mineral. 109, 555-573.
Muir, J.E. and Naldrett, A.J. (1973) A natural occurrence of a two-phase chromium-bearing spinel. Can. Mineral. 11, 930-939.
Onyeagocha, A.C. (1974) Alteration of chromite from the Twin Sisters Dunite, Washington. Am. Mineral. 59, 608-612.
Pan, Y., and Fleet, M.E. (1989) Cr-rich calc-silicates from the Hemlo area, Ontario. Can. Mineral. 27, 565-577.
Paraskevopoulos, G.M. and Economou, M. (1981) Zoned Mn-rich chromite from podiform type chromite ore in serpentinites of northern Greece. Am. Mineral. 66, 1013-1019.
Peacock, S.M. and Norris, P.J. (1989) Metamorphic evolution of the Central Metamorphic Belt, Klamath Province, California: an inverted metamorphic gradient beneath the Trinity peridotite. J. Met. Geol. 7, 191-209.
Pedersen, A.K. (1978) Non-stoichiometric magnesian spinels in shale xenoliths from a native iron-bearing andesite as Asuk, Disko, central west Greenland. Contrib. Mineral. Petrol. 67, 331-340.
Percheck, L., Gerya, T., and Nozhkin, A. (1989) Petrology and retrograde P-T path in granulites of the Kanskaya formation, Yenisey range, Eastern Siberia. J. Met. Geol. 7, 599-617.
Reinecke, T. (1986a) Crystal chemistry and reaction relations of piemontites and thulites from highly oxidized low grade metamorphic rocks at Vitali, Andros Island, Greece. Contrib. Mineral. Petrol. 93, 56-76.
Reinecke, T. (1986b) Phase relationships of sursassite and other Mn-silicates in highly oxidized low-grade, high-pressure metamorphic rocks from Evvia and Andros Islands, Greece. Contrib. Mineral. Petrol. 94, 110-126.
Rice, J.R. (1977) Contact metamorphism of impure dolomitic limestone in the Boulder aureole, Montana. Contrib. Mineral. Petrol. 59, 237-259.
Sack, R.O. and Ghiorso, M.S. (in press) Chromian spinels as petrogenetic indicators: Thermodynamics and petrological applications. Am. Mineral.
Sandiford, M., Neall, F.B., and Powell, R., (1987) Metamorphic evolution of aluminous granulite from Labwor Hills, Uganda. Contrib. Mineral. Petrol. 95, 217-225.
Santosh, M. (1987) Cordierite gneisses of southern Kerala, India: petrology, fluid inclusions and implications for crustal uplift history. Contrib. Mineral. Petrol. 96, 343-356.
Schumacher, J.C. and Robinson, P., 1987, Mineral chemistry and metasomatic growth of aluminous enclaves in gedrite-cordierite-gneiss from southwestern New Hampshire, USA. J. Petrol. 28, 1033-1037.
Schumacher, J. C., Schäfer, K. and Seifert, F. (1987) Lamellar nigerite in Zn-rich spinel from the Falun deposit, Sweden. Contrib. Mineral. Petrol. 95, 182-190.
Selverstone, J. and Stern, C.R. (1983) Petrochemistry and recrystallization history of granulite xenoliths from the Pali-Aike volcanic field, Chile. Am. Mineral. 68, 1102-1112.
Shannon, E.V. (1923) Note on cobaltiferous ghanite from Maryland. Am. Mineral. 8, 147-148.
Shedlock, R.J. and Essene, E.J. (1979) Mineralogy and petrology of a tactite near Helena, Montana. J. Petrol. 20, 71-97.
Shive, P.N., Frost, B.R., and Peretti, A. (1988) The magnetic properties of metaperidotitic rocks as a function of metamorphic grade: Implications for crustal magnetic anomalies. J. Geophys. Res. 93, 12187-12195.
Shulters, J.C. and Bohlen, S.R. (1989) The stability of hercynite and hercynite-gahnite spinels in corundum or quartz-bearing assemblages. J. Petrol. 30, 1017-1031.
Spencer, K.J. and Lindsley, D.H. (1981) A solution model for coexisting iron titanium oxides. Am. Mineral. 66, 1189-1201.
Spry, P.G. (1982) An unusual gahnite-forming reaction, Geco base-metal deposit, Manitouwadge, Ontario. Can. Mineral. 20, 549-553.
Spry, P.G. (1987) Compositional zoning in zincian spinel. Can. Mineral. 25, 97-104.
Spry, P.G. and Scott, S.D. (1986) Zincian spinel and staurolite as guides to ore in the Appalachians and Scandinavian Caledonides. Can. Mineral. 24, 147-163.
Stephenson, N.C.N. and Hensel, H.D. (1982) Amphibolites and related rocks from the Wongwibinda metamorphic comples, northern N.W.S. Australia. Lithos 15, 59-75.
Stoddard, E.F. (1979) Zinc-rich hercynite in high-grade metamorphic rocks: a product of the dehydration of staurolite. Am. Mineral. 64, 736-741.
Stüwe, K. and Powell, R. (1989) Low-pressure granulite facies metamorphism in the Larsemann Hills area, East Antarctica: petrology and tectonic implications for the evolution of the Prydz Bay aea. J. Met.

Geol. 7, 465-483.
Tracy, R.J. (1982) Compositional zoning and inclusions in metamorphic minerals Rev. Miner. 10, 355-397.
Tracy, R.J., Jaffe, H.W. and Robinson, P. (1978) Monticellite marble at Cascade Mountain, Adirondack Mountains, New York. Am. Mineral. 63, 991-999.
Tracy, R.J. and Robinson, P. (1988) Silicate-oxide-sulfide-fluid reactions in granulite-grade pelitic rocks, central Massachusetts. Am. J. Sci. 288-A, 45-74.
Trelor, P.J. (1987) The Cr-minerals of Ourokumpu - their chemistry and significance. J. Petrol. 28, 867-886.
Trommsdorff, V. and Evans, B.W. (1980) Titanian hydroxyl-clinohumite: Formation and breakdown in antigorite rocks (Malenco, Italy). Contrib. Mineral. Petrol. 72, 229-242.
Turnock, A.C. and Eugster, H.P. (1962) Fe-Al oxides: Phase relations below 1000°C. J. Petrol. 3, 533-565.
Van Lamoen, H. (1979) Coronas in olivine gabbros and iron ores from Susimäki and Ruittamaa, Finland. Contrib. Mineral. Petrol. 68, 259-268.
Vielzeuf, D. (1983) The spinel and quartz associations in high grade xenoliths from Tallante (S.E. Spain) and their potential use in geothermometry and geobarometry. Contrib. Mineral. Petrol. 82, 301-311.
Waters, D.J. (1986) Metamorphic history of sapphirine-bearing and related magnesian gneisses from Namaqualand, South Africa. J. Petrol. 27, 541-565.
Waters, D.J. and Moore, J.M. (1985) Kornerupine in Mg-Al - rich gneisses from Namaqualand, South Africa: mineralogy and evidence for late-metamorphic fluid activity. Contrib. Mineral. Petrol. 91, 369-382.
Weiser, T. (1967) Zink und Vanadium führende chromite von Outokumpu/Finland. N. Jb. Mineral. Monat. 234-243.
Williams, M.L. and Grambling, J.A. (1990) Manganses, ferric iron, and the equilibrium between garnet and biotite. Am. Mineral. 75, 886-908.
Windley, B.F., Ackermand, D., and Herd, R.K. (1984) Sapphirine/kornerupine-bearing rocks and crustal uplift history of the Limpopo belt, Southern Africa. Contrib. Mineral. Petrol. 86, 342-358.
Windley, B.F., Herd, R.K., and Ackermand, D. (1989) Geikielite and ilmenite in Archaean meta-ultramafic rocks, Fiskenaesset, West Greenland. Eur. J. Mineral. 1, 427-437.
Wise, W.S. (1959) An occurrence of geikielite. Am. Mineral. 44, 879-882.
Wylie, A.G., Candela, P.A., and Burke, T.M. (1987) Compositional zoning in unusual Zn-rich chromite from the Sykesville district of Maryland and its bearing on the origin of "ferritchromit". Am. Mineral. 72, 413-422.
Zakrzewski, M.A. (1989) Chromian spinels from Kuså, Bergslagen, Sweden. Am. Mineral. 74, 448-455.

Chapter 14

B. Ronald Frost

MAGNETIC PETROLOGY: FACTORS THAT CONTROL THE OCCURRENCE OF MAGNETTE IN CRUSTAL ROCKS

INTRODUCTION

It is a sad truth that all too often scientists are prisoners of the way they look at the world. Petrologists rely heavily on thin sections to understand mineral reactions and too many of them still clump the oxides and sulfides as "opaques" (i.e., minerals one cannot identify in transmitted light microscopy). Economic geologists, who usually study reactions in reflected light often have an equally egregious habit of referring to any silicate phase as gangue (i.e., a mineral that cannot be easily identified in reflected light). Geophysicists and rock magnetists use sophisticated instruments to measure magnetic fields in the Earth's crust or to determine the magnetic properties of rock samples. Although they can determine the mineralogic source of the magnetism (usually magnetite but less commonly pyrrhotite or native metals), they cannot say why some rocks have magnetite and others, of similar composition, do not. Furthermore, because petrologists usually consider the presence or absence of magnetite and other opaque phases in a rock as a secondary question (if they consider it all) geophysicists usually find the petrologic literature of little help in understanding the factors that control the stability of magnetic minerals in crustal rocks. The question of the mineralogic control of rock magnetism has been a growing field of study (e.g. Haggerty, 1978; Wasilewski and Fountain, 1982; Schlinger, 1985; Frost and Shive, 1986; Shive et al., 1988) and it has been recently graced with a formal name: magnetic petrology (Wasilewski and Warner, 1988).

To a great extent, the problems addressed in the field of magnetic petrology are different from those found in typical igneous and metamorphic petrology. Rock magnetists wish to understand the factors that control magnetite abundance in the common rocks in the earth's crust. As likely as not these rocks will lack the low-variance assemblages that petrologists prefer to study. Indeed, many of the rocks most important to the magnetic signature of mountain belts, such as metamorphosed greywackes and quartzofeldspathic gneisses, are petrologic orphans. Because there may be little mineralogic change in such rocks during metamorphism, there is a very meager petrologic literature from which one can draw to determine how the magnetic phases behave in such rocks.

The field of magnetic petrology consists of two aspects: (1) identification of the minerals responsible for the magnetic signature of a rock, and (2) determination of the factors that control the occurrence and abundance of these phases. The first part, which is best termed magnetic mineralogy, is a fairly well established field (see Haggerty, 1976); the second, however, is still in its infancy. The first aspect has been addressed by several chapters in this book. In this chapter I will cover the general factors that control magnetite stability in crustal rocks. Because magnetite is the most abundant magnetic mineral in crustal rocks, understanding its stability in the crust is a major aspect of magnetic petrology. Other magnetic phases, pyrrhotite and awaruite (Ni-Fe alloy) will be mentioned only in passing.

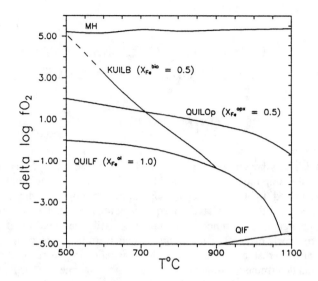

Figure. 1. $\Delta \log fO_2$ - T diagram showing the relative cooling trends for oxide-silicate equilibria. $\Delta \log fO_2$ refers to deviation from the FMQ buffer, MH = hematite-magnetite buffer, QIF = quartz-iron-fayalite buffer, QUILF = quartz-ulvöspinel-ilmenite-fayalite equilibrium, QUILOp = quartz-ulvöspinel-ilmenite-orthopyroxene equilibrium, KUILB = quartz-ulvöspinel-ilmenite-annite equilibrium. See Chapter 1 for balanced reaction describing the buffers and the chapter by Frost and Lindsley in this volume for balanced reactions for the other equilibria.

There are two important aspects to the study of magnetite occurrence in crustal rocks. One concerns the factors that control magnetite composition, including the subsolidus reactions that affect the oxides during cooling of plutonic rocks. Although they are nominally metamorphic reactions, they have been given little attention by metamorphic petrologists (and even less be igneous petrologists) and are but poorly understood. The second concerns the reactions that control the occurrence and abundance of magnetite during metamorphism of rocks of various protoliths.

CONTROLS OF MAGNETITE STABILITY DURING COOLING OF PLUTONIC ROCKS

The most important reason why the compositions of Fe-Ti oxides in plutonic rocks differ from those in volcanic rocks is that Fe-Ti oxides undergo extensive sub-solidus re-equilibration during cooling. Oxide re-equilibration actually involves three different processes: (1) oxide-silicate re-equilibration, (2) oxide-oxide, or interoxide re-equilibration, and (3) intraoxide re-equilibration (Frost et al., 1988).

Oxide-silicate re-equilibration

This process involves the exchange of ferrous iron and, to a lesser extent, ferric iron and titanium as well, between the oxides and coexisting silicates. Cooling paths following oxide-silicate reactions involving quartz, two oxides and either fayalite (QUIIF) or orthopyroxene (QUIIO) will tend to be weakly oxidizing (Fig. 1) (Frost et al., 1988; Lindsley et al., 1990). Oxide-silicate cooling paths involving olivine and pyroxene apparently close at moderately high temperatures; for example Frost et al. (1988) show that cooling paths for a fayalite-two oxide-quartz assemblage in the Skaergaard closed over a temperature range from 700°C to 900°C.

Unlike the trends with olivine and pyroxene, cooling trends involving two oxides, biotite, and K-feldspar (KUIIB) have steeply oxidizing paths (Frost, 1990) (Fig. 1) (see Chapter 12 for further discussion of this equilibrium). Furthermore, because the presence of H_2O fluxes mineral reactions, oxide-silicate equilibria involving biotite do not seem to close until temperatures of 400°C to 500°C are reached. For this reason the magnetite in granitic rocks is commonly nearly pure Fe_3O_4 (e.g. Czamanske et al. 1977)

Oxide-oxide re-equilibration

Oxide-oxide re-equilibration typically affects both titanomagnetite and ilmenite and involves the $Fe^{2+}TiFe^{3+}_{-2}$ exchange operation. This is an expression of the following equilibrium:

$$Fe_2TiO_4 + Fe_2O_3 = FeTiO_3 + Fe_3O_4$$
in magnetite   in ilmenite   ilmenite   magnetite

With falling temperature this reaction proceeds to the right with the result that both magnetite and ilmenite will approach their pure end-member compositions as cooling proceeds. The T-fO_2 path followed during this re-equilibration is dependent on the relative abundance of magnetite and ilmenite in the rock (Fig. 2). In rocks relatively rich in magnetite relative to ilmenite the cooling trend will be sharply reducing, whereas in rocks with only minor magnetite the cooling path will follow the ilmenite isopleths (Frost et al., 1988). This will be accompanied by either constant relative fO_2 or strong relative oxidation, depending on the composition of the primary ilmenite. (see Andersen and Lindsley, 1988, Figs. 2 and 5).

Intraoxide re-equilibration

During intraoxide re-equilibration titanium in titanomagnetite is removed by the reaction to ilmenite. This process produces the typical texture of ilmenite plates within magnetite and has been called oxyexsolution (Buddington and Lindsley, 1964) (see also Chapter 5). This process is modelled by the reaction:

$$6\ Fe_2TiO_4 + O_2 = 6\ FeTiO_3 + 2\ Fe_3O_4$$
in magnetite   ilmenite   magnetite

This reaction proceeds from right to left on cooling. However, as noted in Chapter 1, there must be a source for oxygen in the rock for this reaction to operate. When the petrologist carefully reconstitutes the composition of titanomagnetite to apply the Fe-Ti oxide thermometer the temperature that will be obtained will be the temperature at which oxyexsolution began (Frost and Chacko, 1989).

The T - fO_2 path followed during cooling of an oxide assemblage depends upon many factors - original oxygen fugacity of the assemblage, the relative abundance of titanomagnetite and ilmenite, and composition of the fluid to name a few. Consider a rock that crystallized at relatively oxidized conditions (condition 1, Fig. 3). Such a rock

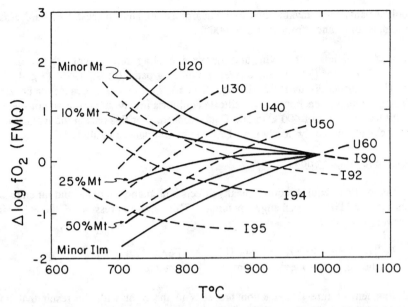

Figure. 2. $\Delta \log fO_2$ - T diagram showing the cooling trends followed during interoxide equilibration of the same initial oxide assemblage but with different proportions of titanomagnetite and ilmenite, assuming that the oxides cooled in a closed system (after Frost et al., 1988).

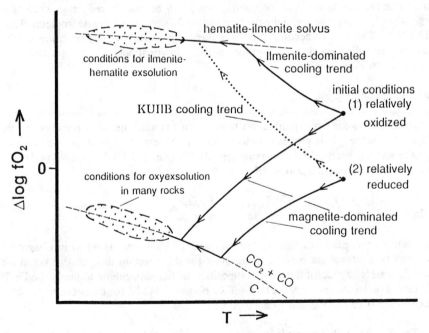

Figure. 3. $\Delta \log fO_2$ - T diagram showing inferred cooling trends for Fe-Ti oxides. In relatively oxidized rocks (condition (1)), an ilmenite-dominated cooling trend could lead to conditions of the hematite-ilmenite solvus. A magnetite-dominated trend in rocks rich in CO_2 can produce relative redution and graphite saturation. In systems containing H_2O re-equilibration between oxides and a hydrous silicate (such at the KUIIB equilibrium) can also lead to relative oxidation. See text for discussion.

may be a granite or a quartz-bearing mafic rock with relatively magnesian pyroxenes (see chapter by Frost and Lindsley, this volume). If the rock contained ilmenite as the major oxide, the interoxide cooling trend will be relatively oxidizing. On cooling the oxygen fugacity may reach that of the ilmenite-hematite solvus, causing hematite to exsolve from the ilmenite. Inspection of the Figures 2 and 5 of Andersen and Lindsley (1988) and of Figure 4 of Ghiorso and Sack (in press) shows that ilmenites with X_{hem} of 0.2 or more can attain oxygen fugacities of the hematite-ilmenite solvus in this manner. The exsolved hemoilmenite from the charnockites described by Bradshaw (1985) may have formed in such a manner. If a rock forming at condition 1 contained titanomagnetite as the major oxide phase, rather than ilmenite, the interoxide cooling trend would be relatively reducing. If a carbonic fluid was present the trend may intersect the graphite-saturation surface, at which point oxyexsolution would occur accompanied by the precipitation of graphite. Precipitation of graphite accompanying oxyexsolution is even more likely in rocks the primary crystallization conditions of which were relatively reduced (e.g. condition 2, Fig. 3). The lower the oxygen fugacity in the primary assemblage of a rock, the more likely an interoxide cooling trend will intersect the graphite saturation surface, even if there is a significant proportion of ilmenite to titanomagnetite. The co-precipitation of graphite accompanying oxyexsolution has been postulated because reconstituted oxide pairs for a number of rock types lie on or close to the graphite-saturated surface. This includes fayalite monzonites (Fuhrman et al., 1988; Kolker and Lindsley, 1989), gabbroic anorthosite (Frost et al., 1989), and granulites (Frost and Chacko, 1989). However, not all oxyexsolved titanomagnetites formed on the graphite-saturated surface. Diamond (1989) described a reconstructed oxyexsolved titanomagnetite that lay at oxygen fugacities more than one log unit above FMQ. The source for oxygen in such oxyexsolution is unlikely to be dissociation of the fluid and is more likely to be reaction with the silicate phases, possibly with associated reduction of ferric iron.

Another possible cooling path is one that is controlled by oxide-silicate reaction, rather that being controlled by interoxide equilibration. In granitoids with a H_2O-rich fluid, the major re-equilibration is likely to be reaction between the oxides, K-feldspar, and biotite (KUIIB, see the chapter by Frost and Lindsley in this volume for further discussion of this equilibrium). This reaction has a relatively oxidizing trend with cooling and can cause rocks of a wide range of primary conditions to be oxidized to the hematite-ilmenite solvus (dotted line, Fig. 3). The exsolved hemoilmenite from metamorphosed granites described by Diamond (1989) could have formed in this manner.

These recrystallization processes have important consequences to petrology and rock magnetism. The complex re-equilibration of oxides means that petrologists must take great care when applying Fe-Ti oxide thermometry to plutonic rocks. Reconstitution of an oxyexsolved titanomagnetite will provide the petrologist with the temperature at which intraoxide re-equilibration began but this may not be the temperature at which the primary assemblage formed (Frost and Chacko, 1989). Re-equilibration processes are important to rock magnetism because they enrich the magnetite in the Fe_3O_4 component. This will increase both the magnetic susceptibility of the spinel phase and the Curie temperature. Because all crustal rocks have been subjected this low-temperature re-equilibration one can model the spinel phase in crustal rocks as being pure magnetite, even in plutonic rocks where the primary high-temperature spinel is postulated to have been rich in Fe_2TiO_4 (Frost and Shive, 1986).

MINERAL EQUILIBRIA EFFECTING MAGNETITE STABILITY DURING METAMORPHISM

The occurrence and abundance of magnetite in metamorphic rocks is a complex function of the composition of the rocks, the stable mineral assemblage that is present, and the temperature and pressure of metamorphism. Because all minerals in a rock will react during metamorphism the compositions of oxides, silicates, and sulfides in a given assemblage are interrelated and are controlled by oxygen fugacity as well as sulfur fugacity of the fluid (Nesbitt, 1986a,b; Frost, 1988). There are three types of equilibria that can affect the abundance and occurrence of magnetite in metamorphic rocks: (1) oxide - silicate equilibria, (2) graphite - bearing equilibria, and (3) sulfide equilibria.

Silicate-oxide and silicate-oxide-carbonate equilibria

Oxidation-reduction equilibria. Oxidation-reduction equilibria are essentially variants of the FMQ buffer:

$$2\ Fe_3O_4 + 3\ SiO_2 = 3\ Fe_2SiO_4 + O_2 \qquad (1)$$
$$\text{in spinel} \quad \text{quartz} \quad \text{in olivine}$$

Reactions such as these may have a significant effect on the oxygen fugacity at which an assemblage equilibrates, but because of mass-balance constraints, they may not produce or consume much magnetite (see Frost, 1982; 1988).

A series of equilibria involving magnetite, native iron, and Fe-silicates has been generated for the system Fe-Si-O-H (Frost, 1979a, 1985, 1988; Engi, 1986) (see Table 1 and Fig. 4). Important features of this diagram can be seen by comparing reaction surfaces (A) and (B). During prograde metamorphism magnetite is a product of the reactions on surface (A), while it is a reactant on surface (B). The production of magnetite requires consumption of free oxygen from the fluid and therefore prograde metamorphism of a rock with a buffering assemblage on equilibrium surface (A) will result in relative reduction. However, because there is a vanishingly small amount of free oxygen in the fluid these reactions can change the oxygen fugacity by many orders of magnitude without producing petrologically significant amounts of magnetite. Conversely, consumption of magnetite by reactions such as those labeled B liberates oxygen. Thus progressive metamorphism of a rock with an assemblage on equilibrium surface (B) will result in relative oxidation.

In magnesium-bearing systems the isobarically univariant curves in Figure 4 will become divariant surfaces that slope to higher oxygen fugacity as the silicates become enriched in Mg (e.g. Lindsley et al., 1990). Consequently it is possible to use Fe/(Fe+Mg) ratio of silicates coexisting with magnetite or hematite to monitor changes in oxygen fugacity (Froese, 1977; Frost, 1982; Nesbitt, 1986b). Even though our knowledge of the energetics of Fe-Mg exchange in most phases is still poor, making it difficult to calculate such buffering surfaces exactly, one can still monitor relatively small changes in oxygen fugacity by comparing the Fe/(Fe + Mg) ratio of the silicates from similar assemblages. One way to do this chemographically is by projecting from Fe-oxide. This technique was used by Frost (1982) to compare the oxygen fugacity at which iron-formations have crystallized. A similar diagram is shown schematically in

Table 1

Equilibria in the System Fe-Si-C-O-H

Phase Compositions
Fayalite (F): Fe_2SiO_4 Iron (I): Fe
Greenalite (Gre): $Fe_3Si_2O_5(OH)_4$ Minnesotaite (Mi): $Fe_3Si_4O_{10}(OH)_2$
Grunerite (Gru): $Fe_7Si_8O_{22}(OH)_2$ Quartz (Q): SiO_2
Hematite (H): Fe_2O_3 Vapor (V): H_2O

Equilibria

1. $2\,M + 3\,Q = 3\,F + O_2$
2. $2\,Gru = 7\,F + 9\,Q + 2\,H_2O$
3. $6\,Gru + 7\,O_2 = 14\,M + 48\,Q + 6\,H_2O$
4. $2\,Gru + 6\,M = 16\,F + 3\,O_2 + 2\,H_2O$
5. $F = 2\,I + Q + O_2$
6. $2\,Gru = 14\,I + 16\,Q + 2\,H_2O + 7\,O_2$
7. $2\,Gru + 18\,I + 9\,O_2 = 16\,F + 2\,H_2O$
8. $7\,Mi = 3\,Gru + 4\,Q + 4\,H_2O$
9. $2\,Mi + O_2 = 2\,M + 8\,Q + 2\,H_2O$
10. $12\,Mi + 2\,M = 6\,Gru + O_2 + 6\,H_2O$
11. $2\,Min = 6\,I + 8\,Q + 3\,O_2 + 2\,H_2O$
12. $4\,Mi + 2\,I + O_2 = 2\,Gru + 2\,H_2O$
13. $9\,Mi + 4\,F = 5\,Gru + 4\,H_2O$
14. $6\,Mi + 10\,M = 24\,F + 6\,H_2O + 5\,O_2$
15. $2\,Mi + 10\,I + 5\,O_2 = 8\,F + 2\,H_2O$
16. $3\,Gre = Mi + 6\,F + 9\,H_2O$
17. $4\,Gre + O_2 = 2\,Mi + 2\,M + 6\,H_2O$
18. $6\,Gre + 2\,M = 12\,F + 12\,H_2O + O_2$
19. $2\,Gre + 2\,I + O_2 = 4\,F + 4\,H_2O$
20. $4\,Gre = 2\,Mi + 6\,I + 6\,H_2O + 3\,O_2$
21. $M = 3\,I + 2\,O_2$
22. $Gre + 2\,Q = Min + H_2O$
23. $2\,Gre + O_2 = 2\,M + 4\,Q + 4\,H_2O$
24. $6\,H = 4\,M + O_2$
25. $4\,Gre + 3\,O_2 = 6\,H + 8\,Q + 8\,H_2O$
26. $2\,Gre = 6\,I + 4\,Q + 4\,H_2O + 3\,O_2$

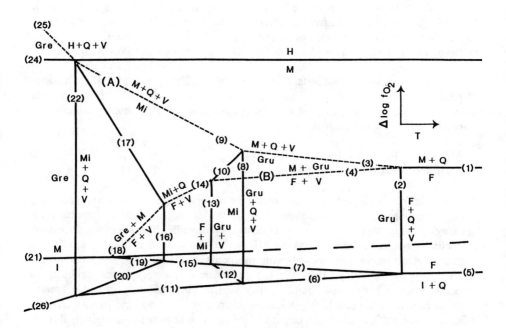

Figure. 4. $\Delta \log fO_2$ - T diagram showing topologic relations in the system Fe-Si-O-H. F = fayalite, Gre = greenalite, Gru = grunerite, H = hematite, I = iron, Mi = minnesotaite, Q = quartz, V = water vapor. Numbers refer to reactions listed on Table 1. Curve A is for assemblages saturated in quartz. Curve B is for assemblages saturated in olivine (modified after Frost, 1988).

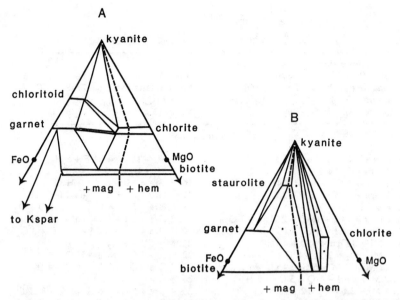

Figure. 5. Chemographic projections from muscovite, quartz, an Fe-oxide, and vapor in the system K_2O-FeO-MgO-Al_2O_3-SiO_2-H_2O. (A) Projection showing expected assemblages in lower amphibolite facies (modified after Thompson, 1972). (B) Phase relations in kyanite-grade pelites from Fernleigh, Ontario. Dots show assemblages encountered by Hounslow and Moore (1976) (after Frost, 1988).

Figure 5 for pelitic schists in upper amphibolite facies (Thompson, 1972). In such a diagram the two-phase tie-lines are isobars of oxygen fugacity. Thus a decrease in Fe/(Fe+Mg) ratio of silicates from a buffering assemblage will reflect an increase in oxygen fugacity. At some value of Fe/(Fe+Mg) hematite will become the stable oxide in place of magnetite. This is shown in Figures 5 A and B by the dashed line.

As noted above, in most instances buffer-type equilibria are not an efficient means of producing or consuming of magnetite. However where the buffer-type equilibria intersect the hematite-magnetite buffer, they can lead to significant production of magnetite by a reactions such as:

$$\text{FeO} + \text{Fe}_2\text{O}_3 = \text{Fe}_3\text{O}_4 \qquad (27)$$
in silicate   hematite   magnetite

This reaction has been postulated as a cause for major changes in magnetite content in weakly metamorphosed iron-formations (Han, 1978). It may also play an important role in changing magnetite abundance in rocks containing ilmenite. The assemblage hematite-magnetite is not restricted to lie on the MH buffer, for with increasing Ti substitution hematite (or ilmenite) can coexist with magnetite over about 8 log units of fO_2 (Buddington and Lindsley, 1964). Thus in ilmenite-bearing systems oxidation or reduction through buffer-type equilibria may lead to observable changes in magnetite abundance through reaction (27). Of course this reaction will become increasingly less efficient at lower fO_2 as the Fe_2O_3 content of the ilmenite becomes progressively diluted.

The reactions between magnetite and an Fe-bearing carbonate (siderite or ankerite), like reactions with Fe-silicates, will buffer the oxygen fugacity of a system. The major difference is that in carbonate-bearing reactions a carbon-bearing fluid species (most likely CO_2) will be present. The behavior of equilibria between siderite (or a siderite-magnesite solution) and magnetite (or hematite) has been discussed in depth by Frost (1979a,b; 1985) who showed the following:

1. Reactions involving Fe,Mg-carbonate, magnetite and quartz assemblages will be relatively oxidizing. At relatively low metamorphic temperatures, such as greenschist and lower amphibolite facies, hematite may be stable with the carbonate rather than magnetite.

2. For most metamorphic conditions the fluid coexisting with magnetite and an Fe-carbonate is likely to be dominantly CO_2 and H_2O, rather than methane, particularly if quartz is in the assemblage.

3. In a buffering assemblage containing magnetite and an Fe-bearing carbonate (be it ankeritic or sideritic), increasing CO_2 in the fluid will increase relative fO_2.

Equilibria involving ferric iron in silicates. In assemblages containing silicates that can accommodate ferric iron, equilibria between Fe oxides and Fe^{3+} in the silicates may play an important role in determining the abundance of magnetite. Such equilibria can operate without any change in the oxidation state of iron and therefore, they may have a major control on the abundance of Fe-oxides in metamorphic rocks. For example it is observed that hematite in contact metamorphosed slates and red sandstones decreases markedly in abundance when biotite appears (Best and Weiss, 1964; Riklin, 1983). Detailed studies by Riklin (1983) show that this decrease is not due to a change in the oxidation state of iron but rather is related to incorporation of Fe^{3+} into biotite. It is likely that reactions involving ferric iron in silicates are also responsible for the changes in magnetite abundance in the Liberty Hill pluton (Speer, 1981). Here magnetite initially increases in abundance with increasing metamorphic grade, probably as a result of the release of ferric iron during breakdown of epidote. At grades above the disappearance of epidote, increasing metamorphic grade leads to a decrease in magnetite abundance possibly because of the solution of ferric iron into biotite which becomes more abundant with increasing grade.

Equilibria involving ferric iron are probably important in mafic rocks as well. Jolly (1980) documented that hematite disappears from metabasalts during the transition from greenschist to amphibolite facies. This is probably caused by Fe^{3+} substitution into hornblende, for it is well known that amphiboles in mafic rocks become more tschermakitic with increasing metamorphic grade (Laird, 1980). Unfortunately the electron microprobe cannot routinely analyze ferric iron and therefore the ferric iron content of many important silicates is poorly constrained. Consequently, even though reactions involving the solution of ferric iron into silicates may play a major role in controlling magnetite abundance during metamorphism, they are very poorly understood.

Graphite-Fluid Equilibria

The equilibrium between graphite and a C-O-H fluid phase is important because it is one of the major processes of reduction in petrology (French, 1966; Ohmoto and Kerrick, 1977; Frost, 1979b; Holloway, 1984). This is a manifestation of the fact that the graphite-saturation surface in the system C-O has a relatively more reducing slope with increasing temperature than do the oxide-silicate or oxide-oxide buffers (Fig. 6). The saturation surface in the C-O system gives the maximum oxygen fugacity at which graphite may occur. The presence of diluting species in the fluid, such as H_2O and CH_4,

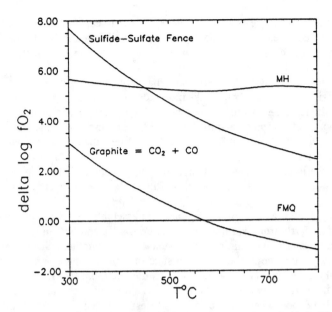

Figure. 6. $\Delta \log fO_2$ - T diagram showing the location of the graphite-saturation surface and of the sulfate-sulfide fence. MH and FMQ refer to buffers discussed in Chapter 1.

will cause the surface to fall to lower fO_2. In pelitic rocks, where the silicate relations are governed by dehydration equilibria, the fluid composition is likely to follow a path on the graphite-saturation surface that is marked by the locus of points where $X(H_2O)$ is at a maximum, i.e., $X(H_2O)^{max}$ (Ohmoto and Kerrick, 1977). This forms a curve that lies one half to one log unit below the graphite-saturation curve in the pure C-O system. Any graphite-bearing rock undergoing metamorphism with concurrent dehydration will follow a buffering path marked by $X(H_2O)^{max}$ rather than by the graphite saturation surface in the pure C-O system. This will lead to a marked relative reduction of oxygen fugacity during metamorphism, the effect of which is to eliminate magnetite from graphitic rocks. With increasing temperature such a process will also cause the fluid phase, which would be initially very rich in H_2O, to become progressively enriched in CO_2 and CH_4 (Ohmoto and Kerrick, 1977; Frost, 1979b).

Most graphitic schists lack magnetite, although the rare assemblage magnetite-ilmenite-graphite has been reported by McLellan (1985). Magnetite is rare in graphitic pelitic rocks for two reasons. One is the strongly reducing trend followed by graphitic rocks during progressive metamorphism, which would cause the magnetite to be reduced to FeO in silicates by the reaction:

$2\ Fe_3O_4\ +\ C\ =\ CO_2\ +\ 6\ FeO$
magnetite graphite   fluid   in silicates

The second reason is the fact that there is a positive correlation between the carbon content of a pelitic rock and the sulfur content (Tracy and Robinson, 1988). In rocks with high sulfur fugacity pyrrhotite or pyrite would be stable in place of magnetite (Nesbitt, 1986a).

Oxide-sulfide equilibria

Phase relations involving sulfides, silicates and oxides are complicated by the fact that the fugacity of sulfur is a function of oxygen fugacity. In metamorphic environments S_2 is only a minor sulfur species. The dominant species are H_2S and SO_2, which are related by the following equilibrium:

$$2 H_2S + 3 O_2 = 2 H_2O + 2 SO_2 \tag{28}$$

We can define the locus of points at which the fugacities of SO_2 and H_2S are equal as the sulfate-sulfide fence. At oxygen fugacities above this fence SO_2 will be the dominant species, while at lower oxygen fugacities, H_2S will dominate. The sulfate-sulfide fence, like the graphite-saturation surface, has a more reducing slope with increasing temperature than do the oxide-oxide or oxide-silicate buffers (Fig. 6). From the stoichiometry of equilibrium (28) it is evident that the location of the sulfate-sulfide fence is dependent on the composition of the attendant fluid. This effect, however, is not large; for example, at 500°C decreasing XH_2O in the fluid from 1.0 to 0.2 will displace the sulfate-sulfide fence downward by only 0.4 log units (Frost, 1988). From Figure 6 it is evident that H_2S will be the dominant sulfur species in all metamorphic rocks, apart from those metamorphosed at high temperatures and under relatively oxidizing conditions or in the presence of a carbonic fluid phase. Thus, although it is entirely valid to write reactions in terms of S_2 when balancing equilibria, when one is considering a process, the reactions must be written with H_2S as the sulfur species.

Recognizing this, the reactions whereby magnetite is converted to pyrite or pyrrhotite in metamorphic rocks can be written as follows:

$$2 Fe_3O_4 + 6 H_2S = 6 FeS + 6 H_2O + O_2 \tag{29}$$

and

$$Fe_3O_4 + 6 H_2S + O_2 = 3 FeS_2 + 6 H_2O \tag{30}$$

These reactions indicate that the reaction of pyrite to magnetite provides a source of oxygen that may be used for oxidation of silicates. In contrast, the reaction of pyrrhotite to magnetite requires a source of oxygen to proceed efficiently. In an assemblage with magnetite, pyrite, and pyrrhotite, the conversion of sulfides to magnetite can proceed as an oxygen-independent reaction:

$$3 FeS_2 + 6 FeS + 12 H_2O = 3 Fe_3O_4 + 12 H_2S \tag{31}$$

It is possible, indeed likely, that magnetite will be produced by the breakdown of Fe-sulfides during metamorphism of pyrite-bearing, graphite-free rocks. Such reactions, however, are not well documented petrologically. The converse reaction, namely the consumption of magnetite in areas of high sulfur fugacity is a well known process that is used to explain aeromagnetic anomalies associated with ore deposits.

Although one can write the reaction of pyrrhotite to pyrite as a simple sulfidation reaction, since the dominent fluid in metamorphic rocks is H_2S and not S_2, to reflect the mass-balance in metamorphic rocks, the reaction should be written:

$$2 \text{ FeS}2 + 2 \text{ H}_2\text{O} = 2 \text{ FeS} + 2 \text{ H}_2\text{S} + \text{O}_2 \tag{32}$$

Reaction (32) indicates that in most metamorphic environments pyrrhotite will be stable in the relatively reducing environments while pyrite will occur in the more oxidizing rocks. This is roughly what is seen in nature. Pyrrhotite is the major Fe-sulfide in metaperidotites and high-temperature metapelites, whereas pyrite is more common in metabasites.

EFFECT OF PROGRADE METAMORPHISM ON VARIOUS PROTOLITHS

During prograde metamorphism of most rocks, the path followed by the fluid composition is essentially determined by the bulk composition of the protolith. The bulk composition will determine the mineral assemblages present at any given point on the P-T path of a rock, and these assemblages will dictate the path followed by the fluids. This concept of mineralogic control of fluid composition is termed buffering (Eugster, 1959) and it is well recognized that in most metamorphic rocks the fluid was internally controlled. Although the buffering path followed by the fluids during prograde metamorphism has not been determined for all rock types, a rough summary of some protoliths is given below.

Metaperidotite and iron-formation

These protoliths can be discussed together because the behavior of both can be closely approximated by the Fe-Mg-Si-C-O-H system (see Fig. 4). Because most iron-formations are saturated with respect to magnetite and quartz, they will lie on surfaces (appropriately displaced because of magnesium substitution) that are equivalent to curve (A) in Figure 4. Progressive metamorphism of these rocks will lead to relative reduction (Frost, 1979a). There is strong evidence (Han, 1978) that the lowest grade iron-formations contain the assemblage greenalite-hematite-quartz (with or without siderite). Buffering by oxide-silicate-quartz equilibria will drive oxygen fugacity to the hematite-magnetite buffer (Figs. 4, 7). At this point hematite can be converted to magnetite by the oxygen-conserving reaction:

$$3 \text{ Fe}_2\text{O}_3 + \text{Fe}_3\text{Si}_2\text{O}_5(\text{OH})_4 = 3 \text{ Fe}_3\text{O}_4 + 2 \text{ SiO}_2 + 2 \text{ H}_2\text{O} \tag{33}$$
$$\text{hematite} \quad \text{in greenalite} \quad \text{magnetite} \quad \text{quartz}$$

Above this intersection the fluid in a hypothetical Mg-free rock will be buffered by the solid curve in Figure 7 until, in the highest grade rocks, the FMQ buffer surface is reached. The presence of Mg in natural systems will cause the buffering surface to be displaced to higher fO_2. In addition, sufficient Mg will stabilize orthopyroxene at the expense of fayalite + quartz. Most iron-formations contain silicates that are only moderately magnesian and, therefore, lie at oxygen fugacities that deviate only slightly from the solid line in Figure 4. Some iron-formations, however, contain silicates that are relatively strongly magnesian, for example the orthopyroxene with $X_{Fe} = 0.23$ reported by Butler (1969). To account for this compositional variation, the fO_2 trajectory on Figure 7 is shown in a stippled pattern with the density of the pattern representing the approximate frequency of occurrence.

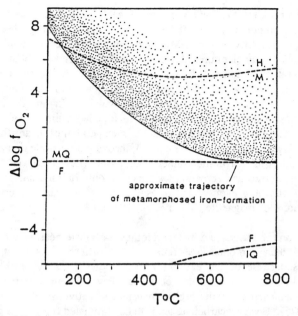

Figure. 7. Inferred T - Δ log fO$_2$ trajectory followed during metamorphism of iron-formations. Density of stippling reflects the frequency of occurrence (after Frost, 1988).

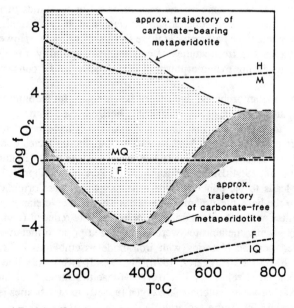

Figure. 8. Inferred T - Δ log fO$_2$ trajectories followed during metamorphism of serpentinites. Area in light stippling refers to carbonate-bearing rocks, that in dark stippling refers to carbonate-free rocks (after Frost, 1985, 1988).

In metaperidotites the lowest oxygen fugacity will occur with the assemblage antigorite-olivine-brucite-magnetite (Frost, 1985). This assemblage lies four to five log units of oxygen fugacity below FMQ and commonly occurs with highly-reduced iron-nickel alloys. At temperatures above the stability of this assemblage, metaperidotites will be olivine-saturated and will lie on displaced equivalents of curve (B) in Figure 4. Prograde metamorphism will lead to relative oxidation along this curve. Because the spinel phase in the highest grade rocks is dominated by Cr and Al rather than Fe^{3+}, oxygen fugacities in these rocks will be one to four log units above FMQ (see Frost, 1985) (Fig. 8). The strongly reducing nature of metaperidotites means that they will also have relatively low sulfur fugacity. As a result, the common sulfide in many serpentinites and metaperidotites is pyrrhotite, rather than pyrite (Eckstrand, 1975; Frost, 1985). Because metaperidotites and serpentinites are commonly rich in magnetite, the magnetic effects of pyrrhotite and even of native Fe-Ni alloys, are vastly overwhelmed by the magnetic signature of magnetite (Shive et al., 1988).

In both iron formations and metaperidotites, carbonate-bearing assemblages will be more oxidized than carbonate-free rocks. This difference is most pronounced in metaperidotites because of the extremely low oxygen fugacity of the carbonate-free assemblages. Thus, talc-magnesite rocks may have hematite, indicating an oxygen fugacity more than ten log units higher than that in native-metal bearing serpentinites (Eckstrand, 1975). Because carbonate tends to be eliminated from these protoliths during prograde metamorphism, the T-fO_2 trends for carbonate-bearing iron-formations and metaperidotites will approach the trends for the carbonate-free assemblages (Fig. 8).

<u>Metabasites</u>

Little work has been done on the oxygen or sulfur fugacities of metabasites. To a large extent this is a consequence of the fact that it is difficult to characterize quantitatively any intensive parameter in these low-variance rocks. However, on the basis of petrologic studies and from studies of magnetic susceptibility a rough picture of the behavior of oxides during metamorphism of mafic rocks can be constructed.

In weakly metamorphosed basalts it is common to recognize both oxidized, red-weathering, and reduced, green-weathering horizons (see Surdam, 1968). However, it is impossible to make such distinctions in rocks that have been metamorphosed to grades above greenschist facies. Mafic rocks in amphibolite facies commonly have the relatively oxidized assemblage ilmenite-hematite ($\pm$ magnetite) (Banno and Kanehira, 1961; Kanehira et al., 1964; Braun and Raith, 1985; Peretti and Köppel, 1986; Laird, pers. comm.). This leads to the conclusion that the rocks that were originally hematite-rich were reduced to the hematite-ilmenite solvus during prograde metamorphism. Although many amphibolites are relatively oxidized, some are quite reduced (Dymek, 1983; Russ-Nabelek, 1989). Many metabasites were not oxidized prior to metamorphism and thus began their metamorphic histories with iron-oxide assemblages indicative of oxygen fugacities near FMQ (see Haggerty, 1976)(Fig. 9). It is possible that these originally reduced rocks stayed at relatively low oxygen fugacities throughout their metamorphic history. The exact process that controls oxygen fugacity in metabasites is not known and is an important topic for future research.

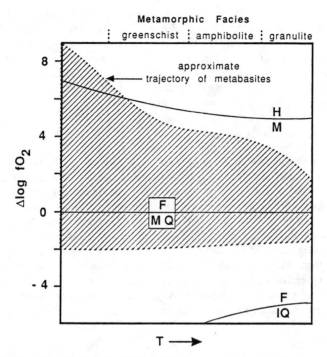

Figure. 9. Inferred T - Δlog fO$_2$ trajectory followed during metamorphism of mafic rocks (altered from Frost, 1988).

Judging from the relatively high oxygen fugacity of most basic schists, one would conclude that sulfur fugacity in these rocks would also be relatively high. This is consistent with the findings of Banno and Kanehira (1961) and Kanehira et al. (1964) who recorded pyrite-chalcopyrite ± bornite as a common assemblage in blueschists and amphibolites of the Sanbagawa belt. Pyrrhotite is thus a far less common phase in metabasites than it is in more reduced rocks such as pelites and serpentinites.

There is ample evidence indicating that the incorporation of ferric iron in amphiboles and in chlorite has important effects on the reactions governing the prograde metamorphism of mafic rocks. Hematite seems to disappear or decrease in abundance with the appearance of hornblende (cf. Banno and Kanehira, 1961; Kanehira et al., 1964; Jolly, 1980). In addition many metabasites are distinctly poor in magnetite, as indicated by their low magnetic susceptibility (Powell, 1970; Williams et al., 1986). This suggests that reactions between Fe-oxides and Fe^{3+} in the silicates, rather than oxidation-reduction equilibria, may be the most important oxide-silicate equilibria in mafic rocks. There are several indications that the Fe^{3+} bound in silicates is released as magnetite during the transition from amphibolite to granulite facies. For example Powell (1970) found mafic granulites to be distinctly more magnetic than amphibolite-grade equivalents. Furthermore, mass-balance calculations (Russ-Nabelek, 1989) indicate that magnetite should be a product phase of hornblende-breakdown reactions. If magnetite is a product of hornblende breakdown, then buffering by the breakdown assemblage should lead to reduction (for example, see curve A in Fig. 4). Indeed, such reduction is indicated for mafic granulites. They generally contain oxide assemblages that equilibrated at oxygen fugacities within a log unit of the FMQ buffer (cf. Price and Wallace, 1976; Bohlen & Essene, 1977; Dufour, 1985), well below those of the hematite-ilmenite solvus (Fig. 6). There are, however, a couple of occurrences of oxidized metabasites (or meta-diorites) where the ilmenite has exsolved hematite on cooling (Dymek, 1983; Bradshaw, 1985). Like many aspects of magnetic petrology, the reason why these rocks are far more oxidized than other high-grade metabasites is not known.

Graphite-free rocks with Fe-Ti oxides

Information to date indicates that in the absence of graphite, the fluid composition of many rocks will be driven to the fO_2 of the hematite-ilmenite solvus. In addition to the metabasites noted above, this assemblage has been found in quartzites (Rumble, 1976), metamorphosed bauxites (Feenstra, 1985), metamorphosed sandstones (Riklin, 1983), and graphite-free pelitic schists (Chinner, 1960; Hounslow and Moore, 1967). It is not known what equilibria control oxygen fugacity in most protoliths, but it is clear that at the lowest grades many of the rocks mentioned above contained hematite. During metamorphism these must have been reduced to the ilmenite-hematite solvus as the hematite was progressively enriched in titanium. This is particularly well demonstrated by the metabauxites studied by Feenstra (1985). These rocks have Ti-poor hematite + rutile at the lowest grades but contain magnetite + ilmenite in upper amphibolite facies.

Pelitic schists

The major factor that determines the path followed by oxygen fugacity during metamorphism of pelitic schists is the abundance of organic matter in the protolith. Those rocks that were originally poor in organic matter evolve to become graphite-free metapelites. In amphibolite facies such rocks generally contain the assemblage ilmenite-hematite (+/- magnetite) (Chinner, 1960; Hounslow and Moore, 1967; Hutcheon, 1979). Originally most of these rocks were probably hematitic. Like hematite-bearing mafic rocks and metabauxites noted above, they were probably reduced to their present assemblage during metamorphism, although the exact equilibria that were responsible have not yet been identified. In general pyrite survives to higher metamorphic grades in graphite-free schists than in graphite-bearing ones (see Hutcheon, 1979). This simply reflects the fact that graphitic schists are more reducing, and hence have a lower sulfur fugacity than non-graphitic schists of the same metamorphic grade.

As would be expected, prograde metamorphism of graphitic schists is accompanied by marked reduction (Fig. 10). This is indicated by the fact that such rocks are distinctly lacking in magnetite and contain ilmenite and rutile as the sole oxides (see Mohr and Newton, 1984). Prograde metamorphism also leads to progressive desulfurization. Low-grade slates commonly contain pyrite, but in graphitic rocks pyrrhotite appears in the lowest amphibolite facies. For example, in the southern Appalachians pyrrhotite appears in graphitic schists at grades slightly below those of the biotite isograd (Carpenter, 1974) and pyrite disappears by the time the staurolite isograd is reached (Nesbitt and Essene, 1983). Pyrite, however, may not disappear at such low grades in all graphitic schists. In some sulfur-rich schists from eastern Maine it persists to upper sillimanite grades (Guidotti, 1970; Tracy and Robinson, 1988) and in high-pressure rocks from Japan it is stable in epidote-amphibolite facies (Kanehira et al., 1964).

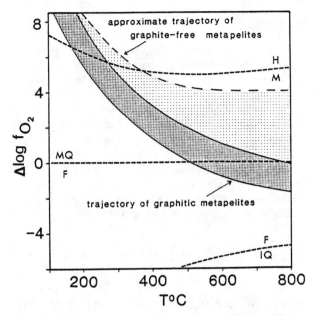

Figure. 10. Inferred T - Δ log fO$_2$ trajectories followed during metamorphism of pelitic rocks. Heavy stippling refers to graphite-bearing pelitic schists, light stippling to graphite-free schists (after Frost, 1988).

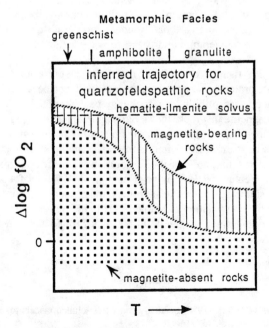

Figure. 11. Inferred T - Δ log fO2 trajectories followed during metamorphism of quartzofeldspathic gneisses. Ruled area refers to magnetite-bearing rocks, stippled area to magnetite free (but probably ilmenite-bearing) rocks.

Quartzofeldspathic gneisses

The buffering path followed during metamorphism of quartzofeldspathic rocks is poorly known. One possible trajectory is shown in Figure 11. Considering that the cooling trend in a few granitoids lies on the hematite-ilmenite solvus, it is assumed in this diagram that this is where many magnetite-bearing quartzofeldspathic gneisses will lie. With increasing temperatures magnetite-bearing granitoids will tend to follow a relatively reducing trend governed by the dehydration of biotite (+ ilmenite) to magnetite + K-feldspar (the KUIlB equilibria). This trend will become much less reducing when biotite begins to react to orthopyroxene and K-feldspar at the beginning of granulite facies. Here oxygen fugacity will be controlled by reaction between orthopyroxene, oxides, and quartz (QUIlOp equilibria, Lindsley et al., 1990). There are many granites that lack magnetite (Ishihara, 1977) and these are constrained to in the stippled area in Figure 11, at oxygen fugacities below those of the magnetite-bearing granitoids.

CONCLUSIONS AND SUGGESTIONS FOR FUTURE WORK

In this chapter I have been able to outline in a very general manner the way that oxygen fugacity and magnetite abundance vary in several rock types as a function of metamorphism. For some rocks, such as metaperidotites and metamorphosed iron-formations, the equilibria that control magnetite abundance and oxygen fugacity are relatively simple and the way they behave with increasing temperature is fairly well understood. In others, such as in non-graphitic pelitic rocks and metabasites, the reactions that control magnetite abundance are barely understood at all and the manner that they behave with changes in temperature and pressure is conjecture at best.

Even when the factors controlling magnetite stability in metamorphism are understood for most protoliths, magnetic petrology will not be a trivial field of study. Although the magnetic susceptibility of a rock may be determined mainly by the stability of coarse-grained magnetite in a rock, the stable magnetic remanence in a rock will reside in the finest-grained, perhaps single domain, magnetite. This fine-grained magnetite may well be a retrogressive feature, as it often is in metaperidotites (Shive et al., 1988), or it may occur as fine plates, that are possibly exsolution features, in pyroxene (Fleet et al., 1980; Schlinger and Veblen, 1989), olivine (Champness, 1970), or feldspar (Geissman et al., 1988). Future studies of magnetic petrology must, therefore consider not only the factors that control the composition of stable Fe-Ti oxide phases in the rock but also the possibility that the fine-grained magnetite in the rock formed as the result of a different process.

REFERENCES

Andersen, D.J. and Lindsley, D.H. (1988) Internally consistent solution models for Fe-Mg-Mn-ti oxides: Fe-Ti oxides. Am. Mineral. 73, 714-726.

Banno, S. and Kanehira, K. (1961) Sulfide and oxide minerals in schists in the Bessi-Ino district, central Shikoku, Japan. Japan. J. Geol. Geogr. 29, 29-44.

Best, M.G. and Weiss, L.E. (1964) Mineralogical relations in some pelitic hornfelses from the

southern Sierra Nevada, California. Am. Mineral., 49, 1240-1266.
Bohlen, S.R. and Essene, E.J. (1977) Feldspar and oxide thermometry of granulites in the Adirondack Highlands. Contrib. Mineral. Petrol. 62, 153-169.
Bradshaw, J.Y. (1985) Geology of the northern Franklin Mountains, northern Fiordland, New Zealand, with emphasis on the origin and evolution of Fiordland granulites. Ph.D. dissertation, University of Otago.
Braun, E. and Raith, M. (1985) Fe-Ti oxides in metamorphic basites from the Eastern Alps, Austria: a contribution to the formation of solid solutions of natural Fe-Ti oxide assemblages. Contrib. Mineral. Petrol. 90, 199-213.
Buddington, A.W. and Lindsley, D.H. (1964) Iron-titanium oxide minerals and synthetic equivalents. J. Petrol. 5, 310-357.
Butler, P. (1969) Mineral compositions and equilibria in the metamorphosed iron formation of the Gagnon region, Quebec, Canada. J. Petrol. 10, 56-101.
Carpenter, R.H. (1974) Pyrrhotite isograd in southeastern Tennessee and southwestern North Carolina. Geol. Soc. Am. Bull., 85, 451-456.
Champness, P.E. (1970) Nucleation and growth of iron oxides in olivines, Mineral. Mag. 37, 790-800.
Chinner, G.A. (1960) Pelitic gneisses with varying ferric/ferrous ratios from Glen Clova, Angus, Scotland. J. Petrol. 1, 178-217.
Czamanske, G.K., Wones, D.R., and Eichelberger, J.C. (1977) Mineralogy and petrology of the intrusive complex of the Pliny Range, New Hampshire. Am. J. Sci. 277, 1073-1123.
Diamond, J.L. (1989) Mineralogical controls on rocks magnetism in the Sierra San Pedro Martir Pluton, Baja California, Mexico. M.S. Thesis, University of Wyoming.
Dufour, E. (1985) Granulite facies metamorphism and retrogressive evolution of the Monts du Lyonnais metabasites (Massif Central, France). Lithos 18, 97-113.
Dymek, R.F. (1983) Fe-Ti oxides in the Malene supracrustals and the occurrence of Nb-rich rutile. Rapp. Grølands geol. Unders. 112, 83-94.
Eckstrand, O.R. (1975) The Dumont Serpentinite: A model for the control of nickeliferous opaque mineral assemblages by alteration reactions in ultramafic rocks. Econ. Geol. 70, 183-201.
Engi, M. (1986) Towards a thermodynamic data base for geochemistry. Consistency and optimal representation of the stability relations in mineral systems. Habilitationsschrift, E. T. H., Zürich.
Eugster, H.P. (1959) Oxidation and Reduction in metamorphism. In P. H. Abelson, ed. Researches in Geochemistry. New York, John Wiley & Sons, 397-426.
Eugster, H.P. and Wones, D.R. (1962) Stability relations of the ferruginous biotite annite. J. Petrol. 3, 82-125.
Feenstra, A. (1985) Metamorphism of bauxites on Noxos, Greece. Ph.D. Dissertation, University of Utrecht.
Fleet, M.E., Bilcox, G.A., and Barnett, R.L. (1980) Oriented magnetite inclusions in pyroxenes from the Grenville province. Can. Mineral. 18,89-99.
French, B.M. (1966) Some geologic implications of equilibrium between graphite and a C-H-O gas at high temperatures and pressures. Rev. Geophysics, 4, 223-253.
Froese, E. (1977) Oxidation and sulphidation reactions. In Greenwood, H. J.,ed. Short Course in Application of Thermodynamics to Petrology and ore Deposits. Mineral. Assoc. Canada. Short Course Handbook 2, 84-98.
Frost, B.R. (1979a) Metamorphism of iron-formation: Parageneses in the System Fe-Si-C-O-H. Econ. Geol. 74, 775-785.
Frost, B.R. (1979b) Mineral equilibria involving mixed volatiles in a C-O-H fluid phase: The stabilities of graphite and siderite. Am. J. Sci. 279, 1033-1059.
Frost, B.R. (1982) Contact metamorphic effects of the Stillwater Complex, Montana: the concordant iron-formation: a discussion of the role of buffering in metamorphism of iron-formation. Am. Mineral. 67, 142-148.
Frost, B.R. (1985) On the stability of sulfides, oxides, and native metals in serpentinite. J. Petrol. 26, 31-63.
Frost, B.R. (1988) A review of graphite-sulfide-oxide-silicate equilibria in metamorphic rocks. Rendiconti Societa Italiana Mineral. Petrol. 43, 25-40.
Frost, B.R. (1990) Biotite crystallization as an oxidation agent in granitic rocks. Geol. Soc. Am. Abstracts with Programs 22, A301.
Frost, B.R. and Chacko, T. (1989) The granulite uncertainty principle: Limitations to thermobarometry in granulites. J. Geology 97, 435-450.

Frost, B.R., Fyfe, W.S., Tazaki, K., and Chan, T. (1989) Grain-boundary graphite in rocks from the Laramie Anorthosite Complex: Implications for lower crustal conductivity. Nature 340, 134-136.

Frost, B.R., Lindsley, D.H., and Andersen, D.J. (1988) Fe-Ti oxide - silicate equilibria: Assemblages with fayalitic olivine. Am. Mineral. 73, 727-740.

Frost, B.R. and Shive, P.N. (1986) Magnetic mineralogy of the lower continental crust. J. Geophys. Res. 91, 6513-6521.

Fuhrman, M.L., Frost, B.R., and Lindsley, D.H. (1988) Crystallization conditions of the Sybille Monzosyenite, Laramie Anorthosite Complex, Wyoming. J. Petrol. 29, 669-729.

Geissman, J.W., Harlen, S.S., and Brearley, A.J. (1988) The physical isolation and identification of carriers of geologically stable remanent magnetization: Paleomagnetic and rock magnetic microanalysis and electron microscopy. Geophys. Res. Lett. 15, 479-482.

Ghiorso, M.S. and Sack, R.O. (in press) Fe-Ti oxide geothermometry: Thermodynamic formulation and the extimation of intensive variables in silicic magmas. Contrib. Mineral. Petrol.

Guidotti, C.V. (1970) The mineralogy and petrology of the transition from lower to upper sillimanite zone in the Oquossoc Area, Maine. J. Petrol. 11, 277-336.

Haggerty, S.E. (1976) Opaque mineral oxides in terrestrial igneous rocks. Rev. Mineral. 3, Hg101-Hg300.

Haggerty, S.E. (1978) Mineralogical constraints on Curie isotherms in deep crustal magnetic anomalies. Geophys. Res. Lett. 5,105-108.

Han, T.M. (1978) Microstructures of magnetite as guides to its origin in some Precambrian iron-formations. Fortschr. Mineral. 56, 105-142.

Holloway, J.R. (1984) Graphite-CH_4-H_2O-CO_2 equilibria at low-grade metamorphic conditions. Geology 12, 455-458.

Hounslow, A.W. and Moore, J.M. (1967) Chemical petrology of Grenville schists near Fernleigh, Ontario. J. Petrol. 8, 1-28.

Ishahira, S. (1977) The magnetite-series and ilmenite-series granitic rocks. Mining Geology 27, 293-305.

Jolly, W.T. (1980) Development and degradation of Archean lavas, Abitibi Area, Canada, in light of major element geochemistry. J. Petrol. 21, 323-363.

Kanehira, K., Banno, S., and Nishida, K. (1964) Sulfide and oxide minerals in some metamorphic terranes in Japan. Japan J. Geol. Geophys. 35, 175-191.

Kolker, A. and Lindsley, D.H. (1989) Geochemical evolution of the Maloin Ranch pluton, Laramie Anorthosite Complex, Wyoming: Petrology and mixing relations. Am. Mineral. 74, 307-324.

Laird, J. (1980) Phase equilibria in mafic schists from Vermont. J. Petrol. 21, 1-37.

Lindsley, D.H., Frost, B.R., Andersen, D.J., and Davidson, P.M. (1990) Fe-TI oxide - silicate equilibria: Assemblages with orthopyroxene. Geochem. Soc. Spec. Pap. 2, 103-119.

McLellan, E. (1985) Metamorphic reactions in the kyanite and sillimanite zones of the Barrovian type area. J. Petrol. 26, 789-818.

Mohr, D.W. and Newton, R.C. (1983) Kyanite-staurolite metamorphism in sulfidic schists of the Anakeesta Formation, Great Smoky Mountains, North Carolina. Am. J. Sci. 283, 97-134.

Nesbitt, B.E. (1986a) Oxide-sulfide-silicate equilibria associated with metamorphosed ore deposits. Part I. Theoretical Considerations. Econ. Geol. 81, 831-840.

Nesbitt, B.E. (1986b) Oxide-sulfide-silicate equilibria associated with metamorphosed ore deposits. Part II. pelitic and felsic terrains. Econ. Geol. 81, 841-856.

Nesbitt, B.E. and Essene, E.J. (1983) Metamorphic volatile equilibria in a portion of the southern Blue Ridge Province. Am. J. Sci. 283, 135-165.

Ohmoto, H. and Kerrick, D.M. (1977) Devolatilization equilibria in graphitic schists. Am. J. Sci. 277, 1013-1044.

Peretti, A. and Köpel, V. (1986) Geochemical and lead-isotope evidence for a mid-ocean ridge type mineralization within a polymetamorphic ophiolite complex (Monte del Forno, N. Italy - Switzerland) Earth Planet. Sci. Lett. 80, 252-264.

Powell, D.W. (1970) Magnetized rocks within the Lewisian of western Scotland and under the Southern Uplands. Scott. J. Geol. 6, 353-369.

Price, R.C. and Wallace, R.C. (1976) The significance of corona textured inclusions from a high pressure fractionated alkalic lava: North Otago, New Zealand. Lithos 9, 319-329.

Riklin, K. (1983) Contact metamorphism of the Permian "red sandstones" in the Adamello area. Mem. Soc. Geol. Ital. 26, 159-169.

Russ-Nabelek, C. (1989) Isochemical contact metamorphism of mafic schist, Laramie Anorthosite Complex,

Wyoming: Amphibole compositions and reactions. Am. Mineral. 74, 530-548.
Schlinger, C.M. (1985) Magnetization of lower crust and interpretation of regional magnetic anomalies: example from Lofoten and Vesterålen, Norway. J. Geophys. Res. 90, 11484-11504.
Schlinger, C.M., and Veblen, D.R. (1989) Magnetism and transmission electron microscopy of Fe-Ti oxides and pyroxenes in a granulite from Lofoten, Norway. J. Geophys. Res. 94, 14009-14026.
Shive, P.N., Frost, B.R., and Peretti, A., 1988, The magnetic properties of metaperidotitic rocks as a function of metamorphic grade: Implications for crustal magnetic anomalies. J. Geophys. Res. 93, 12187-12195.
Speer, J.A. (1981) The nature and magnetic expression of isograds in the contact aureole of the Liberty Hill pluton, South Carolina: Summary. Geol. Soc. Am. Bull. Part 1, 92, 603-609.
Surdam, R.C. (1968) Origin of native copper and hematite in the Karmutsen Group, Vancouver Island, British Columbia. Econ. Geol. 63, 961-966.
Thompson, J.B. (1972) Oxides and sulfides in regional metamorphism of pelitic schists. 24th Int'l. Geol. Congress, Sect. 10, 27-35.
Tracy, R.J. and Robinson, P. (1988) Silicate-oxide-sulfide-fluid in granulite-grade pelitic rocks, central Massachusetts. Am. J. Sci. 288-A, 45-74.
Wasilewski, P. and Fountain, D.M. (1982) The Ivrea Zone as a model for the distribution of magnetization in the continental crust. Geophys. Res. Lett. 9, 333-336.
Wasilewski, P. and Warner, R.D. (1988) Magnetic petrology of deep crustal rocks - Ivrea Zone, Italy. Earth Planet. Sci. Lett. 87, 347-361.
Williams, M.C., Shive, P.N., Fountain, D.M., and Frost, B.R., (1986) Magnetic properties of exposed deep crustal rocks from the Superior Province of Manitoba. Earth Planet. Sci. Lett. 76, 176-184.